# Computational Design of Engineering Materials

Successful computational design of engineering materials requires a combination of multiscale computational methods such as the CALPHAD method, first-principles calculations, phase-field simulation, and finite element analysis, covering the atomic–meso–macro scale ranges. Written jointly by a team of recognized experts in these fields, this book provides unique insights in both the fundamentals and case studies for a variety of materials. The fundamentals of computational thermodynamics, thermophysical properties, first-principles calculations, mesoscale simulation methods, and crystal plasticity finite element method are introduced. The nonspecialist reader with a general science or engineering background should understand these tools deeply enough to consider their applicability and assess the results. In particular, the important role of CALPHAD and its scientific databases in materials design and the integration of simulation tools at different levels are highlighted. Case studies for designing a wide range of materials, including steels, light alloys, superalloys, cemented carbides, hard coating, and energy materials, are demonstrated in detail through a step-by-step methodology. Ancillary materials provide the reader with hands-on experience in simulation tools. This book is intended for professionals in design of engineering materials and other materials, being also an invaluable reference to graduates, undergraduates, researchers, and engineers who use various computational tools in their study, research, and/or development of materials.

**Yong Du** received his PhD from Central South University (CSU) of China in 1992. From 1993 to 2003, he continued his research at the Tokyo Institute of Technology, University of Barcelona, Clausthal University of Technology, University of Vienna, and University of Wisconsin at Madison. Since 2003, Du has been a professor at CSU. His research fields are thermodynamics, thermophysical properties, and materials design. His recognitions and honors include the National Outstanding Youth by the National Natural Science Foundation of China (NSFC), the Cheung Kong Chair Professorship by the Ministry of Education of China, the Leader of the Innovative Research Team by the NFSC, and the Leader of 973 National Basic Research Program of China. Du is an associate editor for both *CALPHAD* and the *Journal of Phase Equilibria and Diffusion*. He has been awarded one First-Class Prize of Hunan Provincial Natural Science Award, one Third-Class Prize of National Natural Science Award of China, and one best paper prize of the Alloy Phase Diagram International Committee (APDIC).

**Rainer Schmid-Fetzer** received his PhD from Clausthal University of Technology (TU Clausthal) in 1977. With background in metallurgy and physics he earned merits in thermodynamics, solidification, interface reactions and applications to designing alloys. His career comprised research at the University of Wisconsin at Madison between 1982 and 1984, as lecturer, and professor at TU Clausthal and also in 1994 at the Microelectronics and Microsystems section of the Daimler Benz Corporate Research Institute, Frankfurt, and in 1997 as Visiting Professor at the University of Wisconsin at Madison. He retired in 2015. Schmid-Fetzer is an associate editor of the *Journal of Phase Equilibria and Diffusion* and a member of the advisory board of the *International Journal of Materials Research* (formerly *Zeitschrift für Metallkunde*). He is a fellow of ASM International since 2003. He has been awarded the most prestigious honors and awards, including the Tammann Award and Werner-Köster Prize from the German Society of Materials, the Hume-Rothery Prize of British IoM3, the William Hume-Rothery Award from the Minerals, Metals, and Materials Society (TMS) of USA, and the Lee Hsun Lecture Award from the Institute of Metal Research of the Chinese Academy of Sciences.

**Jincheng Wang** received his PhD from Northwestern Polytechnical University of China in 2001. He did postdoctoral work at the National Institute for Materials Science of Japan from 2002 to 2004. Since 2005, Wang has been working at the State Key Laboratory of Solidification Processing at the Northwestern Polytechnical University as lecturer, associate professor, and professor. From 2009 to 2010, he worked at the Institute of Space and Astronautical Science of Japan as a visiting professor for six months. His major research fields include solidification; modeling and simulation of microstructure evolution; alloy design; additive manufacturing; and high-entropy alloys. His awards include two Second-Class Prizes of Shaanxi Provincial Natural Science Award.

**Shuhong Liu** received her PhD from Central South University (CSU) in 2010. Then she did postdoctoral work for one year at the Materials Chemistry Institute of RWTH Aachen University. Since 2010, Liu has been working at the Research Institute of Powder Metallurgy of CSU as lecturer, associate professor, and professor. She is an editorial member of the *Journal of Mining and Metallurgy B*. Her awards include the best paper prize of APDIC and the First-Class Prize of Hunan Provincial Natural Science Award. Her research interests include phase diagram, CALPHAD-type calculations of thermodynamic and thermophysical properties, as well as corrosion and precipitation simulation.

**Jianchuan Wang** received his PhD from Central South University (CSU) in 2012, being a visiting PhD student in 2011 at Max-Planck-Institut für Eisenforschung GmbH. Since 2013, he has been working at the State Key Laboratory of Powder Metallurgy of CSU as lecturer and associate professor. His major research focuses on using first-principles calculation and molecular dynamics to study defect properties, diffusion, thermodynamics, interface in condensed matter, especially in hydrogen storage materials, and Li-ion batteries. He has (co)authored more than 60 peer-reviewed publications.

**Zhanpeng Jin** received his master's degree from the Central South Institute of Mining and Metallurgy of China in 1963. He continued his teaching and research there from 1964 until he retired in 2018. From 1978 to 1980, Jin worked at the Royal Institute of Technology of Sweden as a visiting scholar. He was an associate editor of *CALPHAD* and a member of the Asia Pacific Materials Academy. In 2006, Jin became an academician of the Chinese Academy of Sciences. He received one Third-Class Prize of National Natural Science Award of China.

# Computational Design of Engineering Materials

## Fundamentals and Case Studies

YONG DU
*Central South University*

RAINER SCHMID-FETZER
*Clausthal University of Technology*

JINCHENG WANG
*Northwestern Polytechnical University*

SHUHONG LIU
*Central South University*

JIANCHUAN WANG
*Central South University*

ZHANPENG JIN
*Central South University*

CAMBRIDGE
UNIVERSITY PRESS

MATERIALS RESEARCH SOCIETY
*Advancing materials. Improving the quality of life.*

Shaftesbury Road, Cambridge CB2 8EA, United Kingdom

One Liberty Plaza, 20th Floor, New York, NY 10006, USA

477 Williamstown Road, Port Melbourne, VIC 3207, Australia

314–321, 3rd Floor, Plot 3, Splendor Forum, Jasola District Centre, New Delhi – 110025, India

103 Penang Road, #05–06/07, Visioncrest Commercial, Singapore 238467

Cambridge University Press is part of Cambridge University Press & Assessment,
a department of the University of Cambridge.

We share the University's mission to contribute to society through the pursuit of
education, learning and research at the highest international levels of excellence.

www.cambridge.org
Information on this title: www.cambridge.org/9781108494106

DOI: 10.1017/9781108643764

© Cambridge University Press and Assessment 2023

First published 2023

Printed in the United Kingdom by TJ Books Limited, Padstow Cornwall

*A catalogue record for this publication is available from the British Library.*

*A Cataloging-in-Publication data record for this book is available from the Library of Congress.*

ISBN 978-1-108-49410-6 Hardback

Y. Du wishes to dedicate this book to Professor Peiyun Huang, who received his PhD from MIT in 1945 and was one of the two supervisors (along with Professor Zhanpeng Jin) for Du's research.

# Contents

*Colour plates are to be found between pages 460 and 461.*

# Foreword

A revolution has been under way for several decades, transforming materials engineering from a much-discredited slow and costly process of trial-and-error experimental "materials by discovery" to one of true *design* enabled by predictive science inverted to exploit the system of CALPHAD fundamental databases now known as the materials genome. Driven by a systems approach to control of hierarchical microstructure, university research initiated in the 1980s integrated materials science, quantum physics, and continuum mechanics to bring materials into a new age of computational engineering design. Moving beyond the reductionist philosophy of traditional academic research, these efforts tested the accuracy limits of density functional theory (DFT)-based quantum mechanical methods in demonstrating particular utility in the prediction of surface thermodynamics, while continuum micromechanics of heterogeneous systems brought new insights to the unit processes of fracture and fatigue where quantitative structure–property relations had previously been lacking. Paramount to the achievements was a synthetic philosophy that can be traced back to the founding of the international CALPHAD collaboration by the work of Kaufman and Cohen in 1956. Rather than the "calculation of phase diagrams," as implied by the CALPHAD acronym, their work actually entailed the opposite – reducing the information in the Fe–Ni equilibrium diagram to its underlying thermodynamics for the specific purpose of controlling behavior far from equilibrium, as represented by martensitic transformations. It is this recognition of the importance of thermodynamics in defining the forces driving dynamic systems, and the attendant expansion of CALPHAD data to incorporate kinetic parameters, that has given CALPHAD such power in enabling the application of our fundamental knowledge of materials dynamics in a quantitative and system-specific form. Just as the human genome functions as a database directing the assembly of the structures of life, the CALPHAD genome embodies the fundamental parameters driving the dynamic assembly of multiscale materials microstructures – the defining concept of the materials genome metaphor.

Successful demonstrations of efficient CALPHAD-based parametric materials design in the 1990s led to the founding of QuesTek Innovations as the first computational materials design company. This was soon followed by the US Defense Advanced Research Projects Agency–Accelerated Insertion of Materials (DARPA-AIM) initiative aiming to accelerate the full materials development and qualification cycle. Here full simulation of microstructural evolution in complex processing enabled a probabilistic science approach to accurately forecast manufacturing variation with minimal test data. Ultimately, that was applied to full flight qualification of two computationally designed aircraft landing gear steels. Achievements of the DARPA-AIM program were highlighted in a 2004 US National Research Council report *Accelerating Technology Transition*, which outlined a national initiative to promulgate this technology, ultimately undertaken in 2011.

The rare event of a speech by a US president announcing this materials research initiative immediately attracted the attention of apex corporations, which brought their resources to its efficient implementation, enabling for the first time the incorporation of materials design and development into concurrent engineering. Here, a historic milestone was the announcement of four new alloys with the Apple Watch in 2014, designed concurrently with the product in less than two years of acquiring the design technology that enabled it. Further migration of the technology led to Elon Musk's announcement of the novel SpaceX SX500 fire-resistant Superalloy as a vital enabler of the Raptor engine of the Mars Starship, integrated at an early stage of development into a highly accelerated concurrent engineering process. A major innovation in automotive technology was Tesla's rapid development of aluminum structure Giga-Casting technology enabled by a novel Al casting alloy designed through tight control of eutectic phase fractions, also delivered concurrently with its casting pilot plants in under a two-year cycle, now in production in the Tesla model Y. With the efficiency and efficacy of CALPHAD-based materials design clearly established by these major successes, Tesla announced the formation of a materials applications group to accelerate the replacement of legacy alloys with designed alternatives to enable a higher level of full-system optimization. We thus find ourselves at the start of another technology revolution whereby the rest of engineering is being retrained to embrace the new opportunity of materials concurrency.

Written by Rainer Schmid-Fetzer and Yong Du (both world-level outstanding scientists) along with four Chinese scientists, this unique book offers a compendium of the full computational toolset enabling this revolution.

Choose wisely.

G. B. Olson
*Massachusetts Institute of Technology*
*Cambridge, MA*

# Preface

Recently, with the rapid development of computational techniques at different scales and various materials databases, materials design has become a research hotspot in different disciplines, including materials science, metallurgy, physics, chemistry, geology, biotechnology, and more. The most important trend is the integration of multiscale computational techniques for materials design, such as the CALPHAD technique, first-principles calculations in atomistic scale, mesoscale phase-field simulation, and finite element analysis in macroscale. However, most of the relevant books published so far do not reflect this important trend. Moreover, contributors of previously published books have focused on only one or two computational tools and, therefore, could not cover the tools in different scales. Thus, there is a need to publish a new book on this topic.

About half of this book presents for the first time a wide spectrum of various computational methods used in the design of engineering materials. An important feature of this part of the book is the methodology to establish thermodynamic and thermophysical databases for multicomponent and multiphase systems. Such databases are critical for an effective design of various engineering materials, which are usually multicomponent and multiphase alloys. This theoretical part of the book should be very useful for researchers, engineers, and students from materials sciences, metallurgy, physics, mathematics, and chemistry.

The other half of this book features a step-by-step demonstration of the design of engineering materials. This demonstration covers a very wide range of materials, including steels, light alloys, superalloys, cemented carbides, hard coatings, and energy materials.

The major motivation to write this book originated from a long-term cooperation between Professor R. Schmid-Fetzer and Professor Yong Du, which dates back to November 1994, when Dr. Du joined Professor Schmid-Fetzer's group as Alexander von Humboldt Research Fellow. Subsequently, they have established a close collaboration through several channels, such as mutual visits, attending conferences simultaneously, supervising PhD students together, and publishing papers jointly. Through many discussions and their individual experiences, both Professor Schmid-Fetzer and Professor Du have wondered why there is no book on the market that introduces the design of engineering materials via a step-by-step methodology. This book tries to fill that gap.

**Figure 0.1** Five authors (Jianchuan Wang, Jincheng Wang, Rainer Schmid-Fetzer, Yong Du, and Shuhong Liu, from left to right) discussing the overall structure of the book in Changsha on September 21, 2018. A black and white version of this figure will appear in some formats. For the colour version, please refer to the plate section.

We believe that researchers, engineers, and graduate and undergraduate students in materials science and engineering, including ceramics, metallurgy, and chemistry, will find the book to be of great value. Moreover, we feel that even other fields, including computational biomaterials science, where modeling approaches have been used extensively for the research and development of various engineering materials, might substantially benefit from the methods and design methodology presented in this book.

Computational techniques and software have developed rapidly in recent times, and new concepts such as machine learning and artificial intelligence have emerged in the past few years. Consequently, it has been a tremendous challenge to keep the content of the book up to date. In addition, we do not expect the book to be error-free. Your comments and feedback on the book are highly appreciated and will enable us to address any shortcomings through the book's website or during the book's next revision.

# Acknowledgments

The authors are grateful to many colleagues for inspiration, reading some chapters of the book, contributing some text, giving valuable comments, and producing a number of diagrams. Thanks to Professors Weibin Zhang, Keke Chang, and Wei Xiong; Doctors Yuling Liu, Lianchang Qiu, Yafei Pan, Shiyi Wen, Peisheng Wang, Yingbiao Peng, Yuxiang Xu, Haixia Tian, Kaiming Cheng, Cong Zhang, Peng Zhou, Dongdong Zhao, Dandan Liu, Ying Tang, Fan Zhang, Dandan Huang, Huixin Liu, Fangyu Guo, Shaoqing Wang, Chong Chen, Jinghua Xin, Mingjun Yang and Yinping Zeng; as well as our PhD students Xiaoyu Zheng, Peng Deng, Qianhui Min, Bo Jin, Han Li, Qiang Lu, Baixue Bian, Fengyang Gao, Yiqi Guan, Fangfang Zeng, Huaqing Zhang, Qi Huang, Liying Wu, Ya Li, Tong Yang, Shiwei Zhang, and many others.

# 1 Introduction

## Contents

For millennia, the advance of human civilization has been closely linked to materials available from the Stone Age through the Bronze and Iron Ages and into the current Information Age. The Information Age is associated with a revolution of technology resulting from significant advances in Si-based semiconductors and other materials. In particular, since 2018 the world has been significantly influenced and changed by the fifth-generation technology standard for cellular networks (5G). Such a network enables users to connect virtually with almost everyone and everything, delivering data among users and machines with much higher speed, lower latency, more reliability, and better availability compared to 4G networks. 5G networks have been utilizing a wide variety of materials, including metals, ceramics, plastics, composites, and materials with low dielectric, high thermal conductivity, and high electromagnetism. In consequence, the continuing technological advancement in current society is strongly dependent on advanced engineering materials that satisfy the ambitious requirements of new products.

The earliest documented materials were the Cu–Sn alloys in ancient China during the Bronze Age (Chang, 1958) The compositions of the "six alloys" (Cu–Sn alloys with the compositions of 17, 20, 25, 33, 40, and 50 wt.% Sn) were determined purely via trial-and-error experiments to guide the casting of various civil and military tools. It is unbelievable that such an empirical approach has persisted for more than three millennia. With the development of computers as well as software and various databases (such as thermodynamic and diffusion databases), computational design of engineering alloys has demonstrated its significant role in efficient developments of new alloys since Kaufman and Bernstein published the book entitled *Computer Calculations of Phase Diagrams* in 1970 (Kaufman and Bernstein, 1970). The interested reader could refer to the following selection of recent (mainly edited) books for more details in the field of computational design (Ashby et al., 2019; Bozzolo et al., 2007; Da Silva, 2019; Horstemeyer, 2012; LeSar, 2013; Raabe et al., 2004; Saito, 1999; Shin and Saal, 2018).

## 1.1    Definition of a Few Terms Used in Computational Design of Materials

The goal of this book is to introduce the basic methods used in computational design of engineering alloys and to demonstrate several step-by-step case studies for computational design of these alloys. To facilitate reading, a few most frequently used terms are defined in a more precise way before we proceed to discuss computational design of engineering alloys.

A **model** is an idealization of an actual phenomenon, i.e., an approximate description of a phenomenon based on some empirical and/or physically sound reasoning. A model often begins with a set of concepts, and then it is usually transcribed into mathematical equations from which one can calculate some quantities with a desire to describe some phenomena. For example, a thermodynamic model is usually established according to the crystal structure of one phase in order to calculate its Gibbs energy. Thermodynamic models expressed in different mathematical forms contain adjustable parameters that can be optimized to reproduce many kinds of experimental phase diagrams and thermodynamic properties (activity, heat capacity, enthalpy of mixing, and so on) as well as theoretical data such as first-principles computed enthalpy of formation. The main focus of a model is to create an idealization of an actual phenomenon within an accepted accuracy instead of a strictly true fundamental description of the phenomenon. One may argue that any model is only a picture of reality – not reality itself.

**Simulation** is a numerical calculation for a modeled system with respect to external and/or internal fields as well as applied constraints. It requires algorithms based on the models and numerical solution strategies that are the backbone of simulation software. Consequently, the simulations are performed by subjecting models to inputs and constraints for the sake of describing an actual phenomenon, such as the solidification of an alloy. The accuracy of a simulation for an actual phenomenon depends on several factors, such as the adequacy of the model, the accuracy in solving sets of equations numerically, and the reliability of input parameters in the equations.

A **database** is an organized set of data that is stored in a computer and can be accessed, managed, updated, and used in many ways. Various types of databases are reported for the systems of interest, resulting in different definitions for the databases. According to a recent analysis for engineering alloys (Li et al., 2018), three kinds of databases are defined for these alloys: the original technological database, the evaluated technological database, and the scientific database. The first type of database is usually a compilation of one or several typical quantities, such as hardness and toughness during one or several processes, such as homogenization and age hardening. The second type of database is the critically evaluated technological database, eliminating inaccurate values. The last type of database (i.e., the scientific database) is the most important database for engineering alloys. This scientific database is based on physically sound models, and the parameters in the database are obtained by fitting accurate experimental and/or theoretical data for targeted systems. Thermodynamic and thermophysical databases (for example, diffusion coefficient, interfacial energy, and thermal conductivity) are the typical scientific databases.

Armed with powerful materials design software, these databases can be utilized to design alloy composition, optimize the heat treatment schedule, simulate microstructural evolution, and predict mechanical and other properties.

The term **materials design** may have different meanings for different readers. Olson presented a very deep and wide definition for this term (Olson, 1997; 2000). In view of the four cornerstones (i.e., processing, structure, property, and performance) in materials science and engineering, **our definition for materials design** is to establish relationships among these four cornerstones through computationally based approaches implemented with experimental and/or empirical approaches for the sake of yielding materials with the desired sets of properties and performances to meet the needs of users. Materials design will be more powerful when modeling and simulation tools are integrated with experiments, as highly stressed in integrated computational materials engineering (ICME) (ICME, 2008) and the Materials Genome Initiative (MGI) (MGI, 2011), which will be briefly described in the next subsection.

## 1.2    The Past and Present Development of Computational Design of Engineering Materials

The computational design of engineering materials dates back to 1970, when the calculation of phase diagrams (CALPHAD) approach was developed by Kaufman and Bernstein (Kaufman and Bernstein, 1970), who advanced the pioneering work on phase diagram calculations by Van Laar (1908) and Meijering (1950). The major justification for such a statement is that almost all engineering materials are multi-component and multiphase systems, and the CALPHAD approach is the only one that can deal with such complex systems.

However, it should be mentioned that before the birth of the CALPHAD approach, there were a few important milestones for alloy design, such as the Hume-Rothery rule (Hume-Rothery, 1967). This rule utilizes information about atomic size, valence, electronegativity, and crystal structure to predict phase formation, being applicable to both solid substitutional and interstitial solutions. Due to its empirical feature, this rule can only be used as component selection criteria instead of alloy composition optimization, which is usually the first step for computational design of engineering alloys.

Another approach for alloy design is the phase computation (PHACOMP) method, which was developed by Boesch and Slaney (1964). This method utilizes the average number of electron vacancies in the metal d band above the Fermi level to predict phase stability of the harmful topologically close-packed (TCP) phases. One modification for the original PHACOMP method is the so-called d orbital method (Matsugi et al., 1993), in which the d orbital energy level of alloying transition metal and the bond order are used as indicators for the occurrence of TCP phases. Although the electron vacancies in the d orbital method can be obtained through quantum mechanical electronic structure calculations, the average number of electron vacancies and the bond order are only approximate, affecting the validity of the method. One more obvious shortcoming for the PHACOMP method and its other revised form (Morinaga et al., 2003) is that the criterion for judging phase stability is independent of temperature. Due to the preceding drawbacks, the PHACOMP method and its modified form cannot be used for design of multicomponent and multiphase engineering alloys in which temperature dependence of phase stability should be considered.

This is why we consider the aforementioned book entitled *Computer Calculation of Phase Diagrams* by Kaufman and Bernstein (1970) as the outstanding milestone for computational design of engineering alloys. This book leads to the birth and development of the CALPHAD approach. According to Lukas et al. (2007), a comprehensive definition for CALPHAD is that the "CALPHAD method" means the simultaneous use of all available experimental and theoretical data to assess the parameters in Gibbs energy models selected for individual phases. Armed with powerful software tools, such as Pandat (www.computherm.com) and Thermo-Calc (www.thermocalc.com), which are based on the principles and concepts of thermodynamics, this method can be used to calculate phase diagrams and various thermodynamic properties in multi-component systems within all the composition and temperature ranges. The successful use of CALPHAD for materials design relies on reliable thermodynamic databases. Reliability means that the properly selected thermodynamic models and optimized parameters can reproduce both thermodynamic and phase stability experimental data as well as first-principles or other theoretically calculated data. Recently, the CALPHAD method has been broadened to include a range of fundamental phase-level thermophysical properties. In conjunction with other computational methods, such as the phase-field method and the finite element method, the CALPHAD method has shown its importance for process and phase transformation simulations. Due to these unique features, thermodynamics, which is the theoretical basis for CALPHAD, was regarded as the fundamental building block for simulation-supported materials design (McDowell et al., 2010).

For the past two decades, the computational design of engineering materials has focused on the following three aspects: multiscale/multilevel modeling methodologies for more quantitative materials design, more user-friendly simulation software, and high-quality scientific databases (i.e., thermodynamic and thermophysical databases). In order to demonstrate the major aspects for materials design in detail, a flow chart for through-process simulation and experimentation with aluminum alloys during the whole heat treatment schedule is presented in Figure 1.1 (Du et al., 2017). The first-principles method, phase-field method, and finite element method are typical nano-, meso-, and macro-level simulation methods, respectively. The Kampmann–Wagner numerical (KWN) model (Kampmann and Wagner, 1984), which is the basis of the one-dimensional precipitation simulation package, is included in this figure due to its high computational efficiency for design of engineering alloys. The CALPHAD method can cover the micro to meso levels of phase transformation in engineering alloys. As shown in Figure 1.1, these multiscale numerical simulations from nano ($10^{-10}$–$10^{-8}$ m), micro ($10^{-8}$–$10^{-4}$ m), meso ($10^{-4}$–$10^{-2}$ m), to macro ($10^{-2}$–10 m) were utilized to describe multiscale structures and their response to mechanical properties

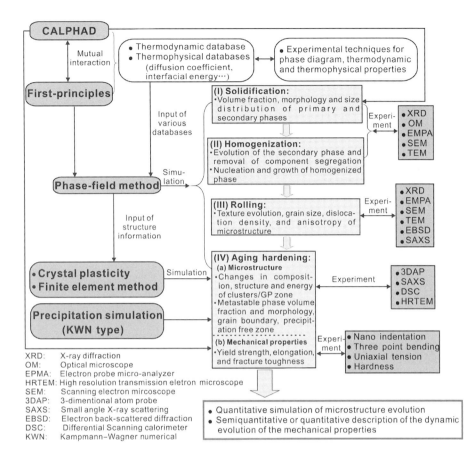

**Figure 1.1** Flow chart for through-process simulation and experiment of Al alloys during solidification, homogenization, rolling, and age strengthening.

during the whole research and development (R&D) process of aluminum alloys. Similarly, time scales can range from femtoseconds of atomic vibrations to decades for the use of products. As a strong tie to multiscale simulations, various materials characterizations are needed to serve as validation for the accuracy of simulations or to perform decisive experiments based on the simulation results. Transmission electron microscope (TEM) / three-dimensional atom probe (3DAP), scanning electron microscope (SEM) / electron probe microanalyzer (EPMA) / electron backscatter diffraction (EBSD), and optical microscope (OM) are typically the nano-, micro-, and meso-level structure characterization methods. Consequently, computational design of engineering materials is often considered the most powerful approach when it is integrated with experiments. In Figure 1.1, thermodynamic and thermophysical databases are also indicated. These scientific databases are key inputs for various simulations based on the CALPHAD, phase-field, and finite element methods, just to mention a few.

Recently, the concurrent design of materials and product was described (McDowell et al., 2010). To reflect this idea, a simple but comprehensive diagram is presented in

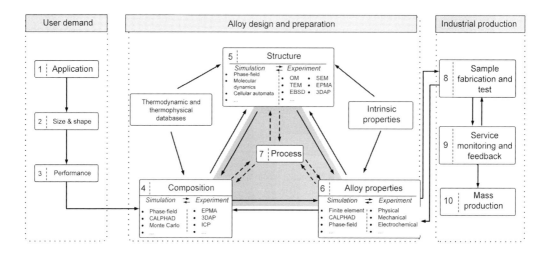

**Figure 1.2** Three stages (user demand, alloy design and preparation, and industrial production) for the development of engineering materials.

Figure 1.2, where three stages (user demand, alloy design and preparation, and industrial production) for development of engineering alloys are shown and linked to each other.

Following the strategies similar to those described in Figures 1.1 and 1.2, many new engineering alloys have been developed recently. By establishing the relationships among processing, structure, properties, and performance by means of the integrated computational materials design (ICMD) technique enhanced with experiment, Olson and his colleagues developed the ultrahigh-strength steel Ferrium S53 in only 8.5 years (McDowell et al., 2010). The development of this steel would need more than 15 years by the traditional and costly experimental approach. Gottstein (2007) introduced the scientific concept Integral Materials Modeling (IMM) to guide the development for a variety of materials, including Al alloys and steels. According to IMM, the properties of terminal materials can be predicted from the information about materials chemistry and various processing conditions. The unique feature for IMM is that the microstructural evolution through the whole processing chain is traced to manipulate the final microstructure needed at the end of the processing chain for the sake of predicting the corresponding properties. Most recently, using a through-process simulation (the CALPHAD and finite element methods) and calculation-guided key experiments, Qiu et al. (2019) developed a high-performance chemical vapor deposition (CVD) hard coating in two years. In their work, the CALPHAD calculations predict the phases and their compositions accurately by means of the computed CVD phase diagrams, where the phase regions are shown for the given temperature and gas mixtures. Finite element simulations predict temperature field, gas distribution field, and the deposition rate of hard coating. The costly experimental work was significantly reduced with these simulations.

In 2008, the promising engineering approach ICME was announced by the National Research Council in the USA. Its goal is to optimize materials, processes, and component design, prior to the fabrication of components, by linking models at multiple length and time scales into a holistic system. The successful application of ICME (ICME, 2013; Schmid-Fetzer, 2015) in industry supports the MGI announced in June 2011 by US President Barack Obama. The goal of MGI is to discover, develop, and manufacture advanced materials at least twice as fast as was possible at the time through the integration of three platforms: high-throughput calculations, high-throughput experiments, and databases. Kaufman and Ågren (2014) mentioned that "a materials genome is a set of information encoded in the language of thermodynamics that serves as a blueprint for a material's structure."

Recently, materials informatics (Rajan, 2008) has shown its role to speed up the discovery of new materials. By using data mining and visual analysis of databases, materials informatics could extract and/or establish quantitative relationships among the four cornerstones of materials science and engineering much faster than the current materials design methods. Future prospects for ICME, MGI, and materials informatics will be addressed in Chapter 13 of the book.

## 1.3 The Structure of the Book

This book is intended to be an introductory as well as a reference book for computational design of various engineering materials. The text of the book is divided into two parts, and each chapter in these two parts has its special focus. The first part, comprising Chapters 1–6, presents broad but not deep introductions to basic computational methods used in materials design. For more detail about these methods, the interested reader could refer to the references recommended in each chapter. Our intent is to provide a concise but sufficient background in the theory of these methods so that the readers of this book can understand case studies of the representative engineering materials. One very short case study is integrated in each chapter of Chapters 2–6, while detailed case studies for selected engineering materials are described in the second part (Chapters 7–12) via a step-by-step strategy.

The balance of this book is well realized by the interplay between theory described in part one and practice demonstrated in part two. It is our hope that this book will (1) help to attract and prepare the next generation of materials design modelers, whether modeling is their principal focus or not; and (2) encourage readers, in particular undergraduates, graduates, and engineers, to apply these methods for materials design.

This first chapter briefly introduces the past and present development of computational design of engineering materials. The scope and structure of the book are also described.

Chapter 2 gives a short introduction to density functional theory (DFT) and molecular dynamics (MD). These methods are typically atomistic simulation methods. The chapter demonstrates how to calculate some basic materials properties through the first-principles method. One case example of material design through atomistic methods is included.

In Chapter 3, firstly a brief introduction to the phase-field method and cellular automaton is presented. These methods are typical mesoscale methods for materials design. Secondly, the integration of phase-field method and/or cellular automaton with CALPHAD, MD, crystal plasticity computations, and machine learning is stressed. One case study for material design mainly based on phase-field simulation and machine learning is addressed.

Chapter 4 presents fundamental concepts in the crystal plasticity and finite element method. Crystal plasticity can be integrated with finite element method to describe the mechanical response of crystalline materials, from single crystals to engineering components. The finite element method is widely used as a macroscale method in engineering. One case study of plastic deformation-induced surface roughening in Al polycrystals will be described by means of the crystal plasticity finite element method.

Chapter 5 presents a fundamental introduction to computational thermodynamics and the CALPHAD method, followed by a strategy to establish a consistent thermodynamic database. Such a database is highly needed for efficient material design of engineering material. A few case studies for Al and Mg alloys design are described using solely thermodynamic databases or extended CALPHAD-type databases. The aspects in these case studies include calculations of both stable and metastable phase diagrams, property diagram calculation, and Scheil solidification simulations, as well as extended simulations of solidification and heat treatment in Al and Mg alloys.

In Chapter 6, a few important thermophysical properties (diffusion coefficient, interfacial energy, viscosity, volume, and thermal conductivity) are very briefly described. The procedure to establish thermophysical databases is also described from a material design point of view. One case study for material design mainly using thermophysical properties is demonstrated.

Part two begins with Chapter 7, demonstrating a step-by-step material design for two representative steels, the S53 ultrahigh-strength and high-corrosion resistance steel as well as the AISI H13 hot-work tool steel. The materials design for these steels based on a hybrid approach of thermodynamic calculations, precipitation simulations, first-principles calculations, and the finite element method will cover the selection of alloy composition, optimal combination of the heat treatment schedule, control of the microstructure, and the correlation of structure properties.

Chapter 8 describes a few case studies for material design of Al and Mg alloys. Two case studies for Al alloy design were described through computations using the thermodynamic database and extended CALPHAD-type databases (atomic mobility and kinetic databases). In the first case study on cast alloy A356, the solidification simulation involving microsegregation modeling is presented. The second case study on wrought alloy 7xxx presents heat treatment simulations based on precipitation kinetics. In the case of Mg alloys, the first two case studies present the simulations of solidification path and T6 heat treatment of AZ series Mg alloys using solely thermodynamic databases. The third case study describes the computational design and development of Mg–Al–Sn and AT72 alloys by means of both thermodynamic database and extended CALPHAD-type databases. The last case study is on biomedical Mg alloy implants, presenting the state-of-the-art bioresorbable Mg alloy stent to cure coronary artery disease, the development of which utilized the CALPHAD method.

In Chapter 9, two case studies for the design of single crystal Ni-based superalloy and Ni–Fe-based superalloy for A-USC are demonstrated. The computational methods used in the case studies are thermodynamic calculations, the property prediction model, the multistart optimization algorithm, and machine learning.

In Chapter 10, three types of cemented carbides (gradient cemented carbide, ultrafine cemented carbide, and WC–Co–NiAl cemented carbide) are designed according to thermodynamic calculations, diffusion modeling, phase-field simulations, and the finite element method. The calculations cover the selection of alloy composition, optimal combination of the heat treatment schedule (such as temperature, time, and atmosphere), control of the microstructure, and the correlation of structure properties. In comparison with the time-consuming and costly experimental approach, these simulation-driven materials designs led to the development of these industrial products in three years.

Chapter 11 demonstrates a step-by-step material design for TiAlN-based CVD and PVD hard coatings. The utilized methods include thermodynamic calculations, diffusion modeling, first-principles calculations, phase-field simulations, computational fluid dynamics, and the physically sound structure-property model. The designed materials have found their industrial applications.

In Chapter 12, two case studies for energy materials (hydrogen storage material and lithium battery) are demonstrated. The dominant methods employed in the two case studies are first-principles calculations and thermodynamic calculations.

In Chapter 13, the main contents of the book are summarized, followed by highlighting computational designs of four other engineering materials not covered in the preceding chapters and discussing future directions and key challenges for materials design.

Appendix A provides a summary of ancillary materials available online at Cambridge University Press. These are pertinent data files and step-by-step instructions for hands-on experience of the reader with the simulation tools and examples discussed in the book.

Appendix B compiles the notation in three tables of symbols for the entire book. For the reader's benefit, the equation numbers where these symbols are defined are given as well as the SI-units.

## References

Ashby, M. F., Shercliff, H., and Cebon, D. (2019) *Materials: Engineering, Science, Processing and Design*, fourth edition. Oxford: Butterworth-Heinemann (Elsevier).

Boesch, W. J., and Slaney, J. S. (1964) Preventing sigma phase embrittlement in nickel base superalloys. *Metal Progress*, 86, 109–111.

Bozzolo, G., Noebe, R. D., and Abel, P. B. (2007) *Applied Computational Materials Modeling: Theory, Simulation and Experiment*. Berlin and Heidelberg: Springer Science & Business Media, LLC.

Chang, T. K. (1958) A critical interpretation of "liu-ch'i'". *Journal of Tsinghua University*, 4(2), 159–166.

Da Silva, L. F. M. (ed), (2019) *Materials Design and Applications II*, volume 98. New York: Springer.

Du, Y., Li, K., Zhao, P. Z, et al. (2017) Integrated computational materials engineering (ICME) for developing aluminum alloys. *Journal of Aeronautical Materials*, 37(1), 1–17.

Gottstein, G. (2007) *Integral Materials Modeling*. Weinheim: Wiley-VCH Verlag GmbH & Co. KGaA.

Horstemeyer, M. F. (2012) *Integrated Computational Materials Engineering (ICME) for Metals: Using Multiscale Modeling to Invigorate Engineering Design with Science*. John Wiley & Sons.

Hume-Rothery, W. (1967) *Phase Stability in Metals and Alloys*. Edited by P. S. Rudman, J. Stringer, and R. I. Jaffee. New York: McGraw-Hill, 3.

ICME (2008) *Integrated Computational Materials Engineering: A Transformational Discipline for Improved Competitiveness and National Security*. Committee on Integrated Computational Materials Engineering and National Materials Advisory Board. Washington: National Academies Press.

ICME (2013) *Integrated Computational Materials Engineering (ICME): Implementing ICME in the Aerospace, Automotive, and Maritime Industries*. Pittsburgh: The Minerals, Metals and Materials Society.

Kampmann, R., and Wagner, R. (1984) Kinetics of precipitation in metastable binary alloys – theory and application to Cu-1.9 at % Ti and Ni-14 at % Al. In Haasen, P., Gerold, V., Wagner, R., and Ashby, M. F. (eds), *Decomposition of Alloys: The Early Stages: Proceedings of the 2nd Acta-Scripta Metallurgica Conference. Sonnenberg, Germany, 19-23/09/1983*. Oxford: Pergamon, 91–103.

Kaufman, L., and Ågren, J. (2014) CALPHAD, first and second generation – birth of the materials genome. *Scripta Materialia*, 45, 3–6.

Kaufman, L., and Bernstein, H. (1970) *Computer Calculation of Phase Diagrams*. New York: Academic Press.

LeSar, R. (2013) *Introduction to Computational Materials Science: Fundamentals to Applications*. Cambridge: Cambridge University Press.

Li, B., Du, Y., Qiu, L. C., et al. (2018) Shallow talk about integrated computational materials engineering and Materials Genome Initiative: Ideas and Practice. *Materials China*, 37(7), 506–525.

Lukas, H. L., Fries, S. G., and Sundman, B. (2007) *Computational Thermodynamics: The Calphad Method*. Cambridge: Cambridge University Press.

Matsugi, K., Murata, Y., Morinaga, M., and Yukawa, N. (1993) An electronic approach to alloy design and its application to Ni-based single-crystal superalloys. *Materials Science and Engineering A*, 172(1), 101–110.

McDowell, D. L., Panchal, J. H., Choi, H. J., Seepersad, C. C., Allen, J. K., and Mistree, F. (2010) *Integrated Design of Multiscale, Multifunctional Materials and Products*. Amsterdam: Elsevier.

Meijering, J. L. (1950) Segregation in regular ternary solutions: Part I. *Philips Research Reports*, 5, 333–356.

MGI (2011) *Materials Genome Initiative for Global Competitiveness*. Washington: National Science and Technology Council, Executive Office of the President, USA.

Morinaga, M., Murata, Y., and Yukawa, H. (2003) An electronic approach to materials design. *Journal of Materials Science and Technology*, 19, 73–76.

Olson, G. B. (1997) Computational design of hierarchically structured materials. *Science*, 277, 1237–1242.

Olson, G. B. (2000) Pathways of discovery: designing a new materials world. *Science*, 288, 993–998.

Qiu, L. C., Du, Y., Wang, S. Q., et al. (2019) Through-process modeling and experimental verification of titanium carbonitride coating prepared by moderate temperature chemical vapor deposition. *Surface & Coating Technology*, 359, 278–288.

Raabe, D., Roters, F., Barlat, F., and Chen, L.-Q. (2004) *Continuum Scale Simulation of Engineering Materials: Fundamentals – Microstructures – Process Applications*. Indianapolis: John Wiley & Sons.

Rajan, K. (2008) Learning from systems biology: an "omics" approach to materials design. *JOM*, 60, 53–55.

Saito, T. (1999) *Computational Materials Design*. Springer Series in Materials Science, 34. Berlin and Heidelberg: Springer-Verlag Berlin Heidelberg GmbH.

Schmid-Fetzer, R. (2015) Progress in thermodynamic database development for ICME of Mg alloys. In Manuel, M. V., Singh, A., Alderman, M., and Neelameggham, N. R. (eds), *Magnesium Technology 2015*, The Minerals, Metals and Materials Society (TMS), Pittsburgh, 283–287.

Shin, D., and Saal, J. (eds) (2018) *Computational Materials System Design*. Cham: Springer.

van Laar, J. J. (1908) Melting or solidification curves in binary system. *Zeitschrift für Physikalische Chemie*, 63, 216–253.

# 2 Fundamentals of Atomistic Simulation Methods

## Contents

Atomistic simulations have grown into powerful and efficient tools for studying various materials, including metallic materials, semiconductors, amorphous alloys, and biomolecules, due to the rapid development of computing processors and parallel processing techniques, as well as the establishment of basic physical and computational theories. The widely used atomistic simulation methods are first-principles calculation, mainly density functional theory (DFT) calculations, and molecular dynamics (MD), among others. The former deals with interacting atoms and electrons based on the laws of quantum mechanics, while the latter considers classical Newtonian mechanics.

First-principles calculation, also known as *ab initio* calculation, means that the calculation does not need any empirical or experimentally determined parameters, and the ground state properties of the material are obtained just from the basic physical constants and atomic configurations. It starts from the idea that all materials consist of atoms, which in turn are made up of a positively charged nucleus and a number of negatively charged electrons. Basically, the physical quantities of materials are determined by the interactions of their constituent nuclei and electrons. If one wants to model material properties accurately, all these interactions should be treated appropriately in calculations. What makes first-principles calculations troublesome is not the difficulty of the physics but rather the complexity in terms of numerical calculations. Thus, first-principles calculations can only deal with dozens to hundreds of atoms.

For MD simulation, the microscopic trajectory of each individual particle in the system is determined by the integration of Newton's equations of motion. If all particles' trajectories are known, in principle everything can be calculated. A key element for MD simulation is the potential function used to describe the interactions of particles in the system. Thus, the accuracy or credibility of MD depends to a large extent on the quality of the potential functions used. Because the potential functions in MD are empirical and relatively simple numerical formulations are employed, MD can be used for large systems, which typically contain thousands to millions of atoms, and runs over a long simulation time for dynamic behaviors. Hence, MD has advantages in both space and time domains for materials simulation.

In this chapter, we briefly present the basis of the first-principles method and MD. We also demonstrate how to use the first-principles method to calculate some basic materials properties widely used in computational materials science owing to their remarkable accuracies in predicting the physical and chemical properties of materials. Lastly, we take the design of Mg–Li alloys for ultra-lightweight application as an example to show the critical role of atomistic simulation methods in materials design.

## 2.1 Density Functional Theory

The first-principles calculation is now a common and significant component of research tools in physics, chemistry, and materials science. The first-principles method, which is based on the principles of quantum mechanics, can be used to investigate various properties of materials, such as structure, thermodynamics, mechanics, spectroscopy, and magnetism, by solving the Schrödinger equation, the Hartree–Fock equation, or the Kohn–Sham equation using different approximations. The current first-principles calculations in solids are always based on the DFT, which has been invented and pioneered by Kohn, Hohenberg, and Sham (Hohenberg and Kohn, 1964; Kohn and Sham, 1965). In this section, we briefly introduce the fundaments of DFT.

### 2.1.1 The Many-Body Schrödinger Equation

In a microscopic picture, any real materials, such as an isolated $H_2O$ molecule or an Al–Cu alloy, are composed of microscopic particles of electrons and nuclei. Their material properties are in essence determined by the interactions between these microscopic particles, and a fundamental description of those interactions requires quantum mechanics. For a quantum system, its status is described by the Schrödinger equation:

$$\hat{H}\Psi = E\Psi. \tag{2.1}$$

In this equation, $\hat{H}$ is the Hamiltonian operator, which is a quantum mechanical operator associated with the energy of the system. The Hamiltonian depends on the physical system. $\Psi$ is the wave function that describes the state of the system, and $E$ the energy eigenvalue. Both $\Psi$ and $E$ comprise the solutions of (2.1). In principle, the wave function contains all the information for a given system.

Real materials are basically many-body systems of nuclei and electrons. The energy of a many-body system contains the kinetic energies of nuclei and electrons and the potential energies due to interactions within the electrons and nuclei. Thus, the Hamiltonian is expressed as follows:

$$\hat{H} = -\sum_I \frac{\hbar^2}{2m_I}\nabla_I^2 - \sum_i \frac{\hbar^2}{2m}\nabla_i^2 + \frac{1}{2}\sum_{I\neq J}\frac{Z_I Z_J e^2}{|\mathbf{R}_I - \mathbf{R}_J|} + \frac{1}{2}\sum_{i\neq j}\frac{e^2}{|\mathbf{r}_i - \mathbf{r}_j|} - \sum_{i,I}\frac{Z_I e^2}{|\mathbf{r}_i - \mathbf{R}_I|}, \tag{2.2}$$

where $e$ is the elementary charge, $Z_I e$ the charge of nuclei $I$, and $m$ and $m_I$ are the masses of electron and nuclei $I$, respectively. Vectors $\mathbf{R}$ and $\mathbf{r}$ are, respectively, the positions of nuclei and electrons. The first and second terms in the right-hand side of (2.2) are the kinetic energies of nuclei and electrons, respectively. The last three terms are potential energies due to the Coulomb interactions between nuclei and electrons.

Clearly, an exact solution of (2.2) is not possible, and thus approximations must be made. Bear in mind that the mass of a proton or neutron is about 1,830 times that of an electron, and the nucleus is much heavier than an electron. Thus, electrons respond much more rapidly to changes in their surroundings than nuclei can. Therefore, we can think that the nuclei are fixed when the electrons move in the potential field of the nuclei. As a result, the motions of the nuclei and the electrons can be separated. This separation of the motions of the nuclei and the electrons into separate mathematical problems is the adiabatic or Born–Oppenheimer approximation. Within the Born–Oppenheimer approximation, the nuclei only provide an external potential for electrons, and one can only deal with electrons. Consequently, the many-body system becomes a many-electron system. If the nuclei are at rest and the potential energy term between the nuclei is not considered, then the Schrödinger equation for the system reads

$$\left(-\sum_i \frac{\hbar^2}{2m}\nabla_i^2 + \frac{1}{2}\sum_{i\neq j}\frac{e^2}{|\mathbf{r}_i - \mathbf{r}_j|} - \sum_{i,I}\frac{Z_I e^2}{|\mathbf{r}_i - \mathbf{R}_I|}\right)\Psi = E\Psi. \tag{2.3}$$

In this stationary case, $\Psi$ is the wave function of electrons, which is a function of each of the spatial coordinates and spin states of each electron, and $E$ is the energy eigenvalue of electrons. For simplicity we will neglect the spins of electrons below.

## 2.1.2    The Hartree Approximation

Even within the Born–Oppenheimer approximation with static nuclei, solving (2.3) remains intractable, since too many degrees of freedom are involved, and the Hamiltonian contains interactions within electrons that make the solving of the Schrödinger equation very costly. The wave function of (2.3) could be solved by taking some approximations. The simplest approach to solve (2.3) is to assume that the many-electron wave function can be expanded as a product of individual electron wave functions (i.e., orbitals): $\Psi = \phi_1(\mathbf{r}_1)\phi_2(\mathbf{r}_2)\ldots\phi_N(\mathbf{r}_N)$. Here $N$ is the number of electrons. This expression for the wave function is known as Hartree approximation and would be appropriate when there is no interaction between electrons. The Hartree approximation neglects the fermionic nature of electrons, that is, the wave function of a many-electron system changes its sign when any two electrons interchange their locations. The antisymmetric nature of the wave function of a many-electron system can be retained by using the Slater determinant.

Applying the variation principle to (2.3), the Hartree equation is obtained:

$$\left(-\frac{\hbar^2}{2m}\nabla_i^2 - \sum_I \frac{Z_I e^2}{|\mathbf{r}_i - \mathbf{R}_I|} + \sum_{j\neq i}\int \phi_j^* \frac{e^2}{|\mathbf{r}_i - \mathbf{r}_j|}\phi_j d\mathbf{r}_j\right)\phi_i = \varepsilon_i \phi_i, \qquad (2.4)$$

where $\varepsilon_i$ is a Lagrange multiplier, which represents the energy contribution of electron $i$. The term $\sum_{i\neq j}\int \phi_j^* \frac{e^2}{|\mathbf{r}_i - \mathbf{r}_j|}\phi_j d\mathbf{r}_j$ represents the potential due to other electrons in which electron $i$ moves. For every electron in the system, there is a single-particle equation like (2.4). Even the wave function is simplified in the Hartree approximation; however, solving for $\phi_i$ is still an extremely difficult task because of the nature of electrons and the number of individual electron wave functions. Firstly, the individual electron wave function $\phi_i(\mathbf{r}_i)$ cannot be found without simultaneously considering the individual electron wave functions associated with all the other electrons $\phi_j(\mathbf{r}_j), j \neq i$. In fact, each orbit $\phi_i(\mathbf{r}_i)$ can be determined if the other orbits $\phi_j(\mathbf{r}_j)$ are known; namely, the equation for one $\phi_i$ is related on all the other $\phi_j$'s. In principle, this can be solved by a self-consistent approach. We first assume a set of initial $\phi_i$'s and use these initial wave functions to construct a single-particle Hamiltonian, which allows one to solve the equations for each new $\phi_i$. Subsequently, we compare the resulting $\phi_i$'s with the original $\phi_i$'s and update them. This cycle is continued until both the output and input $\phi_i$'s are the same up to a tolerance. Another important issue to solve is the expensive calculation time. If we calculate an isolated $H_2O$ molecule, the full wave function is a 54-dimensional function (each electron has three spatial variables, even neglecting electron spin degrees of freedom). If we deal with a transition metal cluster of ~100 atoms, the full wave function requires more than 10,000 dimensions.

## 2.1.3    Kohn–Sham Equation

It is necessary to introduce another approximation and also a simpler picture to deal with a many-electron system. According to the core ideas and shortcomings of the

Hartree method, we hope to find an effective method, which aims at solving a single-electron problem without interaction. At the same time, this method should be more strict and more accurate, which can overcome the deficiency of the Hartree method and reduce computational complexity as much as possible. Hohenberg, Kohn, and Sham have established the DFT and shown how to solve the equation of a many-electron system (2.3) effectively (Hohenberg and Kohn, 1964; Kohn and Sham, 1965). A many-electron system is described as a collection of classical ions and essentially single quantum mechanical particles that reproduce the behaviors of electrons in the DFT.

The basic concept of DFT is that instead of handling the many-body Schrödinger equation of (2.3), which involves the many-body wave function $\Psi(\mathbf{r}_1, \mathbf{r}_2, \ldots, \mathbf{r}_N)$, one just deals with a mathematical problem involving the total electron density $n(\mathbf{r})$. This is a great simplification compared to the Hartree approximation, since the many-body wave function never needs to be specified explicitly. Thus, instead of starting with a drastic approximation for the behaviors of the system, one can develop appropriate single-particle equations in an exact manner and then, if needed, introduce reasonable approximations.

The fundamental basis of DFT is that the ground state energy of an interacting electron system is a functional of the electron density $n(\mathbf{r})$. Within the frame of DFT, the total energy $E[n(\mathbf{r})]$ of electrons in a many-electron system is the sum of three kinds of energy: (1) the kinetic energy of electrons $T[n(\mathbf{r})]$, (2) the electron–electron interaction energy $V_{ee}[n(\mathbf{r})]$, and (3) the interaction energy $E_{ext}[n(\mathbf{r})]$ between electrons and the external potential:

$$E[n(\mathbf{r})] = T[n(\mathbf{r})] + V_{ee}[n(\mathbf{r})] + E_{ext}[n(\mathbf{r})]. \tag{2.5}$$

Note that the total energy can be obtained by energy minimization with respect to $n(\mathbf{r})$, being a minimum for the correct density function $n(\mathbf{r})$.

The basic idea of the Schrödinger equation and the Kohn–Sham equation for solving a many-electron system is schematically shown in Figure 2.1. The Kohn–Sham equation is formally equivalent to the Schrödinger equation in describing a quantum mechanical system. The properties of a quantum mechanical system can be calculated by solving the Schrödinger equation with particle spatial coordinates as variables. A more tractable, formally equivalent way is to solve the DFT Kohn–Sham equation with particle density as a variable. This is detailed in the following paragraphs.

Assuming that there exists a noninteracting electron (fictional electrons) system with the number of electrons and electron density identical to that of an interacting many-electron system, these fictional electrons are called Kohn–Sham particles. The electron density $n(\mathbf{r})$ for such a noninteracting electron system can be written as follows:

$$n(\mathbf{r}) = \sum_{i=1}^{N} |\phi_i(\mathbf{r})|^2, \tag{2.6}$$

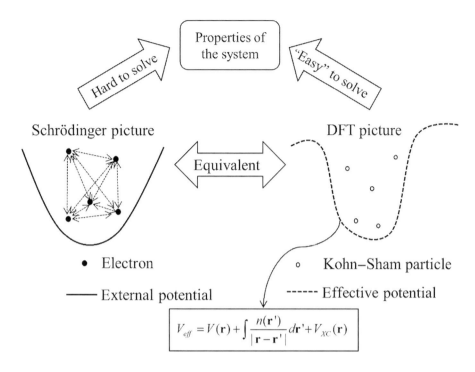

**Figure 2.1** Schematic relation of the Kohn–Sham equation (DFT approach) and the Schrödinger equation (general approach). This figure is adapted from Mattsson et al. (2005) with permission from *IOPscience*.

where $N$ is the number of electrons and $\phi_i(\mathbf{r})$ are single-electron wave functions corresponding to the $N$ lowest occupied states of the noninteracting system.

The kinetic energy of fictional electrons in this lowest state, $T_0$, is given by

$$T_0[n(\mathbf{r})] = \sum_{i=1}^{N} \int \phi_i^*(\mathbf{r}) \left( -\frac{\hbar^2}{2m} \nabla^2 \right) \phi_i(\mathbf{r}) d\mathbf{r}. \tag{2.7}$$

If $V_{ext}(\mathbf{r})$ is a potential describing the interactions between the electrons and the nuclei, then $E_{ext}[n(\mathbf{r})]$ is given by an equation of the following form:

$$E_{ext}[n(\mathbf{r})] = \int n(\mathbf{r}) V_{ext}(\mathbf{r}) d\mathbf{r}. \tag{2.8}$$

The Hartree energy, that is, the classical interaction energy between the charges of density $n(\mathbf{r})$, is defined by the following equation:

$$V_H[n(\mathbf{r})] = \frac{1}{2} \int \int \frac{e^2}{|\mathbf{r} - \mathbf{r}'|} n(\mathbf{r}) n(\mathbf{r}') d\mathbf{r} d\mathbf{r}'. \tag{2.9}$$

Within the DFT, the kinetic energy of a many-electron system is always considered to be the function of electron density of the noninteracting case $T_0[n(\mathbf{r})]$. In consequence, the total energy of a real electron system can be written as follows:

$$E[n(\mathbf{r})] = T_0[n(\mathbf{r})] + V_H[n(\mathbf{r})] + E_{ext}[n(\mathbf{r})] + E_{xc}[n(\mathbf{r})]. \tag{2.10}$$

In (2.10), an energy term $E_{xc}[n(\mathbf{r})]$ appears. This is the so-called exchange-correlation functional and can be defined as follows:

$$E_{xc}[n(\mathbf{r})] = (T[n(\mathbf{r})] - T_0[n(\mathbf{r})]) + (V_{ee}[n(\mathbf{r})] - V_H[n(\mathbf{r})]). \tag{2.11}$$

Here $T[n(\mathbf{r})]$ represents the kinetic energy of a real electron system, and $V_{ee}[n(\mathbf{r})]$ is the interaction potential between electrons, which includes any many-body effects. All the missing terms in many-body interactions and quantum effects are, thus, included in the exchange-correlation density functional.

Applying the variation principle to (2.10) results in a single-particle equation, which is called the Kohn–Sham equation for a fictional system of noninteracting electrons (Kohn and Sham, 1965):

$$\left\{ -\frac{\hbar^2}{2m}\nabla^2 + V_{eff}[n(\mathbf{r})] \right\} \phi_i(\mathbf{r}) = E_i \phi_i(\mathbf{r}). \tag{2.12}$$

$V_{eff}[n(\mathbf{r})]$ is the effective potential, which is expressed by an equation of the following form:

$$V_{eff}[n(\mathbf{r})] = V_{ext}(\mathbf{r}) + V_H[n(\mathbf{r})] + V_{XC}[n(\mathbf{r})]$$
$$= V_{ext}(\mathbf{r}) + e^2 \int d\mathbf{r}' \frac{n(\mathbf{r}')}{|\mathbf{r} - \mathbf{r}'|} + \frac{\delta E_{XC}[n(\mathbf{r})]}{\delta n(\mathbf{r})}. \tag{2.13}$$

Here, the exchange-correlation potential in the Kohn–Sham single-particle equation is defined as follows:

$$V_{XC}(\mathbf{r}) = \frac{\delta E_{XC}[n(\mathbf{r})]}{\delta n(\mathbf{r})}. \tag{2.14}$$

The single-particle states $\phi_i(\mathbf{r})$ are the solutions of the Kohn–Sham equation and are called Kohn–Sham orbitals.

The Kohn–Sham equations are a set of coupled, single-particle equations. So far, the single-electron equation without interaction has been found, and the framework of DFT has been successfully established. Instead of solving a complicated Schrödinger equation for a real electronic system in an external potential, $V(\mathbf{r})$, one would solve a much simpler, but formally equivalent, fictitious system of noninteracting Kohn–Sham particles in an effective potential $V_{eff}[n(\mathbf{r})]$.

The effective potential, which is defined in (2.13), is a function of electron density and in turn depends upon all the single-particle states, as described in (2.6). The equation is solved by iterations until self-consistency is reached, in the same way as solving the Hartree equation. One would first guess the initial electron density, which usually takes superposition from atomic charge densities. From charge density, one can calculate the Hartree potential, the exchange-correlation potential, and the effective potential, then solve the Kohn–Sham equations to obtain single-particle states

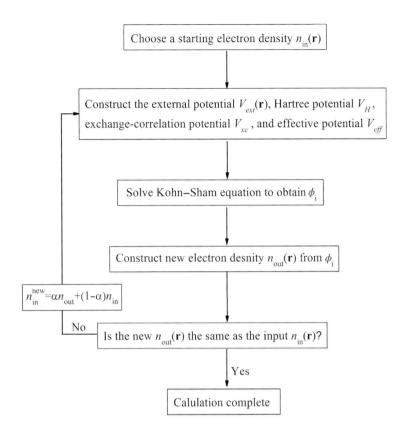

Choose a starting electron density $n_{in}(\mathbf{r})$

Construct the external potential $V_{ext}(\mathbf{r})$, Hartree potential $V_H$, exchange-correlation potential $V_{xc}$, and effective potential $V_{eff}$

Solve Kohn–Sham equation to obtain $\phi_i$

Construct new electron desnity $n_{out}(\mathbf{r})$ from $\phi_i$

$n_{in}^{new}=\alpha n_{out}+(1-\alpha)n_{in}$

No

Is the new $n_{out}(\mathbf{r})$ the same as the input $n_{in}(\mathbf{r})$?

Yes

Calulation complete

**Figure 2.2** Flow chart of the iterative scheme for obtaining the ground state electron density from Kohn–Sham equation (2.12).

$\phi_i(\mathbf{r})$, and, finally, use $\phi_i(\mathbf{r})$ to obtain a new electron density $n(\mathbf{r})$. This loop is repeated until the new electron density (or the new total energy) does not differ much from the former. At this point, the ground state is reached. The iterative process of self-consistently solving the one-electron Kohn–Sham equation to obtain the electron density of a ground state is illustrated in Figure 2.2.

In crystalline solids, the effective potential in the Kohn–Sham equation is a periodic potential $V_{eff}(\mathbf{r}) = V_{eff}(\mathbf{r} + \mathbf{R})$ with $\mathbf{R}$ being the lattice vector of a Bravais lattice. According to the Bloch theorem, the Kohn–Sham wave functions $\phi_i(\mathbf{r})$ can be expanded in plane waves. Solving the Kohn–Sham equation is not the most difficult issue. The more important issue is that the exact form of $V_{XC}$ is unknown. To define the exchange-correlation energy as a functional of charge density, there are two commonly used approximations (Thijssen, 2007).

The first approximation to the exchange-correlation functional is the local density approximation (LDA). The basic idea of LDA is that a general inhomogeneous electronic system is locally homogeneous. Thus, only the electron density at space $\mathbf{r}$ is used to determine the exchange-correlation energy density at that point. The functional is then expressed as

$$E_{XC}^{LDA}[n(\mathbf{r})] = \int n(\mathbf{r}) \varepsilon_{XC}^{LDA}[n(\mathbf{r})] d\mathbf{r}, \tag{2.15}$$

where $\varepsilon_{XC}^{LDA}$ is the local function of the energy density that accounts for the exchange and correlation effect. A common approximation for $\varepsilon_{XC}^{LDA}$ is to use the exchange-correlation energy of a homogeneous electron gas with the same density. Nearly all modern LDA correlation functionals are based on the quantum Monte Carlo data for a homogeneous electron gas (Ceperley and Alder, 1980). LDA works well for systems in which the electron density does not vary too rapidly as it will be exact for a uniform electron gas.

The second approximation of exchange-correlation functional is known as generalized gradient approximation (GGA), which adds the gradient of the density, $|\nabla n(\mathbf{r})|$, as an independent variable:

$$E_{XC}^{GGA}(n) = \int \varepsilon_{XC}^{GGA}[n(\mathbf{r}), \nabla n(\mathbf{r})] n(\mathbf{r}) d\mathbf{r}. \tag{2.16}$$

Generally speaking, GGA functionals can be divided into two categories. One is called "parameter free," and the other is empirical. For the former GGA functionals, the new parameters are determined from known expansion coefficients and other exact theoretical conditions. For the "empirical GGA," the functional parameters are determined by fitting to experimental data or from accurately calculated atomic and molecular properties. The functionals of Perdew–Wang from 1991 (known as PW91) and Perdew, Burke, and Ernzerhof (known as PBE), which are most commonly used in the software of physics and material science, are "parameter free." Most GGAs used in software of quantum chemistry, e.g., Becke, Lee, Parr, and Yang (abbreviated as BLYP), are empirical.

One should note that it is not always the case that GGA is the best choice in first-principles calculations. It is well known that LDA works better than GGA for certain classes of systems and properties, in particular for calculating surface adsorption and nontransition metal oxides. Hence, it is better to perform a test calculation with at least two different types of functionals to get a rough estimation of the accuracy before extensive calculations are done.

## 2.1.4    Pseudopotential Method

To solve the single-particle equation of (2.12), information on ionic potential is needed. Intuitively, a full all-electron Coulomb potential is a choice, but the all-electron potential is inconvenient in first-principles calculation due to the rapidly oscillating wave functions in the ion-core region. A solid is composed of nuclei and electrons, and electrons further fall into two categories, i.e., valence electrons and core electrons. The core electrons are tightly bounded to the nuclei and form ion cores together with the nuclei, and thus a solid can be considered as a collection of ion cores and valence electrons. According to the theory of quantum mechanics, the valence-electron wave functions and core-electron wave functions are orthogonal to each

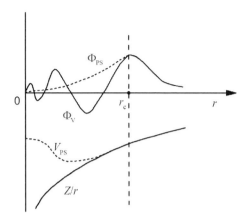

**Figure 2.3** Schematic illustration of all-electron (solid line) and pseudo-electron (dash line) *potentials* ($Z/r$, $V_{PS}$) and their corresponding *wave functions* ($\Phi_V$, $\Phi_{PS}$); $r$ is the radius from the ion-core center. The radius at which all-electron and pseudoelectron values match is designated $r_c$. Outside $r_c$, the pseudopartial waves are equal to the all-electron partial waves. Inside the spheres, the wave functions obtained with the pseudopotential differ from the actual one, and they can be any smooth continuation.

other. All-electron DFT methods treat core and valence electrons on an equal footing. In order to satisfy the orthogonality constraint, all-electron wave functions of valence electrons show rapid oscillations in the core region, as shown in Figure 2.3. Theoretically, one can represent such functions using plane waves, which are generally used as the basis sets of a wave function in a periodical system, but this scheme is impractical since the size of the basis sets would be prohibitive. Thus, it is crucial to replace the full Coulomb potential of the electron–ion interaction by another kind of potential.

Note that the core electrons are almost unaffected when the atoms are gathered to form a solid, thus we are only interested in the valence electrons of atoms. The contribution to the total electron density is the difference between valence and core electrons. The core electrons mainly contribute to the total electron density within the core region. In contrast, the contribution of valence states to the total electron density is negligible in the core region but predominant beyond it. Because of this difference, a highly effective scenario has been developed to separate the two kinds of states. This approach is the so-called pseudopotential method, shown in the form of a dashed line in Figure 2.3, which allows one to take the core electrons out of the picture and at the same time to create a smooth potential for the valence electrons. This means that properties of molecules or solids are calculated based on the idea that the ion cores are not involved in chemical bonding and do not change when structural modifications occur. Pseudo wave functions ideally should have no nodes inside the core region, and thus they only require a small basis set. Replacing explicit treatment of the chemically inert core electrons with an effective core potential reduces the degrees of freedom of a wave function and avoids the numerical challenge of rapid spatial variations in the electronic structure.

## 2.2 Molecular Dynamics

MD is an atomistic simulation method for studying nanoscale and mesoscale phenomena in a wide range of materials, such as alloys, ceramics, and biological molecules. This method typically describes the forces between atoms using interatomic potentials, and the motion of atoms and ions is controlled by Newton's second law of motion. MD is a deterministic technique. By setting an initial set of positions and velocities, the subsequent time evolution of particles' configurations, namely the trajectories of particles, is in principle completely determined. Based on these calculated trajectories of particles, transport properties, structural characteristics, and thermodynamic and reaction pathways can be predicted. Unlike first-principles calculations, which can only deal with hundreds of atoms, typical MD simulations can be performed on systems containing thousands or even millions of atoms, and the simulation times can be from a few picoseconds to dozens of nanoseconds. In MD, the equations of motion for atoms can only be solved numerically, and the equilibrium parameters or transport properties of the system are obtained by statistical methods. In this section, we recall the general concepts in MD simulation. There are some books showing more details of MD (Lesar, 2013; Rapaport, 2004; Thijssen, 2007).

### 2.2.1 The Basic Mechanical Quantities in MD

In classical mechanics, force is the cause of affecting the motion of any objects. In the MD method, force is one of the key physical quantities that determine the movement of every particle in the simulated system. The force $F_i$ acting on particle $i$ is usually derived from the potential $U(\mathbf{r}_1,\mathbf{r}_2,\ldots,\mathbf{r}_2,\ldots,\mathbf{r}_N)$ of the simulated system comprising $N$ particles:

$$\mathbf{F}_i = -\nabla_{ri} U(\mathbf{r}_1, \mathbf{r}_2, \ldots, \mathbf{r}_i, \ldots, \mathbf{r}_N). \tag{2.17}$$

Here $\mathbf{r}_i$ represents the coordinates of atom $i$. The equations of motion for particle $i$ with constant mass $M_i$ follow Newton's second law:

$$M_i \frac{d^2\mathbf{r}_i}{dt^2} = \mathbf{F}_i. \tag{2.18}$$

Note that Newton's second law states force as derivative of momentum. The assumption of constant particle mass in (2.18) is one reason that chemical reactions usually cannot be reproduced. Advanced MD methods, also involving quantum mechanical potentials, are required to this end; see also Section 2.2.6.

In usual MD simulations, based on (2.18), one calculates the velocity of each particle at a given time; subsequently lets every particle move with that velocity a short period of time; and then reevaluates the potential energy, the force, and thus the velocity. Particle collisions are not considered separately. However, the potential rapidly increases between almost colliding particles, thus, avoiding the "collision."

The core of MD method is how to describe the interatomic interactions. The traditional procedure in MD is to determine these interactions, namely interatomic interaction potentials, in advance. The simplest way describing $U$ is to write it as a sum of pairwise interactions $\phi(r_{ij})$, where $r_{ij}$ is the atomic distance:

$$U(r) = \sum_i \sum_{i<j} \phi(|r_{ij}|). \tag{2.19}$$

The clause $i < j$ is to guarantee considering each atom pair only once. The well-known and typical pairwise interaction is the Lennard-Jones (L-J) potential:

$$\phi(r_{ij}) = 4\varepsilon \left[ \left( \frac{\sigma}{r_{ij}} \right)^{12} - \left( \frac{\sigma}{r_{ij}} \right)^{6} \right]. \tag{2.20}$$

This interaction repels at shorter distance, then attracts, and it eventually disappears at a limiting separation $r_c$. The parameter $\varepsilon$ controls the strength of the interaction and $\sigma$ defines a length scale. $\varepsilon$ and $\sigma$ are determined by fitting them to the known physical properties of the materials. The L-J potential function is very simple, and in practice it is suitable for systems containing neutral atoms, such as noble gases, where no electrons are available for binding and atoms are attracted to each other only through the weak van der Waals forces.

The pair potentials have, in principle, an infinite range. In practical simulation, it is customary to set a cutoff radius $r_c$ and disregard the atomic interactions when their distances are larger than $r_c$. The two-body approximation does not work well for metals and semiconductors, thus somewhat complex many-body potentials have been developed and are now commonly used in materials science. One famous many-body potential is the embedded-atom method (EAM) potential:

$$U = \frac{1}{2} \sum_{i=1}^{N} \sum_{j=1, j \neq i}^{N} \phi_{ij}(r_{ij}) + \sum_{i=1}^{N} F_i(\rho_i). \tag{2.21}$$

The EAM potential contains a two-body potential $\phi_{ij}$ and a many-body potential $F_i$. $\phi_{ij}$ is a pair potential that only depends on $r_{ij}$, whereas the embedding energy $F_i$ is a nonlinear function of the electron density $\rho_i$ at the position of atom $i$. $\rho_i$ is constructed by the superposition of contribution from the neighboring atoms of atom $i$ and is a short-ranged function of distance. The idea of EAM is that every atom in a solid is viewed as "impurity" and that this impurity embeds in electron gas, also known as the jellium model.

The Hamiltonian or total energy, which is the sum of potential energy $U$ and kinetic energy $K$, is a conserved quantity in Newtonian dynamics. Usually it is calculated at each time step in order to check whether it is indeed constant with time. In practice, there could be small fluctuations in the total energy. These fluctuations are mainly caused by numerical errors in the time integration.

The instantaneous kinetic energy is given by an equation of the following form:

$$K(t) = \frac{1}{2} \sum_i M_i [v_i(t)]^2, \tag{2.22}$$

where $v_i(t)$ is the instantaneous velocity of particle $i$.

According to the equipartition theorem, each translational degree of freedom gives a universal kinetic energy of $k_B T/2$. Here $k_B$ is the Boltzmann constant and $T$ the absolute temperature. This is the classical statistical estimate, assuming that $k_B T$ is not too small compared to quantum effects. An estimate about the temperature of a $d$-dimensional ($d = 2$ or 3) system at a time point is therefore directly obtained from the average kinetic energy:

$$\frac{d}{2} N k_B T = \sum_{i=1}^{N} \frac{1}{2} \left( M_i v_i^2 \right). \tag{2.23}$$

Then the temperature is given by

$$T = \frac{1}{d N k_B} \sum_{i=1}^{N} \frac{v_i^2}{M_i}. \tag{2.24}$$

The temperature is related to the kinetic energy, which is not a conserved quantity and therefore fluctuates. It is needed to adjust the instantaneous temperature to the desired temperature by rescaling the velocities in the initial steps of a nonequilibration process.

The pressure $p$ cannot be obtained directly. The pressure can be derived from the virial theorem (Rapaport, 2004):

$$pV = N k_B T + \frac{1}{d} \left\langle \sum_{i=1}^{N} \mathbf{r}_i \cdot \mathbf{F}_i \right\rangle. \tag{2.25}$$

The symbol $<>$ means average over several time steps. $d$ is 3 for a three-dimensional system, and 2 for a two-dimensional one. Except for the pressure $p$, all the other quantities are accessible in a simulation.

The atoms of a liquid or a gas move and are subject to displacement. This displacement can be particularly important in the case of a liquid to ensure the fluid properties. Furthermore, the displacement of a single atom does not follow a simple trajectory. A feature of significant interest in liquid system is the mean square displacement (MSD), which contains information on the atomic diffusivity of the particles.

The MSD is defined as follows:

$$MSD(t) = \frac{1}{N} \sum_{i=1}^{N} (\mathbf{r}_i(t) - \mathbf{r}_i(0))^2. \tag{2.26}$$

The summation means average over all the atoms (or all the atoms in a given subclass). For solids, MSD shows a finite value, while for a liquid system, MSD grows linearly with time. In such cases, it is interesting to characterize the behavior of the system compared to the slope of the MSD. The slope of the MSD or so-called diffusion coefficient $D$ is defined as follows:

$$D = \lim_{t \to \infty} \frac{1}{6t} \left\langle |\mathbf{r}_i(t) - \mathbf{r}_i(0)|^2 \right\rangle, \tag{2.27}$$

where $<\ >$ denotes averaging over all atoms of a specific species. It should be noted that in two-dimensional systems the value 6 in denominator is replaced with 4.

The radial distribution function (RDF), $g(r)$, which is also called the pair correlation function or pair distribution function, is a very useful tool to extract the structural properties from numerical simulations. The $g(r)$ represents the probability to find a particle in the shell of the thickness $dr$ at the distance $r$ of another particle. Selecting an atom as the center atom, by dividing the physical space into shells $dr$, it is possible to compute the number of atoms $dn(r)$ at a distance between $r$ and $r + dr$ of this atom:

$$dn(r) = \frac{N}{V} g(r) 4\pi r^2 dr, \tag{2.28}$$

where $N$ represents the total number of atoms, and $V$ the volume; hence $N/V$ is the ordinary atomic density $\rho$ [atoms·m$^{-3}$] of the system. Note that the local density in the differential volume $dV_{shell}$ of shell of thickness $dr$ reads

$$\rho(r) = \frac{dn(r)}{dV_{shell}} = \rho g(r). \tag{2.29}$$

Thus, the RDF can be viewed as the ratio of the local density at a distance $r$ from a reference particle to the ordinary density in the system. In other words, RDF describes how density varies as a function of distance from a reference particle.

By distinguishing the chemical species, it is possible to compute the partial radial distribution functions $g_{\alpha\beta}(r)$:

$$g_{\alpha\beta}(r) = \frac{dn_{\alpha\beta}(r)}{4\pi r^2 dr \rho_a} \tag{2.30}$$

with

$$\rho_\alpha = \frac{N_\alpha}{V}, \tag{2.31}$$

where $N_\alpha$ represents the number of species $\alpha$. These functions give the density probability, $g_{\alpha\beta}(r)$, for an atom of the $\alpha$ species to have a neighbor of the $\beta$ species at a given distance $r$.

The RDF carries information on the structure of the system. The position and magnitude of the peaks reflect the crystal structure or short-range order of the system. Figure 2.4 shows the RDF for a liquid and a crystal phase. For a crystal, the RDF exhibits an infinite number of sharp picks whose separations and heights are characteristic of the lattice structure. For a liquid, $g(r)$ exhibits a major peak close to the average atomic separation of neighboring atoms followed by less pronounced as well as oscillated peaks. At longer distances, $g(r)$ decays steadily to a constant value of 1.

The RDF can serve as a tool for revealing phase structure and phase transition in microstructure evolution. When a simple liquid is supercooled below its freezing temperature, a crystalline solid typically becomes the state of lowest free energy and a first-order crystallization transition may occur. If the liquid is supercooled rapidly enough, crystal nucleation and subsequent crystallization may be suppressed. Now the

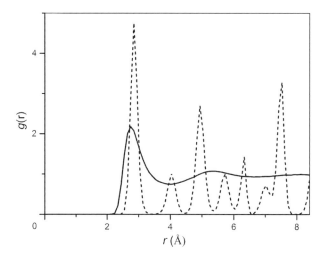

**Figure 2.4** Radial distribution function of a liquid (solid line) and crystal phase (dot line) of a unary substance, schematically.

supercooled liquid instead may undergo, at a characteristic glass transition temperature, a rapid but continuous minor structural change. It distinguishes the "supercooled liquid" from the "glass." The glass is typically an amorphous phase composed of clusters with short-range and medium-range order. To examine the details of the structural relation between medium-range order clusters and crystal nuclei more closely, RDF is often used to analyze the product structure. It is emphasized that crystal nucleation is among the most important processes in the synthesis of materials.

## 2.2.2    Periodic Boundary Conditions

One must distinguish simulation of macroscopic systems from simulation concerning finite objects such as molecules and clusters. For finite systems, the open boundary conditions – i.e., no boundaries at all, just put $N$ particles in space – are credible. But real macroscopic systems have a much larger number of particles (in the order of $10^{23}$) than can be handled in a simulation. The practical MD simulation for macroscopic systems takes place in a container of some kind with hundreds of thousands of atoms in it, which is only a very small fraction of the actual material. No matter how large the simulation-accessible system is, the number of atoms in the simulated system is negligible compared with the number of atoms in a macroscopic piece of matter. Then what should we do at the boundaries of our simulated system? The solution to this issue is to use periodic boundary conditions (PBC), as shown schematically in Figure 2.5. The PBC means that particles are enclosed in a box, and the box and particles are replicated to infinity by rigid translation in all the three Cartesian directions, completely filling the space. These duplicated boxes are referred as images, and accordingly the particles in the duplicated boxes are "image particles." The introduction of periodic boundaries is equivalent to considering an infinite,

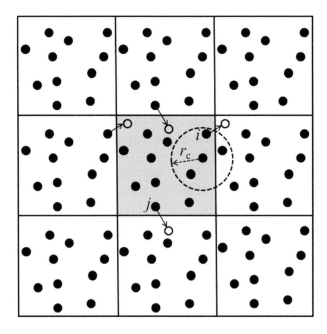

**Figure 2.5** The schematic of PBC in a two-dimensional system. The simulation box is represented by the shaded gray square. Particles in simulation box and the eight image boxes are represented by solid circles. Two particles $i$ and $j$ that will leave the simulation box in the next time step are indicated by their velocity vector. If a particle leaves the simulation box through one face, one of its periodic images will move into it from the opposite face, as shown by the solid arrows and the open circles. The two open circles in the simulation box will manifest themselves in the next time step. In calculating particle interactions within the cutoff range $r_c$, both real and image neighbors are included, as indicated by the dashed circle.

space-filling array of identical copies of the simulation region. Although PBC could mitigate the finite-size effect in the simulated system to some extent, there are two consequences to note about this periodicity.

The first is due to the requirement to conserve the number of particles in the simulation box. Therefore, if an atom moves out the basic simulation box through a bounding face, then it immediately reenters the box through the opposite face. After each integration step, the coordinates must be examined. In other words, if we find that an atom has moved outside the box, we must adjust its coordinates to bring it into the box again. It is rather simple to deal with the "escaped" atoms. For example, assuming the $x$-coordinate is defined to lie between $-L_x/2$ and $L_x/2$, where $L_x$ is the box size in the $x$-direction, if the $x$-position of particle $i$ $r_{ix} \geq L_x/2$, replace it with $r_{ix} - L_x$; otherwise, if $r_{ix} < -L_x/2$, replace it with $r_{ix} + L_x$. Its velocity vector is unchanged.

The second consequence is that atomic interaction may go through box boundaries. Note that the potentials have a cutoff distance $r_c$. If two particles are separated by a distance larger than the cutoff radius $r_c$, they do not interact with each other. Special attention must be paid to the case where atoms are near a boundary. Under this condition, although the distance inside the box between particle $i$ and $j$ is larger than

$r_c$, particle $i$ may interact with one of the images of particle $j$ when the distance between particle $i$ and one of the images of particle $j$ is in the interactive range. In this case, we can use the minimum image criterion: Among all possible images of a particle $j$, select the closest image to account for the interaction with particle $i$, and throw away all the others. Thus, in MD simulation, one would check the effect of periodicity on atomic interactions. Another requirement is that one must make sure that the box size is significantly larger than $2r_c$ along all directions where PBCs are used.

Even with periodic boundaries employed in MD, the finite-size effects still exist. It is necessary to select a suitable size of calculation cell in the MD method. How many atoms in the simulated system are necessary? What is the minimum size of the simulation box? There is no unique answer. The answer depends on both the type of simulated system (i.e., the range of interatomic interaction potential) and the properties of interest. As a minimal requirement, the size should exceed the range of any significant correlations.

## 2.2.3        Time Integration Algorithm

In MD, most properties are obtained by simulating the motions or dynamics of particles numerically. This can be done through computing trajectories, by means of the simultaneous integration of Newton's equations of motion. Time integration algorithms are based on the finite difference method (FDM), where time is discretized on a finite grid. In Chapter 4, a very brief introduction to the fundamentals of the finite element method (FEM) will be presented. Both FEM and FDM are typical numerical methods to solve partial differential equations. Two simple numerical schemes that are widely used in MD are the Verlet and leap-frog methods.

The basic idea of the Verlet method is to make two third-order Taylor expansions for the position variable, one forward and one backward in time. Defining $\mathbf{v}$, $\mathbf{a}$, $\mathbf{b}$, and $\Delta t$ as the velocity, the acceleration, the third derivation of $r$ with respect to time, and the time step, respectively, the expansion reads as follows:

$$\mathbf{r}_i(t + \Delta t) \approx \mathbf{r}_i(t) + \mathbf{v}_i \Delta t + \frac{1}{2}\mathbf{a}_i(t)\Delta t^2 + \frac{1}{6}\mathbf{b}_i(t)\Delta t^3 + O(\Delta t^4) \tag{2.32}$$

$$\mathbf{r}_i(t - \Delta t) \approx \mathbf{r}_i(t) - \mathbf{v}_i \Delta t + \frac{1}{2}\mathbf{a}_i(t)\Delta t^2 - \frac{1}{6}\mathbf{b}_i(t)\Delta t^3 + O(\Delta t^4). \tag{2.33}$$

Adding the two expressions gives the position

$$\mathbf{r}_i(t + \Delta t) = 2\mathbf{r}_i(t) + \mathbf{a}_i(t)\Delta t^2 - \mathbf{r}_i(t - \Delta t) + O(\Delta t^4). \tag{2.34}$$

According to the force law, acceleration $a(t)$ is expressed as follows:

$$\mathbf{a}_i(t) = \frac{\mathbf{F}_i(t)}{m_i} = -\frac{1}{M_i}\nabla U(\mathbf{r}(t)). \tag{2.35}$$

It is obvious from (2.34) and (2.35) that the positions of particles at $t + \Delta t$ can be obtained from the positions at current $t$ and the previous $t - \Delta t$ for the given potential $U$. The truncation error of this algorithm is in the order of $\Delta t^4$. The velocity in the Verlet method is not directly generated in the solution. If it is required (for example, one needs the kinetics energy), it can be obtained from the positions as follows:

$$\mathbf{v}_i(t) = \frac{\mathbf{r}_i(t + \Delta t) - \mathbf{r}_i(t - \Delta t)}{2\Delta t}. \tag{2.36}$$

In the leap-frog method, positions and velocities are evaluated at different times: The positions are calculated at every time step, while the velocities are evaluated in a half time step. The velocity of every particle is constant during this time step. The velocity at the middle step $\mathbf{v}(t + \Delta t/2)$ ought to be a good compromise. The velocity $\mathbf{v}(t + \Delta t/2)$ is obtained from the velocity at time $t - \Delta t/2$ and the acceleration at time $t$:

$$\mathbf{v}_i\left(t + \frac{1}{2}\Delta t\right) = \mathbf{v}_i\left(t - \frac{1}{2}\Delta t\right) + \mathbf{a}_i(t)\Delta t. \tag{2.37}$$

The velocity at $t + \Delta t/2$ is estimated as follows:

$$\frac{\mathbf{r}_i(t + \Delta t) - \mathbf{r}_i(t)}{\Delta t} = \mathbf{v}_i\left(t + \frac{1}{2}\Delta t\right). \tag{2.38}$$

The position at $t + \Delta t$ is obtained accordingly as follows:

$$\mathbf{r}_i(t + \Delta t) = \mathbf{r}_i(t) + \mathbf{v}_i\left(t + \frac{1}{2}\Delta t\right)\Delta t. \tag{2.39}$$

If an estimate for velocity is required to correspond to the time at which coordinates are evaluated, then the following equation can be used:

$$\mathbf{v}_i(t) = \frac{\mathbf{v}_i\left(t + \frac{1}{2}\Delta t\right) + \mathbf{v}_i\left(t - \frac{1}{2}\Delta t\right)}{2}. \tag{2.40}$$

The local errors introduced at each time step due to the truncation of what should really be an infinite series in $t$ are of order $O(\Delta t^4)$ for the coordinates and $O(\Delta t^2)$ for velocities.

Besides these two most commonly used time integration methods, there are other time integration algorithms, such as predictor–corrector methods, the velocity Verlet method, and the Beeman method. For more details on these methods, the interested reader can refer to Rapaport (2004) for a general survey.

## 2.2.4    Ensemble

A thermodynamic system can be described by a number of state variables, such as particle number ($N$), volume ($V$), temperature ($T$), pressure ($p$), and total energy ($E$). Here the symbol $E$ is used instead of $U$ so as to avoid confusion with the potential in (2.17), even though in thermodynamics the equivalent is the internal energy $U$; see Chapter 5. State variables describe the macroscopic state of the system, and they are not all independent but connected by the equation of state for the given material system. For instance, the macroscopic parameters $p$, $T$, $N$, and $V$ have the relation $pV = Nk_BT$ for ideal gas where the particles are noninteracting. This means that we need to fix some macroscopic parameters in MD simulation.

If one simulates a rigid isolated system with fixed particle number $N$, volume $V$ and total energy $E$, this is the "microcanonical ensemble," or $NVE$ ensemble. In the microcanonical ensemble, $N$, $V$, and $E$ are referred as external parameters, while temperature $T$ and pressure $p$ are observables to be calculated. If thermodynamic equilibrium is attained the entropy will be at a maximum. However, MD simulation in $NVE$ ensemble differs from most experimental conditions: temperature and pressure in some cases are invariant. Thus, sometimes one wants to perform a simulation at constant $T$ and/or constant $p$ instead of constant $E$ or constant $V$. In that case, canonical ensemble or isothermal–isobaric ensemble is reasonable. In the "canonical ensemble" ($NVT$ ensemble), $N$, $V$, and $T$ are external parameters and invariant, while total energy $E$ and pressure $p$ are observables. If thermodynamic equilibrium is attained, the Helmholtz energy will be at a minimum. In order to keep the temperature constant, a thermostat, which is an algorithm that adds and removes energy, is required. For simulating a system with constant $N$, $T$, and $p$, one needs the isothermal–isobaric ensemble ($NTP$ ensemble). In the isothermal–isobaric ensemble, one requires an additional barostat, an algorithm that changes volume to keep the pressure constant, besides the thermostat. If thermodynamic equilibrium is attained, the Gibbs energy will be at a minimum. All of the preceding conditions refer to a closed system where $N$ is fixed. In contrast, the "grand canonical ensemble" is actually an open system in equilibrium with a reservoir and not considered here. See also Table 5.1 with the comments on energetic state functions and their natural variables ($S$, $V$, $U$, $T$, $p$).

The interested reader can refer to Lesar (2013) and Thijssen (2007) for more details about $NVE$ and $NPT$ ensembles. In the following subsection, we describe the procedure for MD simulation.

## 2.2.5 Procedure of MD Simulation

The procedure of MD simulation is schematically shown in Figure 2.6. The general steps of MD simulation are as follows:

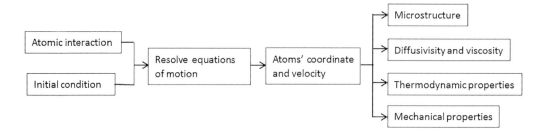

**Figure 2.6** The schematic procedure of MD simulations.

(1) Define the system to be studied, set the boundary conditions, and choose suitable ensemble and atomic interaction potentials, which is $U$ in (2.17).

(2) Set the initial conditions: Establish initial configuration (initial positions of particles) of the system and assign the initial velocity of particles.

(3) Choose algorithms to calculate the forces acting on particles as well as the position and velocity of particles at every simulated time.

(4) After starting from artificial/arbitrary initial values, substantial fluctuations (e.g., of a target temperature) are observed in the subsequent time steps, which should decay approaching a quasistationary state with small fluctuations, also called *equilibrium*.

(5) Obtain equilibrium values of macroscopic parameters and transport properties by means of statistical mechanics.

To perform a simulation, we need to define the simulation box, choose an atomic potential to describe the interaction of particles, and assign the initial positions and velocities of the particles. There are two common ways of setting initial position and velocity of each particle.

The initial position is mainly defined according to experimental data, such as the molecular structure or crystal structure determined by X-ray diffraction or nuclear magnetic resonance spectroscopy. This structure is typically the most stable one at $T = 0$ K with the given potential. Initial velocities may be set as zero or as values from a Maxwell–Boltzmann distribution at a temperature of interest.

The initial positions and velocities can also be chosen from the previous MD run. This is the most commonly used method when studying materials behaviors at high temperatures.

Of course, the state of the initial condition of MD simulation does not correspond to an equilibrium condition. If we are interested in equilibrated state, we must wait for a number of time steps to reach equilibrium before "measuring" observables. Once the simulation is started, equilibrium is usually reached within a time of the order of 1,000–2,000 time steps. A ballpark figure for a time step may be several picoseconds for metal so that equilibrium is reached in nanoseconds. In ab initio molecular dynamics (AIMD), a typical time step is measured in femtoseconds. One simple and direct approach to judge if equilibrium is reached is to check a characteristic observable. The characteristic observable approaches a constant value when thermodynamic equilibrium is reached. For example, the temperature is an invariant in *NVT* ensemble, thus one can monitor the temperature at each time step. If the temperature fluctuates around the setting value with small amplitude, one can say that the system reaches equilibrium, as shown in Figure 2.7.

## 2.2.6    Ab Initio Molecular Dynamics

One very important but most challenging aspect of classical MD simulations is the construction of a suitable potential $U$ in 2.17 to calculate the force. In classical MD simulations, force is computed from empirical potential functions, which have been

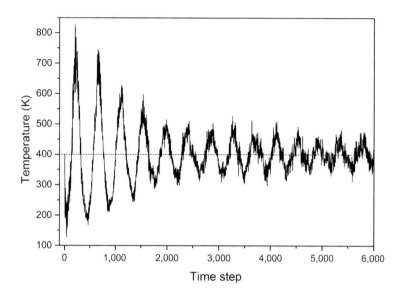

**Figure 2.7** Fluctuation of temperature with steps of time for an MD simulation in *NVT* ensemble. The temperature fluctuates around the nominally fixed temperature (400 K) of the system. For the initial 1,500 time steps (one time step is $10^{-15}$ s), the fluctuation is drastic. After that, the temperature displays a weak fluctuation.

parameterized to reproduce experimental or accurate first-principles calculated data obtained from small model systems. Even though great progress in elaborating empirical potentials has been made, the transferability to chemically complex systems is often restricted. The chemically complex systems are characterized by (i) many different atom or molecule types giving rise to a variety of different interatomic interactions that have to be parameterized, and (ii) the electronic structure causing the chemical bonding pattern to change qualitatively during the course of the simulation. To remedy this situation, a method called as ab initio molecular dynamics has been developed. Figure 2.7 actually results from an AIMD simulation with a time step of 2 fs, i.e., $2 \cdot 10^{-15}$ s.

Essentially AIMD differs from classical MD in two aspects. Firstly, classical MD relies on Newton's equation, while AIMD is based on the quantum Schrödinger equation. Secondly, classical MD depends on semi-empirical atomic potentials that approximate quantum effects, while AIMD is based on the real physical potentials. These differences mean that AIMD does not rely on any adjustable parameter. AIMD falls into two fundamental categories: Born–Oppenheimer molecular dynamics (BOMD) and Car–Parrinello molecular dynamics (CPMD). The BOMD, as the name implies, means that the Born–Oppenheimer approximations are still valid. In the BOMD, the nuclei follow a semiclassical Newton equation. They move according to ionic forces that are calculated within quantum mechanics theory. The electronic wave function is in its ground state for every ionic configuration. This means that one should perform a self-consistent calculation for charge density in every simulated time step. In contrast, for the Car–Parrinello method, the motion of ions and electrons is coupled by explicitly introducing the electronic degrees of freedom as dynamical (fictitious) variables. In CPMD, an explicit electronic minimization at each time step, as done in BOMD, is not needed. The reader can refer to Marx and Hutter (2009) for more detail on AIMD.

AIMD is an advanced simulation technique; its increased accuracy and predictive power are associated with a significant computational cost. Thus, AIMD simulations are limited in the system size and time scale.

## 2.3 Some Quantities Obtained from First-Principles Calculations

In this section, we explain how to calculate structural, thermodynamic, mechanical, and defect properties using the first-principles method. These properties are usually needed for computational design of engineering alloys.

### 2.3.1 Lattice Parameter

In most cases, the physical quantities of crystals in equilibrium conditions are involved. Thus, the first step is to obtain the equilibrium lattice parameter in the state of lowest energy at constant $T$ and $p$. In first-principles calculations, there are two schemes to obtain the equilibrium lattice parameter: direct structure optimization and fitting energy-volume data.

The direct structure optimization means that cell volume and shape are allowed to change during the calculation process; however, the basic crystal structure is given by the user. The equilibrium lattice parameters and atomic coordinates are obtained when the external pressure and force acting on all atoms are close to zero. This method is widely used in first-principles calculations, but the bulk modulus, $B$, cannot be obtained directly.

The lattice parameter can also be obtained through fitting the equation of state (EOS), which shows how pressure $p$, temperature $T$ and volume $V$ are coupled in condensed matters. The widely used EOSs are the Murnaghan EOS, the third-order Birch–Murnaghan isothermal EOS, and the Vinet EOS (Vinet et al., 1986). The Murnaghan EOS reads as follows:

$$E = E_0 + B_0 V_0 \left[ \frac{1}{B_0'(B_0'-1)} \left(\frac{V}{V_0}\right)^{1-B_0'} + \frac{1}{B_0'} \frac{V}{V_0} - \frac{1}{B_0'-1} \right]. \tag{2.41}$$

The third-order Birch–Murnaghan isothermal EOS is as follows:

$$E = E_0 + \frac{9V_0 B_0}{16} \left\{ \left[ \left(\frac{V_0}{V}\right)^{2/3} - 1 \right]^3 B_0' + \left[ \left(\frac{V_0}{V}\right)^{2/3} - 1 \right]^2 \left[ 6 - 4\left(\frac{V_0}{V}\right)^{2/3} \right] \right\}. \tag{2.42}$$

And the Vinet EOS is given by an equation of the following form:

$$E = E_0 + \frac{9V_0B_0}{\eta^2}\left\{1 - e^{\eta(1-X)}[1 - \eta(1 - X)]\right\},\tag{2.43}$$

with

$$\eta = \frac{3}{2}\left(B_0' - 1\right),\tag{2.44}$$

$$X = \left(\frac{V}{V_0}\right)^{1/3}.\tag{2.45}$$

In the preceding equations, $E_0$, $V_0$, $B_0$, and $B_0'$ are the equilibrium energy, the equilibrium volume, the bulk modulus, and the pressure derivation of the bulk modulus, respectively. See also Section 5.3.1.3 in Chapter 5.

In calculations, one needs to calculate energies at several volumes (or lattice parameters), usually three to five volumes around the experimental volume of the crystal. Then a least squares fit of the set $(V_i, E_i)$ should be made to one kind of EOS. To apply EOS to the calculated data, $V_0$, $E_0$, $B_0$, and $B'_0$ are treated as fitting parameters. Figure 2.8 shows the result of Fcc Al. The experimental lattice parameter of Al is $\sim$4.046 Å, corresponding to the volume of 66.43 Å$^3$ per unit cell. Seven volumes around the experimental values are calculated by using DFT with the PBE functional. The dashed line is the result of fitting the data using lattice parameters in the range values 3.7–4.3 Å. This fitted curve predicts that $V_0$ is 66.0113 Å$^3$ (corresponds to lattice parameter of 4.0415 Å) and $B_0$ is 0.4874 eV/Å$^3$ (corresponds to 78 GPa). Note, however, that the experimental relative accuracy of lattice parameter determination can be very high, often better than $10^{-3}$.

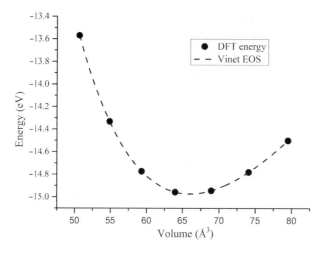

**Figure 2.8** Total energy of Al in the Fcc crystal structure as a function of the volume. Data points are from present DFT calculations at 0 K, and the solid curve is the Vinet equation of state.

## 2.3.2 Thermodynamic Properties above 0 K

Thermodynamics is the cornerstone of materials science and engineering. The first-principles computed thermodynamic properties above 0 K are needed in many simulations for the sake of material design. For example, many metastable phases precipitating during the aging of Al alloys are driven by the Gibbs energies of these phases. It is extremely difficult to obtain the thermodynamic properties for these metastable phases experimentally, thus computational methods play an extremely important role in obtaining thermodynamic properties of metastable phases that are not stable at any temperature or composition of the system.

As an important thermodynamic quantity, the enthalpy of formation at 0 K is given by the energy of reaction, that is, the energy change between compound $(A_xB_y)$ and components (A and B). If the reaction is $xA + yB \rightarrow A_xB_y$, the formation enthalpy $\Delta_f H$ reads as follows:

$$\Delta_f H\left(A_xB_y\right) = E\left(A_xB_y\right) - xE(a) - yE(b), \tag{2.46}$$

where $E$ is the Gibbs energy of the substance, which equals the enthalpy or internal energy, since the entropy of ordered substances is zero at 0 K. In first-principles calculation, it is the DFT energy at 0 K.

The analysis of the ground state energy $E_0$ (at $T = 0$ K) allows us to calculate structural and elastic properties accurately at low temperatures. For thermodynamic properties and phase stability, the Gibbs energy at a finite temperature is needed. At finite temperature, the Gibbs energy mainly includes internal energy and vibrational energy due to atomic vibration around their equilibrium position. In some cases, for example for metals, the contribution of electronic excitations to Gibbs energy cannot be neglected. The thermodynamic properties of crystals are to a large extent determined by phonons, thus the key step for the investigation of thermodynamic properties is to deal with the phonons.

We do not illustrate how to calculate the phonon spectrum here. The reader can refer to van de Walle and Ceder (2002) for detailed accounts. The key quantity to be calculated in order to have access to the thermodynamic properties and the phase stability is the vibrational free energy $F_{vib}$. In the framework of the harmonic approximation, the Helmholtz energy can be expressed as follows:

$$F = E_0 + F_{vib}. \tag{2.47}$$

$E_0$ is the ground state energy, which is easily accessible from the standard DFT calculations. $F_{vib}$ depends on temperature and the frequency of the phonon:

$$F_{vib} = k_B T \sum_{qj} \ln\left\{2\sinh\left[\frac{\hbar\omega_j(\mathbf{q})}{2k_B T}\right]\right\}. \tag{2.48}$$

$\omega_j(\mathbf{q})$ is the frequency of the $j$th phonon mode at wave vector $\mathbf{q}$. Equation (2.48) can also be written as the following integration form:

$$F_{vib} = k_B T \int_0^\infty g(\omega) d\omega \ln \left[ 2 \sinh \left( \frac{\hbar\omega}{2k_B T} \right) \right]. \tag{2.49}$$

Here $g(\omega)$ is the phonon density of state. From $F_{vib}$, the other thermodynamic properties can be obtained. From the Helmholtz energy $F(T,V)$, the Gibbs energy $G(T,p)$ can also be calculated; see Chapter 5 for the reverse transformation.

Vibrational entropy $S_{vib}$ is defined by an equation of the following form:

$$S_{vib} = -\left( \frac{\partial F_{vib}}{\partial T} \right)_V = k_B \sum_{qj} - \ln \left[ 1 - e^{-\hbar\omega_j(\mathbf{q})/k_B T} \right] + \frac{\hbar\omega_j(\mathbf{q})/k_B T}{e^{\hbar\omega_j(\mathbf{q})/k_B T} - 1}. \tag{2.50}$$

The heat capacity per unit cell at a constant volume can be calculated from the following equation:

$$C_V = -T\left( \frac{\partial^2 F_{vib}}{\partial T^2} \right)_V = k_B \sum_{qj} \left[ \frac{\hbar\omega_j(\mathbf{q})}{2k_B T} \right]^2 \frac{1}{\sinh^2 \left[ \frac{\hbar\omega_j(\mathbf{q})}{2k_B T} \right]}. \tag{2.51}$$

The volume thermal expansion coefficient $\alpha_V$ is defined as

$$\alpha_V = \frac{1}{V} \left( \frac{\partial V}{\partial T} \right)_P. \tag{2.52}$$

The heat capacity at constant pressure $C_p$ is obtained by the following equation:

$$C_p = C_V + (\alpha_V)^2 B_0 VT. \tag{2.53}$$

The linear thermal expansion coefficient $\alpha_h$ of the $h$-lattice direction can be written as

$$\alpha_h = \frac{1}{a_h} \frac{\partial a_h}{\partial T}. \tag{2.54}$$

Figure 2.9 shows the calculated heat capacities for Ag by the first-principle approach (Xie et al., 1999). It shows that the calculated heat capacity is in good agreement with available experimental data over a wide range of temperatures.

## 2.3.3     Elastic Properties

In this subsection, we will briefly describe how to calculate basic mechanical properties, including elastic constant, bulk modulus, Young's modulus, and shear modulus.

Basically, there are two ways to compute single-crystal elastic constants in first-principles methods: the energy–strain approach and the stress–strain approach. The energy–strain approach is based on the computed total energies of properly selected strained states of the lattice cell. The stress–strain approach, on the other hand, relies on the stress tensor directly. Here only the energy–strain approach is shown. The underlying concepts for the stress–strain approach are discussed in detail in Le Page and Saxe (2002).

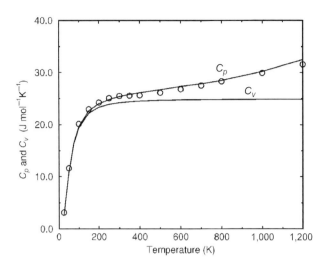

**Figure 2.9** First-principles calculated temperature dependence of heat capacity of Ag at constant pressure ($C_p$) and at constant volume ($C_v$). The circles denote the experimental $C_p$ data. The agreement between theory and experiment is remarkable over a wide range of temperatures. This figure is reproduced from Xie et al. (1999) with the permission from the American Physical Society.

In order to obtain elasti constants, one expands the total energy $E$ of a strained crystal in power series of the strain tensor $e_i$ as

$$E = E_0 + V_0 \sum_{i,j} \sigma_i e_i + \frac{V_0}{2} \sum_{i,j} C_{ij} e_i e_j. \tag{2.55}$$

$V_0$ and $E_0$ are the volume and energy of the undistorted lattice cell, respectively. $\sigma_i = \frac{1}{V_0} \left( \frac{\partial E}{\partial e_i} \right)$ are the components of the stress tensor, and $C_{ij}$ the second-order adiabatic elastic constants (all in Voigt notation: $i, j = 1, 2, 3, 3, 5, 6$, replacement of double indices $11{\rightarrow}1$; $22{\rightarrow}2$; $33{\rightarrow}3$; $12{\rightarrow}6$; $13{\rightarrow}5$; $23{\rightarrow}4$). The maximum number of independent elastic constants is 21. The symmetry of a specified crystal structure reduces the number of independent $C_{ij}$. For a cubic structure, there are the conditions of $C_{11} = C_{22} = C_{33}$, $C_{12} = C_{13} = C_{23}$, $C_{44} = C_{55} = C_{66}$, and all other $C_{ij} = 0$. Thus, the total independent elastic constants are only three for a cubic structure. For a hexagonal phase, there are five independent elastic constants. Compared with the cubic lattice, two independent elastic constants, $C_{13}$ and $C_{33}$, are added for the hexagonal phase in addition to the elastic constants of the cubic structure.

Individual $C_{ij}$ or the combination of them can be determined by computing the total energy of a series of specific strain states. According to (2.55), the elastic energy increment of a solid under strain is represented as follows:

$$\Delta E = \frac{V_0}{2} \sum_{i,j} C_{ij} e_i e_j, \tag{2.56}$$

where $\Delta E$ is the energy increment due to the strain with vector $\mathbf{e} = (e_1, e_2, e_3, e_4, e_5, e_6)$. For instance, $\mathbf{e}(\delta) = (\delta, 0, 0, 0, 0, 0)$ corresponds to simple straining in one direction, and for a cubic crystal, (2.56) is reduced to $E(\delta) = E_0 + (V_0/2)C_{11}\delta^2$, which allows direct computation of $C_{11}$ when the total energy for several values of $\delta$ is known.

For different cells, the deformation is different. For example, for the cubic lattice, by applying the triaxial shear strain $\mathbf{e} = (0, 0, 0, \delta, \delta, \delta)$ to the crystal, $C_{44}$ can be calculated from

$$\Delta E = \frac{3V_0}{2}C_{44}\delta^2. \tag{2.57}$$

For $C_{11}$–$C_{12}$, we apply volume-conserving strain $\left(\delta, \delta, \delta^2/(1 - \delta^2), 0, 0, 0\right)$, which has the advantage that the energy change is an even function of strain $\delta$. Then $\Delta E$ is changed as

$$\Delta E = V_0(C_{11} - C_{12})\delta^2. \tag{2.58}$$

By applying a strain of $\mathbf{e} = (\delta, \delta, \delta, 0, 0, 0)$, the relation for $C_{11} + 2C_{12}$ is obtained:

$$\Delta E = \frac{3}{2}V_0(C_{11} + 2C_{12})\delta^2. \tag{2.59}$$

Alternatively, $C_{11} + 2C_{12}$ can be calculated from bulk modulus:

$$B_0 = \frac{C_{11} + 2C_{12}}{3}. \tag{2.60}$$

Note that the bulk modulus can be obtained by fitting energy-volume relation using one kind of EOS.

In practical calculation, two positive and two negative small strains with the values of less than 1% of the lattice vector are applied. After obtaining five $(\Delta E, \delta)$ points, including the original point, a polynomial of $\Delta E = \Delta E(\delta)$ can be fitted to the data pairs $(\Delta E, \delta)$ using the least squares method.

The elastic constants can be used to judge the mechanical stability of the crystal. A given crystal structure cannot be in a stable or metastable state unless its elastic constants obey certain relationships. For a cubic crystal, the criterion is $C_{11} > C_{12}$, $C_{44} > 0$, and $C_{11} + 2C_{12} > 0$. The lower the symmetry is, the greater the restrictions. For example, the mechanical stability constraints for tetragonal crystal are $C_{11} > C_{12}$; $C_{11} + C_{33} - 2C_{13} > 0$; $C_{11} > 0$; $C_{33} > 0$; $C_{44} > 0$; $C_{66} > 0$; and $2C_{11} + C_{33} + 2C_{12} + 4C_{13} > 0$. A prominent example is pure tungsten, stable in the Bcc structure, Bcc–W. The energy difference to the Fcc–W calculated by DFT is significantly different from the so-called lattice stability obtained by the CALPHAD method; see Chapter 5. This puzzle was resolved by DFT calculations showing that the Fcc structure of tungsten is mechanically unstable and not metastable.

The elastic constants can also be used to obtain bulk modulus $(B)$ and shear modulus $(G)$. Polycrystalline elastic modulus can be estimated from single-crystal elastic constants using Voigt or Reuss homogenization schemes. The Voigt approach gives an upper bound on elastic properties in terms of the uniform strain, while the Reuss scheme gives a lower bound according to the uniform stress. Hill has shown that the actual effective elastic moduli of polycrystalline can be approximated by the arithmetic mean of the two bounds of Voigt and Reuss, which is referred to as the Voigt–Reuss–Hill (VRH) value (Hill, 1952):

$$B = \frac{1}{2}(B_V + B_R), \tag{2.61}$$

$$G = \frac{1}{2}(G_V + G_R). \tag{2.62}$$

Here the subscript $V$ and $R$ represent the Voigt scheme and Reuss scheme, respectively. Based on the shear to bulk modulus ratio $B/G$, the ductility and brittleness can be judged according to Pugh's empirical criterion: the $B/G$ ratio separates ductile ($>1.75$) and brittle ($<1.75$) materials. But there are some exceptions where the Pugh's criterion is invalid for determining the ductility and brittleness of materials, especially for some Laves phases (Long et al., 2016).

The parameters such as Young's modulus $Y$ and Poisson ratio $v$ can also be estimated from the obtained bulk modulus and shear modulus:

$$Y = \frac{9BG}{3B + G} \tag{2.63}$$

$$v = \frac{3B - 2G}{2(3B + G)}. \tag{2.64}$$

## 2.3.4 Defect Properties

Lattice defects, including intrinsic defects and exotic impurities, are common in crystals. It is of fundamental importance to understand the formation of defects in crystals due to their influence on physical properties such as diffusion, electrical conductivity, catalytic properties, and so on. One impressive example is doping in semiconductors, where the incorporation of impurities even in small concentrations determines the electrical conductivity. Point defects also play an important role in atomic transport and mechanical properties of metals. For example, self-diffusion in metals mainly occurs by vacancy-related mechanisms, and segregation of vacancies in the form of pores or even cracks may lead to a degradation of material resistance. One of the most important quantities for defects is the formation energy of the defect (Freysoldt et al., 2014), and this is also the key quantity for vacancy-related properties. In first-principles calculations, the defect is usually modeled in a super-cell, consisting of the defect surrounded by a few dozen to a few hundred atoms of the host material, which is then repeated periodically throughout space. For charged defects, supercell calculations always include a homogeneous compensating background charge.

The defect formation energy is defined as

$$\Delta E_f(X^q) = E(X^q) - E(bulk) - \sum_i n_i \mu_i + q(E_V + \varepsilon_F) + E_{corr}, \tag{2.65}$$

where $\Delta E(X^q)$ and $E(bulk)$ denote the total energies of the supercell containing a defect $X$, such as an interstitial, vacancy, antisite, Frenkel, or other defect pair, in the charged state $q$ and of the defect-free supercell, respectively. $n_i$ represents the number of atoms of type $i$ (host atoms or impurities) that have been added to ($n_i > 0$) or removed from ($n_i < 0$), the supercell upon defect creation, and $\mu_i$ are the corresponding chemical potentials of these species, which represent the energy of the reservoirs with which atoms are being exchanged. $\varepsilon_F$ is the chemical potential of the electrons or Fermi level, which accounts for exchanging electrons with an electron reservoir. $\varepsilon_F$ is conventionally taken with respect to the valence band maximum (VBM) $E_V$ of the perfect lattice, and it can vary from VBM to the conduction band minimum (CBM). The last term of (2.65) is a correction term that accounts for the artificial electrostatic interactions between the defect $X^q$ and its periodic array as well as the compensating background charge.

The slope of the lines in Figure 2.10 corresponds to the charge of the defect. It can be seen obviously that the negatively charged hydrogen vacancy $V_H^-$ and the positively charged hydrogen interstitial $H_i^+$ are the dominant defects due to the lowest formation energy. The neutral and positively charged H vacancy can only be stable within the range $0\sim0.8$ eV above the VBM. As for H interstitial, the neutral state is

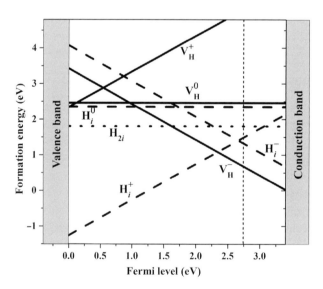

**Figure 2.10** First-principles calculated formation energies of hydrogen-related defects in LiNH₂ as a function of energy position in band gap above VBM. The dashed vertical line denotes the special Fermi-level position due to charge neutrality, where the formation energies of the dominant defects ($V_{Li}^-$ and $I_{Li}^+$, not shown here) are equal. This figure is reproduced from Wang et al. (2011) with the permission from the American Physical Society.

higher in energy than either the positively or the negatively charged one in any position of the band gap, indicating that the neutral interstitial will not be stable and will decay into charged states. Figure 2.10 also shows the formation energy of interstitial neutral hydrogen molecule ($H_{2i}$). The formation energy of ($H_{2i}$) is obviously lower than that of neutral H vacancy and interstitial. Anyway, hydrogen is usually incorporated into $LiNH_2$ in (charged) atomic form because at any position in the band gap the formation energy of either $H_i^+$ or $V_H^-$ is lower than that of ($H_2$)$_i$. These findings indicate that it is possible to control the thermodynamics and kinetics of dehydrogenation reactions of $LiNH_2$ through doping, as some dopants can adjust the position of Fermi level; as a result, the formation energy of H-related charged defects could change.

## 2.4    A Case Study: Design of an Ultra-lightweight Mg–Li Alloys Using First-Principles Calculations

In the last section of this chapter, we take Mg–Li alloys for ultra-lightweight applications as an example to demonstrate how atomistic simulation methods guide material design (Counts et al., 2009, 2010a, b).

Mg is a lightweight metal and has potential application in the field of auto and aerospace industry. However, performing large-scale manufacturing operations by forming Mg and Mg alloys is difficult due to the poor ductility at room temperature, and once formed they often display undesirable mechanical properties. The low room temperature ductility is ascribed to the Hcp crystal structure of Mg. Changing the crystal structure from Hcp to either Bcc or Fcc and simultaneously retaining the advantage of lightweight is one way to address this problem. Li is the lightest metal, and Mg–Li alloys can be stable in Bcc phase at room temperature when Li content is above 30 at.%. Thus, the Mg–Li system may be among the lightest possible metallic alloys and hence could be applied if extreme lightweight is required. Experimental investigation of Mg–Li alloys is expensive and time consuming since both components easily react with air and humidity. The standard procedure in experimental studies includes manufacture of the alloys and then determination of the relevant mechanical and other physical properties of the alloys. A great advantage of computational material science tools is the ability to estimate reasonably certain key mechanical and physical properties prior to focused experimental work. As a result, the computational data can provide valuable information to guide the designing of new alloys.

Figure 2.11 shows the first-principle predicted engineering parameter *B/G* ratio (a measure of ductility) for the polycrystalline single-phase Bcc Mg–Li alloys (Counts et al., 2009). This calculation for the Bcc extends below 30 at.% Li into the Bcc + Hcp region but does not include the Hcp phase. The majority of the Bcc Mg–Li alloys has a *B/G* above 1.75, which means in general this alloy behaves in a ductile manner. Decreasing the Li content to 30–50 at.% leads to a significant decrease for the *B/G* ratio. All the *B/G* values in this composition range are below the ductile–brittle

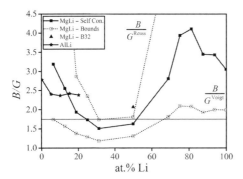

**Figure 2.11** $B/G$ ratio for single-phase Bcc Mg–Li alloys obtained from the Reuss, Voigt, and self-consistent homogenization schemes. The horizontal solid line separates the ductile and brittle regions. Single-phase Fcc Al–Li results are based on self-consistent homogenized data. This figure is reproduced from Counts et al. (2009) with the permission from Elsevier.

transition value, hinting in this composition that the Mg–Li alloys cannot be used in the transportation industry. However, the $B/G$ ratio is sensitive to the atomic arrangement; we can see in Figure 2.11 that for the alloy with 50 at.% Li and Mg, the B32 structure shows about 25% greater ductility than the Bcc (exactly Bcc-B2) structure. For alloys with less than 20 at.% Li, the $B/G$ ratio is in the range of $1.9\sim3.2$, which is in the ductile range. Unfortunately, the Bcc alloys in this composition range have no practical value, as the Hcp + Bcc two-phase region is thermodynamically stable at ambient temperature and pressure in this composition range.

The specific modulus, i.e., the ratio of Young's modulus to mass density ($Y/\rho$, a measure of stiffness per weight), is another important quantity when designing Mg–Li alloys for transportation applications. The specific modulus of Bcc Mg–Li alloys is shown in Figure 2.12. Again, Bcc Mg–Li alloys with less than 20 at.% Li are thermodynamically unstable as single-phase Bcc and therefore have no practical value. Mg alloys with $70\sim100$ at.% Li are both the lightest and least stiff (the Young's modulus is lower than that of other compositions). Consequently, their specific modulus is also on the low end (in the range of $13.9\sim21.6$ MPa m$^3$/kg$^{-1}$). Mg–Li alloys with $30\sim50$ at.% Li have the highest values of specific modulus. Their specific modulus changes from 24.5 MPa m$^3$/kg$^{-1}$ (for B32) to 31.6 MPa m$^3$/kg$^{-1}$ (at 30 at.% Li). The data for Fcc Al–Li alloys are shown for comparison in Figures 2.11 and 2.12. Note that Al–Li alloys are stable as single-phase Fcc only below about 1 at.% Li at room temperature, and at higher composition the intermetallic phase LiAl is formed in the Fcc + LiAl equilibrium.

Figure 2.13a shows the Ashby map of $Y/\rho$ versus $G/B$ of Bcc Mg–Li alloys. It is obvious that the $G/B$ value increases (corresponding to decreasing $B/G$ values) when the specific modulus increases. The Ashby map shows that it is not possible to increase $Y/\rho$ without simultaneously increasing $G/B$ (i.e., brittleness) by changing only the composition or distribution of atoms over the lattice sites of the Mg–Li binary alloys (see data points for Mg$_8$Li$_8$-B2 and Mg$_8$Li$_8$-B32 compounds with the same

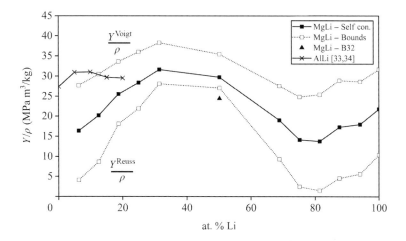

**Figure 2.12** Specific modulus of single-phase Bcc Mg–Li and single-phase Fcc Al–Li. Al–Li results are based on self-consistent homogenized data. Modulus values are representative of polycrystalline materials with a random texture. This figure is reproduced from Counts et al. (2009) with the permission from Elsevier.

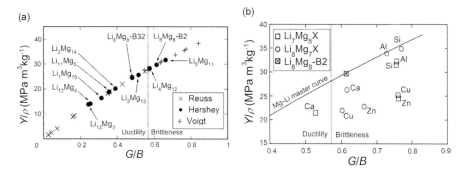

**Figure 2.13** Ashby map of $Y/\rho$ versus $G/B$ of (a) Bcc Mg–Li binary alloys and (b) Bcc MgLi–X ($X$ = Ca, Al, Si, Cu, Zn) compounds. The maps show that it is not possible to increase both $Y/\rho$ and ductility by changing only the composition or local order of a binary alloy. The dotted vertical line (for $G/B – 0.57$) is the threshold separating the ductile and brittle regions. This figure is reproduced from Counts et al. (2010b) with permission from Wiley.

composition but different atomic distributions). Alloys with 30~50 at.% Li may be the optimal alloys in Mg–Li system for ultra-lightweight applications. In an attempt to identify whether alloying with a third element can improve both the specific Young's modulus and ductility of the Mg–Li binary alloys or not, a set of Mg–Li–X ternaries ($X$ = Ca, Al, Si, Cu, Zn) based on stoichiometric Mg–Li with CsCl structure was also studied. The relationship between $Y/\rho$ and $G/B$ of the ternary system is shown in Figure 2.13b. It is shown that none of the studied ternary solutes is able to simultaneously improve both the specific Young's modulus and ductility, and all values are below those of the binary Mg–Li master alloy.

The preceding computational results suggest that it may be impossible to fabricate binary Bcc Mg–Li alloys with simultaneously improved ductility as well as stiffness by changing only the composition or local atomic structure. Even adding alloy elements such as Ca, Al, Si, Cu, and Zn to the $Mg_8Li_8$–B2 structure cannot improve the specific Young's modulus as well as ductility. Hence it is not the best choice for one to carry out experimental investigations on the binary Mg–Li alloys for ultra-lightweight applications. Not only the "positive results" but also this type of "negative result" are extremely useful for guiding necessary, though costly, experimental work in a focused way.

## References

Ceperley, D. M. and Alder, B. J. (1980) Ground state of the electron gas by a stochastic method. *Physical Review Letters*, 45(7), 566–569.

Counts, W. A., Friák, M., Raabe, D., and Neugebauer, J. (2009) Using *ab initio* calculations in designing Bcc Mg–Li alloys for ultra-lightweight applications. *Acta Materialia*, 57(1), 69–76.

Counts, W. A., Friák, M., Raabe, D., and Neugebauer, J. (2010a) *Ab initio* guided design of Bcc ternary Mg–Li–X (X = Ca, Al, Si, Zn, Cu) alloys for ultra-lightweight applications. *Advanced Engineering Materials*, 12(7), 572–576.

Counts, W. A., Friák, M., Raabe, D., and Neugebauer, J. (2010b) Using *ab initio* calculations in designing Bcc MgLi–X alloys for ultra-lightweight applications. *Advanced Engineering Materials*, 12(12), 1198–1205.

Freysoldt, C., Grabowski, B., Hickel, T., et al. (2014) First-principles calculations for point defects in solids. *Reviews of Modern Physics*, 86(1), 253–305.

Hill, R. (1952) The elastic behaviour of a crystalline aggregate. *Proceedings of the Physical Society A*, 65(5), 349–354.

Hohenberg, P. and Kohn, W. (1964) Inhomogeneous electron gas. *Physical Review*, 136(3B), B864–B871.

Kohn, W. and Sham, L. J. (1965) Self-consistent equations including exchange and correlation effects. *Physical Review*, 140(4A), A1133–A1138.

Le Page, Y. and Saxe, P. (2002) Symmetry-general least-squares extraction of elastic data for strained materials from *ab initio* calculations of stress. *Physical Review B: Condensed Matter*, 65(10), 104.

Lesar, R. (2013) *Introduction to Computational Materials Science: Fundamentals to Applications*. Cambridge: Cambridge University Press.

Long, Q., Nie, X., Shang, S.-L., et al. (2016) C15 NbCr2 Laves phase with mechanical properties beyond Pugh's criterion. *Computational Materials Science*, 121, 167–173.

Marx, D. and Hutter, J. (2009) *Ab Initio Molecular Dynamics: Basic Theory and Advanced Methods*. Cambridge: Cambridge University Press.

Mattsson, A. E., Schultz, P. A., Desjarlais, M. P., Mattsson, T. R., and Leung, K. (2005) Designing meaningful density functional theory calculations in materials science – A primer. *Modelling and Simulation in Materials Science and Engineering*, 13(1), R1–R31.

Rapaport, D. C. (2004) *The Art of Molecular Dynamics Simulation*, second edition. Cambridge: Cambridge University Press.

Thijssen, J. (2007) *Computational Physics*, second edition. Cambridge: Cambridge University Press.

van de Walle, A. and Ceder, G. (2002) The effect of lattice vibrations on substitutional alloy thermodynamics. *Reviews of Modern Physics*, 74(1), 11–45.

Vinet, P., Ferrante, J., Smith, J. R., and Rose, J. H. (1986) A universal equation of state for solids. *Journal of Physics C: Solid State Physics*, 19(20), L467–L473.

Wang, J., Du, Y., Xu, H., et al. (2011) Native defects in LiNH2: a first-principles study. *Physical Review B: Condensed Matter*, 84(2), 024107.

Xie, J., de Gironcoli, S., Baroni, S., and Scheffler, M. (1999) First-principles calculation of the thermal properties of silver. *Physical Review B: Condensed Matter*, 59(2), 965–969.

# 3 Fundamentals of Mesoscale Simulation Methods

## Contents

## 3.1    Mesoscale Simulation

The mesoscale, the spatial scale between atomic structures and the engineering continuum, is critical for controlling the macroscopic behaviors and properties of materials. To investigate or predict the macroscopic behaviors and properties, it is necessary to have a deep understanding of the mechanisms controlling the phenomena at this length scale. Probing, characterizing, and ultimately controlling materials at the mesoscale is at the core of materials science and engineering. However, despite the tremendous progress made during the past two decades in this regard, controlling materials at the mesoscale is still extremely challenging, requiring advanced experimental tools and modeling techniques to capture the complex phenomena.

Modeling and simulation play critical roles in all scientific disciplines, providing tools that can accelerate scientific discovery in ways that are not possible with a purely experimental approach. In materials science, for example, modeling and simulation can be employed to understand and improve the applied processing methods, investigate the evolution of the microstructure, determine the structure–property relationships, and ultimately predict the performance (Tonks and Aagesen, 2019). Mesoscale, which often refers to microstructure modeling, is appropriate at intermediate scales between the atomic and continuum scales and can provide a bridge between the atomistic structures and macroscopic properties of materials. Consequently, modeling has become increasingly significant, particularly with the rapid development of high-performance computing technologies. Modeling and simulation at the mesoscale, however, are not easy to implement because the phenomena of interest are neither atomistic (in which the solutions can be obtained by understanding the behaviors of a few to hundreds of atoms or molecules) nor macroscopic (where the continuum properties of the material can be assumed without losing events occurring within the smaller-scale regime) (Mohanty and Ross, 2008).

Recently, many mesoscale simulation methods have been developed and applied in various fields of materials science. Mesoscale simulation methods can be particle based, such as dissipative particle dynamics and Brownian dynamics, or field based, such as phase-field methods, cellular automaton methods, and fluctuating hydrodynamics. Among these mesoscale simulation methods, the phase-field and cellular automaton (CA) methods are extremely popular and powerful for simulating microstructure evolution processes, including phase transformations and grain growth at the mesoscale, particularly for predicting the evolution of the complex three-dimensional (3D) microstructures.

In this chapter, we focus on introducing the fundamentals of typical mesoscale simulation methods, application examples, and integrations with other methods. One case study for material design, mainly using the phase-field method, is briefly presented.

## 3.2     Phase-Field Method

The phase-field (PF) method describes microstructures by a set of spatially dependent field variables from which the spatial distributions of different phases and the boundaries between them can be analyzed. The PF method can handle the evolution of complex microstructures without explicitly tracking the interfaces and deal with multiple physical phenomena simultaneously within an irreversible thermodynamic framework. Although the PF method was originally developed for describing phase transformations, it is also a powerful method for other fields, such as physics, chemistry, and even topology optimization. Therefore, the PF methodology has rapidly gained popularity over the last two decades across numerous fields of materials science (Boettinger et al., 2002; Chen, 2002; Moelans et al., 2008; Provatas and Elder, 2010; Steinbach, 2009; Takaki, 2014).

### 3.2.1     History of the Phase-Field Method

The development of the PF method can be traced back to the middle of the last century. During the early 1950s, Ginzburg and Landau (1950) used an order parameter and its gradient to propose a model for superconductivity, from which the famous time-dependent Ginzburg–Landau dynamics equation was obtained. Twenty years later, a similar type of dynamics equation was proposed independently by Allen and Cahn (1979) to describe the process of phase separation. In 1958, Cahn and Hilliard (1958) developed a thermodynamic model that takes into account the gradient of thermodynamic properties in an inhomogeneous system with a diffusive interface, providing classical nonlinear diffusive dynamics. In 1974, by combining the dynamics of both conserved and nonconserved field variables, Hohenberg and Halperin (1977) proposed "Model C" to describe the critical dynamics of phase transformations. In 1978, to avoid interface tracking, Langer (1986) adapted Model C to produce a PF model that considered the Gibbs–Thomson equation of solidification at solid–liquid interfaces. This was the first time the concept of a "phase-field model" had been proposed.

Throughout the 1980s, no significant progress was made regarding this method except for several studies conducted from the perspective of mathematics. This

(a)

(b)

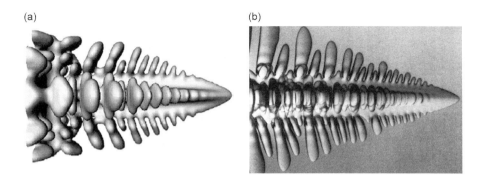

**Figure 3.1** Comparison of the first PF simulation of 3D dendrites morphology obtained by Kobayashi (1994) (a) and experimental observations (b) (Glicksman et al., 1993). The left figure is reproduced from Kobayashi (1994) with permission from Taylor and Francis, and the right one is reproduced from Glicksman et al. (1993) with permission from Elsevier.

situation did not change until Kobayashi's pioneering simulations of 3D dendritic growth in 1994 (Kobayashi, 1993, 1994). As shown in Figure 3.1, the first 3D dendritic structures simulated by Kobayashi using the PF model showed a striking similarity to the real dendrite structures observed in experiments conducted by Glicksman et al. (1993). Kobayashi's study contributed to the rapid development of PF simulations in materials science and engineering. Since then, many PF models have been proposed in various fields, particularly in investigations of phase transformations and grain growth of different materials.

With the PF method, complex microstructural changes are described using PF variables (or order parameters). According to the macroscopic physical interpretations of PF variables, two kinds of PF models have been proposed independently by different communities: one based on Langer's PF model, and the other based on the continuum field model derived by Chen and Wang in 1996 (Chen and Wang, 1996) from the microscopic theory developed by Khachaturyan (1983). In Langer's PF model, PF variables are introduced to avoid interface tracking. With the continuum field model, however, the field variables are well defined physical order parameters. For example, to reflect the crystal symmetry relations between coexisting phases for order/disorder transformations, the long-range order parameter is defined as a field variable. This type of PF model has been widely employed in various solid–solid phase transitions, such as order/disorder transitions and martensitic transformations.

Because the phase field in PF models can be identified as a physical quantity that is coarse grained at mesoscales, we also named the PF models coarse-grained or continuum PF models to distinguish them from two other types of PF models: the phase-field crystal (PFC) (Elder et al., 2007) and microscopic phase-field (MPF) (Khachaturyan, 1983) models, which were proposed to overcome the inability of the coarse-grained PF method to handle structural evolution at the atomic scale. The PFC and MPF models are also briefly introduced in this chapter. For simplicity, PF models in this book refer to continuum or coarse-grained PF models, unless otherwise noted.

## 3.2.2      Principles of the Phase-Field Method

PF models were originally developed to study solidification and other phase transformation processes. This method can deal with multicomponent, multiphase, and multigrain problems in various fields of materials, such as liquid–solid or solid–solid phase transition, grain growth, recrystallization, and fracture. Here, taking liquid–solid transformation as an example, we introduce the fundamentals of the PF method, including the diffuse interface, order parameters, free-energy functional, dynamic equations, and anisotropy.

### 3.2.2.1      Diffuse Interface

Although a rich variety of PF models are available, their commonality is based on the description of a diffuse interface. Consequently, one of the distinguishing characteristics of the PF approach is the diffuse interface between phases, and such an approach is sometimes called diffuse interface models. This concept was first proposed by van der Waals in 1893 and independently 60 years ago by Cahn and Hilliard. In contrast to the conventional sharp interface model (in which properties are discontinuous at the interface), in a diffuse interface model (in which properties evolve continuously at the interface between the equilibrium values in the neighboring phases/grains), field variables that change smoothly from one phase to another are employed to describe interfaces and boundaries, which can overcome the difficulties in solving free boundary or moving boundary problems. As shown in Figure 3.2a, with the PF method, a system of continuous variables is used to describe the microstructure, where the microstructure interfaces have a finite width ($\delta$) over which the variables shift between values. However, for the sharp interface model, as shown in Figure 3.2b, there is a steep change in properties across the interface, and each point in the material is either fully in one phase or fully in another. This discontinuity in the sharp interface model results in different controlling equations for different phases that should be solved together with the initial and boundary conditions to evolve the microstructure (Moelans et al., 2008).

The most important advantage of a diffuse interface model is that it avoids interface tracking, which is obtained by coping smartly with the interfacial energy using a gradient term. Of course, the diffuse interface model also has certain drawbacks. One of the model's disadvantages is that simulations based on this type of model can be very time consuming owing to the steep gradient at the interface.

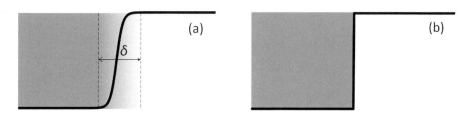

(a)      (b)

**Figure 3.2**  Schematics of (a) diffuse interface and (b) sharp interface.

### 3.2.2.2 Order Parameter

In the PF method, as the key point, interfaces are implicitly described by continuous field variables, which take the equilibrium values in the neighboring bulk phases and change continuously but steeply through the diffuse interface. Concentration and long-range order parameters are typical examples of field variables that characterize the compositional and structural heterogeneities, respectively. According to Landau's theory of phase transitions, the order parameter ($\phi$) is a state variable that is used to distinguish ordered and disordered phases. It can be interpreted as a nonzero average of the local order parameter field, which exhibits spatial variations. Therefore, $\phi$ can be imagined as the spatial average of the local order parameter averaged over the bulk phases (Provatas and Elder, 2010). For a two-phase material, $\phi$ is typically set to be 0 or 1 for the bulk phases; thus, the domain where $0 < \phi < 1$ is the interface.

### 3.2.2.3 Free-Energy Functional

Unlike classical thermodynamics, where the potential state variables are homogeneous throughout a system in equilibrium, in the PF method, the free energy is a function of the continuous field variables and their gradients. Thus, as another important feature of the PF model, the microstructural evolution is defined in terms of the free energy of the entire system but not the individual phases. Similarly, in the CALPHAD method of classical thermodynamics the equilibrium is defined as the global minimum of free energy of the entire system but not the individual phases, though space and time independent. This makes it possible to couple the phase field with other external fields (such as external magnetic or electronic fields) for the sake of providing a complete view of the material behaviors. The free energy ($F$) of the system may have contributions from the bulk free energy ($F_{\text{bulk}}$), interfacial energy ($F_{\text{int}}$), strain energy ($F_{\text{str}}$), and other energies, such as those from magnetic or electrostatic interactions:

$$F = F_{\text{bulk}} + F_{\text{int}} + F_{\text{str}} + \cdots \tag{3.1}$$

Similar to classical thermodynamics, the bulk free energy determines the equilibrium compositions and phase fractions, while the interfacial and strain energies not only influence the compositions and phase fractions, but also determine their morphologies and distributions. If no elastic or magnetic or electric field exists, the total free energy under constant temperature is as follows:

$$F = \int_V \left[ f(c, \phi) + \frac{\varepsilon_c^2}{2} (\nabla c)^2 + \frac{\varepsilon_\phi^2}{2} (\nabla \phi)^2 \right] dV. \tag{3.2}$$

The local free-energy density $f(c, \phi)$ is typically formulated as $f(c, \phi) = f_{\text{chem}} + f_{\text{doub}}$. The gradient energy coefficients ($\varepsilon_c$ and $\varepsilon_\phi$) are always positive because energetically unfavorable gradient terms can give rise to surface tension. $V$ is the volume of the system. The chemical free-energy density ($f_{\text{chem}}$) can be formulated by interpolating the free energies of the bulk phases as $f_{\text{chem}} = p(\phi)f^S + (1 - p(\phi))f^L$, where $f^S$ and $f^L$ are the free-energy densities of the solid phase and liquid phase, respectively, and $p(\phi)$

is a monotonically increasing interpolation function, and $p(\phi) = \phi^2(3 - 2\phi)$ is often used.

The interfacial energy, which is extremely significant in microstructural evolution, is often sensitive to the chemical, mechanical, or physical state of the interface. According to Cahn and Hilliard's description of the free energy of an inhomogeneous system, the interfacial energy ($F_{\text{int}}$) can be defined as follows:

$$F_{\text{int}} = \int_V \left( f_{\text{doub}} + f_{\text{grad}} \right) dV = \int_V \left[ Wg(\phi) + \frac{\varepsilon_c^2}{2}(\nabla c)^2 + \frac{\varepsilon_\phi^2}{2}(\nabla\phi)^2 \right] dV, \tag{3.3}$$

where $f_{\text{grad}} = \frac{\varepsilon_c^2}{2}(\nabla c)^2 + \frac{\varepsilon_\phi^2}{2}(\nabla\phi)^2$. The double-well potential ($f_{\text{doub}}$) has the form $f_{\text{doub}} = Wg(\phi)$, where $W$ denotes the energy barrier height, and the double-well function $g(\phi)$ often uses the form $g(\phi) = \phi^2(1 - \phi)^2$ (Takaki, 2014). This indicates that the interfacial energy consists of two contributions: one from the difference between the PF variables and their equilibrium values at the interface, and the other from the gradient of the PF variables. In the bulk phase, the double-well potential and the gradient terms become 0 because the PF variables are constant. Therefore, the bulk phases have no contribution to the interfacial energy.

The equilibrium profile of the PF variables can be obtained by minimizing the integral in (3.3) for a planar boundary. Subsequently, the specific interfacial energy can be obtained. The interfacial width is controlled by two opposite effects. On one hand, a wider interfacial region has a smaller contribution from the gradient energy term to the interfacial energy. On the other hand, a more diffuse interface region results in a larger contribution from the double-well potential. For some PF models, there is an analytical relation among the gradient energy coefficient, interfacial energy, and interfacial width. However, in most cases, the interfacial energy is determined numerically from (3.3) (Moelans et al., 2008).

### 3.2.2.4    Dynamic Equations

In the PF model, microstructural evolution is described by the temporal and spatial evolution of the PF variables through coupled partial differential equations. For a conserved PF variable, such as a concentration field, it evolves as

$$\frac{\partial c}{\partial t} = \nabla \cdot \left[ M_c \nabla \frac{\delta F}{\delta c} \right], \tag{3.4}$$

where $M_c$ is related to atomic mobility. Note that the term "concentration" is often used improperly (i.e., $\text{mol/m}^3$ or $\text{kg/m}^3$) in the solidification and phase-field community but is synonymous for "content" in at.% or wt.%. Usually, a conserved Gaussian noise term that satisfies the fluctuation-dissipation theorem is added on the right side of (3.4) to describe the effect of thermal fluctuations on microstructural evolution. $\delta F / \delta c$ is a variational derivative of the total free energy of the system $F$ with respect to $c$. Equation (3.4) is the well-known Cahn–Hilliard equation, or Model B as it is often called in condensed matter physics. If the gradient energy related to the concentration field can be neglected (the term $f_{grad}$ in (3.3)), the term $\delta F/\delta c$ is the difference in chemical potentials between the two components, which is essentially the thermodynamic factor in diffusion theory.

For a nonconserved PF variable ($\phi$), the dynamic equation is as follows:

$$\frac{\partial \phi}{\partial t} = -M_\phi \frac{\delta F}{\delta \phi} + \xi_\phi, \tag{3.5}$$

where $M_\phi$ is related to the interface mobility and has the relation $\tau \to 1/M_\phi$ with the kinetic attachment time ($\tau$), and $\xi_\phi$ is a nonconserved Gaussian noise field. In addition, $\delta F/\delta \phi$ (the variational derivative that determines how the free-energy density varies with changes in $\phi$) is the driving force for the rate of change of a nonconserved PF variable, driving the system down gradients in a free-energy landscape.

### 3.2.2.5 Anisotropy

Anisotropy is a direction-dependent characteristic, i.e., properties vary depending on the direction. The atoms in the crystals are arranged in a symmetrical manner, which makes some properties of the crystal anisotropic, such as the density, thermal diffusivity, interfacial energy, interface kinetics, and solute redistribution coefficient. For many materials (including metals), the interfacial energy and the kinetic coefficient are orientation dependent. Owing to its significant impact on phase morphology, anisotropy should be included in models of microstructural evolution. For example, the morphology of growing dendrites is very sensitive to the anisotropy, although the anisotropy is quite small ($\sim 1\%$), particularly for the interfacial energy. Thus, in the form presented in (3.5), the PF model cannot simulate anisotropic growth forms, such as dendrites, because an isotropic surface tension can only lead to isotropic structures. PF models usually introduce anisotropy into the surface energy by making the gradient term coefficient ($\varepsilon$) and the kinetic attachment time of the interface ($\tau$) functions of the angle ($\theta$) of the local interface normal vector ($\vec{n}$) (as shown in Figure 3.3a). Specifically, the gradient term in the free-energy functional and the kinetic attachment time in the PF dynamics become the following:

$$\frac{1}{2}\varepsilon^2 \nabla^2 \phi \to \frac{1}{2}|\tilde{\varepsilon}(\theta)\nabla\phi|^2 \tag{3.6}$$

$$\tau \to \tilde{\tau}(\theta), \tag{3.7}$$

where

$$\theta \equiv \arctan\left(\frac{\partial\phi/\partial y}{\partial\phi/\partial x}\right) \tag{3.8}$$

is the angle between the normal direction of interface and the reference axis in an $x$–$y$ space. A convenient choice for describing the anisotropy is as follows:

$$\tilde{\varepsilon}(\theta) = \varepsilon a(\theta), \quad \tilde{\tau}(\theta) = \tau a^2(\theta), \tag{3.9}$$

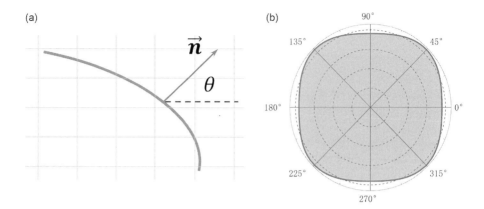

**Figure 3.3** (a) Schematics of local interface normal vector $\vec{n}$ and the angle $\theta$. (b) Polar plot of $a(\theta)$ as a function of $\theta$ when $K = 4$ and $\varepsilon_4 = 0.05$ in which $a(\theta) = 1 + \varepsilon_K \cos{(K\theta)}$.

where the function $a(\theta)$ modulates the anisotropy of the energy and kinetic terms of the interface. For a weak anisotropy in two-dimensional (2D) systems, the following formulation is commonly applied:

$$a(\theta) = 1 + \varepsilon_K \cos{(K\theta)}, \tag{3.10}$$

where $\varepsilon_K$ describes the degree of anisotropy of the interfacial energy (with $\varepsilon_K = 0$ corresponding to the isotropic situation), and K is the order of the anisotropy. For a cubic anisotropy, K is 4, while for a hexagonal one, K is 6. A polar plot of $a(\theta)$ as a function of $\theta$ when $K = 4$ and $\varepsilon_4 = 0.05$ is shown in Figure 3.3b. Moreover, according to the Gibbs–Thomson condition, $\varepsilon_K$ in (3.10) must be smaller than 1/15.

A strongly anisotropic interfacial energy will result in faceted morphologies, such as faceted WC morphology in the WC–Co cemented carbide. Examples of such anisotropy functions include $a(\theta) = 1 + \varepsilon_K(|\sin{(\theta)}| + |\cos{(\theta)}|)$ (resulting in square particles), and $a(\theta) = 1 + \varepsilon_K|\cos{(\theta)}|$ (resulting in plate-like particles) (Moelans et al., 2008).

### 3.2.3 Phase-Field Model for Solidification of a Pure Substance

To construct a PF model, the primary steps typically include selecting the PF variables that fit the problem at hand, constructing the free-energy functional in terms of these variables, and then determining the dynamic equations and model parameters. To handle the complex changes in the interface shape during solidification of a pure substance, the liquid–solid interface is described by a PF variable $\phi$, which usually takes a value from zero (liquid phase) to 1 (solid phase). Then, the free energy of the pure substance during solidification can be expressed as

$$F = \int_V \left[ f(\phi, T) + \frac{1}{2}\varepsilon^2(\nabla\phi)^2 \right] dV, \tag{3.11}$$

where the chemical free energy, $f(\phi, T)$, is defined as

$$f(\phi, T) = p(\phi)f^S(T) + [1 - p(\phi)]f^L(T) + Wg(\phi). \tag{3.12}$$

To avoid mixing the solid and liquid phases, the product $Wg(\phi)$ is introduced as an energy barrier term.

Then, the kinetic evolution equation is as follows:

$$\frac{\partial \phi}{\partial t} = -M_\phi \frac{\delta F}{\delta \phi}, \tag{3.13}$$

where the proportionality factor $M_\phi$ is the interface mobility, which is generally dependent on the field variable. As to the temperature field, a heat-transfer equation reads

$$\frac{\partial T}{\partial t} = \kappa \nabla^2 T + \frac{L}{C_p} \frac{\partial p(\phi)}{\partial t}, \tag{3.14}$$

in which $L$ is the latent heat, $\kappa$ the thermal diffusion coefficient, and $C_p$ the heat capacity. The second term on the right-hand side of (3.14) represents the latent heat generated during solidification (Takaki, 2014).

## 3.2.4 Phase-Field Model for Alloy Solidification

Numerous PF models for alloy solidification have been proposed. Among these models, the Wheeler–Böttinger–McFadden (WBM) and Kim–Kim–Suzuki (KKS) models are two typical models with different mixture compositions defined in the interfacial region.

### 3.2.4.1 WBM Model

Directly extending pure substance solidification to the case of alloys, Wheeler et al. (1993) developed a PF model, which is typically called the classic WBM model, for isothermal solidification of binary alloys. The basic idea of this model is to construct a generalized free-energy functional by weighting two single-phase free energies according to the alloy concentrations. This model is one of the most widely used models for binary alloys, particularly for alloys described by a simple phase diagram with an ideal solution assumption.

In the WBM model, the complete free-energy functional incorporating the bulk and interface effects for a binary alloy with components A and B under isothermal conditions is given through the same basic equation as (3.2):

$$F = \int_V \left[ f(\phi, c) + \frac{1}{2}\varepsilon_c^2(\nabla c)^2 + \frac{1}{2}\varepsilon_\phi^2(\nabla \phi)^2 \right] dV, \tag{3.15}$$

where $c$ is the solute composition (molar fraction), and $\varepsilon_\phi$ and $\varepsilon_c$ are the gradient coefficients of the phase-field and solute composition, respectively. For the ideal solution assumption, a simple choice for the free-energy density $f(\phi, c)$ is given by the following equation:

$$f(\phi, c) = cf_B(\phi) + (1 - c)f_A(\phi) + \frac{RT}{V_m}[c \ln c + (1 - c) \ln (1 - c)], \qquad (3.16)$$

where $R$ is the gas constant, $V_m$ the molar volume, and $f_A(\phi)$ and $f_B(\phi)$ are the free-energy densities of pure $A$ and $B$, respectively. A more complex form of (3.16) might include an excess free energy proportional to $c(1 - c)$ (Wheeler et al., 1993).

According to thermodynamics, the dynamic equations must satisfy the condition that the free energy of the entire system decreases monotonically with time. Under the condition of concentration conservation, minimizing the potential can automatically lead to the distribution of solutes between phases. Thus, the following equations can be derived:

$$\frac{\partial \phi}{\partial t} = -M_\phi \frac{\delta F}{\delta \phi} = -M_\phi \left[ \frac{\partial f}{\partial \phi} - \varepsilon_\phi^2 \nabla^2 \phi \right] \qquad (3.17)$$

$$\frac{\partial c}{\partial t} = \nabla \left( M_c \nabla \frac{\delta F}{\delta c} \right) = \nabla \left[ M_c c(1 - c) \left( \nabla \left( \frac{\partial f}{\partial c} \right) - \varepsilon_c^2 \nabla^2 c \right) \right]. \qquad (3.18)$$

If the gradient energy of the solute can be neglected, the diffusion equation can be derived as follows:

$$\frac{\partial c}{\partial t} = \nabla \left[ M_c c(1 - c) \nabla \left( \frac{\partial f}{\partial c} \right) \right] = \nabla D \nabla c + \nabla M_c c(1 - c) \nabla (f_B - f_A), \qquad (3.19)$$

where $D \equiv M_c RT/V_m$. Compared with the diffusion equation in Fick's approximation, the above equation adds a term proportional to $\nabla(f_B - f_A)$ to account for solute partition in the interface.

To further apply the WBM model under nonisothermal conditions, Warren and Boettinger (1995) further derived the PF equations for the nonisothermal solidification of binary alloys from the entropy functional, which was a breakthrough toward applying the PF model to real materials. Using this model, realistic 2D/3D dendritic growth in binary alloys can be successfully simulated under both isothermal and nonisothermal conditions. Most recently, using the Gibbs energy functions from the CALPHAD energy formalism in place of the Helmholtz free energy density used in WBM model, Rajkumar et al. (2020) modified the governing equations of the WBM model and applied it to Cu–Ni alloys by treating phases as real solutions. This modified model can describe the nonideal behaviors of phases according to the CALPHAD contribution, leading to a more realistic simulation.

### 3.2.4.2    KKS Model

The WBM model, which was derived under consistent thermodynamic conditions, has been widely used. In this model, it is assumed that both the solid and liquid phases have the same composition within the interfacial region. Thus, the interfacial energy depends intrinsically on the local concentration. In 1999, Kim, Kim, and Suzuki

(Kim et al., 1999) proposed a PF model for binary alloys that is called the KKS model. In this model, the interface is defined as a fractionally weighted mixture of solids and liquids. Compared to the WBM model, the KKS model assumes that the solid and liquid phases have different compositions within the interfacial region, which can eliminate the extra energy due to the interpolation of the liquid and solid bulk energies.

Under the constraint of equal chemical potentials, the compositions of the solids and liquids within the interfacial region can be determined by the following equations:

$$c = p(\phi)c_S + [1 - p(\phi)]c_L \tag{3.20}$$

$$f_{c_S}^S(c_S(x, t)) = f_{c_L}^L(c_L(x, t)), \tag{3.21}$$

where $c_S$ and $c_L$ are the local solute compositions of the solid and liquid phases, respectively, while $f_{c_S}^S \equiv df^S(c_S)/dc_S$ and $f_{c_L}^L \equiv df^L(c_L)/dc_L$ are the corresponding chemical potentials. Equation (3.21) does not imply that the entire interfacial region has a constant chemical potential. Only under a state of thermodynamic equilibrium can this chemical potential be constant across the interface.

Then the local free-energy density, $f(c, \phi)$, can be modified as follows:

$$f(c, \phi) = p(\phi)f^S(c_S) + [1 - p(\phi)]f^L(c_L) + Wg(\phi). \tag{3.22}$$

These free-energy densities are dependent on the compositions and can be determined by means of the CALPHAD approach. In the KKS model, the PF dynamic equation and solute diffusion equation are as follows:

$$\frac{\partial \phi}{\partial t} = M_\phi \left( \varepsilon^2 \nabla^2 \phi - \frac{\partial f}{\partial \phi} \right) \tag{3.23}$$

$$\frac{\partial c}{\partial t} = \nabla \cdot \left( \frac{D(\phi)}{f_{cc}} \nabla \left( \frac{\partial f}{\partial c} \right) \right) = \nabla[D(\phi)\nabla c] + \nabla[D(\phi)p'(\phi)(c_L - c_S)\nabla \phi], \tag{3.24}$$

where the solute diffusivity, $D(\phi)$, is dependent on the phase-field variable $\phi$, and $F_{cc}$ is the second derivative of the chemical free-energy density with respect to the solute concentration $c$.

The main difference between the KKS and the WBM models is the definition of the free-energy density within the interfacial region under an equilibrium state (Kim et al., 1999). In the KKS model, the interfacial region at equilibrium is defined as a mixture of a solid and liquid with different compositions. In the WBM model, however, this is defined as a mixture of solid and liquid phases with the same composition. Moreover, in the WBM model, besides an imposed double-well potential $Wg(\phi_0)$, there is an extra double-well potential originating from the definition of the free-energy density at the interface. Compared to the double-well potential $Wg(\phi_0)$, this extra potential can be ignored only in the sharp interface limit where $W \to \infty$ or in an alloy with an extremely small difference between the equilibrium composition of liquid and solid phases $c_L^e - c_S^e$. Nevertheless, with the increase of the interface width or $c_L^e - c_S^e$, the height of this extra potential becomes too large to be ignored. In the KKS model, however, the term in the free energy, $p(\phi_0)f^S(c_S^e) + [1 - p(\phi_0)]f^L(c_L^e)$, corresponds to the common tangent line itself exactly. Under this circumstance, no extra potential exists (Kim et al., 1999).

Although the WBM and KKS models have been widely employed to model the solidification of binary alloys, both models have their own strengths and weaknesses. The WBM model is very straightforward and thus easy to implement. However, because the bulk energy contributes to the interfacial energy, the analytical relationship between the model parameters and the physically measurable values cannot be determined conveniently. Then the parameters are typically determined numerically. Moreover, for given kinetic and thermodynamic properties, this method typically results in a thinner interface, which makes it difficult to choose a wider interface to reduce the computational costs. The most attractive advantage of KKS model is its considerably looser restriction on the interfacial thickness compared with other models. In consequence, it is very suitable for applications at a practical length scale. In this model, the bulk free energy no longer contributes to the interfacial energy. Thus, this kind of relationship between the interfacial energy and the interface thickness can be determined under equilibrium conditions. In addition, increasing the interface width does not change the thermodynamic and kinetic properties. As a result, a larger-scale simulation can be achieved. However, the KKS model introduces two additional degrees of freedom, which increases the computational cost (Moelans et al., 2008).

In addition, if the difference between the diffusion coefficients on both sides of the interface is very large, some artificial nonequilibrium effects, such as extra surface diffusion, interface stretching, and solute trapping, could occur (Karma, 2001). To overcome these artificial effects, Karma (2001) developed a quantitative one-sided PF model for solidification of dilute binary alloys with zero diffusivity in the solid. In Karma's quantitative PF model for solidification of binary alloys, a nonvariational antitrapping current term is added to the diffusion equation to counterbalance the solute trapped in the artificially wide interfaces.

## 3.2.5    Multiphase-Field Model

With the development of the PF method, its basic principles have been used to develop numerous so-called multiphase or multi-order parameter PF models, aimed at describing the polycrystalline, multiphase, or multicomponent phenomena in phase transformations. There are primarily three types of approaches for constructing multi-PF models: the incorporation of multiple order parameters, the introduction of orientational order parameters, and the introduction of multiple phase fields. Chen and Yang (1994) developed a multi-PF model in which the microstructure is described by nonconserved order parameters related to specific phases or crystalline orientations. This multi-PF model has been widely applied in the modeling of solid–solid phase transitions involving multiple grains. Employing the grain orientation as an order parameter, Kobayashi and Warren (Kobayashi et al., 1998; Warren et al., 2003) developed a multi-PF model that allows the simulation of solidification, grain growth, and grain rotation in multiple grain structures. Steinbach et al. (1996) and Kim et al. (2006) extended the single-PF formalism toward multiphase systems and formulated a multi-PF model, in which each phase is identified with an individual PF variable, and the transformation between all two phases is treated with its own characteristics.

In the following, taking polycrystalline solidification as an example, we will briefly introduce the multi-PF model proposed by Li et al. (2012). In the multi-PF model, a system with $p$ coexisting phases is described by $p$ PF variables ($\phi_K$) representing the local phase fractions. Hence, the PF variables must sum to 1 at any point in the system, $\sum_{j=1\ldots p}\phi_j = 1$, and only $p$-1 PF variables are independent. The phase state with $\phi_i = 1 (i \neq 1)$ represents the solid phase but with different crystallographic orientations, whereas $\phi_i = 0 (i \neq 1)$ represents the liquid phase, $\phi_1 = 1$. In the dilute solution limit, the dynamics of the PF variables are controlled by

$$\frac{\partial \phi_i}{\partial t} = -\frac{2}{N}\sum_{j\neq i}^{n} s_i s_j M_{ij}\left(\frac{\delta f}{\delta \phi_i} - \frac{\delta f}{\delta \phi_j} + \Delta g_{ij}\right), \tag{3.25}$$

where $s_i$ is a step function ($s_i = 1$ if $\phi_i > 0$, while $s_i = 0$, otherwise), and

$$\frac{\delta f}{\delta \phi_i} = \begin{cases} \sum_{j\neq 1}^{n}\left(\frac{\varepsilon_{SL}^2}{2}\nabla^2\phi_j + W_{SL}\phi_j\right) & \text{for } i = 1 \\ \frac{\varepsilon_{SL}^2}{2}\nabla^2\phi_1 + W_{SL}\phi_1 + \sum_{j\neq 1, i}^{n}\left(\frac{\varepsilon_{SS}^2}{2}\nabla^2\phi_j + W_{SS}\phi_j\right) & \text{for } i \neq 1 \end{cases} \tag{3.26}$$

$$\Delta g_{ij} = \begin{cases} 0 & \text{for } i \neq 1 \text{ and } j \neq 1 \\ 6\phi_1(1 - \phi_1)\Delta S(T_m - T - m_L c_L) & \text{for } i = 1 \text{ and } j \neq 1 . \\ -6\phi_1(1 - \phi_1)\Delta S(T_m - T - m_L c_L) & \text{for } i \neq 1 \text{ and } j = 1 \end{cases} \tag{3.27}$$

To ensure that the chemical potentials of the solid and liquid are equal, an antitrapping current term is added to the solute diffusion equation as follows:

$$\frac{\partial c}{\partial t} = \nabla\phi_1 D_L\nabla c_L + \nabla\left(\frac{\varepsilon_{SL}}{\sqrt{2W_{SL}}}(c_L - c_S)\sqrt{\phi_1(1 - \phi_1)}\frac{\partial\phi_1}{\partial t}\frac{\nabla\phi_1}{|\nabla\phi_1|}\right), \tag{3.28}$$

where the concentrated mixture of solid and liquid is $c = \phi_1 c_L + (1 - \phi_1)c_S$. The relations among parameters $\varepsilon$ and $W$, with subscripts $SL$ or $SS$, the interfacial energy ($\sigma$), and interface width ($2\delta$) of the solid–liquid or solid–solid interface are as follows:

$$\varepsilon_{SL} = \frac{4}{\pi}\sqrt{\delta_{SL}\sigma_{SL}}; W_{SL} = \frac{2\sigma_{SL}}{\delta_{SL}}; \varepsilon_{SS} = \frac{4}{\pi}\sqrt{\delta_{SS}\sigma_{SS}}; W_{SS} = \frac{2\sigma_{SS}}{\delta_{SS}}. \tag{3.29}$$

The interface mobility ($M_{SL}$) can be determined as follows:

$$M_{SL} = \frac{8\sigma D_L\sqrt{2W_{SL}}}{\pi\Delta S m_L c_L^e(1 - k)\varepsilon_{SL}^3}. \tag{3.30}$$

Most multi-PF models have been applied to grain growth, coarsening, and, more recently, multiphase solidification/precipitation. Some models also incorporate elastic effects to study the role of strain in phase transformations. Other models, particularly those introducing an orientational order parameter, have been used predominantly to examine dendritic solidification and the subsequent formation of a polycrystalline network. In addition, by allowing the description of multiple grains in a microstructure, multi-PF models have been used to consider the occurrence of multiple thermodynamic phases in grades of technical alloys that may have numerous grains, phases, and components. It is therefore predictable that multi-PF models will become increasingly important in the modeling and simulation of microstructural evolution, particularly for technical alloys.

## 3.2.6      Phase-Field Crystal Model

In the coarse-grained PF method, a PF variable that is spatially uniform in bulk phases at equilibrium is used to model the phase transformation. Although the simplicity of this description is advantageous for computational and analytical calculations, this method has deficiencies. The coarse-grained PF model operates at a length scale much larger than the interatomic distances, and therefore cannot describe the atomistic features, including crystal defects. Thus, although the PF method can predict microstructural evolution during phase transformations from a more statistical perspective, it is generally unable to handle structural evolution at atomic scales. In addition, in the coarse-grained PF model, the formulation requires uniform fields that are in equilibrium. Consequently, properties such as elasticity, anisotropy, and grain orientations are not inherent in the physical description for the model (Emmerich et al., 2012; Provatas and Elder, 2010).

As a new paradigm for modeling material behavior, the phase-field crystal (PFC) methodology provides a representation of the atomic density field at the continuum scale, which evolves on diffusive time scales. The PFC model is a simple dynamical density functional theory that automatically incorporates the crystal structure, anisotropy, elasticity, and crystal defects. Thus, it contains considerably richer physics than the coarse-grained PF models, including atomic-scale descriptions of the elasticity, dislocations, grain rotation, and grain boundaries. This allows for the analysis of various phenomena that are unapproachable using the coarse-grained PF method. The PFC model has shown great potential for addressing the shortcomings of the PF method and has emerged as a powerful computational technique that can capture atomic-scale processes at time scales much larger than that of molecular dynamics (MD) approaches.

In addition, owing to the time scale of the PFC model, diffusive behaviors can be studied. Thus, the PFC method is a potential modeling technique that can bridge the gap between the continuum PF technique and atomistic MD simulations. This new method has great potential in modeling issues in materials science, including solidification, precipitation, dislocation motion, dislocation interaction, and various other applications (Emmerich et al., 2012; Provatas and Elder, 2010). As the primary difference between the PFC and PF models, the order parameter is refashioned as temporally coarse but is spatially described by an atomic probability density. This is

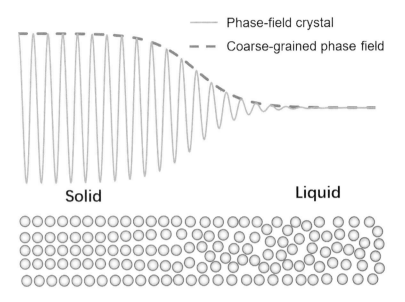

**Figure 3.4** Schematic representation of the field profile in the PFC method and the PF method.

achieved by choosing a free-energy functional that is minimized by an atomic probability density that is spatially periodic for crystalline solids and constant for liquids. As shown in Figure 3.4, the PFC model relinquishes the mesoscale order parameter $\phi(\mathbf{r}, t)$ of the coarse-grained PF method and uses a local time-averaged atomic probability density $\varphi(x)$. For the solid (crystalline) phase, a periodic symmetry of the atomic density, $\varphi(x) = \varphi_0(x) + \xi$, is used, where $\xi$ is the Gaussian fluctuation of the density field, while for the liquid phase, a constant atomic density, $\varphi(x) = \varphi_0(x)$, is applied. This description incorporates the necessary physics to describe phenomena such as surface tension, elasticity, anisotropy, and lattice defects.

Here, we briefly introduce the single-mode PFC model derived from the density functional theory by expanding the direct dual-particle correlation function, $C_2(k)$. The dimensionless free energy can be expressed in terms of the reduced density ($\varphi$) as follows:

$$F = \int d\mathbf{r} \left\{ \frac{\varphi}{2} \left[ a + \lambda \left( q_0^2 + \nabla^2 \right)^2 \right] \varphi + g \frac{\varphi^4}{4} \right\}. \tag{3.31}$$

Here, the order parameter ($\varphi$) represents the reduced atomic number density relative to a reference liquid state, and the parameter $q_0$ is the position of the first peak in the structure factor, $S(k)$, of the liquid phase close to its melting point. The parameters $a$ and $\lambda$ can be obtained from the polynomial approximation of $S(k)$. Similar to the nonconservative order parameters in the PF model, the dynamic equation of the reduced atomic number density is given as follows:

$$\frac{\partial \varphi}{\partial t} = M \nabla^2 \frac{\delta F}{\delta \varphi}. \tag{3.32}$$

By defining $\vec{x} = \vec{r}q_0$, $\psi = \varphi\sqrt{\frac{g}{\lambda q_0^4}}$, $\varepsilon = -\frac{a}{\lambda q_0^4}$, and $\tilde{F} = \frac{g}{\lambda^2 q_0^5}F$, the dimensionless free-energy functional can be rewritten as follows:

$$\tilde{F} = \int d\vec{x} \left\{ \frac{\psi}{2}\left[ -\varepsilon + \left(\nabla^2 + 1\right)^2 \right]\psi + \frac{\psi^4}{4} \right\}. \tag{3.33}$$

In the original PFC model, the conserved dynamic equation has the dimensionless form

$$\frac{\partial \psi}{\partial \tau} = \nabla^2 \left(\delta\tilde{F}/\delta\psi\right) = \nabla^2 \left\{ \left[\varepsilon + \left(\nabla^2 + 1\right)^2\right]\psi + \psi^3 \right\}, \tag{3.34}$$

where the dimensionless time $\tau = \Gamma\lambda q_0^6 t$.

Since the original PFC model was initially proposed, many improvements have been made, such as new models for stabilizing complex crystal structures, improved predictive capabilities, and connections between model parameters and real physical properties. Initially, the original PFC model proposed by Elder et al. (2007) was only suitable for simulating systems with simple crystal structures, such as 2D hexagonal and 3D body-centered cubic (Bcc) lattices. It was later proved that the original PFC model can also stabilize face-centered cubic (Fcc) and hexagonal close-packed (Hcp) structures. The three-mode PFC model can even describe all five Bravais lattices and other more complex structures, including honeycomb and kagome lattices (Guo et al., 2015).

The PFC method is becoming an increasingly comprehensive tool for capturing the phase diagrams, nucleation events, and microstructure evolution of materials. In the past decade, the PFC method has been widely employed to simulate phenomena related to solidification, solid–solid phase transitions, grain boundary migration, and dislocation motion, among others. PFC models have also been extended to binary and multicomponent alloy systems as well as colloids, liquid crystals, quasicrystalline, and ferromagnetic materials. Figure 3.5 shows an example of simulation results using a binary PFC model (Tegze et al., 2009). The simulated dendritic and eutectic solidification structures demonstrate that binary eutectic solidification may also lead to the formation of eutectic colonies.

Another important good point of the PFC model is that its length scale can be extended to the mesoscale by a renormalization group or multiscale expansion technique (Provatas et al., 2007). This is particularly important because of its significant potential to bridge the gap between the PFC models and the coarse-grained PF models. By providing a seamless model that directly illustrates the atomic-scale phenomena during microstructural evolution, PFC-PF integration will be a breakthrough in the development of true multiscale models. As an important step in this integration, the amplitude expansion of a PFC model can be developed. The amplitude models not only can preserve the salient features of the underlying PFC models, but they also can resolve the microstructures at the mesoscale, similar to the coarse-grained PF method. The applicability of this complex amplitude model has been confirmed by simulations of peritectic solidification and grain growth (Tegze et al., 2009).

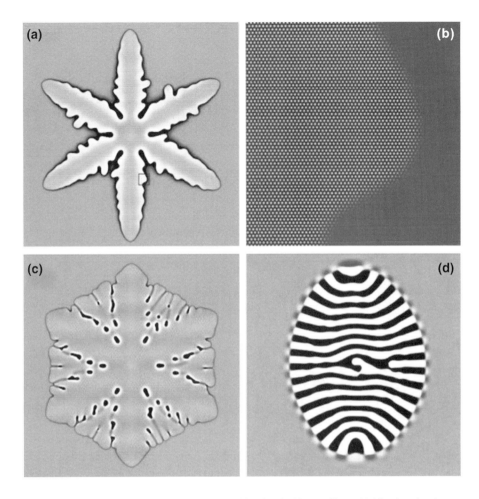

**Figure 3.5** Illustrative PFC simulations for solidification in binary alloys. (a) Number density difference map of a solutal dendrite. (b) Total number density map of the small square area shown in (a). (c) A more compact dendritic structure grown at a higher driving force. (d) Eutectic structure. This figure is reproduced from Tegze et al. (2009) with permission from Elsevier. A black and white version of this figure will appear in some formats. For the colour version, please refer to the plate section.

The PFC methodology is a promising technique that can operate on diffusional time scales and atomic spatial scales, and thus shows great potential in dealing with defects and deformations, particularly in coupled diffusional-displacive processes. However, to make this promising method a material-specific simulation tool, significant developments are still required.

## 3.2.7    Microscopic Phase-Field Models

In addition to the PFC model, the microscopic phase-field (MPF) model is another type of PF model that can handle structural events at atomic scales. This model was first proposed by Khachaturyan and further developed by Poduri and Chen (1998). In essence, the MPF is the discrete lattice form of the Cahn–Hilliard equation at the atomic scales. In 1968, to simulate the phase separation phenomena, Khachaturyan proposed an improved Ginzburg–Landau-type continuum field kinetic model, also known as the microscopic lattice diffusion theory. In this model, the macroscopic kinetic coefficients that appear in the Ginzburg–Landau or Onsager equations can be calculated at the microscopic scale (Khachaturyan, 1983). Using this approach, Poduri and Chen (1998) successfully investigated the kinetics of order–disorder transitions, phase separation, and morphological evolution, as well as the coarsening kinetics of $L1_2$-ordered (Al$_3$Li) precipitates in Al–Li alloys.

In the MPF model, the atomic structure and shapes of the phase can be described by the microscopic field represented with the single-site occupation probability function $\chi(\mathbf{r}, t)$. This function represents the probability of a solute atom being observed at the lattice site $\mathbf{r}$ and time $t$ in real space. It is an average of the occupation number $c(\mathbf{r})$ over the time-dependent ensemble. Here, the occupation number $c(\mathbf{r})$ is described by the following equation:

$$c(\mathbf{r}) = \begin{cases} 1, \text{ if a site } \mathbf{r} \text{ is occupied by a solute atom} \\ 0, \text{ otherwise.} \end{cases} \tag{3.35}$$

The Ginzburg–Landau or Onsager-type kinetic equation is used to describe the dynamic evolution of this microscopic field as follows:

$$\frac{\partial \chi(\mathbf{r}, t)}{\partial t} = \sum_{\mathbf{r}'} L(\mathbf{r} - \mathbf{r}') \frac{\delta F}{\delta \chi(\mathbf{r}', t)}, \tag{3.36}$$

where $F$ is the free-energy functional, and $L(\mathbf{r} - \mathbf{r}')$ is the symmetric microscopic kinetic matrix.

The MPF model can be used to analyze individual dislocation cores through the generalized stacking fault energy function. A unique feature of this model is that it can predict the fundamental properties of individual defects, such as the size, formation energy, and activation energy of the defect nuclei, and the mechanisms of their mutual interactions (Wang and Li, 2010). The model can directly use the physical properties, surface energy, and elastic constants from ab initio and atomistic modeling as inputs. With the MPF model, Qiu et al. (2019) predicted the equilibrium dislocation structures associated with $(1\bar{1}0)$ pure twisted grain boundaries (GBs) and their energies in five different Bcc metals ($\beta$-Ti, Mo, Nb, W, and Ta). The generalized stacking fault energy surfaces and elastic constants obtained from atomistic simulations are employed as inputs to the MPF model. Figure 3.6 shows a comparison of the GB structures among the different Bcc crystals.

The MPF models can be regarded as an application of the coarse-grained PF theory at or near the atomic scale, and thus have significant promise in bridging from first-principle calculation and atomistic modeling to the simulation of partial dislocations on time-dependent deformation phenomena (Louchez et al., 2017). In addition, because it directly uses inputs from ab initio calculations and is independent of the accuracy of the interatomic potential, the MPF model is much more efficient than full

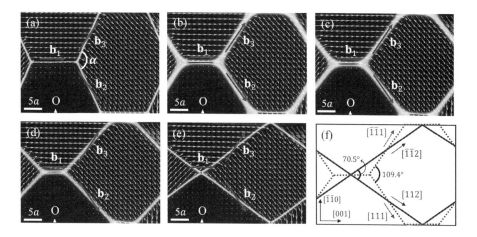

**Figure 3.6** Equilibrium dislocation structures within $(1\bar{1}0)$GB in (a) Nb, (b) Mo, (c) W, (d) Ta, (e) $\beta$ – Ti, together with (f) a schematic drawing of two special states of equilibrium dislocation structures in Bcc crystals. This figure is reproduced from Qiu et al. (2019) with permission from Elsevier. A black and white version of this figure will appear in some formats. For the colour version, please refer to the plate section.

atomistic simulations. However, under the assumption that continuum theory is still applicable at the atomic scale, this model can only address a limited number of dislocations with near-atomic-scale accuracy.

The PF approach has been given increasing prominence in modeling pattern formation and microstructural evolution applied in numerous materials science and engineering applications. As pointed out by Wang and Li (2010), this approach can be applied to single defect levels and coarse-grained levels. When applied at a single-defect scale, the MPF model is a superposition of the Cahn–Hilliard description of chemical inhomogeneity and the Peierls description of displacive inhomogeneity. When applied at the subangstrom scale, the PFC model can describe thermally averaged atomic configurations and motions of defects on diffusional time scales. When applied at the mesoscopic scale, the coarse-grained PF models can predict the microstructural evolution involving both chemical and mechanical interactions through coupled displacive and diffusional mechanisms (Wang and Li, 2010).

## 3.2.8 Applications of the Phase-Field Method

In the past three decades, the PF method has been thoroughly developed and has gradually become a powerful and promising method in computational materials science owing to several important characteristics, such as the diffuse interface; lack of interface tracking; automatic accounting of the interfacial energy contributions; natural consideration of topological singularities; easy coupling with stress, magnetic, and electric fields; and a particular suitability for visualizing microstructural developments. The PF methodology has become one of the most successful techniques for modeling microstructural evolution and has been widely applied to various material processes,

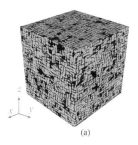
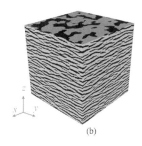
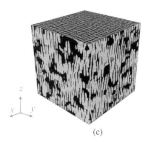

(a)  (b)  (c)

**Figure 3.7** (a) Simulated $\gamma/\gamma'$ microstructures in single-crystal Ni–Al alloys without external stress after 4.7 hours aging at 1,300 K. Rafted $\gamma/\gamma'$ microstructure developed from (a) after 5.67 hours aging under 152 MPa tension along [001]; (b) for negative misfit $(-0.3\%)$; and (c) for positive misfit $(+0.3\%)$. Figure 3.7 is reproduced from Zhou et al. (2010) with permission from Taylor and Francis.

including solidification, solid–solid phase transitions, and other microstructural changes (Dantzig et al., 2013).

One successful application of the PF method to obtain a fundamental understanding of microstructural evolution is the quantitative modeling of $\gamma'$ rafting in Ni-based superalloys. Zhou et al. (2010) performed a 3D PF simulation of $\gamma/\gamma'$ microstructural evolution coupled with plastic deformation. Figure 3.7 shows the simulated rafting microstructures of single-crystal Ni–Al alloys with positive and negative lattice misfits. In addition to the similar N-type and P-type rafting morphologies under uniaxial loading, the simulated microstructures show a remarkable resemblance to those observed during experiments regarding both the microstructure morphologies and time scale. Here, the N-type (P-type) rafting represents that $\gamma'$ precipitates directionally coarsen normal to (parallel to) the direction of external loading. Zhou et al. (2010) also considered the plastic deformation in the $\gamma$-channel, which is described by the local channel dislocation density of a single active sliding system and performed micrometer-level simulations. With the help of the PF method, the interaction between elastic inhomogeneity and channel plasticity can be characterized, thereby providing potential opportunities for new material-design strategies (Zhou et al., 2010).

## 3.3    Cellular Automaton Method

Since the 1990s, the PF method has been introduced to the field of modeling and simulation of microstructural evolution because of its high resolution in the simulation domains of both bulk phases and interfaces. However, the high-quality simulation results come at a cost: The PF method usually requires considerable calculations. Even with currently available computing resources, PF simulations are still computationally expensive for large spatial domains or long temporal ranges, particularly for 3D cases.

The cellular automaton (CA) method is a well-known general-purpose numerical method used to simulate different processes and phenomena. This method offers the possibility of simulating large domains or long temporal ranges with a moderate computational demand. The CA method was originally proposed as an algorithm to describe the discrete space-time evolution of complex systems based on the interaction between the automata themselves and with the defined environment (neighborhood). At present, it has become a discrete model used in many disciplines, such as mathematics, physics, biology, computability theory, and materials science, and has been widely used to model and simulate microstructural evolution.

## 3.3.1    Historical Background

The CA method was introduced by von Neumann at the end of the 1940s (Aspray and Burks, 1987), aiming to reproduce complex physical phenomena with simple rules defined at the microscopic level and simulate self-reproducing Turing automata and population evolution. In 1970, Conway (Allan, 1974) introduced the famous Game of Life, arguably the most popular automaton ever. During the 1980s, Wolfram (1986) extensively studied one-dimensional CA and provided the first qualitative taxonomy of its behavior. This study laid the groundwork for further research and helped to place the growing community of CA researchers on the scientific map. Later applications primarily described the nonlinear dynamic behaviors of fluids and reaction-diffusion systems (Raabe, 2002; Reuther and Rettenmayr, 2014). During the past two decades, CA methods have gained increasing momentum for modeling microstructural evolution, especially since the noticeable success in predicting solidified grain structures in 1990s (Rappaz, 1993).

## 3.3.2    Principles of the Cellular Automaton Method

The CA method is an algorithm describing the spatiotemporal evolution of complex systems in discrete lattices. CA is a dynamic grid model with discrete-time, space and state, spatial interaction, and time causality. Unlike general dynamic models, CA is not determined by strictly defined equations or functions with a solid physical background, but rather by transition rules constructed by the model. In principle, any model that employs this idea can be considered a CA model. Accordingly, the term CA is a general expression for a class of models or a method framework.

In the CA method, the simulation domain is divided into multiple cells that contain all necessary information to describe a given process. Each cell will change its state over time according to the defined rules (including the state of adjacent cells), and the state of a cell depends on the states of its own and its immediate neighbors in the previous time step. Figure 3.8 schematically shows the construction of a CA model for solidification in 2D. Each cell is assigned information about the state (for solidification, the states include a solid phase, a liquid phase, and an interfacial region) and the calculated fields (such as the temperature field, concentration field, and flow field). The change in the cell state is calculated using analytical or probabilistic transition rules, while the field of the cell is calculated using the transport and transformation equations.

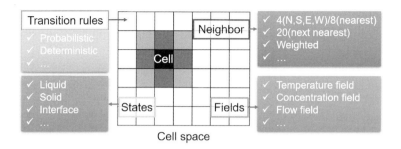

**Figure 3.8** Schematics of construction of a CA model for solidification.

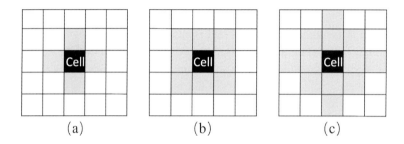

**Figure 3.9** Different types of neighborhood (in gray) in the CA method: (a) the von Neumann neighborhood, (b) the Moore neighborhood, and (c) the extended Moore neighborhood.

In general, the construction of a CA model includes four steps. First, the simulation space is divided into a lattice of cells. In 2D, a square or hexagonal lattice can be applied, whereas in 3D, a cubic lattice is generally adopted. Then, a neighborhood for the cells is defined, which can be modified by introducing weighted neighbors, indicating that their influences on other cells are taken with certain weights. Next, one or more state variables are defined for each cell. Finally, the transition rules are determined for the cells. The state of a cell at the subsequent step is determined by the transition rules according to the current state of the cell and its neighborhood.

Classical cells have classical neighborhood definitions, the most famous being the definitions of von Neumann and Moore neighborhoods on 2D square cells, as shown in Figure 3.9. The von Neumann neighborhood is a region where each cell interacts with only its four horizontal and vertical neighbors, while the Moore neighborhood includes all eight immediately adjacent cells. For the hexagonal cell lattice, only the nearest neighbors are typically used, but defining an environment with more distant neighbors is only impeded by the difficultly in the computational implementation.

### 3.3.3 Classical Cellular Automaton Model

Rappaz (1993) was the first to use the CA method to predict solidified grain structures with arbitrary preferential crystallographic orientations in both 2D and 3D, by considering the nucleation, growth kinetics of dendrite tips, and crystallographic orientations. The primary idea of a CA model for solidification is to describe the evolution of the interface by capturing nearby liquid cells into the solid phase based on some transition rules. When a cell is liquid, the index of the cell state is zero, while when it is solid, the index is an integer representing the crystallographic orientation. The initial position and crystallographic orientation of the crystal grains are randomly selected from numerous cells and a certain number of orientations, respectively. In the classical CA model for grain growth, the transition rules include the mechanisms of heterogeneous nucleation and grain growth. Thus, the evolution of the solidified grain structure is governed by the following two submodels.

### 3.3.3.1 Nucleation Model

A continuous nucleation model is typically used, which considers two different Gaussian distributions to describe heterogeneous nucleation both at the mold–casting interface and in the bulk melts. The relationship between the grain density increase, $d(n)$, and undercooling, $d(\Delta T)$, is described by the Gaussian distribution:

$$\frac{d(n)}{d(\Delta T)} = \frac{n_{max}}{\sqrt{2\pi}\Delta T_\sigma} \exp\left[-\frac{1}{2}\left(\frac{\Delta T - \Delta T_{mn}}{\Delta T_\sigma}\right)^2\right], \tag{3.37}$$

where $\Delta T_{mn}$ is the mean nucleation undercooling (averaged over all liquid cells), $\Delta T_\sigma$ the standard deviation, and $n_{max}$ the maximum density of the nuclei obtained from the integral of this distribution from $\Delta T = 0$ to $\infty$ (Rappaz, 1993). The density of grains $n(\Delta T)$ is determined by the following:

$$n(\Delta T) = \int\limits_0^{\Delta T} \frac{d(n)}{d(\Delta T')} d(\Delta T').$$

When nuclei form on the mold wall or in the melts, the corresponding cell is assumed to have a random crystallographic orientation.

### 3.3.3.2 Grain Growth Model

Initially, since the adopted transition rules are very simple, the CA model for grain growth cannot resolve the details of the dendritic morphologies. To address this shortcoming, a combination of envelopes, which is defined as "smooth surfaces" surrounding all dendritic tips around a grain, is simulated to represent the dendritic structures. This means that the final simulation result contains dendritic grains rather than dendritic crystals. In this model, the kinetic theory of dendritic tip growth is used, and the growth velocity is thus averaged for a given element. In this respect, this model greatly depends on averaged quantities, similar to the traditional continuum deterministic models. However, unlike these traditional models, it can show the grain structures. First, a "nucleation" cell, $v$, is considered, in which a new grain has formed at time $t_n$, as shown in Figure 3.10. The primary [10] and [01] crystallographic directions create angles of $\theta$ and $\theta + \pi/2$, respectively, with respect to the horizontal axis. The temperature of this "nucleation" cell is assumed to be locally uniform, and

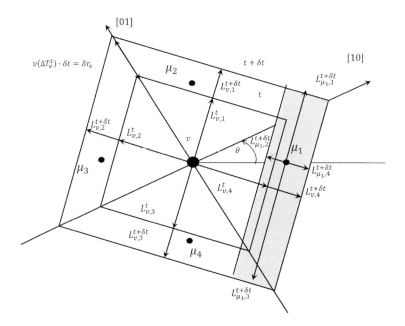

**Figure 3.10** Schematic diagrams illustrating the growth algorithm used in the CA model. This figure is reproduced from Rappaz (1993) with permission from Elsevier.

the "nucleation" cell is surrounded by four nearest-neighbor cells, labeled $\mu_1$, $\mu_2$, $\mu_3$, and $\mu_4$. Assuming that the dendritic structure in the cell $v$ develops into a square envelope, the four semidiagonals of the square $(j = 1, 4)$ correspond to the direction of the dendrite tip. Then the extension $L_{v,j}^t$ can be given as follows:

$$L_{v,j}^t = \int_{t_n}^{t} v(\Delta T)d\tau. \tag{3.38}$$

Here, $v(\Delta T)$ is the growth rate of dendrite tip as a function of undercooling, $\Delta T = T_l - T$, where $T_l$ is the liquidus temperature, and $T$ is the thermal history of the cell given by the temperature field calculation. To improve the computational efficiency, $v(\Delta T)$ is usually related to the grain growth with the local undercooling by a polynomial function based on the Lipton–Glicksman–Kurz (LGK) model:

$$v(\Delta T) = k_1 \Delta T + k_2 \Delta T^2 + k_3 \Delta T^3, \tag{3.39}$$

where $k_1$, $k_2$, and $k_3$ are the fitting coefficients. The implementation details for this method were provided by Rappaz (1993). In addition to the LGK model, the Kurz–Giovanola–Trivedi (KGT) model, extended KGT model, and PF model are also used to establish the dependence of the dendrite tip growth velocity on the local supercooling.

In 1994, Gandin and Rappaz (1994) further proposed a mesoscale model by combining the CA growth model and a finite element heat flow model, which is named as CA-FE model. In the CA-FE model, the explicit temperature is interpolated at a cell location firstly and used in the subsequent CA calculation. Then the nucleation and growth of the grains are simulated using the CA algorithm. The potential of this CA-FE model has been demonstrated by the predictions of typical grain structures formed during the investment and continuous casting processes (Gandin and Rappaz, 1994), which has been adopted in the commercial software ProCast for modeling and simulation of casting process.

## 3.3.4    Modified Cellular Automation Model

To simulate the dendritic structures formed during solidification, Zhu and Hong (2001) proposed a modified CA model based on the classical CA model for grain growth. Unlike the classical CA model, where only the temperature field is incorporated, the modified CA model took into account the effects of both constitutional and curvature undercooling on the equilibrium interface temperature, thus the solute field is also included. Moreover, a coupled transport model can also be incorporated to calculate the moment, heat, and mass transfer from both convection and diffusion. This modified model has been applied in simulating the growth of dendrites in both 2D and 3D, dendritic growth with melt convection, and regular and irregular eutectic solidification. It has also been successfully extended from solidification to solid-state phase transitions.

In the modified CA model, each cell has three types of status (solid, liquid, or interface). To develop a physically sound model for dendritic growth, every cell is characterized by temperature, concentration, solid fraction, and crystallographic orientation, among other factors (Beltran-Sanchez and Stefanescu, 2004). These quantities are used to solve the diffusion equations or to adjust the solid fraction to fulfill the condition of local thermodynamic equilibrium of the interface cells. The calculation procedure for the modified CA model is as follows (Reuther and Rettenmayr, 2014): (1) calculate diffusion in the bulk phase and at the solid–liquid interface; (2) couple the evolution of the solid fraction with the concentration field; (3) determine the phase state of the cell; and (4) calculate the interface geometry. Therefore, choosing different algorithms for each of the four submodels can characterize a CA model of dendritic growth. The nucleation issue in the modified CA model is the same as that in the classical CA model.

### 3.3.4.1    Growth Kinetics and Orientation

According to the solidification theory, the total undercooling at the dendrite tip, $\Delta T$, includes four individual undercooling contributions:

$$\Delta T = \Delta T_c + \Delta T_t + \Delta T_r + \Delta T_k, \tag{3.40}$$

where $\Delta T_c$, $\Delta T_t$, $\Delta T_r$, and $\Delta T_k$ are the constitutional, thermal, curvature, and kinetic undercooling, respectively. Then, based on the KGT or LKT model, the growth kinetics of the dendrite tip can be obtained.

The growth kinetics can also be calculated by solute conservation at the interface:

$$V_n\left(c_L^* - c_S^*\right) = \left[-D_L\left(\frac{\partial c_L}{\partial x} + \frac{\partial c_L}{\partial y}\right) + D_S\left(\frac{\partial c_S}{\partial x} + \frac{\partial c_S}{\partial y}\right)\right] \cdot \vec{n}, \qquad (3.41)$$

where $\vec{n}$ is the interface normal vector; $V_n$ is the interface growth velocity along the normal direction; $c_S\left(c_S^*\right)$ and $c_L\left(c_L^*\right)$ are the concentrations (equilibrium) of the solid and liquid phases, respectively; and $D_S$ and $D_L$ are the solute diffusion coefficients of the solid and liquid, respectively.

### 3.3.4.2    Solute Redistribution

Solute redistribution consists of two steps:

(1) The assumption of local equilibrium is satisfied at the solid–liquid interface as follows:

$$c_S^* = kc_L^*, \qquad (3.42)$$

where $k$ is the solute redistribution coefficient.

(2) During solidification, the solute field is controlled by diffusion in liquids and solids. Thus, diffusion must be simulated within the entire domain. For single-phase solidification, the governing equation for solute redistribution reads

$$\frac{\partial c}{\partial t} = D\nabla^2 c + c(1-k)\frac{\partial f_s}{\partial t}, \qquad (3.43)$$

where $f_s$ is the fraction of the solid phase.

For eutectic/peritectic solidification, the concentration field is governed by

$$\frac{\partial c}{\partial t} = D\nabla^2 c + c(1-k_\alpha)\frac{\partial f_{s,\alpha}}{\partial t} + c(1-k_\beta)\frac{\partial f_{s,\beta}}{\partial t}, \qquad (3.44)$$

where $f_{s,\alpha}$ and $f_{s,\beta}$ are the solid fractions for the $\alpha$ and $\beta$ phases, respectively, and $k_\alpha$ and $k_\beta$ are the partition coefficients for the $\alpha$ and $\beta$ phases, respectively.

### 3.3.4.3    Thermal Transport

For transient heat conduction, the governing equation is

$$\rho C_p \frac{\partial T}{\partial t} = \lambda \nabla^2 T + \rho L \frac{\partial f_s}{\partial t}, \qquad (3.45)$$

where $\rho$ is the density, $C_p$ the heat capacity, $\lambda$ the thermal conductivity, and $L$ the latent heat of solidification.

### 3.3.4.4    Interface Curvature

The dependencies of the interface temperature, $T^*$, and interface concentration, $c_L^*$, with interface curvatures are

$$T^* = T_L + \left(c_L^* - c_0\right)m_L - \Gamma\bar{\kappa}f(\vartheta, \theta) \tag{3.46}$$

$$c_L^* = c_0 + \left(T^* - T_L^e + \Gamma\bar{\kappa}f(\vartheta, \theta)\right)m_L^{-1}, \tag{3.47}$$

where $\Gamma$ is the Gibbs–Thomson coefficient, $T_L^e$ the liquidus temperature at the initial composition $c_0$, and $m_L$ the slope of the liquidus. The average interface curvature $\bar{\kappa}$ for a cell with a solid fraction $f_s$ can be obtained from

$$\bar{\kappa} = \frac{1}{a}\left[1 - 2\left(f_s + \sum_{k=1}^{N}f_s(k)\right)\bigg/(N+1)\right], \tag{3.48}$$

where $N$ is the number of neighboring cells.

The anisotropy of the surface tension $f(\vartheta, \theta)$ can be calculated by

$$f(\vartheta, \theta) = 1 + \varepsilon_K \cos\left[4(\vartheta - \theta)\right], \tag{3.49}$$

with the growth angle $\vartheta = \arccos\left[\dfrac{\partial f_s/\partial x}{\left((\partial f_s/\partial x)^2 + (\partial f_s/\partial y)^2\right)^{1/2}}\right]$, where $\varepsilon_K$ accounts for the degree of anisotropy and $\theta$ represents the preferential crystallographic orientation.

## 3.3.5    Discussion Regarding Grid Anisotropy

Compared with other numerical methods for microstructural evolution, including the front-tracking method, the PF method, and the level-set method, the CA method is relatively easy to implement and requires fewer computing resources. However, the discrete nature of the CA method impacts its accuracy. It has been shown that the CA method tends to bias the results by introducing a lattice-related anisotropy, and therefore lacks the ability to precisely account for the effect of interfacial energy anisotropy, which is extremely important in dendritic growth. Thus, for the simulation results from early CA models, the effect of the artificial anisotropy is very obvious. For example, during the growth of a single dendrite, the growth direction is not sustained, resulting in a slightly curved and asymmetric primary arm. To utilize the high computational efficiency of the CA model while simultaneously realizing an accurate description of microstructural evolution (particularly for dendritic growth), the effect of the grid anisotropy must be reduced considerably (Reuther and Rettenmayr, 2014).

The grid anisotropy superimposes physical anisotropy, thus hindering the interpretation of simulated microstructures. Several algorithms have been proposed to reduce the effect of grid anisotropy. For example, the change in solid fraction can be limited according to the local curvature to avoid sharp tips or valleys. Using this method, the state change of a liquid cell into an interface cell requires a minimum total solid fraction within its neighborhood (Reuther and Rettenmayr, 2014). It was found that the shape of the obtained dendrites depends to a large extent on the minimum solid fraction required. Marek (2013) scaled the calculated interface velocity using a function that depends on the numerical parameters obtained from smoothing the field of phase state of the cell within a given neighborhood. Compared with the traditional CA model, the resulting model can simulate the anisotropic benchmark problem with a greatly reduced grid effect.

To reduce the grid anisotropy, another option is to rotate the grid during the simulation, which can disperse the influence of grid anisotropy in various directions. If the time scale for the rotation is sufficiently small, the error in the anisotropy can be expected to average itself out. Based on grid rotation, (Wei et al. (2011) proposed a random zigzag capture method to reduce the grid anisotropy. This random zigzag capture method can be used to obtain a seaweed dendritic morphology with zero anisotropy strength, indicating that the grid anisotropy was remarkably reduced.

### 3.3.6    Applications of the Cellular Automation Method

Owing to its flexibility in considering various state variables and transformation rules, the CA method has a unique versatility in microstructure simulations. From a fundamental viewpoint, the CA method is desirable for understanding the kinetics and topology changes during microstructural evolution. From a practical viewpoint, based on phenomenological but sound physical foundations, it is desirable for predicting microstructure parameters, such as grain size or texture, which can remarkably influence the mechanical and physical properties of materials. Important areas where microstructure-based CA has been successfully applied include static recrystallization and recovery, dendritic growth, and related nucleation and coarsening phenomena (Lian et al., 2019).

Materials-related applications of the CA method can be found in the field of solidification. Because of the lower computational requirements (memory and time) compared to other numerical methods, the CA method is suitable for studying phase transformation in multicomponent, multiphase, and/or multigrain systems, particularly in 3D. Zhang et al. (2012) developed a 3D CA model to simulate the solidification of an Al–Cu–Mg alloy by coupling the thermodynamic and kinetic calculations. Figure 3.11 shows the simulated microstructural evolution of an Al-3.9 wt.% Cu–0.9 wt.% Mg alloy during directional solidification. Both the steady-state morphologies of the dendrite tip in 3D and a 2D slice are shown. The results showed that some secondary dendrite arms were overgrown and therefore disappeared during the growth process, resulting in an increased secondary dendrite arm spacing.

### 3.4    Other Mesoscale Simulation Methods

Various modeling and simulation approaches have been developed to model material behavior at the mesoscale. In recent years, a large number of numerical models for treating interfaces have been proposed. In addition to the PF and CA methods,

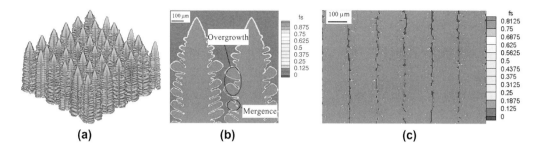

**Figure 3.11** Calculated dendritic microstructure of an Al-3.9 wt.% Cu-0.9 wt.% Mg alloy solidified directionally. The temperature gradient ahead of the S/L interface is 7 K mm$^{-1}$ and the cooling rate is 1 K s$^{-1}$. (a) The 3D dendrite tip morphology; (b) the 2D dendrite tip morphology; (c) the 2D morphology of the dendrites after solidification. This figure is reproduced from Zhang et al. (2012) with permission from Elsevier. A black and white version of this figure will appear in some formats. For the colour version, please refer to the plate section.

a front-tracking method and a level-set method are commonly applied mesoscale simulation approaches, particularly for microstructural evolution.

## 3.4.1 Front-Tracking Method

The front-tracking method usually refers to a family of numerical methods that apply a type of tracking to evolving discontinuities, using surfaces or lower-dimensional manifolds as the computational degrees of freedom. This method has been widely used to model interface evolution processes, such as solidification and multiphase flow. It can reproduce complex dendritic structures during crystal growth in supercooled melts, including tip splitting, solute trapping, side branching, and coarsening. (Juric and Tryggvason, 1996; Li et al., 2003). For the modeling and simulation of solidification, the advantage of the front-tracking method is that it can directly handle the Gibbs–Thomson relationship and energy balance condition. However, the complexity of processing the interface and calculating normal vectors and curvatures limits the applicability of the front-tracking method in single solid-phase systems (Tan and Zabaras, 2007).

With this method, the local mesh must be adjusted to accommodate the calculated interface morphology. Based on this, the front-tracking method is best suited to unstructured grids, similar to the finite volume method (FVM) and finite element method (FEM). The implementation details of these methods vary widely. For the solidification of a pure substance, two boundary conditions must be fulfilled at the solid–liquid interface: the conservation of energy and the Gibbs–Thomson relationship. In most front-tracking methods, the calculation is divided into two steps. First, the interface position is fixed, and the temperature distribution is solved. Second, after the temperature distribution is obtained, the interface position is recalculated based on conservation of the interfacial energy. Using the temperature distribution obtained in the first step, the speed of the local interface is obtained, the interface position is updated, and the calculation is continued.

Unfortunately, although it is possible to satisfy the Stefan condition in a pointwise manner, many current implementations of this condition do not satisfy global energy conservation. These formulations are very sensitive to the mesh size and orientation. Moreover, the front-tracking method must track the moving interface, which makes the calculation cumbersome and thus limits its development.

## 3.4.2    Level-Set Method

The level-set method, which was developed during the 1980s by Osher and Sethian (1988), is a simple but useful method for tracking the evolution of an interface in both 2D and 3D. As a conceptual framework, this method uses level sets as a tool to analyze the surfaces and shapes, which makes it extremely easy to track shapes that change the topology, and thus can handle topological changes of the evolving interface. This method has been designed particularly for issues in which the topology of the evolving interface changes, or sharp corners and cusps are present (Moelans et al., 2008).

In the level-set method, the interface is represented by a contour where the level-set function, $\Phi$, takes a predetermined value (typically zero) (Tan and Zabaras, 2007). In general, the level-set function is a signed function determined by the distance from the nearest interface, and the sign denotes one phase or the other. The level-set function can be obtained by iterating the following equation until convergence is reached:

$$\frac{\partial \Phi}{\partial \tau} = s(\Phi)(1 - |\nabla \Phi|), \tag{3.50}$$

where $s(\Phi)$ is the smoothed sign function, and $\tau$ the iteration time.

The advantage of this method is that it can perform numerical calculations involving curves and surfaces on a fixed Cartesian grid without parameterization. This enables the level-set method to handle problems in which the interface velocity is sensitive to local properties, such as curvature and normal direction. This is a very promising method for interface tracking with low computational costs. Consequently, this method is widely used in various applications, including fluid flows, microstructural evolution, and image processing. In contrast to the PF method, because the asymptotic analysis is unnecessary, this method can cope with the sharp interface front directly and thus has been employed to study the evolution of various microstructures. The disadvantage of this method is that it requires considerable thought to construct an appropriate speed to advance the level-set function. However, the reward is a highly versatile technique that can solve many complex problems.

To predict the microstructural evolution in multicomponent systems with realistic material parameters, Tan and Zabaras (2007) presented a level-set model that combined features of front-tracking and fixed-domain methods to model the dendritic solidification of alloys. Figure 3.12 shows the simulated 3D results of coupled fluid flow for a Ni–Cu binary alloy.

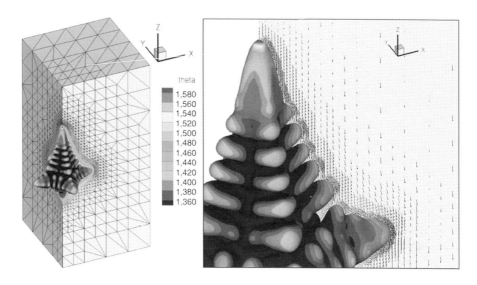

**Figure 3.12** Ni–Cu crystal growth with inlet flow from the top. Left: dendrite colored with interface velocity, mesh colored with temperature; right: flow passing by the upstream and perpendicular stream. This figure is reproduced from Tan and Zabaras (2007) with permission from Elsevier. A black and white version of this figure will appear in some formats. For the colour version, please refer to the plate section.

## 3.4.3 Comparisons among Different Mesoscale Simulation Methods

In recent years, significant progress has been achieved in simulation of microstructural evolution using PF, CA, front-tracking, and level-set methods. These methods have been well established for describing morphological evolution.

Both the front-tracking and level-set methods are based on the tracking or capturing of sharp interfaces. A front-tracking method solves the sharp interface model directly, where the interface must be tracked in every time step according to the conservation boundary conditions. With this method, it is difficult to handle dynamic interfaces undergoing complicated topological changes. At present, this method is seldom used for microstructural evolution, particularly after the development of the PF and CA methods.

The level-set method, as an alternative approach for handling sharp interface fronts directly, is a promising method for tracking interfaces with low computational costs, and it thus is widely used in various fields. This method has many advantages: interface geometries can be easily and accurately computed, the level-set equation has been well studied and can capture topological changes, the intrinsic geometric properties can be easily determined, and the method can be implemented relatively easily by using accurate high-order computational schemes. In addition, the level-set method does not require asymptotic analysis, whereas the PF method requires complex asymptotic analysis to determine the relationship between the measurable physical parameters and the PF model. However, the level-set method typically suffers from mass loss/gain problems owing to numerical dissipation. Consequently, the application of boundary conditions can occasionally be quite cumbersome.

Because the PF method does not require the explicit calculation of the moving interface, it can avoid the numerical difficulties associated with the moving interface. Another advantage is that fully assessed Gibbs energy functions and thermophysical properties obtained from the CALPHAD method can be used, allowing realistic values of these parameters to be used. This is particularly beneficial for multiphase and multicomponent systems. However, this approach is not followed by all PF applications, and simple rough functions are fabricated despite the availability of proper CALPHAD assessments. Users are advised to check this aspect carefully. The primary drawbacks of the PF method are that significant computational effort is required, and many parameters involved in solving the evolution equations are difficult to determine.

Essentially, the huge grids (and thus enormous computing resources) required to simulate realistic complex microstructures (proving both the convergence of the solution and grid independence) are a bottleneck for both the PF and level-set methods.

Compared with most other numerical methods, including the PF and level-set methods, the CA method is relatively easy to implement and requires fewer computing resources, because the defined diffusion equation is solved only in the bulk cells, while the interface mass balance is solved separately in the interface cells (Choudhury et al., 2012). However, there are certain deficiencies in terms of accuracy owing to its discrete nature. In the PF method, the interface geometry is described with high accuracy. In the CA method, however, it can only be described approximately by the coarse-grained solid-fraction field, leading to deviations in the interface shape and growth behavior, though the coarse representation of the interface allows a lower computational cost. Therefore, it is typically found that the quantitative results obtained by the PF method are better than those obtained by the CA method.

The fundamental difference between these mesoscale methods for modeling microstructural evolution is the specific approach used to describe the interface. In the PF method, the interface is represented by the smooth transition of the PF variables, and the interface thickness expands within a few (approximately five to seven) grid points. The interface in the CA method consists of one or more layers in the two-phase region (phase fraction is between 0 and 1), and for the level-set method, the level-set function is employed to capture the evolution of the interface. Hence, to resolve a small morphological length scale, the PF method requires considerably smaller grid spacings compared to the CA and level-set methods (Choudhury et al., 2012; Zaeem et al., 2012).

It should be noted that the aforementioned microstructure modeling methods (PFM, CA, level-set, etc.) are also called direct detailed methods, because these simulation methods can provide detailed descriptions of microstructure evolution during various phase transition processes, including nucleation, growth, and coarsening. As a result, the actual morphologies and corresponding dynamics of the microstructure evolution can be obtained with very high precision. In addition to this kind of direct detailed method, there also exist two other approaches that can describe the evolution of microstructures. One is the internal state variable (ISV) method, which is represented by the Johnson–Mehl–Avrami–Kohnogorov model or the Scheil–Gulliver model. This kind of physics-based method is often called a "micro model" embedded in process-scale simulations to calculate latent heat release, phase size and fraction, recrystallization grain fraction, etc. This method is numerically very effective and easy to set up. The other approach is the frequency distribution function (FDF) method, which is represented by the Kampmann–Wagner numerical (KWN) model. The feature of this method is to provide a statistical description of important microstructural features (such as size distribution) by using the mean-field method. In contrast to the direct detailed method, the FDF method is popular because of its good balance between efficiency and accuracy. In general, the direct detailed method can be used as a computational experiment to verify the reliability of the latter two methods. When it comes to industrial issues, the ISV and FDF methods have good capabilities because they can effectively solve issues such as multiscale, multiparameter, and computational efficiency. For more detailed information about the ISV and FDF methods, interested readers can refer to papers of Du et al. (2017, 2021) and Tang et al. (2018).

## 3.5    Integration of the Phase-Field Method with Other Simulation Approaches

The PF method, along with other mesoscale simulation methods, plays a significant role in multiscale modeling. To fully understand the significant role that mesoscale simulation methods play in the multiscale modeling of materials, strategies are needed to integrate mesoscale simulation methods with methods at other scales. A simple flowchart of this strategy is shown in Figure 3.13. On one hand, the free-energy construction and kinetic coefficients of PF models can be taken from CALPHAD and/or atomistic models. On the other hand, PF modeling can also be used as an input to macroscale modeling to predict mechanical properties and simulate macro processes for the sake of material design. This flowchart also indicates that the mesoscale simulation methods represented by the PF methods are quite significant in bridging the atomistic structures and macroscopic properties of various materials.

### 3.5.1    PF Modeling with Atomic Simulations

One of the primary challenges in multiscale modeling is bridging the gap between atomic and macroscopic methods to ensure that the descriptions at all levels are linked and quantitatively consistent with one another. Atomistic methods, such as the first-principles method, including density functional theory (DFT), and MD, are widely applied in lots of areas of material science to understand material behaviors. However, owing to the small length scale and short time scale of these methods, it is difficult to use an atomistic simulation to establish direct connections to the macroscale.

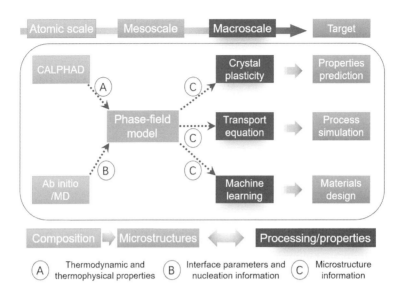

**Figure 3.13** Schematics of multiscale modeling approaches based on the PF method.

Mesoscale modeling and simulations, ranging approximately from 100 nm to 100 μm, can provide a bridge from the atomistic scale to the macroscale.

PF modeling is currently being continuously applied in the fields of materials, physics, and biological sciences and fills the extremely needed gap between the atomic scale and the continuum level of modeling and simulation. There are two ways to couple the PF method with atomic-scale simulation methods, as shown in Figure 3.14. On one hand, simulations at the atomic scale can provide some parameters needed in the PF model. Atomistic modeling is required to predict the key interfacial properties affecting the microstructures, such as the interfacial energy and its anisotropies, to more directly link continuum-scale modeling to experimentally observed changes (Liu and Nie, 2017), allowing for efficient solutions to mesoscopic phenomena, such as dendritic and grain growth, among numerous other aspects. On the other hand, PF simulations can provide data assimilation to verify atomic-scale simulations. However, in most models integrated with the PF method, atomic simulations act as a parameter input.

In principle, almost all parameters in a PF model can be determined from atomistic calculations. The diffusion coefficient, interfacial energy, and kinetic coefficient are examples of parameters required in PF formulations that are difficult to measure. More importantly, the small anisotropy of the last two quantities is essential for morphological evolution and must be included in the PF model. Quantitative characterization of the anisotropy of an interface using atomistic simulations is useful for modeling changes in the dendrite growth orientation. Based on ab initio calculations, mixed-cluster expansion Monte Carlo simulations, and the PF method, Vaithyanathan et al. (2002) proposed a multiscale approach to study the growth and coarsening of the $\theta$-$Al_2Cu$ phase in Al–Cu alloys during aging. From ab initio calculations, they

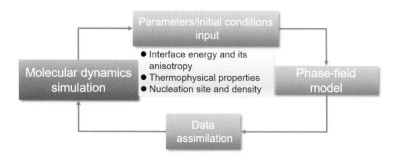

**Figure 3.14** Schematics of the coupling between PF model and molecular dynamics method.

obtained the free-energy density, interfacial energy and its anisotropy, lattice parameters, and elastic constants. Bishop and Carter (2002) proposed a coarse-grained method to relate the atomic structure of grain boundaries calculated from atomistic simulations with the evolution of order parameters across the diffuse interface in the PF model. From atomistic MD simulations, Hoyt and Asta (2002) calculated the interfacial energy and mobility as well as its anisotropy for pure Ni, Au, and Ag, and then input these parameters into a PF model to quantitatively study dendritic growth. They concluded that the necessary parameters for PF modeling derived from atomistic simulations can accurately reproduce the variation in dendrite growth velocity with the measured undercooling in pure Ni. Combining DFT, cluster expansion theory, and potential renormalization theory, Bhattacharyya et al. (2019) derived a relationship between the Gibbs energy and composition, and further proposed a parameter-free PF method. By applying this method to a Ni–Al alloy at 1,027°C, they successfully reproduced the experimentally observed microstructural evolution without any empirical thermodynamic parameters.

To obtain a more accurate and efficient simulation of grain growth, Miyoshi et al. (2018) developed an approach that combines the merits of atomistic and continuum simulations, by converting MD-generated atomic configurations into the multi-PF model. With this method, the nucleation process is simulated by MD, and the microstructure obtained from MD is then input as the initial structures for a multi-PF grain growth simulation, as shown in Figure 3.15. This type of direct bridging method is promising when considering the rapid progress in computer performance.

Not only do coarse-grained models require input parameters from atomic-scale simulations, but also atomistic simulations require many model parameters that must be determined using experimental data or other simulation results to achieve quantitative results. The accuracy of material parameters calculated from MD simulations depends on how accurately the interatomic potential can capture the physics of the bonding in the material. Thus, there is another type of integration between MD simulations and PF modeling. Shibuta et al. (2018) began a study on data assimilation to predict the interfacial parameters for a PF simulation based on the morphology of MD simulations. This can provide a simple method for computing physical quantities in both equilibrium and nonequilibrium states. They further employed the "Ensemble

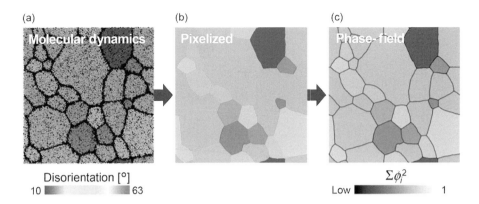

**Figure 3.15** Creation of an initial structure for multi-PF grain growth simulations from MD data. (a) MD-generated solidified microstructure at time $t = 1,000$ ps. (b) Polycrystalline system created by discretizing the atomic configuration shown in (a) into 2D difference grid points. (c) Multi-PF interfacial profiles obtained through the relaxation simulation starting from (b). This figure is reproduced from Miyoshi et al. (2018) with permission from Elsevier. A black and white version of this figure will appear in some formats. For the colour version, please refer to the plate section.

Kalman Filter" method to estimate the input parameters using the microstructures obtained from MD simulations. Thus, the parameters directly associated with the evolution of microstructures can be estimated. The primary drawback of atomistic simulation methods is that they cannot provide material data for multicomponent and multiphase systems. In Chapters 5 and 6, we have described that the CALPHAD approach can provide such data.

## 3.5.2    PF Modeling with Crystal Plasticity

The linkage between microstructures and properties of materials is at the core of material modeling. Crystal plasticity (CP), a physically based plasticity theory, describes the deformation of a material at the microscale. From CP simulations, the mechanical behavior of polycrystalline materials can be predicted based on the interactions among the phases, grains, and subgrains, regardless of the anisotropic elastic and plastic behaviors of these features.

Most existing PF models incorporate the contribution of elastic strain energy into microstructural evolution. However, both experimental and simulation results indicate that in polycrystalline microstructures, the stress may exceed the elastic limit. Therefore, a PF model for modeling microstructural evolution should include the driving force generated not only by the elastic field but also by plastic deformation. Understanding the coupled mechanisms of phase transformations and plasticity promises the ability to design an underlying microstructure that governs macroscopic material behaviors. Irreversible deformations will affect the magnitude of the stored elastic energy and thus the microstructural evolution. Coupling a CP formulation with a PF model will benefit the predictions of both properties and microstructural evolution.

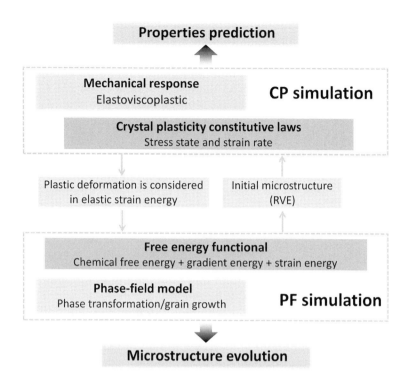

**Figure 3.16** Schematics of the coupling between PF model and crystal plasticity model.

For a better description of integrated models that fully couple the mechanics with microstructure, the coupling strategy between PF models and CP simulations is shown schematically in Figure 3.16. The figure describes how the PF model and mechanical simulation environments interact. With this coupling, the PF method is able to simulate the nucleation and growth of phases and/or grains, whereas the CP method can predict the mechanical properties of the materials under thermomechanical loading conditions. Thus, the contributions of the different mesoscopic microstructures, such as the grains and precipitates, to the mechanical properties can be quantified. It should be noted that when considering problems involving the formation of new grains or new phases during deformation, nucleation models are necessary. Plastic deformation can be introduced into PF models in two ways: explicitly introducing mobile dislocations through continuous fields for each slip system or directly adding a plastic strain field defined at the mesoscale. An example of combining the PF method with the CP model follows.

Cottura et al. (2016) proposed a PF model coupled with a strain gradient CP model based on the dislocation densities to study the rafting behavior of $\gamma'$ precipitates in Ni-based superalloys. Figure 3.17 shows the simulated microstructures in both the elastic (top) and elastoviscoplastic (bottom) cases. The initial microstructure, which is consisted of $\gamma'$ cuboidal precipitates (Figure 3.17, left), was simulated by considering only the elasticity under stress-free conditions. A constant uniaxial tensile stress $(\sigma_a)$ of

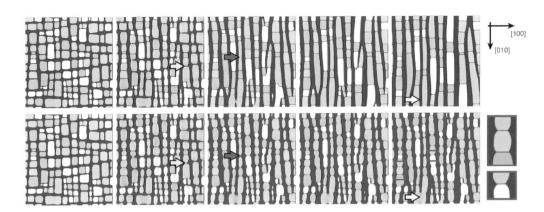

**Figure 3.17** Snapshots of the microstructure (colors indicate $\gamma'$ variants) in a $2.3 \times 2.3$ μm$^2$ periodic box at $t = 0$, 1, 5, 15, and 45 hours (from left to right) in the elastic (top) and elastoviscoplastic (bottom) cases. This figure is reproduced from Cottura et al. (2016) with permission from Elsevier. A black and white version of this figure will appear in some formats. For the colour version, please refer to the plate section.

150 MPa was applied along the [100] direction while maintaining the temperature at 950°C. The simulation results showed that rafting proceeds perpendicular to the [100] loading axis, and plasticity can considerably slow down the evolution of the rafts. This coupled PF and CP model can be used to analyze the microstructural evolution during creep loading.

## 3.5.3    PF Modeling with Macro Transport Equations

Microstructure simulations of the technical materials used during hot working processes, such as casting, forging, and welding, suffer from a strong interdependence between the latent heat release and thermal diffusion at the macroscopic scale. The formation of local microstructures depends on the macroscopic temperature field, which is also closely related to the formation of microstructures. This greatly influences the simulation results of technical castings because the microscale of the phase transition process and the macroscale of heat conduction must be considered simultaneously (Böttger et al., 2006, 2009). A rigorous coupling of both scales is still computationally extremely ambitious.

Figure 3.18 shows the integration strategy between PF modeling and macroscale process modeling. On one hand, PF modeling can supply microstructure information to the transport equations through averaging so that the macroscale process can be simulated. On the other hand, macroscale process modeling through transport equations can also supply information regarding the fields (such as the temperature field, flow field, and stress–strain field) to the PF model.

If the macroscale is the primary focus, a practical and feasible method is to use an extremely simple microstructure model that has sufficient accuracy within the scope of the macroscopic issue. In principle, a dedicated heat flow model for the solidification

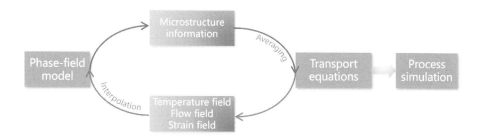

**Figure 3.18** Schematics of the coupling between PF model and macro transport models.

process (considering only thermal diffusion) only requires a proper description of the latent heat caused by the phase transition. However, because of the lack of precise information about the microstructural evolution in most industrial applications, simplified latent heat descriptions based on thermodynamic Scheil modeling are often used in heat flow simulations (Böttger et al., 2009). More elaborate microstructure models can improve heat flow prediction if some microstructure information can be provided that may be of interest to material processing. However, it is possible only if the model is simple enough to be solved simultaneously on all nodes within a reasonable calculation time.

In addition, if the formation of microstructures is the primary focus and this simple microstructure model cannot provide sufficient details, then a more complex model (such as the PF model) will be required. Owing to the high computational cost of these methods, it is impossible to apply them on all nodes of the heat flow grid simultaneously. Thus, in many situations, the microstructure simulation is limited to a very small area. This raises the question of how to obtain latent heat for other nodes, or in what way it can provide practical thermal boundary conditions for complex microstructure modeling (Böttger et al., 2006, 2009).

Laser powder deposition (LPD), a typical additive manufacturing technique, is prominent in the fabrication of high-quality components with various complicated shapes. By coupling the process modeling using transport equations, including the temperature and stress fields, at the macroscale with microstructural evolution modeling using multi-PF modeling at the microscale, Li and Wang (2021) conducted multiscale modeling of the LPD process with the entire thermal history. The macroscale and microstructure simulations were coupled through the transfer of temperature information by a simple bilinear interpolation method in the spatial scale and a linear interpolation method in the time scale. As shown in Figure 3.19, during the four-layer deposition process, the maximum temperature of this bottom area exceeds the liquidus only once, and thus solidification only occurs once. For the top area, however, the temperature exceeds the liquid line twice. Thus, this area will be remelted and resolidified during the second-layer deposition, causing significant microstructural changes. Coupling this multi-PF model with highly efficient macroscale simulations provides a framework for microstructure prediction within the entire thermal history of the additive manufacturing process.

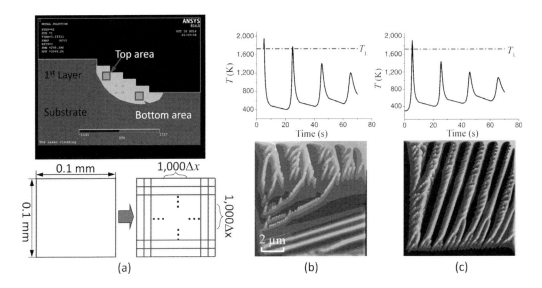

**Figure 3.19** PF modeling of microstructure evolution during additive manufacturing coupling with macro temperature field. (a) Schematics of melting pool and the interpolation, (b) the temperature history and microstructures in the top area, and (c) the temperature history and microstructures in the bottom area (Li and Wang, 2021). A black and white version of this figure will appear in some formats. For the colour version, please refer to the plate section.

## 3.5.4    PF Modeling with CALPHAD

CALPHAD, a methodology introduced by Kaufman and Bernstein in 1970 (Kaufman and Bernstein, 1970), is a phase-based approach to calculate thermodynamics and phase equilibria of a system through a self-consistent framework. This framework can be conveniently extrapolated to multicomponent systems.

To effectively simulate a multicomponent system with the PF method, the input parameters, such as atomic diffusion mobility, interfacial energy, lattice parameters, and elastic constants, should be obtained with high accuracy. In some cases, it is difficult to obtain accurate parameters for complex multicomponent systems. One of the primary obstacles to the application of the PF method to practical multicomponent and/or multiphase systems is that most existing models lack the ability to construct free-energy functionals directly from the real phase diagram and thermodynamic data. As a solution to this issue, coupling the PF method with the CALPHAD approach is helpful for predicting microstructural evolution in multicomponent systems (Kitashima, 2008; Rahnama et al., 2017; Steinbach et al., 2007).

Coupling to CALPHAD thermodynamic and atomic mobility databases can provide realistic driving forces, diffusion potentials, and atomic mobilities at any composition and temperature. This information is a prerequisite for quantitative PF simulations. The thermodynamic data required for multicomponent alloys can be provided by a local linear approximation of the phase diagram or by direct coupling to the thermodynamic data through commercial thermodynamic software, such as

Thermo-Calc, Pandat, and FactSage. Obtaining the driving force from commercial thermodynamic software and databases is powerful for studying various phase transitions. The CALPHAD-type thermodynamic and atomic mobility databases are linked to models through an interface program (such as the TQ interface in Thermo-Calc software) incorporated in the simulation code. Coupling to the thermodynamic datasets allows simulations to be conducted in alloy systems of technical relevance. MICRESS (www.micress.de) is a software used for the simulation of technical alloys because it applies a pragmatic PF model and robust coupling to arbitrary CALPHAD databases.

## 3.5.5    PF Modeling with Machine Learning

To design and manufacture materials more quickly and cost efficiently, the central task of materials science is to improve the ability to model how processes produce material structures, how such structures affect material properties, and how materials should be selected for a given application. Using material data and advanced computer models, the behavior of new materials in specific applications can be realistically simulated. Thus, the lengthy cycles of building and testing can be avoided. In the past 20 years, computational materials science has gradually shifted from purely computational materials to the discovery and design of new materials based on machine learning. This method has been widely applied to numerous problems in material science and engineering. Integrating the PF method with a machine learning approach is quite an interesting topic (Steinmetz et al., 2016).

The integration between a PF simulation and machine learning is illustrated in Figure 3.20. On one hand, the PF simulation results can be used to generate data for the machine learning model. On the other hand, machine learning can also supply some of the parameters required for a PF simulation. For example, PF models of multicomponent alloys are typically coupled with a thermodynamic database, which consumes considerable time to obtain solutions of phase equilibrium equations. To reduce this time while maintaining sufficient accuracy, machine learning can provide an option for determining the input parameters for PF models.

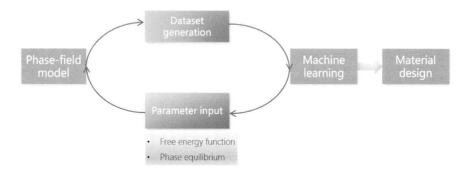

**Figure 3.20** Schematics of the coupling between PF model and machine learning.

Teichert and Garikipati (2019) examined the feasibility of using machine learning techniques to determine equilibrium states in a physical system. Nomoto et al. (2018) employed the linear prediction procedure of a neural network as a machine learning approach to accelerate the thermodynamic calculations in a multi-PF model of the solidification of quinary Fe–Cr–Ni–Mo–C stainless steel. The microstructures generated using parameters from machine learning agree well with those generated from the model that was directly coupled with the CALPHAD database, but the computation was five times faster. Jiang et al. (2019) developed a model to obtain the quasiphase equilibrium based on machine learning. It was demonstrated that this combination is very convenient to acquire quasiphase equilibrium data in PF models for multicomponent alloys with high accuracy and fast speed.

PF simulations have become a valuable tool for exploring the influence of processing parameters on the microstructural evolution of many advanced materials. Using machine learning and other approaches, Yabansu et al. (2017) developed a framework to establish process-structure linkages for ternary eutectic alloys based on the results generated from PF simulations. Figure 3.21 shows the framework used to establish the linkages from simulated microstructures. The workflow is as follows: (1) Identify the process parameters of interest and their ranges, (2) generate a suitable calibration dataset for the PF model, (3) quantify the microstructure through spatial statistics, and (4) project this information onto a reduced-order space to obtain a low-dimensional

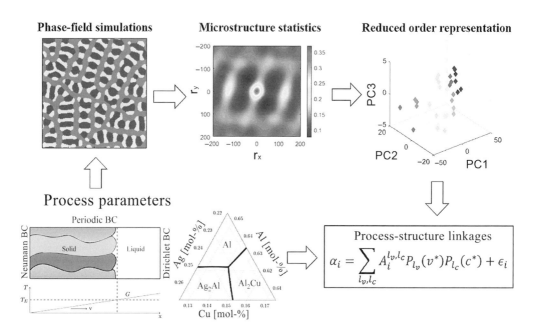

**Figure 3.21** Framework to establish process-structure linkages from microstructures simulated by PF models and machine learning. This figure is reproduced from Yabansu et al. (2017) with permission from Elsevier. A black and white version of this figure will appear in some formats. For the colour version, please refer to the plate section.

representation. Through machine learning and cross-validation techniques, the process space and reduced-order representation of microstructures are mapped to achieve the required process-structure linkages.

## 3.6    A Case Study: Phase-Field Design of High-Energy-Density Polymer Nanocomposites

The PF method is a powerful tool for modeling microstructural evolution, which has been growing in popularity at an ever-increasing rate. Although much of the research using the PF method has focused on simulations of microstructural evolution, it also has potential as a powerful tool for material discovery and plays an important role in computational material design. It is very likely that this method can be used as a design simulation to search for the desired microstructures. Accordingly, a very effective strategy for developing new materials is to elucidate the mechanism of microstructural evolution based on experiments, then model the microstructural evolution using the PF method, and finally screen the most desirable microstructures.

Shen et al. (2018) presented a high-throughput PF design of high-energy-density polymer nanocomposites, which is a good example that demonstrates the potential of the PF method in material design. Taking the polyvinylidene difluoride (PVDF)-BaTiO$_3$ nanocomposite as an example, they first performed a 2D PF simulation to predict the breakdown phase evolution under applied electric fields. Subsequently, from the simulation, they found that the microstructure can remarkably influence the local electric field distribution and thus the breakdown strength. Then they conducted further simulations for various 3D microstructures. In addition to the pure polymer matrix, they further simulated five types of 3D nanocomposites with 10% nanofiller: vertical nanofibers, vertical nanosheets, random nanoparticles, parallel nanofibers, and parallel nanosheets. To screen the optimal microstructure with a high-energy-density

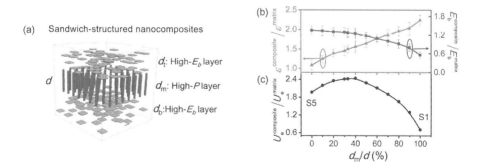

**Figure 3.22** (a) A designed sandwich microstructure based on PF simulations and (b) the corresponding effective dielectric permittivity, $\varepsilon$, and the breakdown strength, $E_b$, and (c) the energy density as a function of the fraction of the middle layer. This figure is reproduced from Shen et al. (2018) with permission from Wiley. A black and white version of this figure will appear in some formats. For the colour version, please refer to the plate section.

using high-throughput computation, they built a microstructure dataset by adjusting the length ratio of nanofillers. The obtained dataset was then employed for microstructure design. To further optimize the designed microstructure, more simulations were conducted by adjusting the volume fraction of nanofiller in the middle layer in the sandwich structure. As shown in Figure 3.22, a sandwich microstructure filled with parallel nanosheets in the upper and lower layers, and with vertical nanofibers in the middle layer, was designed.

This work provides a computational approach for understanding electrostatic breakdown. Moreover, it is expected that this approach can stimulate more experimental efforts to synthesize polymer nanocomposites with novel microstructures and high performance.

## References

Allan, H. (1974) Regular algebra and finite machines by J.H. Conway. *Mathematical Gazette*, 58(405), 243–244.

Allen, S. M., and Cahn, J. W. (1979) A microscopic theory for antiphase boundary motion and its application to antiphase domain coarsening. *Acta Metallurgica*, 27(6), 1085–1095.

Aspray, W., and Burks, A. (1987) *Papers of John von Neumann on Computing and Computer Theory*. Los Angeles: MIT Press.

Beltran-Sanchez, L., and Stefanescu, D. M. (2004) A quantitative dendrite growth model and analysis of stability concepts. *Metallurgical and Materials Transactions A*, 35(8), 2471–2485.

Bhattacharyya, S., Sahara, R., and Ohno, K. (2019) A first-principles phase field method for quantitatively predicting multi-composition phase separation without thermodynamic empirical parameter. *Nature Communications*, 10(1), 3451.

Bishop, C. M., and Carter, W. C. (2002) Relating atomistic grain boundary simulation results to the phase-field model. *Computational Materials Science*, 25(3), 378–386.

Boettinger, W. J., Warren, J. A., Beckermann, C., and Karma, A. (2002) Phase-field simulation of solidification. *Annual Review of Materials Research*, 32(1), 163–194.

Böttger, B., Eiken, J., and Apel, M. (2009) Phase-field simulation of microstructure formation in technical castings – a self-consistent homoenthalpic approach to the micro-macro problem. *Journal of Computational Physics*, 228(18), 6784–6795.

Böttger, B., Eiken, J., and Steinbach, I. (2006) Phase field simulation of equiaxed solidification in technical alloys. *Acta Materialia*, 54(10), 2697–2704.

Cahn, J. W., and Hilliard, J. E. (1958) Free energy of a nonuniform system. I. Interfacial free energy. *Journal of Chemical Physics*, 28(2), 258–267.

Chen, L. Q. (2002) Phase-field models for microstructure evolution. *Annual Review of Materials Research*, 32(1), 113–140.

Chen, L. Q., and Wang, Y. (1996) The continuum field approach to modeling microstructural evolution. *JOM*, 48(12), 13–18.

Chen, L. Q., and Yang, W. (1994) Computer simulation of the domain dynamics of a quenched system with a large number of nonconserved order parameters: the grain-growth kinetics. *Physical Review B*, 50(21), 15752–15756.

Choudhury, A., Reuther, K., Wesner, E., August, A., Nestler, B., and Rettenmayr, M. (2012) Comparison of phase-field and cellular automaton models for dendritic solidification in Al–Cu alloy. *Computational Materials Science*, 55, 263–268.

Cottura, M., Appolaire, B., Finel, A., and Le Bouar, Y. (2016) Coupling the phase field method for diffusive transformations with dislocation density-based crystal plasticity: application to Ni-based superalloys. *Journal of the Mechanics and Physics of Solids*, 94, 473–489.

Dantzig, J. A., Di Napoli, P., Friedli, J., and Rappaz, M. (2013) Dendritic growth morphologies in Al-Zn alloys – Part II: phase-field computations. *Metallurgical and Materials Transactions A*, 44(12), 5532–5543.

Du, Q., Chen, M., and Xie, J. (2021) Modelling grain growth with the generalized Kampmann–Wagner numerical model. *Computational Materials Science*, 186, 110066.

Du, Q., Tang, K., Marioara, C. D., Andersen, S. J., Holmedal, B., and Holmestad, R. (2017) Modeling over-ageing in Al–Mg–Si alloys by a multi-phase CALPHAD-coupled Kampmann–Wagner numerical model. *Acta Materialia*, 122, 178–186.

Elder, K. R., Provatas, N., Berry, J., Stefanovic, P., and Grant, M. (2007) Phase-field crystal modeling and classical density functional theory of freezing. *Physical Review B*, 75(6), 064107.

Emmerich, H., Löwen, H., Wittkowski, R., et al. (2012) Phase-field-crystal models for condensed matter dynamics on atomic length and diffusive time scales: an overview. *Advanced Physics*, 61(6), 665–743.

Gandin, C. A., and Rappaz, M. (1994) A coupled finite element-cellular automaton model for the prediction of dendritic grain structures in solidification processes. *Acta Metallurgica et Materialia*, 42(7), 2233–2246.

Ginzburg, V. L., and Landau, L. D. (1950) On the theory of superconductivity. *Journal of Experimental and Theoretical Physics*, 20(1064).

Glicksman, M. E., Koss, M. B., Hahn, R. C., Rojas, A., Karthikeyan, M., and Winsa, E. A. (1993) The isothermal dendritic growth experiment: scientific status of a USMP – 2 space flight experiment. *Advanced Space Research*, 13(7), 209–213.

Guo, C., Wang, J., Wang, Z., Li, J., Guo, Y., and Tang, S. (2015) Modified phase-field-crystal model for solid-liquid phase transitions. *Physical Review E*, 92(1), 013309.

Hohenberg, P. C., and Halperin, B. I. (1977) Theory of dynamic critical phenomena. *Reviews of Modern Physics*, 49(3), 435–479.

Hoyt, J. J., and Asta, M. (2002) Atomistic computation of liquid diffusivity, solid–liquid interfacial free energy, and kinetic coefficient in Au and Ag. *Physical Review B*, 65(21), 214106.

Jiang, X., Zhang, R., Zhang, C., Yin, H., and Qu, X. (2019) Fast prediction of the quasi phase equilibrium in phase field model for multicomponent alloys based on machine learning method. *CALPHAD*, 66, 101644.

Juric, D., and Tryggvason, G. (1996) A front-tracking method for dendritic solidification. *Journal of Computational Physics*, 123(1), 127–148.

Karma, A. (2001) Phase-field formulation for quantitative modeling of alloy solidification. *Physical Review Letters*, 87(11), 115701.

Kaufman, L., and Bernstein, H. (1970) *Computer Calculation of Phase Diagrams with Special Reference to Refractory Metals*. New York: Academic Press, 334.

Khachaturyan, A. G. (1983) *Theory of Structural Transformations in Solids*. New York: John Wiley & Sons.

Kim, S. G., Kim, D. I., Kim, W. T., and Park, Y. B. (2006) Computer simulations of two-dimensional and three-dimensional ideal grain growth. *Physical Review E*, 74(6), 061605.

Kim, S. G., Kim, W. T., and Suzuki, T. (1999) Phase-field model for binary alloys. *Physical Review E*, 60(6), 7186–7197.

Kitashima, T. (2008) Coupling of the phase-field and CALPHAD methods for predicting multi-component, solid-state phase transformations. *Philosophical Magazine*, 88(11), 1615–1637.

Kobayashi, R. (1993) Modeling and numerical simulations of dendritic crystal growth. *Physica D*, 63(3), 410–423.

Kobayashi, R. (1994) A numerical approach to three-dimensional dendritic solidification. *Experimental Mathematics*, 3(1), 59–81.

Kobayashi, R., Warren, J. A., and Carter, W. C. (1998) Vector-valued phase field model for crystallization and grain boundary formation. *Physica D*, 119(3), 415–423.

Langer, J. S. (1986) Models of pattern formation in first-order phase transitions, in Grinstein, G. and Mazenko, G. (eds), *Directions in Condensed Matter Physics. Series on Directions in Condensed Matter Physics*. Singapore: World Scientific, 165–186.

Li, C. Y., Garimella, S. V., and Simpson, J. E. (2003) Fixed-grid front-tracking algorithm for solidification properties, part I: Method and validation. *Numerical Heat Transfer, Part B*, 43(2), 117–141.

Li, J., Wang, Z., Wang, Y., and Wang, J. (2012) Phase-field study of competitive dendritic growth of converging grains during directional solidification. *Acta Materialia*, 60(4), 1478–1493.

Li, J. J., and Wang, J. C. (2021) Macro-micro coupled simulation of microstructure during laser additive manufacturing process, [Lecture].unpublished.

Lian, Y., Gan, Z., Yu, C., Kats, D., Liu, W. K., and Wagner, G. J. (2019) A cellular automaton finite volume method for microstructure evolution during additive manufacturing. *Materials and Design*, 169, 107672.

Liu, H., and Nie, J. F. (2017) Phase field simulation of microstructures of Mg and Al alloys. *Materials Science and Technology*, 33(18), 2159–2172.

Louchez, M. A., Thuinet, L., Besson, R., and Legris, A. (2017) Microscopic phase-field modeling of Hcp|Fcc interfaces. *Computational Materials Science*, 132, 62–73.

Marek, M. (2013) Grid anisotropy reduction for simulation of growth processes with cellular automaton. *Physica D*, 253, 73–84.

Miyoshi, E., Takaki, T., Shibuta, Y., and Ohno, M. (2018) Bridging molecular dynamics and phase-field methods for grain growth prediction. *Computational Materials Science*, 152, 118–124.

Moelans, N., Blanpain, B., and Wollants, P. (2008) An introduction to phase-field modeling of microstructure evolution. *CALPHAD*, 32(2), 268–294.

Mohanty, S., and Ross, R. B. (2008) Overview of multiscale simulation methods for materials, in Ross, R. B. and Mohanty, S. (eds), *Multiscale Simulation Methods for Nanomaterials*. Hoboken: John Wiley & Sons, Inc., 1–7.

Nomoto, S., Seguwa, M., and Wakameda, H. (2018) Non-equilibrium PF model using thermo-dynamics data estimated by machine learning for additive manufacturing solidification. In *Solid Freeform Fabrication 2018: Proceedings of the 29th Annual International Solid Freeform Fabrication Symposium – an Additive Manufacturing Conference*, 1975–1886.

Osher, S., and Sethian, J. A. (1988) Fronts propagating with curvature-dependent speed: algorithms based on Hamilton–Jacobi formulations. *Journal of Computational Physics*, 79(1), 12–49.

Poduri, R., and Chen, L. Q. (1998) Computer simulation of morphological evolution and coarsening kinetics of $\delta'$ (Al3Li) precipitates in Al–Li alloys. *Acta Materialia*, 46(11), 3915–3928.

Provatas, N., Dantzig, J. A., Athreya, B., et al. (2007) Using the phase-field crystal method in the multi-scale modeling of microstructure evolution. *JOM*, 59(7), 83–90.

Provatas, N., and Elder, K. (2010) *Phase-Field Methods in Materials Science and Engineering*. Weinheim: Wiley–VCH Verlag GmbH & Co. KGaA.

Qiu, D., Zhao, P., Shen, C., et al. (2019) Predicting grain boundary structure and energy in Bcc metals by integrated atomistic and phase-field modeling. *Acta Materialia*, 164, 799–809.

Raabe, D. (2002) Cellular automata in materials science with particular reference to recrystallization simulation. *Annual Review of Material Research*, 32(1), 53–76.

Rahnama, A., Dashwood, R., and Sridhar, S. (2017) A phase-field method coupled with CALPHAD for the simulation of ordered $\kappa$-carbide precipitates in both disordered $\gamma$ and $\alpha$ phases in low density steel. *Computational Materials Science*, 126, 152–159.

Rajkumar, V. B., Du, Y., Zeng, Y., and Tang, S. (2020) Phase-field simulation of solidification microstructure in Ni and Cu–Ni alloy using the Wheeler, Boettinger and McFadden model coupled with the CALPHAD data. *CALPHAD*, 68, 101691.

Rappaz, M. (1993) Modelling of microstructure formation in solidification processes. *Acta Metallurgica et Materialia*, 34(1), 93–124.

Reuther, K., and Rettenmayr, M. (2014) Perspectives for cellular automata for the simulation of dendritic solidification – a review. *Computational Materials Science*, 95, 213–220.

Shen, Z. H., Wang, J. J., Lin, Y., Nan, C. W., Chen, L. Q., and Shen, Y. (2018) High-throughput phase-field design of high-energy-density polymer nanocomposites. *Advanced Materials*, 30(2), 1704380.

Shibuta, Y., Ohno, M., and Takaki, T. (2018) Advent of cross-scale modeling: high-performance computing of solidification and grain growth. *Advanced Theory and Simulations*, 1(9), 1800065.

Steinbach, I. (2009) Phase-field models in materials science. *Modelling and Simulation in Materials Science and Engineering*, 17(7), 073001.

Steinbach, I., Böttger, B., Eiken, J., Warnken, N., and Fries, S. G. (2007) CALPHAD and phase-field modeling: a successful liaison. *Journal of Phase Equilibria and Diffusion*, 28(1), 101–106.

Steinbach, I., Pezzolla, F., Nestler, B., et al. (1996) A phase field concept for multiphase systems. *Physica D*, 94(3), 135–147.

Steinmetz, P., Yabansu, Y. C., Hotzer, J., et al. (2016) Analytics for microstructure datasets produced by phase-field simulations. *Acta Materialia*, 103, 192–203.

Takaki, T. (2014) Phase-field modeling and simulations of dendrite growth. *ISIJ International*, 54(2), 437–444.

Tan, L., and Zabaras, N. (2007) A level set simulation of dendritic solidification of multicomponent alloys. *Journal of Computational Physics*, 221(1), 9–40.

Tang, K., Du, Q., and Li, Y. (2018) Modelling microstructure evolution during casting, homogenization and ageing heat treatment of Al–Mg–Si–Cu–Fe–Mn alloys. *CALPHAD*, 63, 164–184.

Tegze, G., Bansel, G., Tóth, G. I., Pusztai, T., Fan, Z., and Gránásy, L. (2009) Advanced operator splitting-based semi-implicit spectral method to solve the binary phase-field crystal equations with variable coefficients. *Journal of Computational Physics*, 228(5), 1612–1623.

Teichert, G. H., and Garikipati, K. (2019) Machine learning materials physics: surrogate optimization and multi-fidelity algorithms predict precipitate morphology in an alternative to phase field dynamics. *Computer Methods in Applied Mechanics and Engineering*, 344, 666–693.

Tonks, M. R., and Aagesen, L. K. (2019) The phase field method: mesoscale simulation aiding material discovery. *Annual Review of Materials Research*, 49(1), 79–102.

Vaithyanathan, V., Wolverton, C., and Chen, L. Q. (2002) Multiscale modeling of precipitate microstructure evolution. *Physical Review Letters*, 88(12), 125503.

Wang, Y., and Li, J. (2010) Phase field modeling of defects and deformation. *Acta Materialia*, 58(4), 1212–1235.

Warren, J. A., and Boettinger, W. J. (1995) Prediction of dendritic growth and microsegregation patterns in a binary alloy using the phase-field method. *Acta Metallurgica et Materialia*, 43(2), 689–703.

Warren, J. A., Kobayashi, R., Lobkovsky, A. E., and Carter, W. C. (2003) Extending phase field models of solidification to polycrystalline materials. *Acta Materialia*, 51(20), 6035–6058.

Wei, L., Lin, X., Wang, M., and Huang, W. (2011) A cellular automaton model for the solidification of a pure substance. *Applied Physics A*, 103(1), 123–133.

Wheeler, A. A., Boettinger, W. J., and McFadden, G. B. (1993) Phase-field model of solute trapping during solidification. *Physical Review E*, 47(3), 1893–1909.

Wolfram, S. (1986) *Theory and Applications of Cellular Automata (Including Selected Papers 1983–1986) 1*. Singapore: World Scientific.

Yabansu, Y. C., Steinmetz, P., Hötzer, J., Kalidindi, S. R., and Nestler, B. (2017) Extraction of reduced-order process-structure linkages from phase-field simulations. *Acta Materialia*, 124, 182–194.

Zaeem, M. A., Yin, H., and Felicelli, S. D. (2012) Comparison of cellular automaton and phase field models to simulate dendrite growth in hexagonal crystals. *Journal of Materials Science and Technology*, 28(2), 137–146.

Zhang, X., Zhao, J., Jiang, H., and Zhu, M. (2012) A three-dimensional cellular automaton model for dendritic growth in multi-component alloys. *Acta Materialia*, 60(5), 2249–2257.

Zhou, N., Shen, C., Mills, M., and Wang, Y. (2010) Large-scale three-dimensional phase field simulation of $\gamma'$-rafting and creep deformation. *Philosophical Magazine*, 90(1–4), 405–436.

Zhu, M. F., and Hong, C. P. (2001) A modified cellular automaton model for the simulation of dendritic growth in solidification of alloys. *ISIJ International*, 41(5), 436–445.

# 4 Fundamentals of Crystal Plasticity Finite Element Method

## Contents

## 4.1 Crystal Plasticity and Its General Features

In Chapter 3, we described the fundamentals of mesoscale methods (mainly the phase-field and CA methods) for simulation of microstructure. That chapter mentioned that the integration of the phase-field method with crystal plasticity can be used to simulate the mechanical response of crystalline materials in mesoscale quantitatively. In this chapter, the fundamentals of crystal plasticity will be briefly described. Crystal plasticity can be integrated with either the phase-field or finite element method to describe the mechanical response of crystalline materials, from single crystals to engineering components.

Previous approaches (Bishop and Hill, 1951a, b) used for treating anisotropic plasticity provide many enlightening ideas for the numerical calculation of plasticity. But these approaches cannot consider the mechanical interactions in polycrystalline materials and are also not suitable for cases with complex internal or external boundary conditions, such as in-grain or grain cluster mechanical problems and abrupt mechanical transitions at interfaces. In order to remedy the situation, since Peirce et al. (1982) first mentioned the method in the 1980s, the crystal plasticity finite element method (CPFEM) has become one of the main methods to describe the mechanical response of materials from microscopic to macroscopic scales. This method can be utilized to describe the elastic-plastic deformation of anisotropic heterogeneous crystalline materials, being of fundamental importance for both mechanical property predictions and performance simulations.

Nowadays, CPFEM has been utilized to describe various crystal mechanical problems, ranging from orientation stability, crystal deformation, nanoindentation, recrystallization, mesoscale damage prediction, cup drawing, multiscale prediction of rolling texture, and so on (Roters et al., 2010; Zhao et al., 2008). A well-known advantage of CPFEM is its ability to analyze multiparticle problems since the method can consider the differences among individual crystal orientations during the simulation. Another important advantage of CPFEM is its ability to solve crystal mechanics problems with complex internal and external boundary conditions, as it enables the treatment of boundary conditions imposed by intergranular and intracrystalline micromechanical interactions. An additional advantage of CPFEM is its high flexibility to include various constitutive formulations (such as dislocation glide, martensite formulation, mechanical twinning, etc.) for plastic flow and hardening. One more advantage is the numerical aspect of CPFEM. The formulations of CPFEM can be either integrated into commercial finite element codes or implemented into commercial solvers in the form of user-developed subroutines. The last feature is very important since many engineering applications are usually performed by means of commercial software packages.

## 4.2    Basic Concepts and Equations of Continuum Mechanics

In this subsection, basic concepts and equations of continuum mechanics will be briefly introduced for the sake of understanding fundamental equations of the crystal plasticity finite element method. Many monographs have been published on continuum mechanics. For more details about this topic, the interested reader can refer to the books of Borja (2013); Dunne and Petrinic (2005); Raabe et al. (2004).

### 4.2.1    Definition of a Few Basic Terms in Continuum Mechanics

*Object and configuration*: In continuum mechanics, an *object* is regarded as a coherent set of "particles" (or "microclusters"). On one hand, particles should be large enough microscopically so that they contain enough microscopic particles (molecules, atoms, ions, etc.). On the other hand, to ensure that all kinds of physical quantities of such particles have a macroscopic statistical table at any time, they should be small enough macroscopically to be regarded as geometric "points." We call space occupied by each object at any time, together with the corresponding distribution of the position of all its particles in this region, as a *configuration* of this object. To describe the configuration and its change, the coordinate system and the deformation gradient need to be defined, which will be described in Section 4.2.2.

*Stress:* When an object is deformed due to external factors (force, temperature field, etc.), it produces an interactive internal force between various parts of the object in order to resist the effect of this external cause and try to restore the object from the position after deformation to its position before deformation. The internal force per unit area at a certain point in the examined small space is called stress. Those perpendicular to the same slide plane are called normal stress, and those tangent to the same slide plane are named shear stress.

*Strain:* Strain refers to the local relative deformation of a body under the action of external force or nonuniform temperature field.

*Constitutive relation:* The relationship between stress tensor and strain tensor defines the constitutive relation. In general, it refers to a set of correlations that relate the parameters describing the deformation of the continuous medium with those describing the internal force. Specifically, the constitutive relation refers to a set of constitutive equations that relate the strain tensor to the stress tensor. For different materials, there are different constitutive relations under different deformation conditions, which are also known as different constitutive models. CPFEM is one of a few methods to establish the constitutive relations for various materials.

## 4.2.2    Three Coordinate Systems and the Deformation Gradient

Usually there are three coordinate systems used in crystal plasticity.

A *shape coordinate* is based on the physical shape of the object. During the deformation process, the object will be deformed with the change of shape.

A *lattice coordinate* has a coordinate axis fixed locally parallel to the crystal direction. When the existence of point defect is not considered, the nodes of the coordinate network maintain one-to-one correspondence within the lattice points during deformation process. The deformation in the shape and lattice coordinates is consistent only when the crystal defect does not move, which is the usual assumption to calculate the elastic stress by measuring the small deformation of the coordinate system deposited on the surface or embedded in the object.

A *laboratory coordinate* will not deform with the body. The deformation of the shape coordinate system and the lattice coordinate system can be calculated according to their components in the laboratory coordinate system.

To observe the deformation, one can consider a small piece of hypothetical material that has not yet been loaded. Consequently, this hypothetical material is in an undeformed (or initial) configuration (or state), as shown in state A in Figure 4.1.

Material in state A is deformed to state B by an external force. Assuming that the material is subjected to composite tension, rigid body rotation, and displacement, we can measure all quantities relative to the global XYZ axis, commonly referred to as the material coordinate system, or laboratory coordinates. Let us consider an infinitesimal line PQ or vector $\mathbf{dX}$ (or $\mathbf{dx}$ for line P'Q') embedded inside the material in a

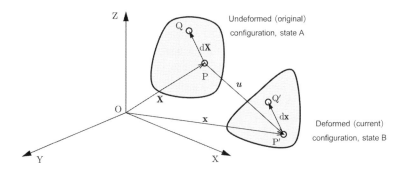

**Figure 4.1** A material element in a reference or undeformed configuration, state A, that deforms into a deformed or current configuration, state B ($\mathbf{x}$ denotes the vector under the current configuration, $\mathbf{X}$ denotes the vector under the reference configuration).

deformable configuration. The line PQ goes from state A to the configuration of state B, shifting the point P by vector $\mathbf{u}$ to the point $P'$. With respect to the reference frame of matter, the point $P'$ is given by the vector $\mathbf{x}$:

$$\mathbf{x} = \mathbf{X} + \mathbf{u}. \tag{4.1}$$

$d\mathbf{X}$ is transformed to its deformed state $d\mathbf{x}$ by the deformation gradient tensor $\mathbf{F}$ according to the finite strain theory:

$$d\mathbf{x} = \mathbf{F}d\mathbf{X}. \tag{4.2}$$

### 4.2.3    Isochoric/Volumetric Split of the Deformation Gradient

Any deformation can be locally decomposed into first, pure volume deformation, and second, isochoric deformation, or isochoric deformation followed by pure volume deformation. To see this, one can note that the deformation gradient can be multiplicatively split as

$$\mathbf{F} = \mathbf{F}_{\mathrm{iso}}\mathbf{F}_{\mathrm{V}} = \mathbf{F}_{\mathrm{V}}\mathbf{F}_{\mathrm{iso}}, \tag{4.3}$$

where

$$\mathbf{F}_{\mathrm{V}} \equiv (\det \mathbf{F})^{1/3}\mathbf{I}, \tag{4.4}$$

is the volumetric component of $\mathbf{F}$ ($\mathbf{I}$ is a second-order unit tensor), and

$$\mathbf{F}_{\mathrm{iso}} \equiv (\det \mathbf{F})^{-1/3}\mathbf{F}, \tag{4.5}$$

is the isochoric component. Note that, by construction, $\mathbf{F}_{\mathrm{V}}$ corresponds to a purely volumetric deformation, and since

$$\det \mathbf{F}_{\mathrm{V}} = \left[(\det \mathbf{F})^{1/3}\right]^{3} \det \mathbf{I} = \det \mathbf{F}, \tag{4.6}$$

$\mathbf{F}_V$ and $\mathbf{F}$ produce the same volume change. The isochoric component represents a deformation with constant volume, that is,

$$\det \mathbf{F}_{\text{iso}} = \left[ (\det \mathbf{F})^{-1/3} \right]^3 \det \mathbf{F} = 1. \tag{4.7}$$

## 4.2.4     Polar Decomposition

The polar decomposition of the deformation gradient is as follows:

$$\mathbf{F} = \mathbf{R}\mathbf{U} = \mathbf{V}\mathbf{R}, \tag{4.8}$$

The local rotation tensor $\mathbf{R}$ is an orthogonal tensor, and the symmetric positive definite tensors $\mathbf{U}$ and $\mathbf{V}$ are the right and left stretch tensors, respectively. They can be related to each other by the following equation due to the orthogonality of $\mathbf{F}$:

$$\mathbf{V} = \mathbf{R}\mathbf{U}\mathbf{R}^{\mathrm{T}}. \tag{4.9}$$

Figure 4.2 shows a simple illustration of the polar decomposition of the deformation gradient. The stretch tensors $\mathbf{U}$ and $\mathbf{V}$ can be expressed as follows:

$$\mathbf{U} = \sqrt{\mathbf{C}}, \ \mathbf{V} = \sqrt{\mathbf{B}}, \tag{4.10}$$

where $\mathbf{C}$ and $\mathbf{B}$ represent the right and left Cauchy–Green strain tensors, respectively, which are defined as follows:

$$\mathbf{C} = \mathbf{U}^2 = \mathbf{F}^{\mathrm{T}}\mathbf{F}, \ \mathbf{B} = \mathbf{V}^2 = \mathbf{F}\mathbf{F}^{\mathrm{T}}. \tag{4.11}$$

## 4.2.5     Eulerian and Lagrangian Finite Strain Tensors

The Eulerian and Lagrangian finite strain tensors $\mathbf{E}$ and $\mathbf{E}^*$ are symmetric tensors that are defined by the deformation gradient $\mathbf{F}$:

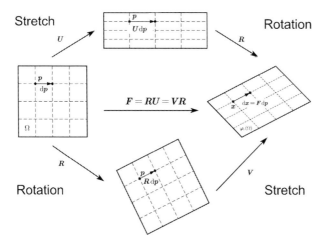

Stretch

Rotation

Rotation

$F = RU = VR$

Stretch

**Figure 4.2** Polar decomposition of the deformation gradient tensor $\mathbf{F}$.

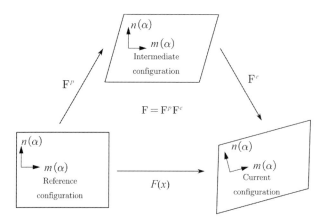

**Figure 4.3** Schematic diagram of the plastic/elastic action of the deformation gradient tensor **F**.

$$E = \frac{1}{2}\left(F^T F - I\right), \tag{4.12}$$

$$E^* = \frac{1}{2}\left(I - F^{-T}F^{-1}\right). \tag{4.13}$$

This deformation can be divided into two parts (Lee and Liu, 1967; Rice, 1971), as shown in Figure 4.3:

$$F = F^e F^p, \tag{4.14}$$

where $F^e$ is the elastic deformation component caused by the reversible response of lattice to external load, and $F^p$ is the plastic deformation, which is the irreversible deformation that still exists after all external forces causing deformation are eliminated.

## 4.3     Mechanical Constitutive Laws of Crystal Plasticity

Since the introduction of computational plasticity in 1950s, the proposed mechanical constitutive laws of empirical elastoplasticity formula (Rice, 1971) have evolved into multiscale internal variable plastic models based on physics, including varieties of direction-dependent and size-dependent effects as well as interface mechanisms (Arsenlis and Parks, 1999; Roters et al., 2010). Basically, the currently advanced theories of crystal plasticity have originated from several classical models, and these basic models are very briefly introduced as follows. For these models, the basic aim is to obtain the plastic strain increment, $\Delta\varepsilon$, due to crystal slip:

$$\Delta\varepsilon = \sum_{\alpha}\dot{\gamma}^{\alpha}\Delta t s^{\alpha} \otimes m^{\alpha}, \tag{4.15}$$

where $\dot{\gamma}^{\alpha}$ denotes the slip rates on each slip system $\alpha$, $\Delta t$ is the time step, $s^{\alpha}$ the slip direction, and $m^{\alpha}$ the normal direction of the slip plane.

In general, the two main types of constitutive models are *phenomenological* and *dislocation-based* models. The latter is the most important one in the framework of microstructure-based constitutive models, which include interface models and displacive transformation models (martensite formulation, mechanical twinning, and so on). Critical resolved shear stress (CRSS) $\tau_c^\alpha$ was used as a state variable on the individual slip system $\alpha$ in most phenomenological constitutive models. Therefore, the slip rate $\dot{\gamma}^\alpha$ is a function of the resolved shear stress,

$$\tau^\alpha = 0.5\mathbb{C}\left[F_e^T F_e - I\right] : m^\alpha \otimes s^\alpha, \tag{4.16}$$

where $\mathbb{C}$ is fourth-order elastic tensor. The slip rate $\dot{\gamma}^\alpha$ depends on the resolved and critical resolved shear stress:

$$\dot{\gamma}^\alpha = f\left(\tau^\alpha, \tau_c^\alpha\right). \tag{4.17}$$

Currently, the phenomenological crystal plasticity models are widely used. However, these models' disadvantage is that the material state can only be described by the CRSS $\tau_c^\alpha$ but not by the lattice defects. The lattice defects, such as dislocations and twins, are important internal variables that can describe the details of plastic deformation. The interested reader can refer to Arsenlis and Parks (1999) and Roters et al. (2010) for more details about the phenomenological crystal plasticity models.

## 4.3.1 Dislocation-Based Constitutive Models

In plastic mechanics, dislocation is the major carrier of plastic deformation, and there is no doubt that dislocation density is the most important internal variable. The constitutive models based on dislocation density can capture the physical behavior of strain hardening accurately at the microscale. By complementing the constitutive model into the finite element code, the macroscopic nonuniform and direction-dependent plastic deformation as well as microscopic dislocation evolution can be simulated simultaneously.

When the dislocation density is selected as an internal variable, its evolution law is usually expressed as the rate equation including dislocation multiplication and annihilation terms. Based on the kinetics of dislocation reactions, the evolution law of static dislocation density has been established. These dislocation responses include the actionless status of moving dislocations caused by dislocation locking, the formation of dipoles, and the dynamic recovery caused by dislocation climbing. However, the number of moving dislocations and the average slip velocity affect the plastic deformation rate, and usually both change at the same time. As a result, it is a formidable challenge to use a simple model to describe the evolution behavior of moving dislocation density. Therefore, if there is no further treatment, it is difficult to obtain an ideal multiplication term in the evolution law of moving dislocations. If the plastic deformation caused by an applied stress is maximum or the external stress caused by a given plastic deformation rate is minimum, the aforementioned problems could be solved.

At the beginning of polycrystalline plastic deformation, dislocations first occur on the slip system where the local shear in the grains with the optimal orientation is the largest. When moving dislocations encounter obstacles such as grain boundaries, they could accumulate in the front of the grain boundary and cause stress concentration on the grain boundary. Such an accumulation of dislocation will increase the applied stress field along the grain boundary. The local distribution of dislocations determines the local stress. In the continuum mechanics of crystal plasticity, the crystal plasticity model uses phenomenological statistics and even empirical formulas to describe the internal dislocation mechanism. Therefore, in this situation the understanding of these microplasticity effects is not deep. However, homogenization, which refers to the transition between the microscale and the macroscale (Arsenlis and Parks, 1999; Roters et al., 2010), is suggested under large plastic strain conditions, and the slip activation process can be captured by long-range stress. This means that the dislocation mechanism can be homogenized in the form of dislocation density, and the dislocation mechanism can be introduced into the theory of crystal plasticity as a rate-dependent constitutive equation.

Many classical models of crystal plasticity have been discussed by Roters et al. (2010). The usual way to introduce dislocations into the crystal plasticity model is to convert strain gradients into geometrically necessary dislocations by using Nye's dislocation tensor $\mathbf{\Lambda}$:

$$\mathbf{\Lambda} = \frac{1}{b}\mathrm{Curl}\ \mathbf{F}^p = -\frac{1}{b}\left[\nabla_x \times (\mathbf{F}^p)^T\right]^T, \tag{4.18}$$

where $\nabla_x = \partial/\partial x$ is the Laplace operator, which is defined as the derivative of the reference coordinate system, and $b$ refers to the magnitude of the Burgers vector.

## 4.3.2　Constitutive Models for Displacive Transformation

The previous section focused on dislocations as carriers of plastic shear. However, transformation-induced plasticity (TRIP) steel, austenite steel, brass, and shape memory alloy are all deformed by dislocation and displacement mechanisms (Roters et al., 2010). These mechanisms are characterized by the nondiffusive collective motion of atomic clusters, in which each atom moves only a short distance relative to its neighbors. This transformation produces shear kinematics similar to dislocation motion.

Much work (Lan et al., 2005; Salem et al., 2005; Thamburaja and Anand, 2001) has focused on the combination of martensite formation (or transformation) and mechanical twinning mechanisms within the framework of CPFEM. The martensitic transformation changes the lattice structure of the crystal, and the changes in shape usually involve the changes in the volume of a unit cell (volume expansion or contraction). Mechanical twinning is carried out through a shear mechanism that reorients the affected volume to a mirror direction relative to the surrounding matrix. For more details about constitutive models with displacive transformation, the interested reader could refer to Roters et al. (2011).

## 4.4        Brief Introduction to the Finite Element Method

The main focus of the crystal plasticity method is to predict the mechanical behavior of aggregates of grains. Generally speaking, the grain size is a few orders of magnitude smaller than the product scale. Therefore, the use of the crystal plasticity method alone cannot simulate the mechanical behavior of these engineering structures. On the other hand, the finite element method (FEM) has been the most popular simulation tool in structural mechanics since Courant performed the first FEM simulation in 1943 (Courant, 1943). Numerous publications about FEM are available. The interested reader can refer to Fish and Belytschko (2007), Logan (2017), and Roters et al. (2011) for exhaustive descriptions of the principles and applications of FEM. Within the framework of this chapter, FEM is considered one method to solve non-linear partial differential equations. Other commonly employed methods include the fast Fourier transform (FFT) method (Lebensohn et al., 2012), finite difference method (Boole, 1880), finite volume method (Versteeg and Malalasekera, 2007), boundary element method (Ali and Rajakumar, 2002), discrete element method (Munjiza, 2004), and smooth particle hydrodynamics (Hoover, 2006). The pertinent nonlinear partial differential equations associated with crystal plasticity have been briefly described in Section 4.3.

FEM usually utilizes the principle of virtual work to derive the fundamental equations associated with the targeted mechanical response. The principle of virtual work is described as follows. When a deformable object in equilibrium is subjected to any virtual displacement, and the object has a coordinated deformation, the virtual work of the external force on the object is equal to the virtual strain energy of the internal stress (Logan, 2017). Figure 4.4 schematically demonstrates the principle of virtual work. There is a deformable body with surface $S$ and volume $V$ in equilibrium under loads and constraints. The total force accelerating the body can comprise body forces per unit mass, g, and tractions (i.e., specific stresses) acting on its surface. The traction force **t** per unit deformed area with normal **n** in the current configuration is determined by the Cauchy stress **σ**. The action of the load produces an actual displacement field. Under the action of this load, the work associated with load and displacement must be equal to the strain energy computed from the stresses and strains

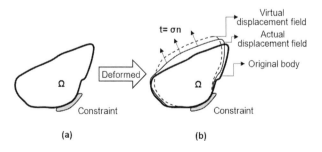

**Figure 4.4** The principle of virtual work on deformable body in equilibrium: (a) original position; (b) equilibrium position and application of a virtual displacement field.

over the domain due to the deformation. Now we assume a virtual arbitrary and smooth displacement field, as shown in Figure 4.4. During the application of the virtual displacement field, the applied load will do additional virtual work. The internal virtual strain energy caused by virtual displacement and virtual strains is equal to external virtual work. Accordingly, the principle of virtual work can be described as follows:

$$\int_S \mathbf{t} \cdot \delta \mathbf{v} dS + \int_V \rho \mathbf{g} \cdot \delta \mathbf{v} dV = \int_V \boldsymbol{\sigma} \cdot \mathrm{grad} \delta \mathbf{v} dV, \tag{4.19}$$

in which $S$ and $V$ denote surface and volume of a deformable body, respectively; $\mathbf{t}$ is traction force per unit deformed area; $\delta \mathbf{v}$ an arbitrary vector-valued field function; $\rho$ the density (mass per volume); $\mathbf{g}$ the body forces per unit mass; and $\boldsymbol{\sigma}$ the Cauchy stress.

In (4.19), the left-hand side is external work done by the surface and body force during a virtual smooth displacement field, while the right-hand side is the internal work associated with the virtual strain field.

Note that the test function $\delta \mathbf{v}$ is arbitrary. Therefore, the integrands can be regarded as the nature of mechanical work if we choose $\delta \mathbf{v}$ to be a displacement. This equation can be further rewritten using $\delta \varepsilon = (\mathrm{grad} \delta \mathbf{v})_{\mathrm{sym}}$ (Roters et al., 2011).

FEM considers the solution domain to consist of many small interconnected subdomains called finite elements, assuming a suitable approximate solution for an individual element, and then deriving the general satisfaction condition for solving this domain, thus getting the solution to the problem. Because it is difficult to obtain accurate solutions for most practical issues, FEM not only has high accuracy, but also can adapt to various complex shapes or structures. Consequently, it becomes an effective engineering analysis tool. Consider the domain discretized with two-dimensional elements (quadrilaterals or triangles), as shown in Figure 4.5. It discretizes the continuous system into a combination of a set of elements and uses the approximation function of the hypothesis in each element to slice the unknown domain function. The approximation function is usually an unknown domain function, and its derivative is expressed in the numerical interpolation function of each node of the element. Consequently, a continuous system with infinite degrees of freedom becomes a discrete finite number of degrees of freedom system.

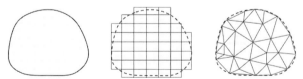

Arbitrary object   Quadrangular discretization   Triangular discretization

**Figure 4.5** Discretization of arbitrary object.

Under the preceding circumstance, the discretization approaches can solve (4.19) by introducing a large set of element shape functions $N_a^e$. As a result, the integrals in (4.19) are split into sums of integrals over the elements:

$$\sum_{e=1}^{N_e} \left[ \int_{S_e} tN_a^e dS_e + \int_{V_e} \rho g N_a^e dV_e \right] = \sum_{e=1}^{N_e} \left[ \int_{V_e} \sigma \mathrm{grad} N_a^e dV_e \right]. \qquad (4.20)$$

In this equation, $N_e$ denotes total number of elements. Because FEM usually employs the elements with simple regular shapes, the right-hand side is associated with the vector of all nodal displacements and a global stiffness matrix. In consequence, the problem can be easily solved by numerical methods.

FFT-based methodology can be also used for prediction of the textures and micromechanical field in polycrystals plastic deformation. When treating a model with periodic boundary conditions, this full-field crystal plasticity fast Fourier transform (CPFFT) approach is a very efficient alternative to the CPFEM due to the repeated use of the FFT algorithm and avoiding the large number of variables in the calculation by FEM.

## 4.5 Software and Procedure for the Crystal Plasticity Finite Element Simulation

### 4.5.1 Brief Introduction to Several FEM Software Packages

The numerous software packages available not only meet the requirement of simulation-based science and engineering but also advance the development of finite element method. Figure 4.6 shows the major commercial FEM software used for different applications, and CAE stands for computational aided engineering. In the

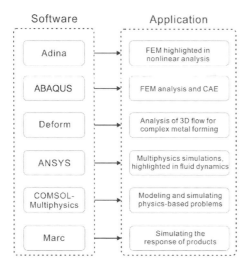

**Figure 4.6** FEM software packages and their pertinent application focus.

following, the major features of these representative commercial FEM software packages are briefly underlined.

*Adina* (www.adina.com) software was developed in 1970s based on finite element theory. The numerical solutions of solid mechanics, structural mechanics, and temperature field problems can be achieved by solving the mechanical linear and nonlinear equations. As a mechanics-based calculation software, Adina has a history of nearly 50 years. Solution modules for structures, computational fluid dynamics (CFD), electromagnetism, thermodynamics, etc., have been gradually developed. Adina is powerful in nonlinear analysis for dynamic analysis such as impacting, crash, buckling, and metal forming.

The *ABAQUS* Unified FEA (finite element analysis) product suite (www.3ds.com/products-services/simulia/products/abaqus) provides powerful and complete solutions for both regular and sophisticated engineering problems covering various industrial applications. With ABAQUS, engineers in the automotive industry can use common model data structures and integrated solver techniques to consider vehicle loads, dynamic vibration, multibody systems, collision, nonlinear static, thermal coupling, and acoustic structural coupling.

The *Deform* (Design Environment for Forming) code is an engineering software developed by Scientific Forming Technologies Corporation (SFTC, USA) (www.deform.com). Metal forming, heat treatment, machining, and mechanical joining processes can be analyzed by designers using Deform. The solver is based on rigid or visco-rigid plastic FEM, which is very efficient in dealing with large plastic deformation problems, where the elastic deformation can be neglected.

*ANSYS* (www.ansys.com) is a general finite element software that combines structure, heat, fluid, electromagnetism and acoustics as an entity. It is widely used in industrial and scientific communities covering nuclear industry, including, petrochemical industry, machinery manufacturing, aerospace, defense, automotive, energy, electronics, civil engineering, biomedicine, hydraulic engineering and household appliances, etc. ANSYS provides various functions, such as highly non-linear structural analysis, electromagnetic analysis, CFD, design optimization, contact analysis, adaptive meshing and extension of macro command based on ANSYS parametric design language.

*COMSOL Multiphysics* (www.comsol.com) is a general simulation software for modeling devices, designs, and processes in engineering, manufacturing, and scientific research fields. Besides for user's projects, the multi-physics can be used to turn models into simulation applications and digital twins that might be used by other design teams, customers, manufacturing departments, and laboratories.

*Marc* (www.mscsoftware.com/product/marc) is an effective, multipurpose, nonlinear finite element analysis code to model the behavior of a product under dynamic, static, and multiphysics loading situations accurately. Marc can be used to simulate nonlinear behavior and transient environmental conditions for a variety of materials. It is ideal for solving complex design problems.

Currently, there are only a few *CPFEM software packages* that are publicly available mainly for fundamental research. There is still a long way to go for its engineering application. The important features for three representative CPFEM software packages are briefly described as follows.

*Düsseldorf Advanced Material Simulation Kit (DAMASK)* (https://damask.mpie.de/; Roters et al., 2019) is used for simulating multiphysics crystal plasticity, thermal, and damage phenomena in the scale from single crystal to component. The internal module structure of DAMASK directly follows the hierarchical structure inherent in the continuum description. The top layer in DAMASK deals with the division of field values specified at the material points between the underlying microstructure components and the subsequent homogenization of the constitutive response of each component. The response of individual microstructure features is described at the intermediate level according to the basic constitutive laws of elasticity, plasticity, phase transformation, damage, and other coupled multiphysical processes. Different constitutive laws can be implemented to provide the response at the lowest level through changing internal state variables. Various CP models have been implemented in DAMASK.

*PRISMS-Plasticity* (https://github.com/prisms-center/plasticity) is an open-source parallel 3D CPFEM software package (Yaghoobi et al., 2019). This toolkit develops an efficient rate-independent crystal plasticity algorithm considering deformation twins, and the program has a good parallelism, which can handle large-scale calculation problems with up to millions of elements.

*MOOSE* (Multiphysics Object-Oriented Simulation Environment) (https://mooseframework.org/) (Bishop and Hill, 1951a; Tonks et al., 2012; Williamson et al., 2012) is a framework for solving engineering problems of multiphysics field coupling. Its design specification, which adopts the object-oriented programming paradigm, is very easy to expand and maintain, and hides the computational mathematical problems behind the physics problems as much as possible, such as adaptive grid algorithm, parallel computing technology, and so on. On the basis of MOOSE, one can concentrate on the exploration of the key theories of CPFEM without spending too much effort on numerical calculations.

## 4.5.2  Procedure for the Crystal Plasticity Finite Element Method

Despite the numerous commercial software packages used for FEM and CPFEM simulations, the general procedures for their utilizations are similar for both simulations. The basic steps for solving a finite element problem include three stages, which are briefly described as follows.

**Step 1**: Preprocessing stage.

In the first step, the problem and solution domain are defined and prepared to be solved. According to the targeted problem, the geometric model is established and the material properties are assigned to the geometric model.

The solution domain is discretized into an assembly of finite elements with different sizes and shapes, which are connected to each other. The smaller the elements, the better the approximation degree of the discrete domain, and the more accurate the calculation result, but the calculation cost will increase. The discretization of the solution domain is one of the core technologies for the finite element method. To ensure the convergence of problem solving, element selection should follow certain principles. For example, the shape of the element should be ruled as good. Otherwise, the deformation will lead to not only low precision, but also the high risk of missing the rank, which will result in failure to solve.

After the loading and boundary conditions are applied to the relevant nodes, the specific application issues can be represented by a set of differential equations, and these equations are transformed into an equivalent functional form. That enables the problem to be solved using numerical techniques.

**Step 2**: Solution stage.

The total matrix equation (joint equations) of the discrete domain is formed by the total assembly. The final assembly is performed at the adjacent unit node, and the continuities of the state variables and their derivatives (if possible) are established at nodes. The finite element method ultimately leads to several equations, which can be solved through numerical methods such as Gaussian elimination, Gauss–Seidel iteration or Cholesky square root methods and so on.

**Step 3**: Post processing stage.

The result of the solution is an approximation of the state variables at the nodes, and they are presented in the form of stress plots, deformed geometry and listing of nodal displacements. These data can be presented in contours or exported for further data processing.

In the next section, the procedure associated with CPFEM will be briefly described using one case study of plastic deformation-induced surface roughening in Al polycrystals.

## 4.6    A Case Study: Plastic Deformation-Induced Surface Roughening in Al Polycrystals

The result of reliable simulation method must be close to the experimental observation as much as possible. The mechanical and microstructure characteristics of a CPFEM simulation can be verified through experiments. The measurement of mechanical properties can be monitored from the aspects of force, elastic stiffness, stress, stress–strain curve, etc. The number of microstructural observations can be mapped according to crystal structure, grain shape, dislocation structure and density, internal stress, surface roughness, etc. Many comparisons between CPFEM and experiments can be made on different scales one by one (Roters et al., 2010). The case study described in this section is a good example of combining experiments with the CPFEM method.

Zhao et al. (2008) studied the plastic strain localization and deformation-induced surface roughness of aluminum polycrystals composed of a small amount of coarse grains. A dog bone specimen analyzed by electron backscatter diffraction (EBSD) indicates plastic deformation under uniaxial tensile load. It is assumed that the grain boundaries on the surface are always perpendicular to the sample surface so that the two-dimensional simulation can achieve the required information.

The solution stage for the targeted issue is briefly described as follows. During the process of deformation, the history of strain localization, surface roughening, microstructure, and intragranular fragmentation were recorded. Using the CPFEM model, the corresponding one-to-one high-resolution simulation was carried out. The results first show that the grain topology and microstructure have an important influence on the source of strain heterogeneity. In addition, the results show that the final surface roughness is not only related to the macroscopic strain localization, but also to the interaction within the grains. Finally, the activation of the sliding system was detected in detail by using the slip lines observed on the sample surface. In the analysis, the ability of the CPFEM model to capture the effect of surface roughness, strain localization related to orientation, and the details of activation mode of slip system in grains are paid special attention. It has been observed that the crystal particles that stretch over the entire width of the dog's bones are very soft. Due to a single isolated soft grain, the lack of dislocation barrier provided by the grain boundary promotes strain localization. Due to the obvious decrease of sample thickness, especially in the soft zone of the sample, serious surface roughness appears in the lower-right corner of Figure 4.7. The main focus is to show how the CPFEM predicts the roughness and

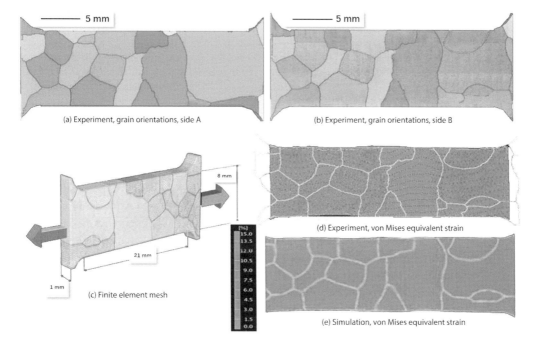

**Figure 4.7** Plastic strain localization and deformation-induced surface roughening in a 3D aluminum polycrystal consisting of a few coarse grains (Roters et al., 2010; Zhao et al., 2008). (a) and (b) are the two sides of the view of the sample. (c) is the finite element model of uniaxial tension after mesh division, which is displayed based on the view of (b). (d) and (e) are the stress distribution obtained by experiment and simulation, respectively. The figures are reproduced from Roters et al. (2010); Zhao et al. (2008) with permission from Elsevier. A black and white version of this figure will appear in some formats. For the colour version, please refer to the plate section.

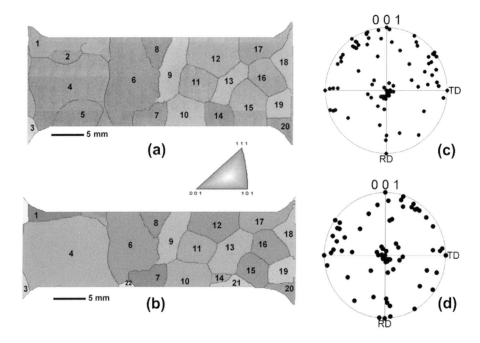

**Figure 4.8** Surface grain profiles and orientation distribution obtained by EBSD (Zhao et al., 2008). Grain shape on top (a) and bottom (b) surface and overall orientation distributions (c, d) on these two surfaces (a, b), respectively. The figures are reproduced from Zhao et al. (2008) with permission from Elsevier. A black and white version of this figure will appear in some formats. For the colour version, please refer to the plate section.

strain localization's history in order to provide more and clearer information related to the microstructure and intragranular fragmentation.

According to the crystallographic information provided in Figure 4.8, the deformation results of Figure 4.7 are obtained by means of the crystal plastic finite element method. At low strain levels, the appearance of surface slip lines is usually caused by slip systems with the largest Schmid factor. Due to the continuity of the crystal plasticity finite element model, in addition to the maximum Schmid factor, additional slip systems need to be activated. The increased strain will produce laminated plate structure, which is beyond the capability of CPFEM.

Similar studies on mechanical unevenness and texture evolution of coarse-grained samples have been reported elsewhere (Beaudoin et al., 1996; Sachtleber et al., 2002). The common point of these works is that they show obvious orientation fracture and strain localization according to the grain orientation.

In fact, using reasonable computational costs and open-source or commercial solvers, the CPFEM can deal with complex boundary conditions, various deformation mechanisms and their interactions, interface characteristics, and slip localization details. It is an effective modeling platform for microscopic and macroscopic mechanical problems with degrees of freedom for shear. Currently, only a few of these microstructure features (second phase fractions, precipitation morphology, and so on) have been introduced into CPFEM. Multiscale simulation is still a serioius challenge to CPFEM.

# References

Ali, A., and Rajakumar, C. h. (2002) *The Boundary Element Method: Applications in Sound and Vibration*. London: Taylor & Francis.

Arsenlis, A., and Parks, D. M. (1999) Crystallographic aspects of geometrically-necessary and statistically-stored dislocation density. *Acta Materialia*, 47(5), 1597–1611.

Beaudoin, A. J., Mecking, H., and Kocks, U. F. (1996) Development of localized orientation gradients in Fcc polycrystals. *Philosophical Magazine A*, 73(6), 1503–1517.

Bishop, J. F. W., and Hill, R. (1951a) CXXVIII. A theoretical derivation of the plastic properties of a polycrystalline face-centred metal. *Philosophical Magazine Series 7*, 42 (334), 1298–1307.

Bishop, J. F. W., and Hill, R. (1951b) XLVI. A theory of the plastic distortion of a polycrystalline aggregate under combined stresses. *Philosophical Magazine Series 7*, 42(327), 414–427.

Boole, G. (1880) *A Treatise on the Calculus of Finite Differences*. London: Cambridge University Press.

Borja, R. I. (2013) *Plasticity: Modeling & Computation*. Berlin, Heidelberg: Springer.

Courant, R. (1943) Variational methods for the solution of problems of equilibrium and vibrations. *Bulletin of the American Mathematical Society*, 49(1), 1–24.

Dunne, F., and Petrinic, N. (2005) *Introduction to Computational Plasticity*. Oxford: Oxford University Press.

Fish, J., and Belytschko, T. (2007) *A First Course in Finite Elements*. Chichester: John Wiley & Sons, Ltd.

Hoover, W. G. (2006) *Smooth Particle Applied Mechanics: The State of the Art*. River Edge: World Scientific Publishing Co. Pte. Ltd.

Lan, Y. J., Xiao, N. M., Li, D. Z., and Li, Y. Y. (2005) Mesoscale simulation of deformed austenite decomposition into ferrite by coupling a cellular automaton method with a crystal plasticity finite element model. *Acta Materialia*, 53(4), 991–1003.

Lebensohn, R. A., Kanjarla, A. K., and Eisenlohr, P. (2012) An elasto-viscoplastic formulation based on fast Fourier transforms for the prediction of micromechanical fields in polycrystalline materials. *International Journal of Plasticity*, 32–33, 59–69.

Lee, E, H , and Liu, D. T. (1967) Finite-strain elastic – plastic theory with application to PlaneWave analysis. *Journal of Applied Physics*, 38(1), 19–27.

Logan, D. L. (2017) *A First Course in the Finite Element Method*, sixth edition. Boston: Cengage Learning.

Munjiza, A. (2004) *The Combined Finite-Discrete Element Method*. Hoboken: Wiley.

Peirce, D., Asaro, R. J., and Needleman, A. (1982) An analysis of nonuniform and localized deformation in ductile single crystals. *Acta Materialia*, 30(6), 1087–1119.

Raabe, D., Roters, F., Barlat, F., and Chen, L. Q. (2004) *Continuum Scale Simulation of Engineering Materials: Fundamentals – Microstructures – Process Applications*. Weinheim: Wiley-VCH Verlag GmbH & Co. KGaA.

Rice, J. R. (1971) Inelastic constitutive relations for solids: an internal-variable theory and its application to metal plasticity. *Journal of the Mechanics and Physics of Solids*, 19(6), 433–455.

Roters, F., Diehl, M., Shanthraj, P., et al. (2019) DAMASK – the Düsseldorf Advanced Material Simulation Kit for modeling multi-physics crystal plasticity, thermal, and damage phenomena from the single crystal up to the component scale. *Computational Materials Science*, 158, 420–478.

Roters, F., Eisenlohr, P., Bieler, T. R., and Raabe, D. (2011) *Crystal Plasticity Finite Element Methods: In Materials Science and Engineering*. Weinheim: Wiley-VCH Verlag GmbH & Co. KGaA.

Roters, F., Eisenlohr, P., Hantcherli, L., Tjahjanto, D. D., Bieler, T. R., and Raabe, D. (2010) Overview of constitutive laws, kinematics, homogenization and multiscale methods in crystal plasticity finite-element modeling: theory, experiments, applications. *Acta Materialia*, 58(4), 1152–1211.

Sachtleber, M., Zhao, Z., and Raabe, D. (2002) Experimental investigation of plastic grain interaction. *Materials Science and Engineering A*, 336(1–2), 81–87.

Salem, A. A., Kalidindi, S. R., and Semiatin, S. L. (2005) Strain hardening due to deformation twinning in $\alpha$-titanium: constitutive relations and crystal-plasticity modeling. *Acta Materialia*, 53(12), 3495–3502.

Thamburaja, P., and Anand, L. (2001) Polycrystalline shape-memory materials: effect of crystallographic texture. *Journal of the Mechanics and Physics of Solids*, 49(4), 709–737.

Tonks, M. R., Gaston, D., Millett, P. C., Andrs, D., and Talbot, P. (2012) An object-oriented finite element framework for multiphysics phase field simulations. *Computational Materials Science*, 51(1), 20–29.

Versteeg, H. K., and Malalasekera, W. (2007) *An Introduction to Computational Fluid Dynamics: The Finite Volume Method*, second edition. London: Pearson Education Limited.

Williamson, R. L., Hales, J. D., Novascone, S. R., et al. (2012) Multidimensional multiphysics simulation of nuclear fuel behavior. *Journal of Nuclear Materials*, 423(1–3), 149–163.

Yaghoobi, M., Ganesan, S., Sundar, S., et al. (2019) PRISMS-Plasticity: an open-source crystal plasticity finite element software. *Computational Materials Science*, 169, 109078.

Zhao, Z., Ramesh, M., Raabe, D., Cuitiño, A. M., and Radovitzky, R. (2008) Investigation of three-dimensional aspects of grain-scale plastic surface deformation of an aluminum oligocrystal. *International Journal of Plasticity*, 24(12), 2278–2297.

# 5 Fundamentals of Computational Thermodynamics and the CALPHAD Method

## Contents

## 5.1    Introduction

The laws of thermodynamics are the scientific foundation for major branches of engineering. *Engineering thermodynamics* deals essentially with questions of heat generation, heat transfer, power-heat conversion, and the corresponding plant design and process engineering. *Materials thermodynamics*, which is in our focus, on the other hand, follows the groundbreaking work *On the Equilibrium of Heterogeneous Substances* by J. Willard Gibbs (Gibbs, 1874; 1878), where these laws are applied to chemical reactions of substances and between various phases, their stability, and transitions. From this basis, the field of chemical thermodynamics, also named thermochemistry or simply thermodynamics, evolved with mathematical descriptions of chemical, physical, electrical, magnetic, and other phenomena. The application to real-world engineering materials was made possible with new computational techniques and the development of the CALPHAD approach. This is why this chapter is specified as "*computational* thermodynamics." The CALPHAD method is recognized as a cornerstone in the modern field of integrated computational materials engineering (ICME) (Xiong and Olson, 2016) and fundamental to the research and development of materials.

This chapter is intended as a practical guide. Hence, on the one hand, the reader may start without prior knowledge of the subject and explanations are focused on real alloy examples. On the other hand, precision in the definitions of terms and equations is pursued and potential traps are indicated that are hard to find in publications or even in software and database documentations, which may also be useful for advanced readers.

Thus, the laws of thermodynamics and the general background of materials thermodynamics are not treated here. They are found in the excellent book by Chang and Oates (2010), by Liu and Wang (2016) and with more focus on phase diagrams by Pelton (2014). Here we focus directly on the CALPHAD method and its thermodynamic basis with the crucial concept of "phase." It is shown why the Gibbs energy is the key thermodynamic function and how it is modeled for very simple and some more complex phases. These mathematical descriptions are implemented in powerful software codes of commercial packages and partly in open-source codes. The intention of Sections 5.2 and 5.3 is to provide a fundamental understanding of the concepts behind CALPHAD by explaining it with examples.

The use of any CALPHAD software depends on the availability and reliability of the thermodynamic database, comprising the thermodynamic parameters for the pertinent material system. The development of such databases is only recommended for experts, and the book of Lukas et al. (2007) gives a good introduction. Current database development also integrates first-principles calculations and CALPHAD modeling with machine learning (Liu, 2018). In this chapter, the focus is on reasonable application of CALPHAD software and databases to engineering materials. It is essential to understand the basics and limitations of databases, as outlined in Section 5.4.

The application of the CALPHAD method in the computational design of alloys and their processing is shown for real-world examples at two levels. At the first level, as indicated in Section 5.5, solely thermodynamic CALPHAD databases are required. Starting with simple examples, the section shows which type of calculations have proved most useful to guide such design. At the second level, as shown in Section 5.6, applications using extended CALPHAD-type databases are outlined for the examples of solidification and heat treatment processes. This involves databases extended to comprise some kinetic and thermophysical data of the alloy system in addition to the thermodynamic database. It also involves the use of advanced CALPHAD-type software packages, such as PanSolidification (Zhang et al., 2020) and PanPrecipitation (Cao et al., 2009). Examples are given for casting and heat treatment processes. Finally, one case study on the design of Al alloys with improved hot cracking resistance is presented. The enormous benefit of these methods by saving time and costs through guidance of large-scale experimental efforts is also demonstrated in many case studies in Chapters 7–12 of this book.

Ancillary material for this chapter is available from Cambridge University Press at www.cambridge.org. These are pertinent data files for the PanSolidification and PanPrecipitation modules of the Pandat software package (Cao et al., 2009; Chen et al., 2002). They are designed to run not only with the licensed Pandat software but also with the Pandat trial/education version that can be downloaded free of charge from CompuTherm's website (https://computherm.com/). The intention is to provide the reader with hands-on experience in solidification simulation (with microsegregation modeling) and heat treatment simulation (with precipitation modeling). A step-by-step instruction to perform the simulations related to Figures 5.23 and 5.24 is given in a readme file. Additional ancillary material is given in Chapter 6 for PanPrecipitation and for both simulation modules in Chapter 8.

## 5.2 Overview of the CALPHAD Method

### 5.2.1 Origins and Development of the CALPHAD Method

The acronym CALPHAD originally stood for **CAL**culation of **PHA**se **D**iagrams, and this method was developed in the early 1970s. The major protagonist was Larry Kaufman, the first author of the book Kaufman and Bernstein (1970), who organized together with some colleagues informal meetings on the subject. Eventually the journal CALPHAD was founded, published since 1977, and the international

CALPHAD conference series started, held each year in a different country, forming the lively network of the CALPHAD community. The journal's subtitle, *Computer Coupling of Phase Diagrams and Thermochemistry*, well describes the original CALPHAD method. A brief history of CALPHAD was given by Spencer (2008).

This approach is detailed in the context of phase transformations in the scholarly book of Hillert (2008), already briefly including subjects such as interfaces and kinetics of transport processes. Advanced methods for calculating more reliable multi-component phase diagrams for technological applications were presented by Chang et al. (2004). Some extended applications are highlighted by Liu (2020) and (Costa e Silva et al., 2007), while Liu et al. (2020) present related calculation of thermophysical properties.

Today, CALPHAD stands for more than calculation of phase diagrams. The method is extended to simulation techniques applied to multicomponent and multi-phase materials in the fields of diffusion, other kinetic processes, thermophysical properties, as well as mechanical properties. The method also involves the development of extended databases for such kinetic and other thermophysical materials properties. This is done using also the CALPHAD assessment technique and building up on the thermodynamic databases, including input from experimental as well as first-principles methods. This great leap forward to a second generation of CALPHAD is providing a key foundation for the materials genome (Kaufman and Ågren, 2014) and ICME (Kattner and Seifert, 2010; Xiong and Olson, 2016). It is also recognized as basis for a variety of other software solutions in ICME (Schmitz and Prahl, 2016).

## 5.2.2    Principles of the CALPHAD Method

Why is the Gibbs energy the key thermodynamic function in the CALPHAD method? That is understood by inspecting Table 5.1, a selection of five important thermodynamic state functions and other variables such as absolute temperature $T$, pressure $p$, and volume $V$. "Natural variables" are special in that the total differential of the corresponding state function depends only on their changes. For the internal energy $U$ of the system, these are entropy, $S$, and $V$. Thus, with $S$ and $V$ constant we have $dU = 0$, defining the equilibrium state at the minimum of $U$. However, there is no laboratory device to keep the entropy constant. That is different for the Helmholtz

**Table 5.1** Important thermodynamic state functions and variables of a single-component system at a fixed amount.

| State function | Symbol definition | Natural state variables | Total differential |
|---|---|---|---|
| Internal energy | $U$ | $S, V$ | $dU = T\,dS - p\,dV$ |
| Entropy | $S$ | $U, V$ | $dS = (1/T)\,dU + (p/T)\,dV$ |
| Helmholtz (free) energy | $F = U - TS$ | $T, V$ | $dF = -S\,dT - p\,dV$ |
| Enthalpy | $H = U + pV$ | $S, p$ | $dH = T\,dS + V\,dp$ |
| Gibbs energy | $G = H - TS$ | $T, p$ | $dG = -S\,dT + V\,dp$ |

energy, $F$, also called free energy, since $T$ and $V$ can be controlled in a furnace with the substance in a strictly rigid container. Thus, the equilibrium is attained at minimum of $F$. That minimum requirement at equilibrium is valid for all examples in Table 5.1, except for the entropy, which attains a maximum for $U$ and $V$ constant. For the majority of laboratory experiments or real materials processing, the obvious choice is to control $T$ in a furnace and $p$ either by ambient pressure or in an autoclave or vacuum equipment. Thus, the Gibbs energy is the most practical choice with $dG = -S\,dT + V\,dp = 0$ at fixed $T$ and $p$. The thermodynamic equilibrium state is then determined by finding the global minimum of $G$.

Once the mathematical function of $G(T, p)$ of the substance is determined, Table 5.2 shows that this implies the complete information on all variables and state functions. All the other state functions and in fact all thermodynamic functions (properties) are uniquely determined from $G$ and its partial derivatives, as detailed in Table 5.2. By knowing $G$, we also know $F$ and can calculate the equilibrium state also at fixed $T$ and $V$.

Additional variables are the amount $n$ of the system or, in multicomponent systems, the amounts of each component $i$, $n_i$. Note that the mol is a tricky unit. One mol equals $N_A$ countable items, where $N_A$ is Avogadro's number. This unit only makes sense if the item is well defined. This is clear if one only deals with

**Table 5.2** Derivation of all thermodynamic quantities from known Gibbs energy, $G$.

| Thermodynamic function | Symbol | Derivation from $G = G(T, p)$ | SI unit |
|---|---|---|---|
| Internal energy | $U$ | $U = G - T\dfrac{\partial G}{\partial T} - p\dfrac{\partial G}{\partial p}$ | J |
| Entropy | $S$ | $S = -\dfrac{\partial G}{\partial T}$ | $\mathrm{J\,K^{-1}}$ |
| Helmholtz (free) energy | $F$ | $F = G - p\dfrac{\partial G}{\partial p}$ | J |
| Enthalpy | $H$ | $H = G - T\dfrac{\partial G}{\partial T} = -T^2\dfrac{\partial(G/T)}{\partial T}$ | J |
| Volume | $V$ | $V = \dfrac{\partial G}{\partial p}$ | $\mathrm{m^3}\ (= \mathrm{J\,Pa^{-1}})$ |
| Molar volume | $V_m$ | $V_m = \dfrac{\partial G_m}{\partial p}$ | $\mathrm{m^3\,mol^{-1}}$ |
| Heat capacity | $C_{p(v)}$ | $C_p = -T\dfrac{\partial^2 G}{\partial T^2} = \dfrac{\partial H}{\partial T}$ $C_v = C_p - \left(\dfrac{\partial p}{\partial T}\right)_v \left(\dfrac{\partial V}{\partial T}\right)_p$ | $\mathrm{J\,K^{-1}}$ |
| Molar heat capacity | $c_p$ | $c_p = \dfrac{C_p}{\mathrm{amount}(n)}$ | $\mathrm{J\,mol^{-1}\,K^{-1}}$ |
| Specific heat | $c_p$ | $c_p = \dfrac{C_p}{\mathrm{mass}}$ | $\mathrm{J\,kg^{-1}\,K^{-1}}$ |
| Chemical potential | $\mu_i$ | $\mu_i = \left(\dfrac{\partial G}{\partial n_i}\right)_{T,p,n_{j\neq i}}$ | $\mathrm{J\,mol^{-1}}$ |

stoichiometric items, as in chemistry. However, in alloy science, with liquid and solid solution phases, stoichiometric and nonstoichiometric intermetallic phases all together, the unit "mol" is not clear at all. Therefore, one should specify the amount in "mol of atoms" or mol-atoms, meaning that the amount is determined by 1 mol of (different) elemental components that may be distributed to various compounds or solution phases in our multiphase system. For example, 1 mol-atoms of cementite $=$ 1 mol-atoms of $Fe_3C =$ 1 mol-atoms of $Fe_{0.75}C_{0.25} = (1/4)$ mol of $Fe_3C$, whereas "1 mol of cementite" is unclear. Unfortunately, the old unit "g-atom" for mol-atoms is no longer used in the SI system.

By fixing all of the amounts, $n_i$, the system is considered "closed" and the materials balance becomes valid. The "state point" is then defined by this chosen set of variables $(T, p, n_i)$. It is also possible to fix the chemical potentials, $\mu_j$, of some components $(j)$ and at least one amount $(i)$ – that is, an "open" system. For example, this is applied if our alloy (condensed system) equilibrates with a gas phase, such as in carburization or nitridation processes of steel. The chemical potentials, $\mu_j$, are then imposed onto the "open" system by a separate "reservoir" of infinite amount. The materials balance is only valid for the first $l$ components with given $n_i$ and $1 \leq i \leq l$, but it is not valid for components with given $\mu_i$ and $(l+1) \leq j \leq k$ of the $k$-component system. That also defines a unique "state point" by the chosen set of variables $(T, p, n_i, \mu_j)$.

Four different snapshots of a system in an approach to equilibrium are sketched in Figure 5.1. The abscissa may be envisaged as a path describing different constitutions of the system. That raises a fundamental question: How can we find the minimum of $G(T, p, n_i)$ at a state point given by fixed values $T$ and $p$ and $n_i$ ? If we fix all variables of any function, there is no room for change. The standard experiment for this state point is to put all the components with fixed amounts $n_i$ in a container with a movable piston to control $p$ and to control $T$ in our furnace. However, usually an internal variable or degree of freedom exists for the system. Nature provides this freedom by distributing the components onto different phases, if available. If not, we have indeed $G(T, p, n_i) =$ constant, and there is no need to search for a minimum.

The solid phases of the elements are always denoted by their crystal structure, such as Fcc (face-centered cubic), Bcc (body-centered cubic), or Hcp (hexagonal close packed). For example, we put 1 mol of aluminum atoms in our container at 1,073 K and $p = 1$ bar. Now a magician could place all atoms in one of the three phases of

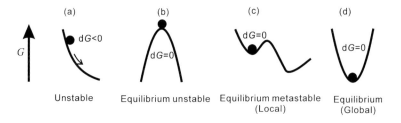

**Figure 5.1** Conceptual illustration of equilibrium states with infinitesimal change of the Gibbs energy of the system.

aluminum (solid Fcc, Liquid, or Gas), and we calculate $G$ for these three different constitutions of the system to find $G(\text{Fcc}) = -46{,}555$ J, $G(\text{Liquid}) = -48{,}143$ J, and $G(\text{Gas}) = +106{,}584$ J. Nature decides to place all atoms in the liquid phase at this state point in thermodynamic equilibrium. Once we know the functions $G(T, \text{phase})$ for the three phases in the entire temperature range at 1 bar, we can immediately calculate the equilibrium constitution of aluminum by finding the minimum of $G(T, \text{phase})$ at any temperature: Fcc $(T < 933$ K$)$, Liquid $(933$ K $< T < 2{,}793$ K$)$, and Gas $(T > 2{,}793$ K$)$.

In addition to these three phases known to be stable at least at some temperature at 1 bar, we can also ask the magician to place all atoms in the hexagonal Hcp crystal structure at 1,073 K and find $G(\text{Hcp}) = -43{,}006$ J. Even though pure Al is never stable in an Hcp structure at 1 bar, the quantity $G(\text{Hcp})$ is of crucial importance to the CALPHAD method. The difference $G(\text{Hcp}) - G(\text{Fcc})$ defines the "lattice stability" of Al(Hcp) referred to the stable Al(Fcc). This concept was the breakthrough to tackle binary or multicomponent systems. Then the functions $G(T, p, x_i, \text{phase})$ are required for each phase in the entire range of state points $(T, p, x_i)$, where $x_i$ is the mole fraction of the component $(i)$, $x_i = n_i / \sum n_i$.

For example, in the binary system Al–Mg, shown also later, extended, though limited, solid solution ranges are stable between pure stable Al (Fcc) and pure stable Mg (Hcp). The Mg atoms dissolve in the Fcc phase and Al atoms dissolve in the Hcp phase. The term "phase" is more general than the term "structure"; sometimes the terms are used synonymously for solid crystalline or amorphous phases. Now the functions $G(T, x_{\text{Mg}}, \text{Fcc})$ and $G(T, x_{\text{Mg}}, \text{Hcp})$ must be given for $0 \leq x_{\text{Mg}} \leq 1$ beyond their stable ranges, up to their "metastable endmember." For the preceding example, we have $G(1{,}073$ K, $x_{\text{Mg}} = 0$, Hcp$) = -43{,}006$ J.

In contrast to Figure 5.1, we conclude that for our example of pure Al, the variation of $G$ at a fixed state point is not continuous but rather stepwise downward the values of $G(\text{phase})$, depending on the manifestation of the Al atoms in the considered phases. For the example of a binary system A–B, there are more options. Shown in Figure 5.2

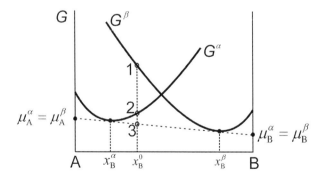

**Figure 5.2** Sketch of the Gibbs energy curves of two phases $\alpha$ and $\beta$ in the binary A–B system at constant $p$ and $T$, demonstrating how the Gibbs energy of the system is minimized in three steps by choice of the constitution toward the $\alpha + \beta$ phase equilibrium.

are the $G$-curves of the two phases $\alpha$ and $\beta$ assumed to exist, not plotted up to their metastable endmembers. The entire diagram is valid for fixed values of $T$ and $p$, and we also fix the composition of the system to the value $x_B^0$; thus, the state point $(T, p, x_B^0)$ is fixed. The degree of freedom in the system is given by different distribution of atoms to the available phases under the constraint of materials balance of A and B atoms. That is set by the value of $x_B^0$ and the generally assumed total amount of $n_A + n_B = 1$ mol-atoms. The values of the Gibbs energy functions of the phases, $G^\alpha$ and $G^\beta$, also refer to 1 mol-atoms. More details for these functional descriptions are given later in Section 5.3.2.

For this state point, our magician could place all atoms in the $\beta$ phase and we have $G(\text{system}) = G^\beta(x_B^0)$ at point 1. With all atoms in the $\alpha$ phase, we obtain the lower value of $G(\text{system}) = G^\alpha(x_B^0)$ at point 2. More freedom is given if some atoms are placed in the $\alpha$ and the rest in the $\beta$ phase. We need to distinguish the composition of the system $x_B^0$ from the compositions of the phases $x_B^\alpha$ and $x_B^\beta$. The Gibbs energy is additive from these parts (phases) of the system, so we have $G(\text{system}) = f^\alpha G^\alpha(x_B^\alpha) + f^\beta G^\beta(x_B^\beta)$. Here $f^\alpha$ and $f^\beta$ are the molar phase fractions that are determined by the chosen phase compositions and the materials balance. The solution of the set of linear equations $x_i^0 = f^\alpha x_i^\alpha + f^\beta x_i^\beta$ (for $i =$ A and B) is the famous *lever rule*.

In the graph of Figure 5.2, this means that for any chosen values of $x_B^\alpha$ and $x_B^\beta$ one may plot a straight line from $G^\alpha(x_B^\alpha)$ to $G^\beta(x_B^\beta)$, and the intersection of this line with the vertical line at $x_B^0$ determines the value of $G(\text{system})$. Obviously, the global minimum of $G(\text{system})$ is attained at point 3, where this straight line forms the common tangent to the curves $G^\alpha$ and $G^\beta$, the scenario plotted in Figure 5.2. It is emphasized that this method to find the system's equilibrium state by minimizing $G$ (system) does not make any use of the chemical potentials. In the phase $\alpha$, these are given at any composition $x_B^\alpha$ by plotting a tangent to $G^\alpha(x_B^\alpha)$ and finding its intersections with the ordinates at pure A for $\mu_A^\alpha(x_B^\alpha)$ and at pure B for $\mu_B^\alpha(x_B^\alpha)$. Inspection of Figure 5.2 reveals that the tangents plotted to find $\mu_i$ in the two equilibrated phases overlap to a common tangent resulting in the "equality of chemical potentials," more precisely the set of two equations $\mu_A^\alpha(x_B^\alpha) = \mu_A^\beta(x_B^\beta)$ and $\mu_B^\alpha(x_B^\alpha) = \mu_B^\beta(x_B^\beta)$. This alternative condition for thermodynamic equilibrium is derived in many textbooks from the condition $dG(\text{system}) = 0$. It is almost equivalent to the condition of the absolute, or global, minimum of $G(\text{system})$, which is the more general condition and explained later in this chapter.

For the simple example in Figure 5.2 with concave Gibbs energy curves $G^\varphi$, i.e., with strictly positive curvature $\partial^2 G^\varphi / \partial x^2$, these equilibrium conditions are completely equivalent. This is the *stable* equilibrium shown in Figure 5.1d. Any deviation of $x_B^\alpha$ and $x_B^\beta$ from their equilibrium values corresponds to an *unstable* state; see Figure 5.1a. However, one of the $G^\varphi$-curves may exhibit a "bump" with regions of $\partial^2 G^\varphi / \partial x^2 < 0$. That is not so rare: One example is a (spinodal) miscibility gap, such as in liquid Al–Bi alloys. Then the "equality of chemical potentials" will have multiple solutions that may be *metastable* (Figure 5.1c) or *unstable* (Figure 5.1b), but only one unique *stable* solution (Figure 5.1d). The only way to safely find it is the search for the global minimum of *G*(system), as demonstrated clearly by Chang et al. (2004). Only one unique *stable* solution exists at a given state point for any set of (parameterized) functions $G^\varphi$ for each phase $\varphi$ of the system, which is the system's thermodynamic database. It was shown for numerous examples (Chang et al., 2004) that the published phase diagrams obtained by the so-called first-generation calculation software did not find this stable equilibrium. Significantly and qualitatively different phase diagrams are obtained by using the same thermodynamic database with a second-generation CALPHAD software, which ensures finding the global minimum of *G*(system) (Cao et al., 2009; Chang et al., 2004; Chen et al., 2002).

A simple flow chart of the CALPHAD method is given in Figure 5.3. The first area is the "thermodynamic modeling," requiring extensive training and experience. It comprises both the educated selection of appropriate models for each phase reflecting structural peculiarities and the assessment of the parameters in these models. The latter part is often called "thermodynamic optimization." The key point is that all available thermodynamic and phase equilibrium data of the system are assessed jointly. The parameter determination is often supported by software codes implemented in CALPHAD software to assist by a least square fit of "calculated" to "accepted" data of the system. Most of the expertise and effort are required for the assessment of the

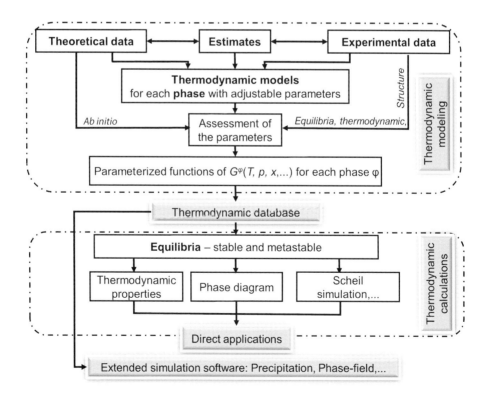

**Figure 5.3** Diagram illustrating the CALPHAD method.

"accepted" data by considering the reliabilities of different – often conflicting – experimental thermodynamic and phase equilibrium data as well as theoretical (e.g., ab initio) or estimated thermodynamic data. The result is boiled down in the set of parameterized functions of $G^\varphi(T, p, x, \ldots)$ for each phase $\varphi$. This set is the "thermodynamic description" of the system, and the set of parameters – linked to the selected functions – constitute the "thermodynamic database."

In the second area, "thermodynamic calculations" are performed using CALPHAD software. Given the database, this requires less training compared to the first area. The equilibrium calculation is the core task of the software. The meaning of stable/metastable equilibria in applications is generally used in context of phase availability (nucleation); see Section 5.5.2. That is different from the aspect of artifacts discussed earlier. Such artifacts due to algorithms trapped in metastable/unstable equilibria should be ruled out by the use of advanced CALPHAD software automatically finding the global minimum of $G$(system) (Cao et al., 2009; Chen et al., 2002). It is the beauty of the CALPHAD method that all thermodynamic properties (enthalpies, activities, etc.) as well as phase diagrams are calculated in a self-consistent manner from the assessed set of $G^\varphi$ functions in the thermodynamic description. These direct applications also comprise the Scheil solidification simulations of alloys.

In advanced applications, extended simulation software also benefits from the same thermodynamic database. This ensures consistent thermodynamic driving force calculations for various kinetic processes and the development of mobility databases for diffusion simulation with a minimum number of additional materials parameters. In a next step, kinetic-thermophysical databases can be developed with only a few additional materials parameters complementing the well-established CALPHAD databases for thermodynamic equilibrium and mobility of species in phases. Examples are outlined in Section 5.6 for microsegregation and precipitation simulation, in Chapter 3 for phase-field simulation, in Chapter 6 for diffusion simulation, and for other ICME software by Schmitz and Prahl (2016).

## 5.2.3　Overview of Commercial and Open-Source CALPHAD-Based Software

Table 5.3 lists some current software tools available for CALPHAD-type calculations, and more information can be found on the corresponding websites. The major

**Table 5.3** Summary of some software tools for CALPHAD-type calculations.

| Name of software, website | Type | Published reference |
|---|---|---|
| Thermo-Calc, www.thermocalc.com | Commercial | Andersson et al. (2002) |
| Pandat, www.computherm.com | Commercial | Cao et al. (2009); Chen et al. (2002) |
| FactSage, www.factsage.com | Commercial | Bale et al. (2009) |
| OpenCalphad, www.opencalphad.com | Open source | Sundman et al. (2015) |
| MatCalc, www.matcalc.at | Commercial | Povoden-Karadeniz et al. (2013) |
| JMatPro, www.sentesoftware.co.uk | Commercial | Saunders et al. (2003) |
| PyCalphad, https://pycalphad.org | Open source | Otis and Liu (2017) |

software packages, Thermo-Calc, Pandat, and FactSage, are all commercial and require a license. However, trial versions may be available free of charge, such as for the Pandat package, which is restricted to binary and ternary systems instead of an unlimited number of components.

In the 1970s, two of the early CALPHAD-type software codes written in Fortran became popular and were freely distributed, even before their journal publications, and prompted progress in the community. One was SOLGASMIX by Gunnar Eriksson (Eriksson, 1975) to calculate the equilibria of multicomponent multiphase systems composed of stoichiometric solid phases and the multispecies gas phase. The other was the "Lukas programs" by Hans Leo Lukas, who focused on the condensed solution phases and compounds, their phase diagrams (BINFKT, etc.) and the parameter optimization (BINGSS, etc.) (Kattner and Seifert, 2010; Lukas et al., 1977). A fundamental work of Hillert (1981) laid out clearly the advantages of the mathematical strategy to calculate the equilibria in the space of mole fractions $x_i$ and using the two material balance constraints to set up the Lagrange function from the molar Gibbs energies of the phases. The equilibrium is calculated as the zero of the gradient of the Lagrange function, which is an efficient way to obtain the local equilibrium of the system's $G$ and the basis of the Thermo-Calc software.

All of these software tools require a thermodynamic database. Its availability and reliability are crucial for the quality of results as outlined in Section 5.4. Extended simulation software, described in Section 5.2.2 and Figure 5.3, is also linked to or integrated in some of the software given in Table 5.3. For example, unidirectional multiphase diffusion is tackled by the software DICTRA, described in detail in Chapter 6, and linked to the Thermo-Calc software. For a similar scope, the PanDiffusion module is integrated in Pandat. For kinetic simulation of precipitation from a solid-state matrix, the software TC-PRISMA is an add-on to Thermo-Calc. With a similar scope, the PanPrecipitation module is integrated into Pandat and application examples are given in Section 5.6.2. For solidification modeling, the PanSolidification module is integrated into Pandat, and application examples are given in Section 5.6.3. More information can be found on the websites listed in Table 5.3.

## 5.3 Thermodynamic Modeling of Gibbs Energy

CALPHAD is a phase-based method, and the modeling aims at a set of functions $G^\varphi(T, p, x)$ for each phase $\varphi$. In the following, we go step by step, starting with $G^\varphi(T, p)$ for phases of fixed composition, unary, or stoichiometric phases. Solution phases with variable $x$ in $G^\varphi(T, p, x)$ are treated after that.

### 5.3.1 Phases with Fixed Composition

#### 5.3.1.1 Pure Elements
The Gibbs energy functions of each element are given relative to the Standard Element Reference (SER) defined at 298.15 K and 1 bar. For each element, a unique phase is

stable at these conditions, such as Fcc for Al, and that is the SER state with the only exception taken for phosphorous. The white form of P is chosen as the reference phase, by convention, because characterization of the more stable red P is difficult (Dinsdale, 1991); see more details in Liang and Schmid-Fetzer (2013, 2014). In the SER state, the enthalpy is by definition zero while the entropy is given by its well-defined absolute value for that phase. The preceding statements are the basis of the widely accepted Scientific Group Thermodata Europe (SGTE) unary database (Dinsdale, 1991), available in slightly updated electronic form also as SGTE-5.0 (downloaded in 2021, version 5.0, dated June 2, 2009, www.sgte.net/en/free-pure-substance-database).

The molar Gibbs energy function of the pure element $i$ in any phase $\varphi$ is generally described by an equation of the following form in the SGTE database:

$$G_i^{0,\varphi}(T) = G_i^{\varphi}(T) - H_i^{SER} = a + bT + cT \ln T + dT^2 + eT^3 + fT^{-1} + gT^7 + hT^{-9},$$
(5.1a)

where $H_i^{SER}$ is the molar enthalpy of the element $i$ at 298.15 K and 1 bar in its *SER* state, and $T$ is the absolute temperature. The function is not only given for a stable temperature range of the phase but also for (metastable) extrapolations. The last two terms in (5.1a) are used only outside the stable range to avoid artifacts. The basic principle is that the $c_p$ functions derived from (5.1a) coincide for all condensed phases at high and low temperatures. Thus $c_p$ of solid phases above the melting point approach the values of $c_p$(Liquid) by using the term $hT^{-9}$ in the high-temperature range. Also, $c_p$(Liquid) below the melting point approaches the values of $c_p$(solid) at 298.15 K due to the term $gT^7$. Otherwise, artificial restabilization of a solid at high temperatures or a liquid at low temperatures may occur due to large differences in $c_p$ (Kauzmann's paradox).

Other formulations of $G_i^{0,\varphi}(T)$ may also be given, for example to extend the description far below $T = 298.15$ K. For the example of stoichiometric InN and GaN, these are given based on the modified Debye or Einstein functions for $T < 298.15$ K (Onderka et al., 2002; Unland et al., 2003), also useful for the determination of absolute entropy at 298.15 K.

A more practical approach is given by including two additional terms to (5.1a):

$$G_i^{0,\varphi}(T) = G_i^{\varphi}(T) - H_i^{SER}$$
$$= a + bT + cT \ln T + dT^2 + eT^3 + fT^{-1} + gT^7 + hT^{-9} + iT^4 + j \ln T.$$
(5.1b)

The corresponding $c_p$ function is given by Wang et al. (2013):

$$c_p = -c - 2dT - 6eT^2 - 2fT^{-2} - 42gT^6 - 90hT^{-10} - 12iT^3 + jT^{-1}. \quad (5.2)$$

This equation is proven to describe the experimental data of all four stoichiometric Li–Si compounds, as well as Li(Bcc) and Si(diamond), from 0 K up to very high temperature, enabling the calculation of the Li–Si phase diagram in that range (Wang et al., 2013). The parameters for $T < 298.15$ K are also given in various temperature ranges. Parameters $c - j$ are first determined from the experimental data of $c_p$, range by range, stepping down to lower $T$, and using parameter $c$ to ensure continuity of $c_p$. Subsequently, the parameters $a$ and $b$ are determined from a continuity constraint for enthalpy and entropy, and thus the continuous function $G_i^{0,\varphi}(T)$ for each phase with continuous first and second derivatives with respect to $T$ is obtained for $T > 0$ K up to the top-end temperature. However, the third derivative is not continuous and kinks occur in $c_p(T)$.

Efforts are ongoing to improve the descriptions of unary data and especially other "endmembers of solutions." These are generally metastable, such as the Al(Hcp) discussed in Section 5.2.2, and of crucial importance for the consistent construction of multicomponent databases using the very same set of the lattice stabilities not only for the unaries but for all phases in the lower-component subsystems. In 1995, the first "Ringberg workshop" (Sundman and Aldinger, 1995) was held with 45 invited participants with a background in physics, chemistry, metallurgy, geology, and computer software discussing in a closed meeting for five days in the Ringberg castle these aspects in six workgroups. Their six reports are published in the CALPHAD journal (Ågren et al., 1995; Aldinger et al., 1995; Chang et al., 1995; Chase et al., 1995; de Fontaine et al., 1995; Spencer et al., 1995). The already included input from first-principles calculations were extended in a later Ringberg workshop in 2013 (Palumbo et al., 2014), and also various options for removing the kinks in $c_p(T)$ were discussed. A method to prevent artificial restabilization of crystalline solid phases in multicomponent systems far above their melting temperature was proposed by Sundman et al. (2020) and revised by Schmid-Fetzer (2022). Despite all efforts, the original SGTE unary database (Dinsdale, 1991) is still the predominantly used one because (i) many multicomponent databases – built upon it – would need significant revision if changing the unary basis, and (ii) because it is still the most comprehensive published unary database.

The Gibbs energy function of element $i$ in its *SER* phase, i.e., $\varphi = SER$, is often denoted as GHSER$_i$. Then

$$\text{GHSER}_i = G_i^{0,SER}(T). \tag{5.3}$$

It is emphasized that in the present notation, the superscript zero in $G_i^{0,\varphi}(T)$ already defines the SER reference, which avoids the distracting replication of subtracting the terms $H_i^{SER}$ in any other phase, for example in stoichiometric compounds. Note that the enthalpy calculated from $H_i^{0,\varphi} = G_i^{0,\varphi} - T\partial G_i^{0,\varphi}/\partial T$ at 298.15 K is zero only for $\varphi = SER$, $H_{Al}^{0,Fcc}(298.15 \text{ K}) = 0$, not for all other phases, e.g., $H_{Al}^{0,Liquid}(298.15 \text{ K}) = +11,005$ J/mol and $H_{Al}^{0,Hcp}(298.15 \text{ K}) = +5,481$ J/mol. No value is known for $H_i^{SER}$ or assigned to it; the point is simply that the same quantity is subtracted in (5.1) for all phases $\varphi$ of this element.

## 5.3.1.2 Magnetic Contribution

Magnetic ordering significantly impacts phase stability and is, for example, the only reason why Fe(Bcc) is not only stable as $\gamma$Fe at high $T$ but stable again as $\alpha$Fe at room temperature. The magnetic contribution to the Gibbs energy, denoted as $G_{mag}^\varphi(T)$, is defined as additive term in (5.4):

$$G_i^{0,\varphi}(T) = G_{i,\text{para}}^{0,\varphi}(T) + G_{i,\text{mag}}^{\varphi}(T), \tag{5.4}$$

where $G_{i,\text{para}}^{0,\varphi}(T)$ is the Gibbs energy expression of a nonmagnetic (paramagnetic) lattice contribution assessed as extrapolation down from very high temperature as given in (5.1).

The magnetic contribution accepted by SGTE (Dinsdale, 1991) is based on a model presented by Inden at a CALPHAD conference in 1976 and later published (Inden, 1981). The explicit expression given by Hillert and Jarl (1978) for $G_{\text{mag}}^{\varphi}(T)$ is as follows:

$$G_{i,\text{mag}}^{\varphi}(T) = RT \cdot g(\tau) \cdot \ln(\beta_0 + 1) \tag{5.5a}$$

$$\tau = \frac{T}{T^*}, \tag{5.5b}$$

where $T^*$ denotes the Curie temperature, $T_C$, for ferromagnetic materials, and the Néel temperature, $T_N$, for antiferromagnetic materials. The average magnetic moment per atom (in Bohr magnetons) is denoted by $\beta_0$. The expression for $g(\tau)$ is given by a truncated power series:

$$\text{For } \tau \leq 1 : g(\tau) = 1 - \frac{1}{D}\left[\frac{79\tau^{-1}}{140p} + \frac{474}{497}\left(\frac{1}{p} - 1\right)\left(\frac{\tau^3}{6} + \frac{\tau^9}{135} + \frac{\tau^{15}}{600}\right)\right] \tag{5.6a}$$

$$\text{For } \tau > 1 : g(\tau) = -\frac{1}{D}\left[\frac{\tau^{-5}}{10} + \frac{\tau^{-15}}{315} + \frac{\tau^{-25}}{1{,}500}\right] \tag{5.6b}$$

$$D = \frac{518}{1{,}125} + \frac{11{,}692}{15{,}975}\left(\frac{1}{p} - 1\right). \tag{5.6c}$$

The value for $p$ is structure dependent, $p = 0.40$ for Bcc_A2 and $p = 0.28$ for other common phases such as Fcc_A1. Thus, the two constant material parameters $T^*$ and $\beta_0$ are sufficient to determine the magnetic contribution for the phase.

The magnetic contributions to enthalpy, entropy and $c_{p,\text{mag}}$, are obtained consistently from the generic temperature derivatives of $G_{i,\text{mag}}^{\varphi}(T)$ as given in Table 5.2 and explicitly in Dinsdale (1991).

For the example of pure nickel in the Fcc phase, the paramagnetic contribution in J/mol is given as

$$G_{\text{Ni,para}}^{0,\text{Fcc}}(T) = -5{,}179.159 + 117.854\,T - 22.096\,T\,\ln(T)$$
$$- 4.8407 \cdot 10^{-3}T^2, \qquad 298 < T < 1{,}728 \text{ K},$$

$$G_{\text{Ni,para}}^{0,\text{Fcc}}(T) = -27{,}840.62 + 279.134977\,T - 43.1\,T\,\ln(T)$$
$$+ 1{,}127.54 \cdot 10^{28}T^{-9}, \qquad 1{,}728 < T < 3{,}000 \text{ K}, \tag{5.7}$$

while $G_{\text{Ni,mag}}^{\text{Fcc}}(T)$ is defined by the parameters $T_C = 631$ K and $\beta_0 = 0.52$ in (5.5) according to the SGTE data (Dinsdale, 1991).

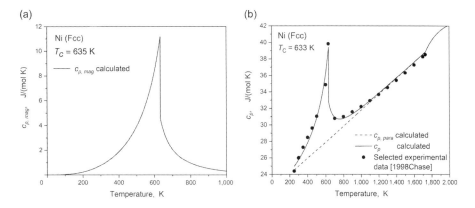

**Figure 5.4** (a) Magnetic contribution to the molar heat capacity of Fcc nickel, $c_{p,\mathrm{mag}}(T)$, calculated from the SGTE data with $T_C = 633$ K and $\beta_0 = 0.52$ (Dinsdale, 1991). (b) Molar heat capacity of Fcc nickel, $c_p(T)$, with selected experimental data (Chase, 1998), and the paramagnetic lattice contribution, $c_{p,\mathrm{para}}(T)$, calculated from the SGTE data (Dinsdale, 1991).

The quality of the temperature dependence of a Gibbs energy function is generally best compared to experimental data by inspection of the heat capacity. The molar heat capacity contributions calculated from $G_{\mathrm{Ni}}^{0,\mathrm{Fcc}}(T)$ are given in Figure 5.4a for the magnetic part and in Figure 5.4b for the paramagnetic part and the total values. It is noted that experimental data are directly available only for the total function, $c_p(T)$, the summation of the lattice and magnetic contributions. The data points in Figure 5.4b are the assessed NIST-JANAF data tabulated by Chase (1998), and they are almost identical to the detailed assessment of Desai (1987), who also shows the wealth of original experimental data in comparison.

Note the kink in $c_p(T)$ of Ni(Fcc) in 5.4b at 1,728 K, the melting point of Ni, which is due to the range change in (5.7) as discussed in the previous section.

Instead of the truncated power series in (5.6), two simple exponential functions for $c_{p,\mathrm{mag}}(\tau)$ were suggested by Chuang et al. (1985) that describe the experimental data for Ni, Co, and Fe very well. The advantage of splitting lattice and magnetic contributions in (5.4) is not just the better description of a lambda-shaped $c_p(T)$, it is also significant in multicomponent solution phases. The paramagnetic base function can simply be extended as generally detailed in Section 5.3.2.1, and the additive composition dependence of $G_{\mathrm{mag}}^{\varphi}(T, x_i)$ is simply expressed by (5.5) with composition dependent $\tau = \tau(x_i) = T/T^*(x_i)$ and $\beta_0 = \beta_0(x_i)$ as detailed in Section 5.3.2.6.

### 5.3.1.3    Pressure Contribution

For gas phases, the pressure contribution in $G^{\varphi}(T, p)$ is given later in Section 5.3.2.2. For condensed phases, a parametric form of the Murnaghan equation (Fernández Guillermet et al., 1985) is used in the SGTE data:

$$G_{\mathrm{press}} = \frac{A \exp\left(A_0 T + A_1 T^2/2 + A_2 T^3/3\right)}{\left(K_0 + K_1 T + K_2 T^2\right)(n-1)} \left\{\left[1 + np\left(K_0 + K_1 T + K_2 T^2\right)\right]^{1-1/n} - 1\right\}.$$

$$(5.8)$$

Since the molar volume of any phase is $V_m(T, p) = (\partial G_m/\partial p)_T$, the parameter $A$ is equivalent to the value of molar volume at temperature of 0 K and pressure of 0 Pa, $A = V_m(0, 0)$. This value is hardly changed between zero and 1 bar ($10^5$ Pa). The material parameters $A_0$, $A_1$, $A_2$ and $K_0$, $K_1$, $K_2$ are given for some elements, such as C, Co, Cr, Fe, Mo, Ni, and Zr in the SGTE database (Dinsdale, 1991). They describe the coefficients of the temperature polynomials of the thermal expansivity $\alpha$ and the isothermal compressibility $\kappa$ (inverse of bulk modulus $B$) of the phase at the selected reference pressure of $p_0 = 0$, respectively. This is shown in (5.9) and (5.10):

$$\alpha(T, p) = \frac{1}{V_m}\left(\frac{\partial V_m}{\partial T}\right)_p \tag{5.9a}$$

$$\alpha(T, p_0) = A_0 + A_1 T + A_2 T^2 \tag{5.9b}$$

$$\kappa(T, p) = \frac{1}{B} = -\frac{1}{V_m}\left(\frac{\partial V_m}{\partial p}\right)_T \tag{5.10a}$$

$$\kappa(T, p_0) = K_0 + K_1 T + K_2 T^2. \tag{5.10b}$$

The parameter $n$ in (5.8) is pressure independent but might be temperature dependent, describing the linear dependence of the bulk modulus $B = B(T, p)$ on pressure of $p$, which was assumed by Fernández Guillermet et al. (1985) and accepted in Dinsdale (1991):

$$B(T, p) = B(T, 0) + n\, p. \tag{5.11}$$

A different formulation of $G_{\text{press}}$ is given by Lu et al. (2005b) and is shown to be superior also at very high pressure for the example of pure copper (Huang et al., 2015). In this alternative formulation of $G_{\text{press}}$ (Lu et al., 2005b), the reference pressure of $p_0 = 1$ bar is selected:

$$G_{\text{press}} = c(T)\left\{\frac{1}{\kappa(T,\ p)} - \frac{1}{\kappa(T,\ p_0)}\right\}. \tag{5.12}$$

The material parameters for the considered phase need also to be determined in a CALPHAD assessment. They are denoted as $V_0$, $V_A$, $V_K$, and $V_C$. Experimental data different from 1 bar are only needed for $V_C$. $V_0$ is equal to the constant molar volume at reference temperature, generally $T_0 = 298.15$ K, and pressure of $p_0 = 1$ bar:

$$V_0 = V_m(T_0, p_0). \tag{5.13}$$

$V_A$, $V_K$, and $V_C$ are functions of temperature only, taken as simple polynomials. The temperature dependence of molar volume is determined by $V_A$ and the compressibility by $V_K$:

$$V_m(T, p_0) = V_0 \exp(V_A) \tag{5.14}$$

$$V_K = \kappa(T, p_0). \tag{5.15}$$

These dependencies at ambient pressure are discussed in more detail in Lu et al., 2005a, c). The high-pressure fitting parameter is $V_C = c(T)$, as given in (5.12), and it is in relation with the molar volume:

$$V_m(T, p) = V_m(T, p_0) + c(T) \ln \left\{ \frac{\kappa(T, p)}{\kappa(T, p_0)} \right\}. \tag{5.16}$$

In fact, (5.16) is the basis of this model, accepted from an empirical logarithmic relation between the molar volume $V_m(T, p)$ and the bulk modulus $B(T,p)$ by Lu et al. (2005b) instead of (5.14). At this point, with the known parameters $V_0$, $V_A$, and $V_K$, all at $p_0 = 1$ bar, the last parameter $V_C$, expressed as a polynomial $c(T)$, is assessed by comparison (least square fit) from a set of experimental or first-principles $V_m(T, p)$ data at high pressure to the values calculated from the model. However, $V_m(T, p)$ cannot be directly calculated from (5.16) because $\kappa(T, p)$ is not known. It is a major contribution in the work of Lu et al. (2005b) to provide a numerical solution for this problem using the exponential integral function $Ei(z)$ and its inverse, $Ei^{-1}(z)$, as follows:

$$V_m(T, p) = c(T) Ei^{-1} \left[ Ei \left\{ \frac{V_m(T, p_0)}{c(T)} \right\} + (p - p_0)\kappa(T, p_0) \exp \left\{ -\frac{V_m(T, p_0)}{c(T)} \right\} \right]. \tag{5.17}$$

This equation can be implemented in a software code together with generic numerical data for $Ei(z)$ and $Ei^{-1}(z)$, as done in Thermo-Calc (Andersson et al., 2002). That provides the numerical solution for a CALPHAD assessment of the last parameter $c(T)$, and, thus, the function $G_{press}(T, p)$ from (5.12).

Based on any of the preceding descriptions of $G_{press}$ in (5.8) or (5.12), the molar Gibbs energy description is extended as follows:

$$G_i^{0,\varphi}(T, p) = G_i^{0,\varphi}(T, p_0) + G_{press}, \tag{5.18}$$

where $G_i^{0,\varphi}(T, p_0)$ is the Gibbs energy function at the reference pressure $p_0$ of zero or 1 bar given in (5.1).

As an application, the $p$-$T$ phase diagram of pure iron calculated from the SGTE data (Dinsdale, 1991) is shown in logarithmic pressure scale to reveal the gas and condensed phases in Figure 5.5. As another example, using the $G_{press}$ model from Lu et al. (2005b), the $p$-$T$ phase diagram of pure copper up to very high pressure (1 GPa $= 10^4$ bar) is given in Figure 5.6 (Huang et al., 2015).

### 5.3.1.4    Stoichiometric Compounds

Stoichiometric compounds with exactly fixed composition are an approximation since thermodynamics requires a finite solution range. This may be very important, for example to model the electronic properties of semiconductor CdTe even though the maximum solution range is only from 49.9995 to 50.01 at.% Te in the binary Cd–Te system (Chen et al., 1998). Therefore, it depends on the application if we consider solid CdTe simplified as stoichiometric or not. Two reference states can be chosen to

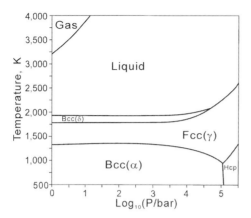

**Figure 5.5** Calculated p-T phase diagram of pure iron. Reproduced from Schmid-Fetzer (2014), Open Access.

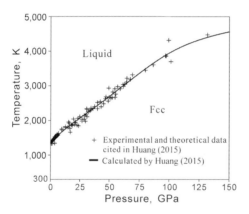

**Figure 5.6** Calculated p-T phase diagram of pure copper. Reproduced from Huang et al. (2015) with permission from Elsevier.

describe the Gibbs energy of a strictly stoichiometric compound, the "absolute reference SER" or "floating reference at T."

The *absolute reference SER* is given for the example of a binary compound $M_xN_y$, where $x$ and $y$ are the stoichiometric factors:

$$G^{M_xN_y}(T) = A^* + B \cdot T + C \cdot T \cdot \ln(T) + D \cdot T^2 + E \cdot T^3 + F \cdot T^{-1} + \cdots, \quad (5.19a)$$

where

$$A^* = A + xG_M^{0,SER}(298.15\ \text{K}) + yG_N^{0,SER}(298.15\ \text{K}) \quad (5.19b)$$

Equation (5.19a) defines the "molar" Gibbs energy of compound $M_xN_y$ referred to the SER of the elemental components. One may define $x$ and $y$ under the constraint of $x + y = 1$ and get $G(M_xN_y)$ per "mol of atoms." One may equally well define $x$ and $y$

as integer or noninteger numbers and get $G(M_xN_y)$ per "mol of $M_xN_y$." This must be explicitly stated; therefore, no subscript "m" is used in (5.19a) because the term "molar" can be quite confusing.

The $c_p(T)$ function corresponding to $G(M_xN_y)$ is already given explicitly in (5.2). Therefore, the parameters $C, D, E, F$ are determined if data for the molar heat capacity of $M_xN_y$ are available. The enthalpy function of the compound is then given, with the only unknown parameter $A^*$, as

$$H^{M_xN_y}(T) = A^* - C \cdot T - D \cdot T^2 - 2E \cdot T^3 + 2F \cdot T^{-1} + \cdots. \qquad (5.20)$$

Note that $H_i^{SER}$ does not appear in (5.20) because it is included in (5.19b) by the definition of $G_i^{0,SER}(T)$. Thus, $H^{M_xN_y}(T)$ is already given in the $SER$ reference state and $H^{M_xN_y}(298.15\text{K})$ equals the standard enthalpy of formation of compound $M_xN_y$, which is generally obtained by calorimetric measurement or approximately by first-principles calculations or estimates. This determines the parameter $A^*$.

The entropy function of the compound is given, with the only unknown parameter $B$, as

$$S^{M_xN_y}(T) = -B - C \cdot (1 + \ln(T)) - 2D \cdot T - 3E \cdot T^2 + F \cdot T^{-2} + \cdots. \qquad (5.21)$$

The value of the absolute entropy $S^{M_xN_y}(298.15 \text{ K})$ is given by integration of $c_p(T)/T$ from 0 K to 298.15 K. For compounds, these data are often available experimentally and also quite precisely by first-principle calculations. That determines the last parameter $B$.

The *floating reference* is given by (5.22):

$$G^{M_xN_y}(T) = xG_M^{0,SER}(T) + yG_N^{0,SER}(T) + A_{x,y} + B_{x,y}T. \qquad (5.22)$$

In this choice, $G(M_xN_y)$ is referred at any temperature to the $G$-functions of the elements in the phase stable at SER. This implies the Neumann–Kopp approximation of $c_p(T)$, additive from the elements, as easily seen from the second derivative and Table 5.2. This modeling is inferior but easier compared to (5.19) since $c_p$ data are not needed and only the two parameters, $A_{x,y}$ and $-B_{x,y}$, need to be determined, corresponding to the enthalpy and entropy of formation of the compound. However, artificial kinks in $c_p(T)$ of the compound at transition temperature of pure elements will occur, most distracting for compounds with much higher melting temperatures than the elements. In that case, a recommended option is to first estimate the missing $c_p$ data with the Neumann–Kopp approximation to fit a simple smooth $c_p(T)$ function to them – or any better estimate – and use that function to model the compound with absolute reference. This way, at least the artificial kinks in $c_p(T)$ of the compound can be avoided. However, the use of proper experimental or ab initio $c_p$ data is always the first choice.

The thermodynamic modeling of the quaternary Q phase in Al–Cu–Mg–Si alloys is given as an example for both methods. In the work of Löffler et al. (2012), the composition of Q is determined with the stoichiometry $Al_{17}Cu_9Mg_{45}Si_{29}$, and the Gibbs energy in J/mol-atoms is given with the *floating reference* without knowledge of the $c_p$ data of the compound:

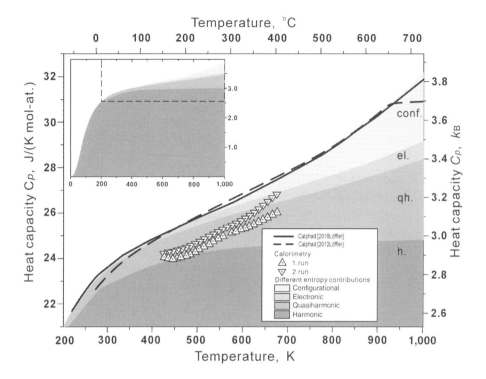

**Figure 5.7** Molar heat capacity of the quaternary compound Al17Cu9Mg45Si29 (Q) in J/(K mol-atoms) and additional scale in multiples of Boltzmann constant, $k_B$. Adapted from Löffler et al. (2016) with permission from Springer Nature.

$$G^Q_{\text{Al:Cu:Mg:Si}} = -45380 + 23.6 \cdot T + 0.17 G^{0,\text{Fcc}}_{\text{Al}} + 0.09 G^{0,\text{Fcc}}_{\text{Cu}} + 0.45 G^{0,\text{Hcp}}_{\text{Mg}}$$
$$+ 0.29 G^{0,\text{Diamond}}_{\text{Si}}. \tag{5.23}$$

In a subsequent work of Löffler et al. (2016), detailed ab initio calculations with density functional theory (DFT) were performed to determine $c_p(T)$ from $0 - 1,000$ K, and the result is given in Figure 5.7. The differently gray-shaded regions also demonstrate that all additive contributions to $c_p(T)$ are important. These are the harmonic (h.), quasi-harmonic (qh.), electronic (e.), and configurational site mixing (conf.) contributions. The solid line from 225 to 1,000 K in Figure 5.7 is the $c_p(T)$ from the CALPHAD assessment based on the complete ab initio data and the corresponding Gibbs energy with *absolute reference* is given, again in in J/mol-atoms, as

$$G^Q_{\text{Al:Cu:Mg:Si}} = -26,450 + 150.329907 \cdot T - 26.2975036 \cdot T \cdot \ln(T) + 0.0023266157 \cdot T^2$$
$$+ 96,177.2212 \cdot T^{-1} - 1.745888853 \cdot 10^{-6} \cdot T^3, \quad 225\,\text{K} < T < 1,000\,\text{K}. \tag{5.24a}$$

The dashed line in Figure 5.7 is the $c_p(T)$ from the Neumann–Kopp approximation used in (5.23). It shows the artificial kinks at the melting points of Mg (923 K) and Al (933 K).

The calorimetric $c_p$ results indicate the same slope as the complete ab initio data but are systematically slightly lower than predicted by theory. This effect is attributed at least partly to a heat loss, mainly due to the brittle structure of the Q phase material and the resulting inefficient thermal contact (Löffler et al., 2016). Therefore, the DFT data were considered superior and used to assess (5.24a).

The peritectic melting temperature of the Q phase has been measured as 981 K (708°C) (Löffler et al., 2016). The applied DFT approach cannot predict the high-temperature limit of the solid phase, therefore a finite terminal value of $c_p \approx 40 \, \text{J}/(\text{K mol-atoms})$ at 5,500 K is assumed and a $c_p$-function approaching that value is used for the temperature range above 1,000 K. The four parameters of the corresponding Gibbs energy for the second temperature range are assessed from this terminal value and the continuity constraints for $G$ and its first three derivatives at 1,000 K (Löffler et al., 2016):

$$
\begin{aligned}
G^Q_{\text{Al:Cu:Mg:Si}} &= -47,422.1991 + 264.37234 \cdot T - 40.2772 \cdot T \cdot \ln(T) \\
&\quad + 4.175 \cdot 10^6 T^{-1}, \qquad T > 1,000 \text{ K.}
\end{aligned}
\tag{5.24b}
$$

The $c_p$ function derived from (5.24b) is as follows:

$$
c_p^Q = +40.2772 - 8.35 \cdot 10^6 T^{-2}, \qquad T > 1,000 \text{ K.} \tag{5.25}
$$

The molar heat capacity assessed in the thermodynamic description of (5.24) and (5.25) evolves smoothly without any kink from 225 K to temperatures above 6,000 K. This procedure may be suggested as best practice.

## 5.3.2 Solution Phases

### 5.3.2.1 Substitutional Solution

Solution phases mean all kinds of phases that can vary in composition. The simplest one is the *substitutional solution*, where all kinds of atoms mix on a single lattice. This is generally used for terminal solutions, emerging from the solid elements and also for common liquid solutions, assuming a quasilattice. For a multicomponent solution phase $\varphi$ with $c$ components, the following equation is used for the Gibbs energy per mol-atoms, where $R$ is the gas constant:

$$
G^\varphi = \sum_{i=1}^{c} x_i G_i^{0,\varphi} + RT \sum_{i=1}^{c} x_i \ln x_i + {}^E G^{bin,\varphi} + {}^E G^{tern,\varphi}. \tag{5.26}
$$

The first term describes the mechanical mixture of all components ($i$) in proportion of their mole fractions, $x_i$. The second term is the contribution from the configurational entropy under the assumption of random mixing on the lattice, so-called *ideal mixing* if the last two terms are zero. Technically, the sum of the first two terms is the ideal *substitutional* solution. The third term ${}^E G^{bin,\varphi}$ describes the contributions of all binary interactions to the "excess Gibbs energy of mixing." The most common polynomial description is the Redlich–Kister formulation (Lukas et al., 2007; Redlich and Kister, 1948):

$$^{E}G^{bin,\varphi} = \sum_{i=1}^{c-1} \sum_{j>i}^{c} x_i x_j \sum_{v=0}^{n} L_{i,j}^{v,\varphi} (x_i - x_j)^{v}. \tag{5.27}$$

$L_{i,j}^{v,\varphi}$ is the Redlich–Kister parameter representing the interaction of order $v$ between elements ($i$) and ($j$) in phase $\varphi$. It is usually expressed with linear temperature dependence as $L_{i,j}^{v,\varphi} = A_{i,j}^{v,\varphi} + B_{i,j}^{v,\varphi} \cdot T$. The solution is called "regular" if only the paramters $L_{i,j}^{0,\varphi}$ for $v = 0$ are used. If only $A_{i,j}^{0,\varphi} \neq 0$, the solution is "strictly regular."

Even if the last term in (5.26) with ternary interactions is set to zero, it is very often found that the thermodynamics of the multicomponent phase $\varphi$ with $c \geq 3$ components is already well described. This finding is most important for applications because it means that the properties (phase diagram, etc.) of the $c$-component system can be predicted from the thermodynamic descriptions of the binaries. Of course, the condition must be fulfilled for all phases in the system. If intermetallic binary phases exist, it must be known if and how they extend by solubility of a third component. Occasionally stable ternary solid phases exist that are not predictable from the binaries. These are the main reasons why a proper CALPHAD assessment is also necessary for the ternary (sub)systems.

The so-called "extrapolation from the binary systems" refers to the aforementioned conditions for solution phases and is equivalent to the "Muggianu method" as discussed with the two other symmetrical methods ("Kohler" and "Colinet") by Hillert (2008). Hillert has also shown why the polynomial formulation of (5.27) (Redlich–Kister / Muggianu) is superior compared to other polynomials for predicting the properties of higher-order solutions.

Occasionally the $G^{\varphi}$ of the solution phase must be corrected for interactions occurring only in the ternary (sub)systems by the following term:

$$^{E}G^{tern,\varphi} = \sum_{i=1}^{c-2} \sum_{j>i}^{c-1} \sum_{k>j}^{c} x_i x_j x_k \left\{ L_{ijk}^{1,\varphi} (x_i + \delta_{ijk}) + L_{ijk}^{2,\varphi} (x_j + \delta_{ijk}) + L_{ijk}^{3,\varphi} (x_k + \delta_{ijk}) \right\},$$

$$\tag{5.28a}$$

where

$$\delta_{ijk} = (1 - x_i - x_j - x_k)/3. \tag{5.28b}$$

In a ternary system, ($c = 3$) $\delta_{ijk} = 0$. In a quaternary or higher system ($c > 3$), the same term $\delta_{ijk} \neq 0$. For a ternary system, this simplifies to

$$^{E}G^{tern,\varphi} = x_1 x_2 x_3 \left\{ L_{123}^{1,\varphi} x_1 + L_{123}^{2,\varphi} x_2 + L_{123}^{3,\varphi} x_3 \right\}. \tag{5.29}$$

Note that "one" (symmetric) ternary interaction parameter means that all three $L_{123}$ parameters are equal. Janz and Schmid-Fetzer (2005) discusses the impact on the activities in a ternary system and also points out the different numbering in the widely used TDB format of databases.

A binary example follows that also highlights the crucial importance of the chosen (metastable) endmember data. We pick up the discussion on $G(T, x_{Mg}, Hcp)$

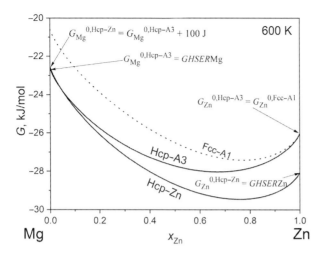

**Figure 5.8** Gibbs energy of two different Hcp solution phases in the Mg–Zn system at 600 K demonstrating the impact of endmember values. The curve for the Fcc–A1 phase is shown in dot for comparison. Adapted from Schmid-Fetzer and Hallstedt (2012) with permission from Elsevier.

in Section 5.2.2, but now in the binary Mg–Zn system, where both elements are stable as Hcp. In this system, the largest solid solubility of Zn in the (Mg) phase is almost 3 at.% Zn while the solid solubility of Mg in the (Zn) phase is negligible with 0.2 at.% Mg. In common notation, the round brackets distinguish a solid solution phase emerging from an element, (Zn), from the element, Zn. The first assessment of the Mg–Zn system by Agarwal et al. (1992) treated both solid phases as the same Hcp-A3. However, in the final version of the public COST507 database compilation the same authors reevaluated the system by including the nonstoichiometry of the Laves-C14 phase MgZn$_2$. Unfortunately, one may say, they also introduced a separate Hcp–Zn phase (Agarwal et al., 1998). This was later disputed by Schmid-Fetzer and Hallstedt (2012). Figure 5.8 demonstrates these choices by the Gibbs energy curves at 600 K.

The same value of GHSER$_{Zn}$ for stable Zn was assigned to the endmember of the new phase, $G_{Zn}^{0,Hcp-Zn}$. For $G_{Zn}^{0,Hcp-A3}$ the value of metastable Zn(Fcc) was chosen (Agarwal et al., 1998; Dinsdale, 1991):

$$G_{Zn}^{0,Hcp-A3} = 2{,}970 - 1.57T + \text{GHSER}_{Zn}. \qquad (5.30)$$

On the Mg side, the same value of GHSER$_{Mg}$ for stable Mg crystal was kept for the $G_{Mg}^{0,Hcp-A3}$. However, the introduction of the additional solution phase Hcp–Zn demands also an endmember at pure Mg, which must be metastable, and it was chosen 100 J above the stable $G_{Mg}^{0,hcp-A3}$. As a consequence of these choices, the Gibbs energy function of the (Mg) solid solution must be changed to maintain more or less the same Liquid + (Mg) phase equilibria assuming the same $G^{\text{Liquid}}$. The revised parameters are as follows (Agarwal et al., 1998):

$$L_{Mg,Zn}^{0,Hcp-A3} = -3{,}056.82 + 5.63801T \tag{5.31a}$$

$$L_{Mg,Zn}^{1,Hcp-A3} = -3{,}127.26 + 5.65563T. \tag{5.31b}$$

In contrast to the preceding revision, the original assessment with only one Hcp phase connecting the stable endmembers had evaluated the following (Agarwal et al., 1992):

$$L_{Mg,Zn}^{0,Hcp} = -1{,}600.77 + 7.62441T \tag{5.32a}$$

$$L_{Mg,Zn}^{1,Hcp} = -3{,}823.03 + 8.02575T. \tag{5.32b}$$

These significantly different parameters are also visible in the different curvature of the two $G$-curves for the (Mg) solid solution on the Mg-rich side in Figure 5.8. Some reasons why it is suggested to not introduce the separate Hcp–Zn phase are given by Schmid-Fetzer and Hallstedt (2012). However, the discussion is ongoing (Dinsdale et al., 2021).

The main reason to show this example in Figure 5.8 is to emphasize that interaction parameters, such as in (5.31) and (5.32), are useful only if linked to consistent endmembers. For many less common crystal structures, the corresponding metastable endmembers are not well defined. This impedes the construction of multicomponent databases, where it is crucial to use exactly the same endmembers. For example, many rare earth metals crystallize in the Dhcp structure, such as La, Nd, and Pr, where Dhcp is the SER phase. Consequently, for systems La-X with X = Al, Ca, Mg, and Y, the metastable endmember functions $G_X^{0,Dhcp}(T)$ are required to describe the solution of element X in the stable phase La(Dhcp). However, these are not given in systematic unary compilations of SGTE (Dinsdale, 1991) or the updated electronic version SGTE-5.0. It is crucial to assign the same functions $G_X^{0,Dhcp}(T) - G_X^{0,SER}(T)$ for all these cases to maintain consistency. Compared to guessing an arbitrary positive value, such as +500 J or +5,000 J, it is recommended to use a value obtained from DFT calculations, at least the approximate value for the ground state difference, $E_X^{0,Dhcp}(0\ \text{K}) - E_X^{0,SER}(0\ \text{K})$.

Solution phases modeled with the first two Redlich–Kister parameters $L_{i,j}^{V,\varphi}$, such as in (5.31) and (5.32), are most common and called "subregular." Sometimes the third parameter of order $v = 2$, "subsubregular," is used as in the Al–Mg example later in this section. The use of more parameters ($v > 2$) is generally disapproved of since such modeling degenerates to an excessive fitting procedure (Chang and Oates, 2010). Strong disagreement over reliable experimental data and a subsubregular description should rather be taken as an indication that a more appropriate model type, closer to the special interactions or structure of the phase, should be chosen. Examples are a significant short-range order, treated in Sections 5.3.2.3 and 5.3.2.4, or a more elaborate sublattice model, especially for intermetallic solution phases.

Two thermodynamic mixing properties are more important because they are accessible to experimental determination: The enthalpy of mixing, $\Delta_{mix}H$ (measured by calorimetry), and the chemical activities, $a_i$, (by partial pressure or electromotive force [EMF] measurements). The chemical activity $a_i^\varphi(x, T)$ is a property of component $i$ in phase $\varphi$ with a reference state and is defined by

$$RT \ln a_i^\varphi = \mu_i - G_i^{0,\varphi}. \tag{5.33}$$

This is the "natural" or "Raoult" reference state where $a_i^\varphi = 1$ at pure $i$, $x_i = 1$, in the same phase $\varphi$. Any reference state can be chosen but must be specified. The chemical potential $\mu_i$ is always the same and refers to SER as chosen for the pure elements. For example, in liquid Fe–C alloys the Raoult reference is inconvenient because the Gibbs energy of pure liquid carbon, $G_C^{0,\mathrm{Liquid}}$, is not well known. By contrast $G_C^{0,\mathrm{Graphite}}$ is very well known, and it is convenient to choose graphite as the reference state so that the activity $a_C^{\mathrm{Liquid/Graphite}} = 1$ if the liquid alloy is in equilibrium with graphite. That is achieved with the following definition:

$$RT \ln a_C^{\mathrm{Liquid/Graphite}} = \mu_C - G_C^{0,\mathrm{Graphite}}. \tag{5.34a}$$

The chemical potential $\mu_i$ does not depend on the *reference* state, but is determined by the *standard* state, chosen as SER in Section 5.3.1.1. Therefore, the activities with different reference states are related by

$$\mu_C = G_C^{0,\mathrm{Graphite}} + RT \ln a_C^{\mathrm{Liquid/Graphite}} = G_C^{0,\mathrm{Liquid}} + RT \ln a_C^{\mathrm{Liquid}}. \tag{5.34b}$$

There is no need to specify $\mu_C$ in a phase because in equilibrium $\mu_C$ has the same value in all phases. The conversion factor between the activities in different reference states is obtained from (5.34b):

$$a_C^{\mathrm{Liquid/Graphite}} = \exp\left\{\frac{G_C^{0,\mathrm{Liquid}} - G_C^{0,\mathrm{Graphite}}}{RT}\right\} a_C^{\mathrm{Liquid}}. \tag{5.34c}$$

The conversion factor depends only on temperature, not on alloy composition. In this example, it is greater than 1, in agreement with the observation of graphite saturation at $x_C = 0.21$ and $1{,}600°C$, as seen later in the Fe–C phase diagram. It is emphasized that the careful notation including the reference state is indispensable. A careless notation of just "$a_C$" or "$a_C^{\mathrm{Liquid}}$" is unclear and not recommended.

All thermodynamic properties of phase $\varphi$, including the chemical activities and enthalpy of mixing, are consistently obtained in the CALPHAD approach from the assessed $G^\varphi$. For example, in the Al–Mg system the liquid phase is completely described by three Redlich–Kister parameters: $L_{\mathrm{Al,Mg}}^{0,\mathrm{Liquid}} = -12{,}000.0 + 8.566 \cdot T$, $L_{\mathrm{Al,Mg}}^{1,\mathrm{Liquid}} = 1{,}894.0 - 3.000 \cdot T$, and $L_{\mathrm{Al,Mg}}^{2,\mathrm{Liquid}} = 2{,}000.0$ in J mol$^{-1}$ (Liang et al., 1998a; Saunders, 1990). The calculated activities plotted in Figure 5.9 indicate essentially a negative deviation from Raoult's law, which is shown as straight lines for the ideal mixing where all $L_{i,j}^{\nu,\varphi} = 0$. Experimental data, available for $a_{\mathrm{Mg}}^{\mathrm{Liquid}}$ only, are compared with the calculation by Saunders (1990). In pre-CALPHAD times, $a_{\mathrm{Mg}}^{\mathrm{Liquid}}(x)$ was often modeled directly and $a_{\mathrm{Al}}^{\mathrm{Liquid}}(x)$ was obtained by application of the Gibbs–Duhem relation. However, that is obsolete by modeling of the integral quantity $G^{\mathrm{Liquid}}$ (Chang and Oates, 2010).

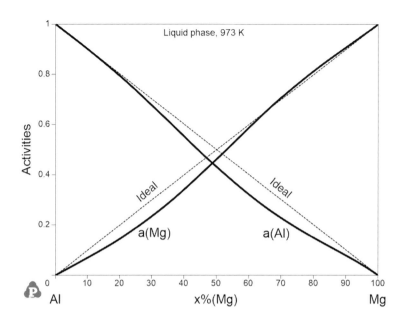

**Figure 5.9** Activities of Al and Mg in liquid Al–Mg solution, $a_i^{\text{Liquid}}$, calculated at 973 K using the data in Saunders (1990); reference states are the pure liquid components. Activities for the ideal solution are shown dashed.

Figure 5.10 shows the calculated enthalpy of mixing, which is the next experimentally accessible quantity, and a comparison is found in Saunders (1990). Both the integral quantity $\Delta_{mix}H$ and the two partial enthalpies of mixing are obtained consistently from $G^{\text{Liquid}}$. The integral curve $\Delta_{mix}H$ is always intersected at its minimum value (–3,015 J/mol in this case) by the partial enthalpies of the components. In addition to the activity and enthalpy data, the determination of $G^{\text{Liquid}}$ is mainly based on the agreement with the phase equilibria in the complete assessment by Liang et al. (1998a) and Saunders (1990).

### 5.3.2.2    Gas Phase

A gas phase composed of only the elemental components, such as in the Ar–He system, is a rare exception. Usually, additional molecules are formed in a reactive gas mixture that is then composed of so-called *species* with a total number $s$ for all the gas species. For example, $s = 7$, species (H, $H_2$, $H_2O$, $H_2O_2$, $O_3$, $O_2$, O) may be considered in the H–O system with $c = 2$ components. When the gas phase is described as an ideal gas mixture of these species, its Gibbs energy per total mole of species $n$ (mol of formula) is given by the following expression:

$$G^{\text{Gas}}(T, p, y_i) = \sum_{i=1}^{s} y_i \left[ G_i^{0,\text{Gas}}(T) + RT \ln (y_i) \right] + RT \ln \left( \frac{p}{p_0} \right), \qquad (5.35)$$

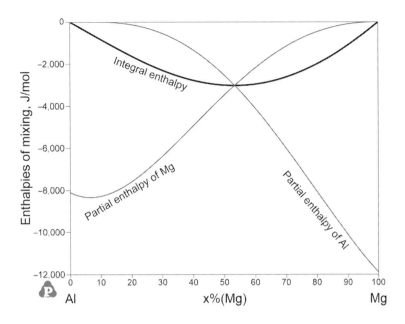

**Figure 5.10** Integral and partial enthalpies of mixing in liquid Al–Mg solution, calculated using the data in Saunders (1990). Reference states are the pure liquid components.

where $y_i = (n_i/n)$ is the molar fraction of the species $i$ in the gas, $p$ is the total pressure, and $p_0$ the reference pressure, generally $p_0 = 1$ bar $= 10^5$ Pa. Note that in a reactive gas mixture composed of $c$ components, we have $s > c$. For some important gaseous components stable at SER, the SGTE data are given for the more stable molecule, such as $O_2$ and $N_2$, thus $GHSER_O - 0.5G_{O_2}^{0,Gas} = 0.5\left(G_{O_2}^{Gas} - H_{O_2}^{SER}\right)$ is given per mol of atoms. From the linearly independent set of components, often a very large number of additional species are formed. For each of these species formed from the component(s), the molar Gibbs energy, $G_i^{0,Gas}(T)$, is a thermodynamic parameter of the gas mixture that needs assessment in analogy to the stoichiometric compounds. These data are found for $p_0 = 1$ bar for example in the NIST-JANAF tables (Chase, 1998).

For $s > c$, the gas mixture has an internal degree of freedom. While the state point is given by fixing $T$, $p$, and the overall composition of the system, counted by atomic fractions $x_i$, the Gibbs energy $G^{Gas}(T, p, x_i)$ is still undetermined. The mole number of species $n_i$ can be varied under the constraint of conserving the number of atoms of each element in the material balance. For each set of $(n_i, i = 1$ to $s)$, a different value of $G^{Gas}$ may be obtained from (5.35) with only one global minimum. This set of $(n_i)$ defines the equilibrium constitution of the reactive gas (i.e., the total mole number $n = \sum n_i$, and the set of equilibrium species fractions $(y_i)$).

Any modern CALPHAD software, as well as the early SOLGASMIX (Eriksson, 1975), determines this equilibrium by numerical methods. This is much more efficient compared to the traditional way of setting up reaction equations and applying the mass action law. Even a case as simple as the H–O system with given $(x_O = 1 - x_H)$ and considering only the species ($H_2$, $H_2O$, $O_2$) involves the solution of a nonlinear equation to find the equilibrium. By adding monatomic oxygen, which actually becomes dominant at high $T$, two reaction equations must be set up for the species ($H_2$, $H_2O$, $O_2$, O) that has not even included the ozone, $O_3$. This comparison of methods is detailed for the S–O system in Chang and Oates (2010).

At very high pressure, the assumption of ideal mixing is not valid and the gas phase must be described as "real gas." For the same mixture of species, the fugacity of the gas, $f$, replaces $p$ in the Gibbs energy function of (5.35):

$$G^{\text{Gas}}(T, p, y_i) = \sum_{i=1}^{s} y_i \left[ G_i^{0,\text{Gas}}(T) + RT \ln (y_i) \right] + RT \ln \left( \frac{f}{p_0} \right). \qquad (5.36)$$

The fugacity is a function of $p$ and $T$ with the low-pressure limit $f(p, T) \to p$. An explicit equation following the general approach of virial coefficients is given for nitrogen by Unland et al. (2003) and applied in the Ga–N system. That enables to use standard CALPHAD software, such as Pandat or Thermo-Calc, for equilibrium calculations because the function $f = f(p, T)$ can be flexibly implemented in the standard TDB database format. A strong impact of the nonideal gas treatment on the phase stability diagram, $p_{N2}$ versus $1/T$, is demonstrated for the Ga–N (Unland et al., 2003) and In–N (Onderka et al., 2002) systems with a high relevance to the crystal growth of the semiconductors GaN and InN from the melt. This approach was extended to more metal-nitrogen systems (Ma et al., 2005).

It was also shown by Unland et al. (2003) that the simple correlation approach for the fugacity suggested by Tsonopoulos (1974) fails for the relevant high-pressure range and most drastically at lower temperatures. It is unfortunate that the software FactSage only provides a hard-wired switch from "ideal" to "real gas" that uses only that simple correlation (Tsonopoulos, 1974).

### 5.3.2.3     Associate Solution

The basic approximation of (5.26) is the description of configurational entropy under the assumption of random mixing. Even for a strictly regular solution, such as (5.27), it should be expected that the local environment of nearest neighbors may deviate from randomness due to the difference in interaction energy between like and unlike neighbors, which is expressed by a constant value of $L_{i,j}^{0,\varphi} = A_{i,j}^{0,\varphi}$. In fact, the temperature-dependent part, $B_{i,j}^{0,\varphi} T$, generates the "excess entropy of mixing" that adds to the "ideal entropy of mixing" $(-R\sum x_i \ln x_i)$. However, in case of very strong short-range order (SRO), the total entropy of mixing must go to essentially zero at the corresponding perfectly ordered composition of the alloy. Additional Redlich–Kister parameters are not useful to compensate the wrong basic assumption of random mixing in that case. Thus, more physics-based approaches are needed. Models where the atoms are distributed on sublattices to describe both SRO and long-range order (LRO) are discussed by Hillert (2008) and Chang and Oates (2010).

In this section, a practical approach is presented to describe the chemical ordering in a liquid phase by the formation of molecular-like clusters of atoms or "associates" (Schmid and Chang, 1985; Sommer, 1982). For example, in this associate solution model, clusters such as $AB_2$ are assumed to form in a binary A–B liquid resulting in the picture of three species (A, B, $AB_2$) mixing in a quasiternary system. There is no topological constraint of a lattice or sublattices for the configurational entropy. The general equation for the Gibbs energy, $G^L$, per total mole of species (mol of formula) is given as follows:

$$G^L(T, y_i) = \sum_{i=1}^{s} y_i \cdot G_i^{0,L}(T) + RT \sum_{i=1}^{s} y_i \ln y_i + \sum_{i=1}^{s-1} \sum_{j>i}^{s} y_i y_j \sum_{v=0}^{n} L_{i,j}^{v,L} (y_i - y_j)^v,$$

(5.37)

where $y_i$ is the molar fraction of species $i$ or $j$ in the liquid and $s$ is the total number of species. The interactions between the species are described in the third term by a Redlich–Kister formulation with parameters $L_{i,j}^{v,L}$. Comparison with (5.35) reveals that the associate solution model reduces to that of an ideal reactive gas mixture at $p = p_0$ if all $L_{i,j}^{v,\varphi} = 0$. In a complete analogy, the main thermodynamic parameters of this liquid associate solution are the Gibbs energies of formation of the associates from the pure liquid elements, $G_i^{0,L}(T)$, for $i = (c+1)$ to $i = s$, where $c$ is the number of elemental components. These are generally used to describe the strong attractive interactions between the elements of the associate, while both repulsive and attractive interactions between any species can be described by the Redlich–Kister parameters.

As an example, we consider the Ca–Sn system (Ohno et al., 2006a), where the binary liquid phase is modeled with the associate $Ca_2Sn$. The total number of moles in the liquid phase $n = n_{Ca} + n_{Sn} + n_{Ca_2Sn}$ results from an internal Gibbs energy minimization since we have one internal degree of freedom. That is because the Gibbs energy function is given explicitly only in the space of species fractions with a higher dimensionality ($y_i$, $s = 3$) than the atomic fractions ($x_i$, $c = 2$). The equilibrium is found numerically by CALPHAD software. In the first sum of (5.37), the parameters, $G_{Ca}^{0,L}(T)$ and $G_{Sn}^{0,L}(T)$, are the Gibbs energies of pure liquid Ca and Sn, which are taken from the SGTE compilation. The Gibbs energy of the assumed $Ca_2Sn$ associate is $G_{Ca_2Sn}^{0,L}(T)$:

$$G_{Ca_2Sn}^{0,L}(T) = 2 \cdot G_{Ca}^{0,L}(T) + G_{Sn}^{0,L}(T) + A_{Ca_2Sn}^L + B_{Ca_2Sn}^L \cdot T.$$

(5.38)

The key parameters of the model are the enthalpy of formation, $A_{Ca_2Sn}^L = -174{,}137 \left(J\, mol^{-1}\right)$, and the entropy of formation, $-B_{Ca_2Sn}^L = -21.0469$ $\left(J\, mol^{-1}K^{-1}\right)$, of the associate determined by Ohno et al. (2006a).

The second term in (5.37) represents the ideal entropy of mixing of species in the quasiternary Ca–Sn–$Ca_2Sn$ system. The third term is the excess Gibbs energy in this quasiternary, and these are determined for the Sn–$Ca_2Sn$ interaction, $L_{Sn,Ca_2Sn}^{0,L} = -110{,}358 + 3.178 \cdot T \left(J\, mol^{-1}\right)$ and $L_{Sn,Ca_2Sn}^{1,L} = -10{,}605 \left(J\, mol^{-1}\right)$. The equilibrium distribution of species in the liquid phase at 1,073 K is shown in

Figure 5.11a, and it is noted that at any point their sum is $y_{Ca} + y_{Sn} + y_{Ca_2Sn} = 1$. At this temperature, five solid intermetallic phases are stable. For the sake of clarity, all solid phases are suspended in the calculation to clearly show the appearance of the associate and the monatomic species.

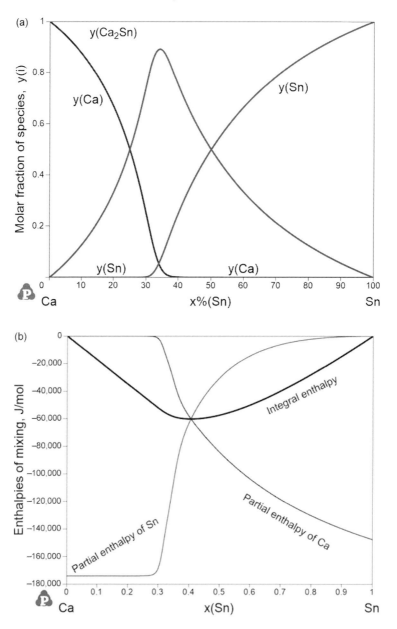

**Figure 5.11** (a) The distribution of species in liquid Ca–Sn solution calculated at 1,073 K using the data (model I) of Ohno et al. (2006a). (b) The integral and partial enthalpies of mixing in liquid Ca–Sn solution calculated at 1,073 K using the data (model I) of Ohno et al. (2006a). The reference states are pure liquid Ca and Sn.

Another peculiarity of SRO is the typical *V*-shape of the integral enthalpy of mixing shown in Figure 5.11(b) with a minimum value of –60,100 J/mol at about 40 at.% Sn. Consequently, *both* partial curves are flattening out in the dilute ranges. That is in contrast to the more rounded bell shape for the integral enthalpy shown in Figure 5.10 for the Al–Mg liquid phase modeled as a substitutional solution. The minimum value of –3,015 J/mol, much closer to zero, also indicates much smaller attractive interactions of unlike neighbors in Al–Mg compared to Ca–Sn liquid alloys. For a regular solution, the perfect bell shape of the integral curve with simple parabolic shape of both partial curves would be obtained.

The associate solution enables modeling of the continuous transition from liquid metal to liquid oxide, including the well-known spinodal miscibility gap and the stabilization of liquid oxide at high oxygen pressure. That was first demonstrated with the consistent modeling of all phases in the Cu–O system by Schmid (1983) using the model (Cu, O, $Cu_2O$) for the liquid phase. Data for the required endmember of pure liquid oxygen at high *T* were also derived (Schmid, 1983) and subsequently accepted in the SGTE database (Dinsdale, 1991).

The associate solution model is also very powerful for consistent thermodynamic modeling of multicomponent liquid slag composed of oxides even with additional sulfides and fluorides. Moreover, modeling of viscosity based on the distribution of associates at equilibrium is possible (Müller et al., 2018). The principle is seen in Figure 5.11a for the distribution of the three species with the maximum value of $y_{Ca_2Sn} = 0.89$ at 34 at.% Sn. In slags, strong ionic bonds are generally dominant between cations $\left(Fe^{2+}, Fe^{3+}, Cu^+, Cu^{2+}, Ca^+, \ldots\right)$ and anions $\left(O^{2-}, S^{2-}, \ldots\right)$. The associate solution model enables a simple distinction of clusters containing cations of different valencies, such as $Fe^{2+}$ and $Fe^{3+}$ in FeO and $Fe_2O_3$, without the need to include the ionicity explicitly in the model.

### 5.3.2.4    Quasichemical Model

The quasichemical model to describe SRO in a liquid solution is based on the concept of nearest-neighbor bond energies. In a binary solution, atoms A and B are distributed over the sites of a quasilattice. The following pair exchange reaction between bonds is considered:

$$(A - A) + (B - B) = 2(A - B) \tag{5.39}$$

The nonconfigurational Gibbs energy change, $\Delta g_{AB}$, in J per mol formula of reaction of (5.39), is the important interaction parameter of the model. The number of bonds of each kind, specifically their fractions $p_{AA}$, $p_{BB}$, and $p_{AB}$, are now the key variables of the model.

The $(A - B)$ and $(B - A)$ bonds are treated generally as different by (Hillert, 2008) but assumed to be always identical by (Pelton et al., 2000), as in (5.39); that gives rise to a slight difference in the following equations.

One may first assume that all atoms have the same number of nearest neighbors, $z$ (coordination number), in the quasilattice. Assuming a random distribution of *atoms* it was shown that the model reduces to a strictly regular solution, (5.27), with parameter $A_{A,B}^0 = z \cdot \Delta g_{AB}/2$ (Hillert, 2008).

The most general definition of the Gibbs energy of mixing of a solution phase, $\Delta_{mix}G = \Delta_{mix}H - T\,\Delta_{mix}S$, is given by the term added to the mechanical mixture of components for the binary example:

$$G^\varphi = \sum_{i=1}^{2} x_i G_i^{0,\varphi} + \Delta_{mix}G^\varphi. \tag{5.40}$$

In this model, we have

$$\Delta_{mix}G = (z \cdot \Delta g_{AB}/2)p_{AB} - T \cdot \Delta_{mix}S. \tag{5.41}$$

The exchange energy $\Delta g_{AB}$ should have an impact on the bond fractions. For example, a negative value of $\Delta g_{AB}$ would favor arrangements with increased $p_{AB}$ at the cost of like bonds. Thus, one may improve the model by assuming a random distribution of bonds, for deriving $\Delta_{mix}S$. That leads to one form of the quasichemical model, and the internal equilibrium of bond fractions is given by Hillert (2008):

$$\frac{p_{AB}^2}{p_{AA}p_{BB}} = \exp\left(-\frac{\Delta g_{AB}}{RT}\right). \tag{5.42}$$

The similarity of (5.42) to a mass action law of chemical reactions gives rise to the name quasichemical model. It was pointed out by Hillert (2008) that for large values of $\Delta g_{AB}$ the physically correct entropy is only obtained for $z = 2$, corresponding to a one-dimensional lattice where all atoms are arranged in a string (Ising model). However, all realistic coordination numbers are much larger, such as $z = 8$ for Bcc or $z = 12$ for Fcc quasilattices.

Notwithstanding this drawback, the model can be formally applied for values of $z \neq 2$. In the modified quasichemical model (MQM) (Pelton et al., 2000), it is set up with more parameters, such as different coordination numbers for the A and B atoms, $z_A$ and $z_B$, which are also composition dependent, and also the exchange energies are made composition dependent. The MQM is extended to multicomponent solutions where, in addition, different interpolation schemes are introduced (symmetric and asymmetric) to determine the exchange energies at the multicomponent composition from the values in the binary subsystems (Pelton and Chartrand, 2001).

It is interesting to compare the basic equations of $G^L$ for the MQM and the associate solution. The $G^L$ of the associate solution in (5.37) is only expressed by the variables $y_i$, the molar fraction of species, which are implicitly related to the atomic fractions $x_i$ by the material balance. However, the MQM is expressed explicitly in (5.40) by the "atomic" fractions $x_i$, and also by the pair fractions $p_{ij}$ in (5.41).

It is emphasized that the MQM is applied mainly to liquid oxide and salt solutions where the components (A and B in (5.39), (5.41), and (5.42)) are mainly nondissociating molecules, such as $Al_2O_3$, $FeO$, and $Fe_2O_3$ in the $Al_2O_3$–$FeO$–$Fe_2O_3$ system. The MQM is further modified by considering two sublattices (Chartrand and Pelton, 2001) and in a last modification to include first- and second-nearest-neighbor SRO on the two sublattices (Pelton et al., 2001). A large database comprising oxide and salt systems was developed by that group in the MQM framework, which is directly embedded in the FactSage software (Table 5.3) (Bale et al., 2009).

### 5.3.2.5 Comparison of Models for Ordered Liquid Solutions

In summary, one may conclude that both the MQM and the associate solution are parameterized models suitable for describing (strong) SRO in multicomponent liquid solutions. The (partially) ionic solution (Hillert, 2001) is the third model in that category. These are *pragmatic models*, and not stringent pictures of a physical reality. They have all been applied frequently and successfully in CALPHAD assessments of multicomponent systems where the liquid phase exhibits strong SRO and cannot be described as (disordered) substitutional solution with (5.26). Other models used for metallurgical slags are briefly mentioned by Sundman et al. (2018).

The thermodynamic model for ionic melts is formulated in the concept of compound energy formalism (CEF) by setting up a (partially) ionic solution (Hillert, 2001). It is based on the assumption of two sublattices, one for cations and the other for anions. However, unlike ionic crystals, variable tendency for ionization complicates the model because the stoichiometric coefficients (sizes) of the sublattices are not constant for ionic liquids. The condition of electroneutrality demands that the stoichiometry varies with composition if there are cations or anions of different valencies (Hillert, 2001), e.g., for $Fe^{2+}$, $Fe^{3+}$, $Cu^+$, $Cu^{2+}$, mentioned previously. For example, the liquid phase of the $ZrO_2$–$TiO_2$–$Al_2O_3$ system was described by the two-sublattice ionic model $\left(Al^{+3}, Ti^{+2}, Ti^{+3}, Zr^{+4}\right)_P \left(O^{-2}, Va, AlO_{1.5}, O, TiO_2\right)_Q$ with stoichiometries $P$ and $Q$ depending on the occupancy of the sublattices (Ilatovskaia et al., 2017).

The actual equations of $G^L$ for the MQM are not given here; they are quite complex, especially with their different options to proceed beyond binary solutions. On the other hand, that enables a close fitting if one component is chemically very different from the other two (e.g., $SiO_2$–CaO–MgO, S–Fe–Cu) in which case the asymmetric interpolation is used, not the symmetric one. The decision making is exemplified for a five-component system (Pelton and Chartrand, 2001). The currently complete details of the MQM are given in Chartrand and Pelton (2001); Pelton and Chartrand (2001); Pelton et al. (2001, 2000). The critique of the MQM as not being realistic begins with the problems for $z = 2$, pointed out earlier (Hillert, 2008). It is further detailed by Hillert et al. (2009), ending up with an attempt to correct the quasichemical model, especially for the wrong predictions of negative configurational entropy.

In contrast to the MQM, there is only one equation of $G^L$ for the multicomponent associate solution, given in (5.37), which is thus much easier to handle. The critique of the associate solution as not being realistic raised by Pelton et al. (2000) concerns two problems, explained for the simple case of three species (A, B, $AB_2$) as follows.

First, "the associate model does not reduce to the ideal substitutional solution if $G^{0,L}_{A_2B}(T) = 0$ because the number of associates is only zero for $G^{0,L}_{A_2B}(T) \to +\infty$, ... the entropy paradox"; by contrast, the MQM reduces to the ideal solution if $\Delta g_{AB} = 0$ (Pelton et al., 2000, p. 655). However, the present author considers this statement as misperception due to the entirely different definitions of these two "interaction parameters." That is easily demonstrated because we agree that the associate solution model reduces to that of the ideal gas at $p = p_0$ if all $L_{i,j}^{V,\varphi} = 0$. Look at the example of gas in the H-O system with species (H, O, $H_2$, $H_2O$, $H_2O_2$, $O_3$, $O_2$) at $p = 1$ bar. In equilibrium at 300 K, the majority species $H_2$, $H_2O$, and $O_2$ occur in molar fractions according to the fixed overall composition, while the fractions of minority species (H, $H_2O_2$, $O_3$, O) are $y_i \ll 10^{-5}$. At very high $T$ (e.g., 10,000 K), the equilibrium shifts indeed to that of the monatomic ideal solution with new majority species H and O while the fractions of the new minority species ($H_2$, $H_2O$, $H_2O_2$, $O_3$, $O_2$), or "associates," are now $y_i < 10^{-5}$ for any chosen overall composition. That result is due to the correct thermodynamic assessment of the Gibbs energies of formation of all these molecules from the atoms, as in the following example:

$$2\,H(g) + O(g) = H_2O(g); \quad \Delta G(\text{formation}) = G_{H2O}^{0,\text{Gas}}(T). \qquad (5.43)$$

In fact, $G_{H_2O}^{0,\text{Gas}}(300\,\text{K}) = -0.87\,\text{MJ/mol}$, $G_{H_2O}^{0,\text{Gas}}(3{,}900\,\text{K}) = 0$, and $G_{H_2O}^{0,\text{Gas}}(10{,}000\,\text{K}) = +1.5\,\text{MJ/mol}$. For all these "associates" ($H_2$, $H_2O$, $H_2O_2$, $O_3$, $O_2$), one finds that $\Delta G(\text{formation}) \to +\infty$ for high $T$, which is physically correct, instead of an assumed $\Delta G(\text{formation}) = 0$. All of these data and equilibrium calculations are obtained by the software FactSage 8.0 with the embedded database FactPS (Bale et al., 2009).

Therefore, the associate solution model correctly reduces to the ideal solution at high $T$ under that constraint. For the example of Ca–Sn, the assessed parameter $\Delta G^L(Ca_2Sn) = -174{,}137 + 21.0469T\,(\text{J/mol})$ (Ohno et al., 2006a) already fulfills the theoretical constraint $\Delta G(\text{formation}) \to +\infty$ for high $T$. That is because the entropy of formation, $\Delta S^L(Ca_2Sn) = -21.0469\,\text{Jmol}^{-1}\text{K}^{-1}$, is correctly assigned a negative value. In fact, the activities of the components calculated at 10,000 K obey almost perfectly Raoult's law, $a_i \approx x_i$, of the ideal solution. Otherwise, one may simply add an additional high-temperature range for the $\Delta G(\text{formation}, T)$ of the associates.

The second problem raised by Pelton et al. (2000), also mentioned by Sundman et al. (2018), is a valid theoretical point. It concerns the possibly wrong approach to Raoult's law for strong association due to a glitch in configurational entropy. For the example (Ca, Sn, $Ca_2Sn$) shown in Figure 5.11a, the solution consists mainly of Sn and $Ca_2Sn$ near pure Sn. Thus, Ca atoms enter the dilute solution from pure Sn mainly paired, and therefore the slope of the activity of Sn differs from unity at pure Sn. That is, $da_{Sn}/dx_{Sn} > 1$ for $x_{Sn} \to 1$ whereas a slope of unity is approached for any disordered solution after (5.26). However, the effect vanishes with increasing temperature and decreasing degree of association, as noted previously. Already at 1,073 K the effect is very small, with $da_{Sn}/dx_{Sn} = 1.0005$ for $x_{Sn} \to 1$, and may be disregarded in practical comparison to experimental activity data. Near pure Ca, the effect does not occur at all because $da_{Ca}/dx_{Ca} = 1$ for $x_{Ca} \to 1$ at any temperature.

### 5.3.2.6    Sublattice Model for Solid Solution Phases

LRO can only be described by using two or more sublattices, meaning that element A prefers one sublattice and B the other. The extreme case was already given by stoichiometric compounds in Section 5.3.1.4, which are technically also "sublattice models," since the atoms (or ions) are on fixed positions in the crystal structure on sublattices. For example, the binary compound CdTe can be simply described as the two-sublattice model $(Cd)_1(Te)_1$ if the minute deviation from stoichiometry is not of interest; see Section 5.3.1.4. One simple way to describe the Cd deficiency in the solution range, experimentally determined up to 50.01 at.% Te, is the introduction of vacancies, Va, on the "Cd sublattice" in the model $(\mathbf{Cd},Va)_1(\mathbf{Te})_1$. The term *constituent* is established for any species on a sublattice in the framework of CEF (Hillert, 2001), and it is helpful to type the majority constituent in boldface. However, to describe both the solution range and the electronic properties of semiconductor CdTe, more ambitious sublattice models are required, and three of them have been tested (Chen et al., 1998). The simplest one is

$$\left(\mathbf{Cd}, Va, Va^{-2}\right)_1 \left(\mathbf{Te}, Va, Va^{+2}\right)_1 \left(\mathbf{Va}\right)_1 \left(Va, e^{-}\right)_1 \left(Va, h^{+}\right)_1. \tag{5.44}$$

The third sublattice provides the interstitial site, empty in this model, and the last two dummy sublattices include the electrons, $e^-$, in the conduction band, and the holes, $h^+$, in the valence band.

A key issue for all sublattice models is the description of the mixing entropy, and often random mixing of constituents on each sublattice is assumed, the so-called Bragg–Williams (BW) approximation (Chang and Oates, 2010), which is also the basis of the CEF (Hillert, 2001; Hillert and Staffansson, 1970; Sundman and Ågren, 1981). This was applied for (5.44) and will be used also for the rest of this section. Since the general equations are found in these references, we will focus on real examples, starting with the simple cases (CEF without internal degree of freedom) and then proceed to the more complex cases. The aim is to provide a simple approach and understanding of the CEF and to indicate some potential traps.

In the first category of sublattice models, each constituent may mix only on one sublattice, such as in the type $(A,B)_p(A)_q$ or $(A,B)_p(C,D)_q$, as opposed to $(A,B)(A,C)$ discussed later. The stoichiometries $p$ and $q$ are constants, and one mole of formula unit contains $p + q$ moles of sites (atoms in this case). The occupancy of each sublattice, $s$, is given by the *site fraction*, $y_j^{(s)}$, of constituents $j$. Thus, $y_A^{(1)}$ varies from 0 to 1, while in the second sublattice $y_A^{(2)} = 1$ in the model $(A,B)_p(A)_q$, and $y_A^{(2)} = 0$ in the model $(A,B)_p(C,D)_q$. For each sublattice $(s)$, one always gets the following:

$$\sum_j y_j^{(s)} = 1. \tag{5.45}$$

The Gibbs energy of the phase is always given explicitly in the space of variables $y_j^{(s)}$ and in J per mole of formula unit, as shown later. The current values of all $y_j^{(s)}$ determine the constitution of the phase. The overall composition of the phase is given by the molar (atomic) fractions of elemental components $i$, $x_i$. The simplicity of this first category of sublattice models is due to the fact that the $y_j^{(s)}$ are explicit functions of $x_i$ according to the materials balance. That is not the case for the model $(A,B)_p(A,C)_q$, where A mixes on both sublattices. In that case, for given values of $x_i$, there is one internal degree of freedom to vary $y_A^{(s)}$ under the constraint $p\, y_A^{(1)} + q\, y_A^{(2)} = x_A(p+q)$. The equilibrium value of $y_j^{(s)}$ is then determined by minimization of the Gibbs energy. The model of (5.44) also belongs to that second category because Va mix on two sublattices with other species.

Two important real examples for the first category are taken from steel: the solutions of carbon in the solid phases of Fe with Fcc and Bcc structure in the Fe–C system. Both phases are modeled as

$$(\mathbf{Fe})_p (\mathbf{Va},\mathbf{C})_q, \tag{5.46a}$$

since C occupies the interstitial sites only. The selection of $p$ and $q$ is based on the crystal structures, $(\mathbf{Fe})_1(\mathbf{Va},\mathbf{C})_1$ for the austenite phase (Fcc) and $(\mathbf{Fe})_1(\mathbf{Va},\mathbf{C})_3$ for the ferrite phase (Bcc). Now the amounts of sites and atoms must be distinguished. For the formula unit $(\mathbf{Fe})_p(\mathbf{Va},\mathbf{C})_q$, the total number of sites per formula unit is constant, $p+q$, whereas the number of *atoms* per formula unit is variable, $p + q\left(1 - y_{Va}^{(2)}\right)$. The content of *atoms*, $x_i$, is generally determined from the $y_j^{(s)}$, one balance equation for each atom:

$$p\, y_{Fe}^{(1)} + q\, y_{Fe}^{(2)} = x_{Fe}\left[p + q\left(1 - y_{Va}^{(2)}\right)\right] \tag{5.46b}$$

$$p\, y_{C}^{(1)} + q\, y_{C}^{(2)} = x_{C}\left[p + q\left(1 - y_{Va}^{(2)}\right)\right]. \tag{5.46c}$$

For the first category, this system of equations can be solved for the variables, $y_j^{(s)}$. Because of $y_{Fe}^{(1)} = 1$, $y_C^{(1)} = 0$, $y_{Fe}^{(2)} = 0$, and $1 - y_{Va}^{(2)} = y_C^{(2)}$, we get the following from (5.46c):

$$y_C^{(2)} = \frac{p\ x_C}{q\ (1 - x_C)}. \tag{5.46d}$$

Equation (5.46b) is redundant for this model. However, it would be used if we had Fe also placed on the interstitial site in a model $(\mathbf{Fe})_p(\mathbf{Va},\mathbf{C},\mathbf{Fe})_q$, with $1 - y_{Va}^{(2)} = y_C^{(2)} + y_{Fe}^{(2)}$, to determine also $y_{Fe}^{(2)}$ directly from $x_C$. That model also belongs to the first category with no internal degree of freedom because none of the three constituents mixes on more than one sublattice.

From (5.46d), we learn that the composition range for the chosen model $(\mathbf{Fe})_p(\mathbf{Va},\mathbf{C})_q$ is limited by the complete filling of interstitial sites, $y_C^{(2)} = 1$, corresponding to the limit $x_C = 0.5$ for the Fcc austenite phase and $x_C = 0.75$ for the Bcc ferrite phase. Therefore, the reference endmember Gibbs energies are not set at pure carbon but at these intermediate compositions for $y_C^{(2)} = 1$. The other endmember Gibbs energies are simply taken at $y_C^{(2)} = 0$ from pure Fe in the Fcc or Bcc phase.

### Magnetic Contribution

Any magnetic contribution in solutions is always treated by separation, as detailed in Section 5.3.1.2 for the elements. Therefore, the Gibbs energy is given for the para-magnetic lattice contribution in terms of the site fractions, $y_j^{(s)}$ in the CEF. The magnetic contribution is added as a function of the site fractions, $y_i$, by inserting the material parameters $T^*(y_i)$ and $\beta_0(y_i)$ in extension of (5.4):

$$G^\varphi\left(T, y_j^{(s)}\right) = G_{para}^\varphi\left(T, y_j^{(s)}\right) + G_{mag}^\varphi\left(T, y_j^{(s)}\right). \tag{5.47}$$

In the case of Fe, the magnetic contribution is strong for $\varphi =$ Bcc. However no composition dependence is assessed for the parameters $T_C$ and $\beta_0$ of Bcc or Fcc in the Fe–C system (Gustafson, 1985). That is different for the Fe–Ni system with a strong composition dependence of $T_C$ and $\beta_0$ for $\varphi =$ Fcc and Bcc (Cacciamani et al., 2010; Chuang et al., 1986; Xiong et al., 2011).

Note the little trap in the unary SGTE magnetic data for pure Fe(Fcc). In the print version, the physically correct data for the Néel temperature $T_N = +67$ K and $\beta_0 = +0.7$ for this antiferromagnetic Fe(Fcc) phase are given, similar to $T_N = +80$ K (Chuang et al., 1986). In fact, the calculated $c_p$ of Fe(Fcc) displays a small peak at 67 K. The baseline is unreal with $c_p < 0$ below 74 K due to the extrapolation out of range. More importantly, the stabilization of Fe(Fcc) due to antiferromagnetic ordering begins only well below about 150 K with increasing $c_{p,mag}$ shown by a slight difference of $c_p$ from the $c_{p,para}$. Compared to the much larger effect shown in Figure 5.4 for Ni(Fcc), and similar for Fe(Bcc), the magnetic effect on Fe(Fcc) well above 67 K is entirely negligible, also because of the small $\beta_0 = +0.7$. These data, however, are displayed differently in the electronic update of unary SGTE-5.0, where negative values for the Curie (not Néel) temperature of $T_C = -201$ K and $\beta_0 = -2.10$ are given for Fe(Fcc). The values relate to the metastable magnetic end-member data for the ferromagnetic Fcc phase from Ni to pure Fe.

That is explained in the next subsection for the austenite phase (Fcc) in the Fe–Ni–C system, modeled as $(Fe,Ni)_1(Va,C)_1$, which is treated with the ferro-/paramagnetic transition originating from the stable $T_C(Fcc,Ni) = 633$ K. The composition dependence of $T_C(Fcc,y_{Ni})$ is given as

$$T_C^{Fcc} = y_{Fe}^{(1)} TC_{Fe:Va}^{Fcc} + y_{Ni}^{(1)} TC_{Ni:Va}^{Fcc} + y_{Fe}^{(1)} y_{Ni}^{(1)} \cdot L_{Fe,Ni:Va}^{TC,Fcc}. \tag{5.48}$$

The general subscript notation in CEF is that the *colon* separates the sublattices. Thus "Ni:Va" means that the first sublattice is filled with Ni and the second with Va at pure Ni, $(Ni)_1(Va)_1$. The *comma* separates the interacting constituents on one sublattice while the other sublattice(s) are filled with one constituent, as in "Fe,Ni:Va". The interaction parameter, $L_{Fe,Ni:Va}^{TC,Fcc}$, describes the deviation from the linear interpolation between the endmembers in (5.48) in the form of a Redlich–Kister expansion. In the present example, we have

$$T_C^{Fcc} = y_{Fe}^{(1)}(-201) + y_{Ni}^{(1)}(633) + y_{Fe}^{(1)} y_{Ni}^{(1)} \cdot \left[2,200 - 700\left(y_{Fe}^{(1)} - y_{Ni}^{(1)}\right) - 800\left(y_{Fe}^{(1)} - y_{Ni}^{(1)}\right)^2\right]$$

$$(\text{for } 0.108 \leq y_{Ni} \leq 1, \text{ in } K). \tag{5.49a}$$

This equation is valid only in the range of positive $T_C(Fcc)$. The approach was developed in the Fe–Cr binary, where the Bcc phase is a stable antiferromagnetic phase at pure Cr and reliable data are available for $T_N(\text{Bcc},y_{Cr})$. Following the suggestion of Weiss and Tauer (1956), the phase changes from ferromagnetic to antiferromagnetic state at 0 K at the composition where $T_C = T_N = 0$, where $\beta_0$ should also be zero. Negative values should be assigned to the antiferromagnetic parameters, $T_N$ and $\beta_0$, in the composition dependence and the result be divided by a constant factor, which is $-3$ for the Fcc phase and $-1$ for the Bcc phase. For Bcc(Fe–Cr), Hertzman and Sundman (1982) employed this approach and could assess $T_C(\text{Bcc},y_{Cr})$ and $T_N(\text{Bcc},y_{Cr})$ by a single smooth function in the form of (5.48). The experimental data for $\beta_0(\text{Bcc},y_{Cr})$ could be described with the same function type (Hertzman and Sundman, 1982). Therefore, the function of $T_C(\text{Fcc},y_{Ni})$ describes the Néel temperature $T_N(\text{Fcc},y_j)$ by (5.49a) multiplied by $(-1/3)$, resulting in $T_N = +67\,\text{K}$ at pure Fe:

$$T_N^{\text{Fcc}} = \left\{ y_{Fe}^{(1)}(-201) + y_{Ni}^{(1)}(633) + y_{Fe}^{(1)} y_{Ni}^{(1)} \cdot \left[ 2{,}200 - 700 \left( y_{Fe}^{(1)} - y_{Ni}^{(1)} \right) - 800 \left( y_{Fe}^{(1)} - y_{Ni}^{(1)} \right)^2 \right] \right\}$$

$$\cdot \left\{ -\frac{1}{3} \right\} \qquad (\text{for } 0 \le y_{Ni} \le 0.108, \text{ in K}). \tag{5.49b}$$

The function of $\beta_0(\text{Fcc}, y_{Ni})$ is given in complete analogy. The magnetic contribution $G_{\text{mag}}^{\varphi}$ in (5.47) is then given directly from (5.5) by inserting the magnetic parameter $T^*$ as a function of $y_j^{(s)}$ as given in (5.48) or (5.49). The splitting of $T^*$ into $T_C$ or $T_N$ in each phase is handled inside the CALPHAD software if negative values appear. The second magnetic parameter, $\beta_0$, is inserted analogously as function of $y_j^{(s)}$.

*Lattice Contribution*
Let us continue with the paramagnetic Gibbs energy function, $G_{\text{para}}^{\varphi}$, for the austenite (Fcc) or ferrite (Bcc) phases in the Fe–Ni–C system, modeled as $(\text{Fe},\text{Ni})_p(\text{Va},\text{C})_q$.

$$G_{\text{para}}^{\varphi}\left(T, y_j^{(s)}\right) = y_{Fe}^{(1)} y_{Va}^{(2)} G_{Fe:Va}^{\varphi} + y_{Fe}^{(1)} y_{C}^{(2)} G_{Fe:C}^{\varphi} + y_{Ni}^{(1)} y_{Va}^{(2)} G_{Ni:Va}^{\varphi} + y_{Ni}^{(1)} y_{C}^{(2)} G_{Ni:C}^{\varphi}$$

$$+ pRT \left( y_{Fe}^{(1)} \ln y_{Fe}^{(1)} + y_{Ni}^{(1)} \ln y_{Ni}^{(1)} \right) + qRT \left( y_{Va}^{(2)} \ln y_{Va}^{(2)} + y_{C}^{(2)} \ln y_{C}^{(2)} \right)$$

$$+ y_{Fe}^{(1)} \cdot y_{Va}^{(2)} y_{C}^{(2)} L_{Fe:Va,C}^{\varphi} + y_{Ni}^{(1)} \cdot y_{Va}^{(2)} y_{C}^{(2)} L_{Ni:Va,C}^{\varphi}$$

$$+ y_{Fe}^{(1)} y_{Ni}^{(1)} \cdot y_{Va}^{(2)} L_{Fe,Ni:Va}^{\varphi} + y_{Fe}^{(1)} y_{Ni}^{(1)} \cdot y_{C}^{(2)} L_{Fe,Ni:C}^{\varphi}. \tag{5.50}$$

The first line in (5.50) is the weighted average of $G_{i:j}^{\varphi}$, the Gibbs energy of each "compound" per one mole of formula unit representing each endmember, the so-called surface of reference. $G_{Fe:Va}^{\varphi}$ is given by $G_{Fe}^{0,\varphi}(T)$ as defined in (5.1) with proper reference to SER. However, $G_{Fe:C}^{\varphi}$ is the key binary Fe–C parameter to be assessed for $\varphi = \text{Bcc}$ or Fcc. The $G_{i:j}^{\varphi}(T)$ describe the interactions of $i$ and $j$ between different sublattices, which are, in many sublattice models, nearest-neighbor bonds. The second line in (5.50) is the entropy of mixing according to the Bragg–Williams approximation. The third line is the excess term describing the interactions of $i$ and $j$ on the second sublattice, depending on how the first sublattice is occupied. The fourth line is the analog excess term for the interactions of $i$ and $j$ on the first sublattice. The $L_{i:j,k}^{\varphi}$ and $L_{i,j:k}^{\varphi}$ are (hopefully small) Redlich–Kister parameters following the notation introduced in (5.48) for the interactions on the second and first sublattice, respectively. They may be expanded in analogy to (5.27). In the present example, that is only required for the binary Fe–Ni parameters for Bcc and Fcc (Cacciamani et al., 2010):

$$L_{\text{Fe,Ni:Va}}^{\varphi} = L_{\text{Fe,Ni:Va}}^{0,\varphi} + L_{\text{Fe,Ni:Va}}^{1,\varphi}\left(y_{\text{Fe}}^{(1)} - y_{\text{Ni}}^{(1)}\right) + L_{\text{Fe,Ni:Va}}^{2,\varphi}\left(y_{\text{Fe}}^{(1)} - y_{\text{Ni}}^{(1)}\right)^2 \tag{5.51}$$

Equation (5.51) can be directly inserted into (5.50), where the other three $L$ parameters are not composition dependent.

This model $(\text{Fe,Ni})_p(\textbf{Va},\text{C})_q$ also belongs to the first category because all $y_i^{(s)}$ are explicit functions of $x_i$. The site fraction $y_{\text{C}}^{(2)}$ is calculated from (5.46d) and the site fractions of Fe and Ni are given by $y_i^{(1)} = x_j/(1 - x_{\text{C}})$. Note that in the binary Fe–C system this reduces to $y_{\text{Fe}}^{(1)} = 1$ as shown for the model in (5.46a).

*Mol-Formula versus Mol-Atoms*

Another trap to avoid is imminent in the unit "mol," explained in Section 5.2.2 for the example of cementite. The frequently used $G$-$x$ plots, as in Figure 5.2, require that the $G^{\varphi}$-functions of all phases relate to the same amount of 1 mol-atoms. Stoichiometric compounds are easily normalized, and substitutional solutions fulfill that by definition in (5.26). For associate solution, all the $G_i^{0,L}(T)$ in (5.37) must refer to 1 mol-atoms, i.e., only those of associates must be normalized, e.g., $G_{\text{Ca}_2\text{Sn}}^{0,L}(T)/3$. For sublattice models, specifically any models with vacancies, the normalization is composition dependent. For the example of Bcc and Fcc in the Fe–Ni–C system, which is modeled as $(\text{Fe,Ni})_p(\textbf{Va},\text{C})_q$, $G^{\varphi}$ from (5.47) and (5.50) is given in J/(mol-formula) and $G^{\varphi}/\left[p + q\left(1 - y_{\text{Va}}^{(2)}\right)\right]$ is given in J/(mol-atoms).

As another example, the $\text{Li}_x\text{Mg}_2\text{Si}$ phase in the Mg–Si–Li system was modeled as interstitial solid solution, $(\textbf{Mg})_2(\textbf{Si})_1(\textbf{Va},\text{Li})_1$, since the ternary solubility of this compound was experimentally detected from the binary $\text{Mg}_2\text{Si}$ directed toward pure Li (Kevorkov et al., 2004). The assessed value of $G^{\text{Mg}_2\text{Si}}$ is known from the binary stoichiometric $\text{Mg}_2\text{Si}$. However, that is not the value for the endmember in the CEF of $(\textbf{Mg})_2(\textbf{Si})_1(\textbf{Va},\text{Li})_1$ at the very same composition in the Mg–Si binary. The correct endmember value is

$$G_{\text{Mg:Si:Va}}^{0,\text{LixMg}_2\text{Si}} = \frac{3}{4}G^{\text{Mg}_2\text{Si}}. \tag{5.52}$$

The factor 3/4 accounts for the fact that the model $(\textbf{Mg})_2(\textbf{Si})_1(\textbf{Va})_1$ has (three atoms/ four sites) in contrast to the stoichiometric model $(\textbf{Mg})_2(\textbf{Si})_1$ with (three atoms/ three sites) for the same binary $\text{Mg}_2\text{Si}$ phase. One mol-formula of $(\textbf{Mg})_2(\textbf{Si})_1(\textbf{Va},\text{Li})_1$ always has 4 mole of sites (Kevorkov et al., 2004). A notation with different subscripts, $G_{\text{M}}$ and $G_{\text{m}}$, was suggested to distinguish the different basis, with "M" indicating a property of the phase per mole formula unit and "m" for a property per mole of components (i.e., nonvacancies, such as atoms, ions, compounds, etc.) (Sundman et al., 2018).

*CEF with Internal Degree of Freedom and Ordering*

The second category shows internal degree of freedom(s) in CEF due to mixing on more than one sublattice and will be explained for two examples from the Cu–Zn binary (Liang et al., 2015), considering also the extension into the Al–Cu–Zn ternary (Liang and Schmid-Fetzer, 2016). The binary Cu–Zn $\gamma$ phase with prototype of $Cu_5Zn_8$ crystallizes in the structure of cI52 with space group *I-43m*. According to that structure and equivalent lattice sites, the best model for binary $\gamma$-$Cu_5Zn_8$ should be $(Zn)_4(Cu)_4(Cu,Zn)_6(Cu,Zn)_{12}$. This phase unites in complete solution with the disordered high-temperature $\gamma$ phase in the Al- Cu system (Liang and Schmid-Fetzer, 2015). The model reflecting the crystal structure of the ternary Al-Cu-Zn $\gamma$ phase perfectly is $(Al,Cu,Zn)_4(Cu)_4(Cu,Zn)_6(Al,Cu,Zn)_{12}$ (Ansara et al., 2000; Liang and Schmid-Fetzer, 2016). However, it involves $3 \times 2 \times 3 = 18$ endmember parameters, plus possible interaction parameters. To avoid this complexity, it was suggested to combine the two $(Al,Cu,Zn)$ sublattices and simplify them to $(Cu)_4(Cu,Zn)_6(Al,Cu,Zn)_{16}$. That reduces to $(Cu)_4(Cu,Zn)_6(Cu,Zn)_{16}$ for the Cu–Zn binary system containing four endmembers (Liang et al., 2015). Balancing the pursuit for perfect reflection of the crystal structure with the need to avoid overparameterized models is a difficult but very important decision. It is not possible to construct the thermodynamic database of a multicomponent system starting systematically from the binaries if the sublattice model of the same phase is set up differently in the binaries.

The second example from the Cu–Zn binary is the Bcc-based intermetallic phase $\beta$, also known as brass with manifold applications. This phase shows an ordering reaction on cooling from $\beta$ (Bcc) to $\beta'$(Bcc_B2) with the ordered CsCl structure; the transition temperature of about 460°C depends on composition (Liang et al., 2015). The disordered Bcc (A2) $\beta$ phase is modeled as a substitutional solution phase, $(Cu,Zn)_1$, and $G^\beta$ is described by (5.26). The model of the ordered $\beta'$ (B2) phase with $(Cu,Zn)_{0.5}(Cu,Zn)_{0.5}$ maintains the number of atoms given by the $\beta$ phase, $(Cu,Zn)_1$. Ordering occurs when $y_j^{(1)}$ start to deviate from $y_j^{(2)}$, where the two sublattices distinguish the edge and center atoms of Bcc. A unique function, $G^{\beta'}$, is used to describe the Gibbs energies of both phases, $\beta$ and $\beta'$, as follows:

$$G^{\beta'} = G^\beta(x_i) + G^{ord,B2}\left(y_i^{(1)}, y_i^{(2)}\right) - G^{ord,B2}\left(y_i^{(1)} = x_i, y_i^{(2)} = x_i\right). \tag{5.53}$$

It is seen that the second and third terms cancel if $y_j^{(1)} = y_j^{(2)}$, in the disordered state, so that $G^\beta = G^{\beta'}$. This function difference, denoted as $\Delta G^{ord,B2}$, represents the change in the Gibbs energy of $\beta$ due to ordering when forming the B2 phase $\beta'$. It vanishes, $\Delta G^{ord,B2} = 0$, when $y_j^{(1)} = y_j^{(2)} = x_j$. According to the sublattice model $(Cu,Zn)_{0.5}(Cu,Zn)_{0.5}$, the ordering term is

$$G^{ord,B2}\left(y_i^{(1)}, y_i^{(2)}\right) = \sum_i \sum_{j \neq i} y_i^{(1)} y_j^{(2)} G_{i:j}^{ord,B2}$$

$$+ 0.5RT\left(\sum_i y_i^{(1)} \ln y_i^{(1)} + \sum_i y_i^{(2)} \ln y_i^{(2)}\right) + G^{ex,B2}, \tag{5.54}$$

in which the excess term is given as

$$G^{ex,B2} = y_{Cu}^{(1)} y_{Zn}^{(1)} \sum_i y_i^{(2)} L_{Cu,Zn:i}^{ord,B2} + y_{Cu}^{(2)} y_{Zn}^{(2)} \sum_i y_i^{(1)} L_{i:Cu,Zn}^{ord,B2}. \tag{5.55}$$

This approach in (5.53)–(5.55) follows the works of Ansara, Dupin, and Sundman (Ansara et al., 1997; Dupin and Ansara, 1999), and is also given in Lukas et al. (2007). However, the notation in the present equations is more specific as detailed in Liang et al. (2015). That is important, as we explain later. The sum of the first and last terms in (5.54), i.e., without the entropy term, equals $\Delta G_m^\circ(y_i^I, y_i^{II})$ in the notation of Dupin and Ansara (1999). However, in the double summation of (5.54), the exclusion $j \neq i$ is not given in Ansara et al. (1997), thus a parameter $G_{i:i}^{0,ord,B2}$ may occur. The exclusion $j \neq i$ is explicitly only given in Dupin and Ansara (1999), where the equations are set up differently with the entropy part in (5.54) taken out to define $\Delta G_m^\circ$. It is mandatory to set $G_{Cu:Cu}^{0,ord,B2} = G_{Zn:Zn}^{0,ord,B2} = 0$ in order to comply with the general proper set of (5.53)–(5.55) with $j \neq i$ because current software packages do not automatically exclude $j = i$ in the summation.

The advantage of writing the order/disorder model equations in the way of (5.53)–(5.55) is that the parameters of the disordered phase in $G^\beta$ may be chosen without any constraint. The parameters of $G^{ord,B2}$, however, must comply with additional relations (Ansara et al., 1997; Dupin and Ansara, 1999) to enable disordering by obtaining $\Delta G^{ord,B2} = 0$ at high temperature. These relations also allow to decrease the number of adjustable parameters in $G^{ord,B2}$.

### First- and Second-Order Transitions

In the example of Cu–Zn, the $\beta/\beta'$ (or A2/B2) transition is of *second order*, which produces a single transition line as opposed to a two-phase region in the Cu–Zn *T-x* phase diagram. The second-order transition is defined by a discontinuity in the second derivative of $G$, $\partial^2 G/\partial T^2$, while the first derivative, $\partial G/\partial T$, is continuous. That is equivalent to a discontinuity in $c_p(T)$ while $H(T)$ is continuous, such as in the magnetic transition of an element; see Figure 5.4 for nickel. In a *first-order* transition, the discontinuity occurs already in the first derivative of $G$, while $G(T)$ is continuous, such as in the melting of an element, where the discontinuity of $H(T)$ describes the enthalpy of melting.

The transition type may change from first to second order. For the example of austenite phase (Fcc) in the Fe–Ni system, as discussed earlier, the para-/ferromagnetic transition originates from $T_C$(Fcc, Ni) at 633 K and continues with decreasing $x_{Ni}$ into the binary Fcc phase with a maximum of $T_C$, initially as a second-order transition without a two-phase region. However, a tricritical point is predicted from the thermodynamic calculation at $x_{Ni} = 0.475$ and $T_C = 735$ K, where the single $T_C$(Fcc) curve splits from the second-order into a first-order transition (Chuang et al., 1986). For $x_{Ni} < 0.475$, a two-phase region $Fcc_1 + Fcc_2$ develops between paramagnetic $Fcc_1$ and ferromagnetic $Fcc_2$, and the impact on heat treatment in that range and on Invar alloys was discussed (Chuang et al., 1986). More examples are given in a recent review of the modeling of ordered phases (Sundman et al., 2018).

*Vacancies in Sublattice Models*

Two types of vacancies should be distinguished in the CEF, namely stoichiometric and thermal vacancies. In the model $(A)_1(Va,C)_1$, only *stoichiometric vacancies* exist, as in Fe–C austenite, and $y_i^{(s)}$ is determined from $x_i$ (no internal degree of freedom). In the model $(A,Va)_1$, only *thermal vacancies* exist, and, despite fixed $x_A = 1$, $y_{Va}^{(1)}$ must be determined from the Gibbs energy minimization (one internal degree of freedom). The relations between the two types are discussed in detail for the model $(A,Va)_1(Va,C)_1$, where both types of vacancies can occur in the enlightening work of Ågren and Hillert (2019).

Of major concern is the "empty endmember compound," $(Va)_1(Va)_1$, and the assignment of $G_{Va:Va}$; see (5.50) and (5.54). In previous work, the value $G_{Va:Va} = 0$ has been assigned frequently, for example in the modeling of Al–Ni Bcc and Bcc_B2 phases with the model $(Al,Ni,Va)_1$ and $(Al,Ni,Va)_{0.5}(Al,Ni,Va)_{0.5}$ in the approach of (5.53) (Dupin and Ansara, 1999; Dupin et al., 2001). However, setting $G_{Va:Va} = 0$ generally creates a serious problem because the equilibrium condition for the always nonconserved Va is $\mu_{Va} = \partial G/\partial n_{Va} = 0$; see Table 5.2. As shown in detail by Oates et al. (2008), the result is that the system always empties itself at equilibrium, $y_{Va} = 1$ for the former and $y_{Va}^{(1)} = y_{Va}^{(2)} = 1$ for the latter model, unless large positive Redlich–Kister $L$-parameters provide a spinodal miscibility gap. The very same problem occurs if the Bcc_B2 phase is modeled as a separate phase $(Al,Va)_1(Ni, Va)_1$ without the ordering contribution in (5.53). Actually, there is a footnote in Dupin and Ansara (1999, p. 81) that "for numerical reasons, it seems necessary to change $G_{Va:Va}$ from 0 to a small positive value, as $10^{-10}$" to avoid this effect.

Setting $G_{Va:Va} = 0$ or $10^{-10}$ J/mol-formula together with large positive interaction parameters $L \gg 0$, such as the typically used $L_{Al,Va:k}^{ord,B2} = L_{k:Al,Va}^{ord,B2} = +100$ kJ/mol $-$ formula($k$ = Al, Ni, Va) (Dupin and Ansara, 1999), however, produces the intended equilibrium only with the first generation of CALPHAD software that determines the *local* equilibrium of $G$ and if appropriate starting values of the phase constitution with all $y_i^{(s)}$ provided. The reason is that three equilibrium solutions of $\mu_{Va} = 0$ exist at different vacancy concentrations (Oates et al., 2008). That different phase diagrams are produced from the same published dataset was demonstrated for many examples by Chang et al., 2004) using the software Pandat (Chen et al., 2002) that automatically determines the *global* minimum of $G$ without any user-provided starting values.

The only physically correct solution, if the Va is introduced on all sublattices in CEF just like other constituents, is to set $G_{Va:Va} \gg 0$, which respects the equilibrium condition with a unique solution of $\mu_{Va} = 0$. That was shown by Oates et al. (2008), who selected $G_{Va:Va} = 60$ kJ/mol -formula for the B2-AlNi phase, which also reduces the number of necessary $L$ parameters. It was also suggested that the choice of the large positive value of $G_{Va:Va}$ may be different for different "empty" host lattices, such as Bcc and Fcc. The minimum value of the endmember to prevent a second stable state at $y_{Va} = 1$ was determined as $G_{Va:Va} \approx 0.38\, RT$, neglecting $L$ parameters (Ågren and Hillert, 2019). The limit is smaller for the model $(A,Va)$, $G_{Va} \approx 0.19\, RT$ and increases with the number of sublattices.

In conclusion, for the nonexpert user, intending to use a certain – possibly older – database, we suggest checking that no smaller value or even $G_{Va:Va} = 0$ is used for any phase modeled as $(A,...,Va)(B,...,Va)$ in CEF. Be careful about "forgotten" parameters: If the TDB dataset does not include the parameter $G_{Va:Va}$, detected as "G (phase,VA:VA;0)," it may be set to zero automatically depending on the software. The Pandat software will give an error message to the user in that case.

## 5.4 Establishment of Thermodynamic CALPHAD Databases

A key feature of CALPHAD databases is the systematic establishment from unary $\rightarrow$ binary $\rightarrow$ ternary $\rightarrow$ ... $\rightarrow$ multicomponent systems. The two major benefits of the CALPHAD method focus on different system sizes. On the one hand, the deep understanding between thermodynamic properties and phase equilibria or transitions is quantitatively worked out mainly in binary and ternary systems. These formerly separated classes of data are combined into a consistent picture that also exploits the strong thermodynamic rules governing the interdependence of all data. On the other hand, the powerful applications in computational materials design are due to the quantitative simulations and predictions mainly in multicomponents, but also in ternary and binary systems. The reliability of these simulations and predictions depends mainly on the quality of the CALPHAD dataset or database.

This *thermodynamic description* of the system comprises, for each phase $\varphi$, both the flag to apply the chosen Gibbs energy model, programmed in the software, and the pertinent thermodynamic model parameters, together forming the function $G^\varphi$. It is created in a "CALPHAD assessment." For small – binary or ternary – systems, the resulting electronic file may be called the thermodynamic "dataset," which should be distinguished from a comprehensive thermodynamic "database" for large systems.

With growing database size, the key issues arising are *consistency, coherency*, and *quality assurance*. These issues also concern extension, maintenance, and updating of the database (Gröbner and Schmid-Fetzer, 2013). *Consistency* demands that, generally, if any newly assessed ternary system is incorporated in the database it may need adjustment to the phase models of the binary systems in the database. Conversely, if any binary system is updated, the check of any higher-order system and probably an adjustment are required, especially for the ternary systems. *Coherency* needs very careful consideration because binary intermetallic phases may be coherent in ternary or higher-order systems if crystallizing in the same structure.

The key issue of *quality assurance* involves addressing four distinct questions concerning the thermodynamic descriptions, also for binary and ternary systems. It should be verified if these descriptions are *correct, reasonable, accurate*, and *safe*, as detailed by Schmid-Fetzer et al. (2007). Assume that a nonexpert user has access to a report with a calculated stable phase diagram, thermodynamic properties, and the accompanying dataset. The test for *correctness* is the most stringent and important one, performed with a reasonable amount of effort, by calculation to check if the intended stable phase diagram and thermodynamic properties are reproduced. Otherwise, this thermodynamic description is denoted as *wrong*.

The test for *reasonability* is important because there are examples of phase diagrams that appear to be reasonable, but on closer inspection exhibit unrealistic thermodynamic features. First of all, the absolute entropies of all solid phases at 298.15 K, $S^0_{298}$, should be checked to identify if these data are not too far away from a linear combination of the $S^0_{298}$ values of the elemental components. For example, in one assessment of the Al–B system the calculated phase diagrams looks acceptable; however, the compound $AlB_2$ was modeled with impossible negative value $S^0_{298}(AlB_2) = -1.9 \, \text{Jmol}^{-1} \, \text{K}^{-1}$. That may happen if the floating reference state is used and the entropy of formation is assessed too negatively; see Section 5.3.1.4. Reassessment of the Al–B system solves these problems (Mirković et al., 2004). It is also recommended to plot the calculated $T - x_B$ phase diagram also with the variables $T - \mu_B$. This chemical potential phase diagram reveals more information from the slope of the two-phase lines. The reasonability of the chosen model type also should be checked. More methods are detailed in Schmid-Fetzer et al. (2007).

The *accuracy* is mainly determined by comparison between the phase diagram calculated from the given dataset and reliable experimental data. In addition, the calculated key thermodynamic data should be compared to experimental data, and in the case of $\Delta H$ and $c_p(T)$ of solid phases, to reliable ab initio data. The judgment of the reliability of reported data requires profound knowledge of the experimental and ab initio methods. Assessing a reliable (sub)set of reported or estimated data should be the major part of any CALPHAD assessment. If an assessor compares reported data and those calculated from the current dataset, it is highly recommended not to react to disagreement by simply increasing the number of parameters; see Section 5.3.2.1. One should first question if error bars involved in the experimental and ab initio methods are realistic and then question the applicability of the chosen model type for $G^\varphi$. One should also not simply look to the minimized value of the "sum of square of errors" between the calculated and "experimental" data provided by software, such as PanOptimizer in Pandat or Parrot in Thermo-Calc, during an optimization run of parameters. These are only supporting tools in the iterative improvement of an optimized parameter set; they must be commanded by experienced expert users and critical manual judgment of overall accuracy. CALPHAD assessment is more than just a fitting exercise.

Finally, the *safety* of a multicomponent thermodynamic database involves checks going beyond the individual assessment of stable binary and ternary subsystems, as detailed for the close proximity of stable and metastable phase boundaries in Schmid-Fetzer et al. (2007). Here we look at it from a user perspective with the simple example of the Y–Zr binary system. For example, the 27-component light alloy database COST507 created by a compilation of many binary and some ternary assessments was shared as open source (Ansara et al., 1998). It also comprises the elements Al–Si–Y–Zr and the calculated Y–Zr phase diagram shows essentially the complete solubility in the phases Hcp, Bcc, and liquid emerging from the stable elements. A truly novice user may take that for granted, but a look in the documentation of the database reveals that the Y–Zr binary system was not assessed, so it was just displayed with ideal solution in all phases, an "extrapolation from the unaries."

Now consider Y–Zr in a less obvious example, starting with an Si–Y–Zr dataset built from all three correctly assessed binary systems, where the Y–Zr phase diagram shows the demixing of both Hcp and Bcc phases correctly. Now the component Al is added with all three correctly assessed binary systems (Al–X, X = Si, Y, Zr) to create the Al–Si–Y–Zr database. The Y–Zr phase diagram calculated from the new database shows a huge intermediate region of Fcc phase existence. How is this serious artifact possible if all the six binary systems had been correctly assessed? The reason is that the Fcc phase was not stable in the Si–Y–Zr dataset but must be introduced with all components as the endmember of the Al–X Fcc solution phase. However, the $L$ parameters describing the interactions of the three endmember components in Fcc were forgotten and rightly set to zero in any CALPHAD software, assuming an intended ideal solution. This ideal Fcc solution is stabilized at higher $T$ and prevails over the demixing Hcp and Bcc phases, which produces the serious artifact. An experienced assessor would have realized that the ideal interaction of Y and Zr in the Fcc phase is a very bad assumption and repulsive interactions must exist. One must introduce a sufficiently large positive $L_{Y,Zr}$ interaction parameter for the Fcc phase, not smaller as evidenced for the demixing Hcp and Bcc phases. A typical estimated value is $L_{Y,Zr}^{Fcc} = 80 \cdot T$, which eliminates the artifact. Interestingly, the artifact is not noticed in the Si–Y and Si–Zr phase diagrams because very stable intermetallic phases prevent any artificial occurrence of Fcc phase.

The conclusion is that – if in doubt on the safety of a database – any user is advised to perform the simple test for *correctness*, detailed in the preceding, by at least calculating all the pertinent binary phase diagrams. In comparison to reported diagrams, it should be kept in mind that good reasons may have led to a simplification of complex intermetallic phases to avoid overparameterization (see Section 5.3.2.6), if the detail of that phase is not in the focus of a dedicated alloy database with many components.

The preceding simple examples are intended to make the reader aware of the fact that CALPHAD assessment and construction of thermodynamic databases require extensive training and experience – much more than applying CALPHAD calculations with a given and proper database. Compilation of a database from "published parameters of systems in the literature" offers manifold traps and hidden errors. For a start, the study of the scholarly book by Lukas et al. (2007) and cooperation with experienced groups in the field is recommended that can be met at the annual CALPHAD conferences.

Large open-source thermodynamic databases are rare, and the most important one is the unary SGTE database, which is the basis of all relevant multicomponent databases. Some differences are noticed between the print version (Dinsdale, 1991) and electronic update version SGTE-5.0; see Section 5.3.1. Such differences may impede the combination of binary assessments to a multicomponent system, requiring reassessment. For some elements, such as for Bi, Gd, and Y, even the stable phases are given with different GHSER functions in both versions. Unfortunately, it is often not explicitly specified in published assessments if the print or electronic update version was used. Thermodynamic datasets for binary or sometimes ternary systems are often freely available in electronic form from the authors of assessments or even as direct download from the CALPHAD and some other journal's websites as supplementary files to a publication.

Large (generally more than 10 components) thermodynamic databases are focused on dedicated alloy systems and confined to application ranges. Examples are databases for Al-based alloys and Mg-based alloys (Chen et al., 2018; Schmid-Fetzer and Zhang, 2018). All of the more recent ones are encrypted to protect the considerable effort necessary by teams of experts to create and maintain them. Each of them can only be used with the licensed software used for the encryption, such as Thermo-Calc, Pandat, or FactSage; see Table 5.3. Databases are available for multicomponent base-metal alloys, such as Fe, Al, Mg, Ni, Co, Cu, Mo, Nb, Ti, and TiAl alloys, for special alloys, such as noble metal, solder, bulk metallic glasses, high-entropy alloys, oxide, and slags, to name some. More information on the large number of dedicated CALPHAD databases available nowadays for all important inorganic engineering material classes are given on the websites listed in Table 5.3.

A major distinction is to be noted on database strategies, which are either *system* oriented or *phase* oriented. This distinction is the same after encryption of the open databases. A system-oriented database comprises all phase descriptions of the A-B-C, ... target system in a single consistent database. The dominant open file format is the TDB text file used by both Pandat and Thermo-Calc software. As opposed to that, phase-oriented database(s) collect all phase descriptions containing any of the A-B-C-... target components from separate databases. The prominent example is the FactSage database system, integrated into the FactSage software, that has different databases for the solution phases ($^*$.SDC) and for the compound phases ($^*$.CDB). On the one hand, that allows much flexibility for experts in compiling some desired phases. On the other hand, it inevitably leads to phase duplications, e.g., solution phases that may overlap in composition ranges, but are assessed only for specific regions, and compounds that overlap with solution phases. The user has to decide which phase to use before starting any calculation. That requires expertise on the material system. In a system-oriented database, by contrast, all these decisions have been made by the expert(s) creating the database, and the user can begin the calculations directly.

Extended CALPHAD-type databases going beyond the solely thermodynamic ones, discussed so far, have emerged in recent years. All of those are amendments that build on the assessed compilation of $G^\varphi$ functions. One example for additional phase-based properties is the description of molar volume and thermal expansion, suggested by Hallstedt et al. (2007) and Lu et al. (2005a, c). The pressure dependence is relevant only above about 1 kbar (0.1 GPa), as detailed in Section 5.3.1.3, and for very high pressure as suggested by Lu et al. (2005b). Amendment of the database by functions of $V_m^\varphi(T)$ provides the additional information on density and volume fractions of phases during thermodynamic calculations, important for solidification shrinkage and other processes. Another example is the development of mobility databases for diffusion simulation mentioned in the introduction and discussed in detail in Chapter 6 on thermophysical properties. The assessment technique for these and other phase-based properties follows also the CALPHAD method (Campbell et al., 2014). Both the parameters on molar volume and mobility are either directly integrated in the TDB text file format or taken out and provided with the same syntax as separate text files to amend the phase descriptions and can be read by both Pandat and Thermo-Calc software. Further database extensions to include kinetic material parameter information necessary for dedicated simulation of solidification and precipitation, as detailed in Sections 5.6 and 5.7, are compiled in separate files in different formats. For example, the Extensible Markup Language (XML) is used for such extended files and batch files in the Pandat software.

## 5.5    Alloy Design Applications Using Solely Thermodynamic CALPHAD Databases

### 5.5.1    Overview

Focused computational design is proven to be strongly supported for all classes of engineering materials for which a thermodynamic CALPHAD database is available that can be used with a major software package; see Table 5.3. The term "alloy" in the following examples is to be taken in a broader sense, also comprising nonmetallic materials, such as ceramics.

Alloy design and process optimization are closely related, and here the materials-centered perspective is taken, as will be discussed later in Chapter 13, as one alternative. That is, we start with the key variables available to the engineer by variation of composition and processing of the alloy, aiming at prediction of resulting structure and properties of the material. In applications, this is done mainly by calculations with individual alloy composition as the key input, and phase formation is predicted as a function of temperature. That can be done under equilibrium conditions or special nonequilibrium, so-called Scheil conditions, for solidification; see Section 5.5.3.

The types of key results are compiled in Table 5.4 for the example of designing cast and heat-treated alloys. These are so-called 1D calculations, where only one state variable varies, for example $T$ at fixed composition, and the result can be displayed in an easily readable property diagram; see Section 5.5.2.3. The impact of composition variation is screened by a certain number of such 1D calculations. Literally thousands of them can be performed in high-throughput calculations (HTC), such as implemented in Pandat package, and the case study in Section 5.7 provides an example. Calculated phase diagrams provide an overview of phase formation and materials compatibility. It is generally recommended to start with that for the material system under consideration, as detailed in the following two subsections. Only the most frequently applied results are listed in Table 5.4.

While only the thermodynamic database of the alloy system is required for all applications in this section, extended applications are exemplified if extended databases involving kinetic parameters are available in Section 5.6.

**Table 5.4** Summary of key results from widely used CALPHAD-type software calculations for the design of cast and heat-treated alloys. Only the thermodynamic database of the alloy system is required.

| Process | Method | Result | Detail of results |
|---|---|---|---|
| Casting | Scheil simulation | Microstructure constituents | • Phase sequence during solidification, primary and secondary phases<br>• Phase fractions<br>• Segregation in phase composition |
| Casting & heat treatment | Scheil simulation | Scheil solidus | • Freezing range for relatively rapid solidification<br>• Incipient melting temperature, first-stage heat treatment temperature limit |
| Casting & heat treatment | Equilibrium calculation | Equilibrium solidus | • Freezing range for relatively slow solidification<br>• Final stage heat treatment temperature limit |
| Heat treatment | Equilibrium calculation | Microstructure constituents | • Final transformed phases with phase fractions and phase compositions depending on heat treatment temperature |
| Property design | Equilibrium calculation | Thermodynamic properties | • Enthalpy exchange, energy storage materials, and hazard safety of batteries<br>• Chemical potentials, partial pressures, EMF |
| General | Phase diagram calculation | Overview of phase formation in equilibrium | • Range of existence of phases in 2D space of state variables (e.g., $T–x$)<br>• Phase transitions, reactions, and transformations<br>• Stable and metastable phase diagrams<br>• Materials compatibility |

## 5.5.2  Applications Based on Equilibrium Calculations

Basic CALPHAD equilibrium calculations may be classified as 0D, 1D, and 2D. The 0D or point calculation is the basic determination of the thermodynamic equilibrium properties at a fixed state point, as explained in Section 5.2.2. Let us assume that the proper software always determines the global minimum of $G$(system) without the artifacts described before due to local minima of $G$(system). The 0D calculation result is the *equilibrium constitution* of the system, expressed by the *phase fractions*, $f^\varphi$, for each phase $\varphi$ and the *phase composition/constitution*, given by $(x_i^\varphi)$ for solution or $\left(y_i^{(s),\varphi}\right)$ for sublattice phases. For all phases not in equilibrium at that state point, $f^\varphi = 0$ holds true. For all phases additionally the driving force, $DF^\varphi$, is obtained, which is a quantitative measure in J/mol-atoms for the distance from equilibrium (Hillert, 2008). For phases prevailing at equilibrium, $DF^\varphi = 0$; for all other phases, $DF^\varphi < 0$. Of course, all other thermodynamic properties of the equilibrium phases are also obtained, such as the determinant and eigenvalues of the Hessian of the Gibbs energy of a phase.

In the applications presented in this section, the meaning of *metastable* equilibrium is generally used in the context of phase availability. In a *stable* equilibrium calculation, all known phases (modeled in the database) compete for the global minimum of $G$(system). Sometimes sluggish formation is known by experience for one or more phases, possibly due to nucleation or growth problems, and these phases are not expected to form in the real process time under consideration. Complete thermodynamic equilibrium is then an

unsound approximation to capture the real process. In a *metastable* equilibrium calculation, these phases are deliberately excluded (suspended) from the phase competition,
and the global minimum of $G$(system) is determined among the remaining phases. This
determines quantitatively the (re-)distribution of components and phase formation in a
much better approximation of that specific real process. In this case, one obtains
$DF^\varphi > 0$ for the suspended phase(s), which is essential information to compare with
the nucleation/growth barrier for that phase. This definition of stable versus metastable
calculations is maintained for the rest of this chapter.

The 1D or line calculation is simply a series of point calculations along a defined
line between two state points. In addition, if a phase transformation occurs, the
transformation boundary point will be also calculated. For example, the system
composition may be fixed and the temperature varied, or a linear combination in
system composition may be varied at a fixed temperature. The 1D calculation is the
workhorse of CALPHAD application calculations because any of the resulting equilibrium property of the 0D calculations in the series may be displayed in a graph with
the "1D-line variable" as abscissa, a *property diagram*. For example, the phase
fractions, phase compositions, activities, etc., may be plotted against temperature.

The 2D or *phase diagram section* calculation is basically different because phase
boundaries are traced in the plane section spanned by two state variables. The section
is always defined by three noncollinear state points, taken as top-left, origin, and
bottom-right corners of the resulting phase diagram. For example, in Figure 5.5, the
unary Fe phase diagram, the state space is defined by the variables $(T, p)$, and the three
state points are (4,000 K, $10^0$ bar), (500 K, $10^0$ bar), and (500 K, $10^{5.5}$ bar). In a closed
binary system at $p = 1$ bar, the state space is defined by $(x_1, x_2, T)$, with $x_1 + x_2 = 1$.
The three state points (1, 0, 700°C), (1, 0, 100°C), and (0, 1, 100°C) define the section
of the common Al–Mg "$T - x$" phase diagram in Figure 5.12.

In a closed ternary system, the state space is defined by $(x_1, x_2, x_3, T, [p])$, and three
state points such as (0, 0, 1, 700°C), (1, 0, 0, 700°C), and (0, 1, 0, 700°C) define the
common isothermal phase diagram section at 700°C in the $1 - 2 - 3$ ternary system.
The value for $p$ is omitted if none of the phases in the database have a $G(p)$
dependence, e.g. for the common case of condensed phases. Similarly, any vertical
section or "$T - x$" isopleth is defined by this simple strategy, which is implemented in
the graphical user interface (GUI) of Pandat software for all 2D sections from unary to
multicomponent systems.

Note that each point in the 2D phase diagram has the meaning of a state point, and
the phases in equilibrium may be indicated by the label of the adjoining region
confined by phase boundaries. In contrast, only the curves have a meaning in a
property diagram resulting from a 1D calculation, and the areas have no meaning
(Schmid-Fetzer, 2014).

After the 2D section calculation is completed, all other thermodynamic properties
along the traced phase boundaries are still stored in the computer memory for some
CALPHAD software. That enables the user to plot the already calculated phase
diagram with alternative variables. Thus, after the 2D calculation of the $T - x_{Mg}$
phase diagram in Figure 5.12 and its plot in the default option, one may directly plot

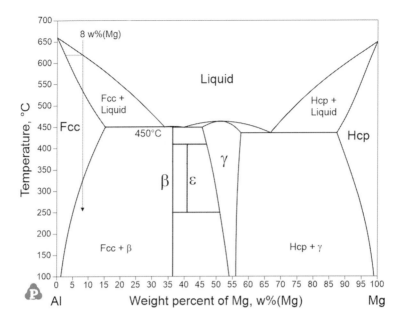

**Figure 5.12** Calculated Al–Mg phase diagram using the thermodynamic data of Liang et al. (1998b); the alloy composition Al–8Mg (wt.%) is indicated.

the chemical potential phase diagram with the variables $T - \mu_B$. However, care must be taken because not all combinations of thermodynamic variables produce a true phase diagram, as shown by Ågren and Schmid-Fetzer (2014).

### 5.5.2.1  Stable Phase Diagram Calculations

The calculated phase diagram first of all reveals the range of existence of stable phases in the 2D space of state variables as shown in the $T - x$ phase diagram of the Al–Mg system in Figure 5.12 with wt.% scale. As an alternative to the molar fraction $x_i$, the weight or mass fraction $w_i$ is often used for applied systems. Also, the phase transitions, reactions, and transformations are displayed, and more information can be read. For example, what is the *materials compatibility* in a sandwich of pure Al and Mg heated to 500°C? The phase diagram reveals that the liquid phase will form at the interface. That is not expected without knowing the phase diagram because the melting points of the initial Al and Mg are some 150°C higher than 500°C. The wealth of information obtained from the most important phase diagrams is explained in Schmid-Fetzer (2014), including examples from Mg-rich alloys, while just two examples for Al-rich alloys are briefly given here.

The vertical arrow in Figure 5.12 for the fixed alloy composition of Al–8Mg (wt.%), $w_{Mg} = 0.08 = 8\%$, indicates the phase regions passed through during solidification and solid-state transformations. These are further explained for this alloy in Section 5.5.2.3 with reference to this diagram. The *tie line* in a phase diagram connects two phases in equilibrium, such as the dotted line connecting liquid at $w_{Mg} = 8\%$ with

Fcc at $w_{Mg} = 2.6\%$ at $619°C$ in the Fcc + Liquid two-phase region. This tie line also defines the *distribution coefficient*, $k_0$, the ratio of the "concentrations" of solid to liquid in equilibrium, commonly used in solidification applications. Often the term *concentration*, unit $[mol/m^3]$ or $[kg/m^3]$, is wrongly used for *contents*, unit $[mol/mol]$ or $[kg/kg]$, such as $x_{Mg}$ or $w_{Mg}$. In the preceding example, we have $k_0(Mg) = w_{Mg}(Fcc)/w_{Mg}(Liquid)$ and $k_0(Mg) = 0.3$ at $619°C$. The definition of $k_0(i)$ assumes a limited concentration of the *solute*, here $i = Mg$, in the liquid and solid phases of a *solvent*, here Al. Applied to binary Cu–Ni alloys with only one Liquid + Fcc phase region, connecting the melting point of Cu with the higher one of Ni, both elements can be treated as solutes with $k_0(Ni) > 1$ and $k_0(Cu) < 1$.

A tie line is always an *isotherm*. However, in ternary or higher systems, the isothermal lines connecting phase boundaries in a $T - x$ phase diagram section, so-called *vertical sections* or *isopleths*, are only by exception also tie lines (Schmid-Fetzer, 2014).

It is instructive to first view any process in the phase diagram to open the perspective for a wider range of composition and temperature. For example, Figure 5.12 reveals that there is some room for variation in the alloy composition since up to $w_{Mg} = 15\%$, the same phase regions are passed. Beyond that limit, no single-phase Fcc phase exists in equilibrium and the eutectic equilibrium Liquid = Fcc + $\beta$ occurs at $450°C$ with liquid composition of $w_{Mg} = 34\%$.

The industrial relevance of the Al–Mg system is that the Al alloy 5xxx series is based on it, often with minor Mn additions. For example, Al alloy 5083 with nominal Al–4.4Mg–0.7Mn (wt.%) has the highest strength of non-heat-treated Al alloys and high-corrosion resistance as needed for marine applications. For simplicity and clearance, the main features are explained on the present example of binary Al–8Mg alloy later in Section 5.5.2.3.

For ternary Al alloys, the example of Al–Mg–Zn is chosen with relevance in the 7xxx series of Al–Zn–(Mg, Cu) alloys, which can be precipitation hardened to very high strength. The 2D phase diagram section of Al–Mg–Zn is not shown here to save space, and it has been explained in detail also with comments on reading ternary isothermal sections and vertical sections together with the concept of zero phase fraction (ZPF) lines in Schmid-Fetzer (2014). The ZPF line of a phase $\varphi$ in a 2D section is defined by the series of state points where phase $\varphi$ is at the edge of equilibrium, but with $f^\varphi = 0$. This is a helpful tool for reading and understanding phase diagram sections of ternary and multicomponent systems. The calculation of any 2D section is in fact done by the leading software simply by tracing the ZPF line of each phase $\varphi$ individually, and their superposition provides all the phase boundaries in the 2D phase diagram section for any number of components.

Figure 5.13 does not show a 2D section but the *projection* of the equilibria of the liquid phase for the Al–Mg–Zn system, or *liquidus surface*. It shows the composition regions of different *primary phases* during solidification. For example, the Al-rich Fcc solution phase is the first solid phase to crystallize upon cooling from a wide range of Al-rich liquid alloys. The isotherms mark the temperature at which solidification starts. The thick lines separating adjoining regions denote liquid phase compositions

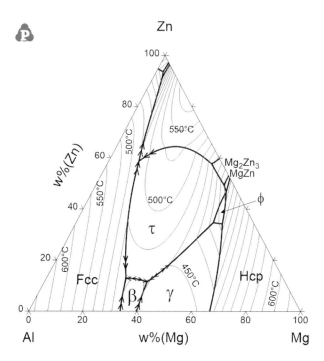

**Figure 5.13** Liquidus projection of the Al–Mg–Zn system calculated using the thermodynamic data of PanMg based on Liang et al. (1998b). The double-arrows indicate decreasing temperature along the monovariant lines, shown in the Al-rich region only.

in equilibrium with the two solid phases along the monovariant temperature. For example, the line originating at 66Al–34Mg (wt.%) and 450°C from the Al–Mg binary describes the Liquid + Fcc + β equilibrium that terminates at the invariant ternary eutectic equilibrium Liquid = Fcc + β + τ at 447°C with liquid composition 57Al–30Mg–13Zn (wt.%).

The composition of the solid phases is not displayed in Figure 5.13. The β phase, often denoted as $Mg_2Al_3$, is modeled as $(Mg)_{89}(Al,Zn)_{140}$. Therefore, its stoichiometry in the binary is more Al-rich, $Mg_2Al_{3.146}$. Moreover, it dissolves Zn by substitution of Al, so its composition in the ternary varies along a straight line at constant 38.86 at.% Mg. Note that here this line is not parallel to the Al–Zn edge because Figure 5.13 is plotted on wt.% scale. In this scale, the composition of β phase $(Mg)_{89}(Al,Zn)_{140}$ also varies along a straight line, but is tilted in the ternary, from 63.6Al–36.4Mg (wt.%) to 80.9Zn–19.1Mg (wt.%). Additional relevant information can be obtained from this liquidus projection as explained for the Mg–Al–Zn system in Schmid-Fetzer (2014).

The *solidification path* of an alloy describes the complete sequence of phase formation from a single-phase liquid alloy to the final solidified phase assembly. It ranges from the liquidus to the solidus temperature, while the liquid phase composition generally varies for binary or multicomponent alloys. In context with a liquidus projection, this variation of the residual liquid composition is most instructive.

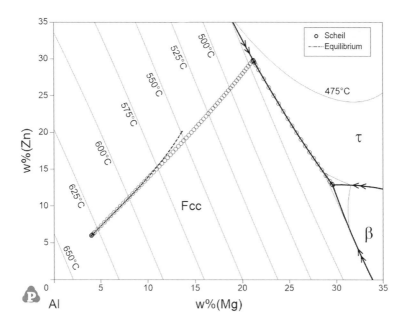

**Figure 5.14** Al-rich corner of Al–Mg–Zn liquidus projection from Figure 5.13; superimposed are the solidification paths of alloy Al–4Mg–6Zn (wt.%) under equilibrium and Scheil conditions.

In Figure 5.14, the solidification path for the ternary example alloy Al–4Mg–6Zn (wt.%) is superimposed on the Al-rich part of the liquidus projection of the Al–Mg–Zn system from Figure 5.13. Because of the restricted composition range, it is highly recommended to use the rectangular scale and not some part of the triangle of Figure 5.13. In these simple Cartesian coordinates, each point is plotted with Mg content as abscissa and Zn content as ordinate, and the Al content adds up to 100%. The equilibrium solidification path of alloy Al–4Mg–6Zn starts at this composition and 630°C with $f^{\text{Liquid}} = 1$ and moves down in temperature on the Fcc primary field of the liquidus surface along a slightly curved line, shown dash-dotted in Figure 5.14. It ends at 549°C with $f^{\text{Liquid}} = 0$. This last droplet of liquid at 66Al–14Mg–20Zn is in equilibrium with the completely solidified Fcc, $f^{\text{Fcc}} = 1$, of the original alloy composition 90Al–4Mg–6Zn. During that solidification, the Fcc phase moves along the solidus surface, not shown here. It starts at 96.5Al–1.2Mg–2.3Zn with the first precipitating Fcc crystal. Significant solid-state diffusion is required in this *equilibrium* solidification to avoid segregation along the path to attain the homogeneous 90Al–4Mg–6Zn composition in the final Fcc phase.

Also shown in Figure 5.14 is the *Scheil* solidification path, where it is assumed that any solid-state diffusion is completely blocked. That path continues along a less curved line in *primary* solidification until it hits the Liquid + Fcc + $\tau$ monovariant line at 470°C. There it bends and in the *secondary* solidification both Fcc and $\tau$ crystallize until the path hits the ternary eutectic. This invariant eutectic reaction Liquid → Fcc + $\beta$ + $\tau$ at the constant temperature of 447°C terminates solidification.

The Scheil conditions, explained in more detail in Section 5.5.2.3, provide one extreme scenario, and the real path is expected between this and the equilibrium path, which is often called the "lever rule solidification path." For many Al and Mg alloys, the formation of the as-cast constitution is often better predicted approximately by the Scheil conditions. More examples are given for the Al–8Mg and other Al alloys in Sections 5.5.2.3 and 5.5.3.

### 5.5.2.2    Metastable Phase Diagram Calculations

The metastable phase diagram with the highest industrial relevance is found in the iron–carbon system forming the basis of steel and cast iron. We start with the stable Fe–C phase diagram, as shown in Figure 5.15a, which is calculated using the thermodynamic description of Chen et al. (2019). The C-rich part above 30 at.% C is cut off in the diagram; it contains only the continuation of the two-phase regions with graphite at 100% C, where the single-phase graphite with negligible Fe-solubility exists. Therefore, the tie lines simply extend to pure carbon from the three phase boundaries, Liquid/(Liquid + graphite), $\gamma/(\gamma + \text{graphite})$, and $\alpha/(\alpha + \text{graphite})$. With this in mind, the phase diagram is simple to read with the eutectic of Liquid $= \gamma + \text{graphite}$ at 1,153.5°C, the eutectoid of $\gamma = \alpha + \text{graphite}$ at 738.0°C and the peritectic of Liquid $+ \delta = \gamma$ at 1,489°C. Unfortunately, most textbooks show the diagram without this explicit information on graphite and moreover in a merged diagram with the "metastable Fe–C" phase boundaries superimposed as some dashed lines. That is difficult to understand for nonexperts.

We look at the metastable Fe–C phase diagram in the separate Figure 5.15b. The meaning of metastability was explained earlier for the 0D calculation. In the Fe–C system, the phase graphite is known to form retarded due to slow nucleation kinetics and, under cooling rates in typical casting conditions of cast iron, does not form at all. In the as-cast microstructure, another C-bearing phase is found, the $Fe_3C$, called cementite, which is the "next" stable phase. This real process is best pictured by suspending the stable phase graphite in the phase diagram calculation, resulting in the metastable Fe–C phase diagram in Figure 5.15b. Technically, the pure carbon in the diamond phase must also be suspended. Comparison of both Fe–C diagrams in Figure 5.15 reveals that the stable eutectic is now replaced by the metastable eutectic, Liquid $= \gamma + Fe_3C$ at 1,148.8°C. Considering a so-called *hypoeutectic* alloy with less than 16.9 at.% Fe, which enters the eutectic from the Liquid $+ \gamma$ region, we see that only a small supercooling of 4.7 K is sufficient to overrun the stable eutectic and enable the formation of $Fe_3C$. For a *hypereutectic* alloy with more than 16.9 at.% Fe, the necessary supercooling is larger and increases with C content, as seen from the difference of the stable liquidus line of graphite and the metastable one of $Fe_3C$. For not too large supercooling, the graphite formation is still hindered even at the low cooling rates of sand casting, and thus only $Fe_3C$ is formed. The metastable eutectic, Liquid $\rightarrow \gamma + Fe_3C$, forms a particular microstructure, so-called *ledeburite* as found in alloys denoted as *white iron*. The microstructure formed at the metastable eutectoid, $\gamma \rightarrow \alpha + Fe_3C$, is denoted as *pearlite*. By heat treatment of any alloy containing $Fe_3C$, for example at 1,000°C, the equilibrium phase graphite forms by decomposition of

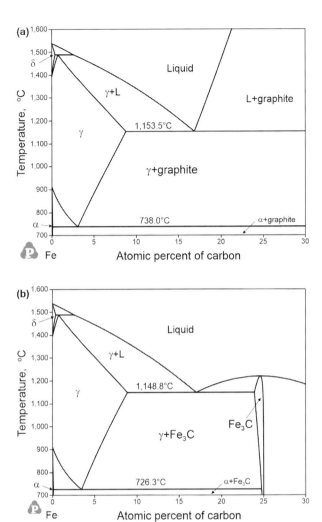

**Figure 5.15** The Fe–C phase diagrams in the Fe-rich region: (a) stable phase diagram (graphite formation), (b) metastable phase diagram (Fe$_3$C formation), calculated from the thermodynamic dataset in Chen et al. (2019).

Fe$_3$C. The formation of graphite in so-called *gray iron* cast alloys is generally favored by addition of silicon in ternary Fe–C–Si alloys and is explained by comparison of the stable versus metastable ternary Fe–C–Si phase diagram (Schmid-Fetzer, 2014).

Note that the peritectic equilibrium, Liquid $+ \delta = \gamma$ at 1,489°C, is exactly the same in both Figures 5.15a and 5.15b. Simply speaking, the C atoms dissolved in liquid, $\delta$, and $\gamma$ phases do "not know" if they would precipitate at higher concentration as graphite or cementite beyond the solubility limits. This and some more aspects of the stable and metastable phase diagrams are explained by Schmid-Fetzer (2014), where an earlier Fe–C dataset was used (Gustafson, 1985), resulting in small differences to the phase diagrams in Figure 5.15.

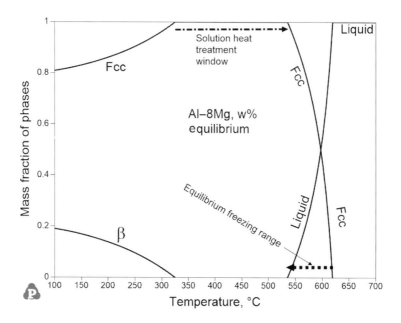

**Figure 5.16** Phase fractions developing in equilibrium for the fixed alloy composition Al–8 wt.% Mg (see Figure 5.14) calculated using the thermodynamic data of Liang et al. (1998b).

### 5.5.2.3    Property Diagram Calculations

Property diagrams result from a 1D or line calculation as defined previously, such as the phase fraction chart in Figure 5.16. It shows the calculated equilibrium phase fractions for the alloy Al–8Mg (wt.%). Phase fractions are dimensionless. However, different values are obtained if molar [mol/mol] or mass [kg/kg] fractions are used. When reporting quantitative data, it is necessary to specify the type, as done in Figure 5.16. Volume fractions of phases can be calculated only from an extended CALPHAD dataset including the molar volume data of all phases.

The phase fraction chart is wrongly denoted in some publications as a phase diagram. However, the phase fraction chart is related to the phase diagram. The vertical arrow in Figure 5.12 passes through four different phase regions, clearly telling the type of phases and the phase fraction could be obtained by graphical application of the lever rule (Schmid-Fetzer, 2014). This procedure can be tedious, and for a multicomponent system it is only possible on a phase diagram section where all tie lines are in the plane of the section. For example, the phase boundary Liquid/(Fcc + Liquid), the *liquidus line*, is intersected by the vertical arrow at 619°C and 92Al–8Mg, the *liquidus point* of alloy Al–8Mg. The first Fcc crystal precipitates with composition 97.4Al–2.6Mg as shown by the dotted tie line at 619°C in Figure 5.12. The same liquidus point is found in Figure 5.16 with $f^{\mathrm{Liquid}} = 1$ and $f^{\mathrm{Fcc}} = 0$ at 619°C. During solidification, the fraction of liquid shrinks while that of Fcc grows until the solidus point is reached at 535°C with $f^{\mathrm{Liquid}} = 0$ and $f^{\mathrm{Fcc}} = 1$. Here the tie line connects liquid(78Al–22Mg) and Fcc(92Al–8Mg). In this *equilibrium freezing range,* the composition variation of the phases could be read from the phase diagram by plotting the tie line at each temperature or, more simply, by plotting it from the 1D calculation of alloy Al–8Mg. These and other property data are also stored in the computer memory for some CALPHAD software after the 1D calculation and can be plotted without performing a new 1D calculation for the same alloy. The beauty of the CALPHAD calculation is that the same type of easily readable phase fraction, phase composition, or other chart is obtained also from the 1D calculation of a complex multicomponent alloy.

Below the solidus point, the single-phase Fcc region with $f^{\text{Fcc}} = 1$ extends down to 325°C, the *solvus point* of alloy Al–8Mg, where Fcc equilibrates with the phase β ($\text{Mg}_2\text{Al}_3$). The temperature range from solidus to solvus is the *solution heat treatment window*, which is wide for this alloy. Phases collected during nonequilibrium solidification in the as-cast alloy may be removed by long enough heat treatment in this range. Below the solvus temperature, the β phase precipitates from Fcc in a solid-state reaction and the equilibrium phase fractions change accordingly in Figure 5.16.

Another example of a property diagram is shown in Figure 5.17. It is obtained from a 1D calculation from pure In to pure Li at 900 K (Zhou et al., 2020). The system In–Li is relevant for the design of Li–ion battery (LIB) materials. For simplicity, only the calculation at 900 K is shown, which is just in the complete single-phase liquid region as seen from the phase diagram. The additional ones at lower temperatures are given by Zhou et al. (2020). For battery design, the open-circuit voltage (OCV) is the most interesting thermodynamic equilibrium property, related to the chemical potential or activity defined in (5.33). The OCV is the same as the EMF, defined as $E$ in the Nernst equation:

$$\mu_{\text{Li}}^{\text{Liquid}} - G_{\text{Li}}^{0,\text{Liquid}} = -1 \cdot F \cdot E, \tag{5.56a}$$

where "1" is the number of electrons given by the $\text{Li}^+$ ion and $F$ is Faraday's constant. In the present case, this means that $E$ is the OCV measured between the electrodes of

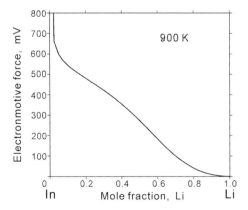

**Figure 5.17** Calculated EMF values at 900 K for liquid In–Li alloys with reference to pure liquid Li at the same temperature. Adapted from Zhou et al. (2020) with permission from Elsevier.

an electrochemical cell where the working electrode is the liquid In–Li alloy, separated from the reference electrode (pure liquid Li) by an electrolyte that only $Li^+$ ions can pass, as detailed for the Si–Li system by Liang et al. (2017). They also showed that it is useful to plot the phase diagram of the alloy system as $T - \mu_{Li}$ graph with an alternative $T - E$ scale. If the alloy is also pure Li, then $E = 0$, as in Figure 5.17. Moreover, $E \rightarrow \infty$ if the alloy is pure In, as also seen from (5.33) with $a_{Li}^{Liquid} = 0$. A reference state must be given for $E$ to be useful, just as discussed for the activity. In real Li–ion battery technology, pure Li is not used as an anode, but another host material, a still mainly graphite-based solid solution, stores the Li via intercalation mechanism (McDowell et al., 2013). The EMF obtained in thermodynamic experiments, such as In–Li, usually at higher temperature, is converted by knowing the EMF of the (graphite)–Li cell and taking the difference of (5.56a) and (5.56b):

$$\mu_{Li}^{Graphite} - G_{Li}^{0,Liquid} = -1 \cdot F \cdot E. \tag{5.56b}$$

Many other CALPHAD applications are based on 1D calculations and diagrams for other properties, such as enthalpy exchange for energy storage materials and hazard safety of batteries, or partial pressures (activities) for material processing with reactive gases, such as nitridation or carburization, where the carbon activity is also the instructive state variable in phase diagrams. Another example is the successful prediction of the single- and multiphase regions for the equiatomic five-component alloy CoCrFeMoNi, a so-called high-entropy alloy (HEA), from the phase fraction CALPHAD calculation (Saal et al., 2018).

## 5.5.3    Applications Based on Scheil Simulations

The Scheil simulation is a key CALPHAD method applied to alloy casting but also important for heat treatment, as summarized in Table 5.4. The Scheil approximation, occasionally called the Scheil–Gulliver approximation, was originally provided with an explicit equation for binary alloys (Scheil, 1942) and earlier without that equation by Gulliver (1913). This approximation comprises five assumptions: (i) local equilibrium at the liquid/solid interface; (ii) perfect mixing in the liquid phase; (iii) solid-state diffusion being completely blocked during and after solidification; (iv) a constant distribution coefficient, $k_0$, of the solute; and (v) only a single solid phase that crystallizes. See Section 5.5.2.1 for the definition of $k_0$, and the comment on the concentration variables. Under these five assumptions, the Scheil equation predicts $c_s$, the concentration of (dilute) solute in the single-phase "solid" as function of a fraction solid, $f^s$, for a given alloy concentration of $c_0$ (Schmid-Fetzer, 2014):

$$c_s = c_0 k_0 (1 - f^s)^{(k_0 - 1)}. \tag{5.57}$$

However, this classic Scheil equation, found in many textbooks on solidification, cannot be applied to real-world scenarios of multicomponent alloys and more than one solid phase precipitating. This is sometimes claimed differently, such as for Al–Si–Ni alloys (Jung et al., 2019). Two reasons are opposing this: (1) Even if "ternary values of $k_0$" are used, the tie lines will not stay in plane during this ternary primary solidification, violating assumption (i); and (2) any secondary solidification path involving additional phases, such as $Al_3Ni$ and (Si) for Al–Si–Ni alloys, can never be covered by (5.57).

The CALPHAD approach solves the problem numerically by retaining the basic Scheil assumptions (i)–(iii) but releasing assumptions (iv) and (v). As shown in Section 5.5.2.1 for the example tie line in Figure 5.12, the concept of constant distribution coefficient, $k_0$, assumes a "linearized" phase diagram and dilute solvent. More importantly, the step toward multicomponent alloys, not covered by the Scheil equation, is successfully done in CALPHAD software without using the concept of "distribution coefficient(s)" or any assumption on these "$k$"-values. A stepwise solidification of a thin (infinitesimal) layer of solid phase(s) from the current local equilibrium at the liquid–solid(s) interface is simulated in a 0D calculation step. The residual liquid is reduced in amount and changed in composition accordingly. For the Al–8Mg example in Figure 5.12, the residual liquid is enriched in Mg. That is repeated in the next step starting with the homogeneous residual liquid. There is no assumption on the number of phases crystallizing in the 0D step, and even primary intermetallic phases may occur.

The sequential (infinitesimal) layers are frozen in and do not change in composition during the simulation, reflecting the blocked solid-state diffusion. The result approximates the as-cast constitution of any alloy with the sequence of phase formation, their fractions, and the composition gradients (segregation) in the solid phases, mainly in the solute base metal, as a function of either temperature or total fraction solid under the approximations (i)–(iii). The main limitations of this kind of Scheil solidification simulation are discussed in Schmid-Fetzer (2014).

Figure 5.18 shows the result of this Scheil simulation for the binary alloy Al–8Mg (wt.%). For comparison, the result of the equilibrium solidification of the same alloy from Figure 5.16 is also superimposed as dashed lines. Exactly the same liquidus point is obtained because the Scheil simulation does not involve any supercooling due to nucleation problems of the solid phase. However, with growing fraction of Fcc (or "number of thin Fcc layers solidified"), the impeded solute redistribution in the solid phase results in a deviation. The difference shown in the (mass) fraction solid, $f^{fcc}$, may be understood by considering the first two steps simplified with rounded values and large temperature steps of a few K. The phase fraction of initial crystal, just saturated at the liquidus point with $w_{Mg}^{Fcc} = 2.6\%$, is still zero. Assume that the first solid layer forms with $w_{Mg}^{Fcc} = 2.8\%$ and the second with $w_{Mg}^{Fcc} = 3.0\%$. The second one is in local equilibrium with Liquid at $w_{Mg}^{Liq} = 9.3\%$. For simplicity, assume the average – or total – solid composition is $w_{Mg}^{Fcc-total} = 2.9\%$ The $f^{Fcc-total}$ at this point is calculated by the lever rule:

$$f^{Fcc-total} = (9.3 - 8.0)\%/(9.3 - 2.9)\% = 0.203. \tag{5.58a}$$

If we look at the equilibrium solidification at the same temperature, we have also $w_{Mg}^{Liq} = 9.3\%$; however, the "solid layers" are now fully homogenized by solid-state diffusion, having also reached $w_{Mg}^{Fcc} = 3.0\%$. This is now not only the local but also the *complete* equilibrium of the alloy at this point, and the Fcc fraction is

$$f^{Fcc} = (9.3 - 8.0)\%/(9.3 - 3.0)\% = 0.206 \qquad (5.58b)$$

Therefore, $f^{Fcc}$(Equilibrium solidification) $> f^{Fcc-total}$(Scheil solidification) and vice versa for the fraction of residual liquid phase, exactly as shown in Figure 5.18.

Now we see clearly the distinction between Scheil and equilibrium solidification: Under Scheil conditions, the lever rule is also applied but to the total (average) solid composition, which differs from the solid composition at the interface to liquid. This should be kept in mind when using the designation *lever rule solidification* for the equilibrium solidification.

At the equilibrium solidus temperature of 535°C, $f^{Liquid}$(Equilibrium) $= 0$, the total solid composition has not yet reached the alloy composition and a significant amount of residual liquid exists, $f^{Liquid}$(Scheil) $= 0.21$. The Scheil solidification path proceeds until the eutectic reaction Liquid $\rightarrow$ Fcc $+ \beta$ at 450°C terminates solidification where $f^{Liquid}$(Scheil) drops from 0.1 to zero; see also Figure 5.12. The phase fractions in Figure 5.18 are plotted down to an arbitrary 420°C to show the eutectic fractions better. The final $f^{Fcc-total}$(Scheil) $= 0.094$ comprises both the $f^{Fcc-total}$(Scheil, primary) $= 0.0906$ and the additional Fcc in the eutectic.

The Scheil solidification path is distinct from the equilibrium path by three main points: (1) occurrence of the eutectic in the as-cast microstructure, (2) much wider

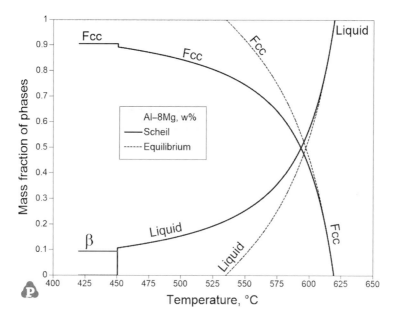

**Figure 5.18** Phase fractions developing in Scheil solidification simulation for the fixed alloy composition Al–8 wt.% Mg calculated using the thermodynamic data of Liang et al. (1998b). The equilibrium fractions from Figure 5.16 are superimposed as dashed lines.

freezing range of 169 K compared to 84 K, and (3) segregation profile of Mg in Fcc from $w_{Mg} = 2.6\%$ (in the core of an Fcc grain) to $w_{Mg} = 15\%$ (at the border of the Fcc grain to the eutectic microstructure). At this point, $w_{Mg}^{Fcc-total} = 5\%$, and the "missing 3 wt.% Mg" are found in the eutectic (Fcc $+ \beta$).

This nonequilibrium occurrence of $\beta$ phase in the eutectic is often detrimental to the mechanical alloy properties because the crystallization of $\beta$ from the liquid phase at high $T$ generally results in coarse microstructure. This may be healed by heating the alloy in the solution heat treatment window shown in Figure 5.16. The $\beta$ phase may be completely dissolved, and single-phase Fcc is obtained after long enough heat treatment. The alloy may then be quenched to shock-freeze in the Fcc state and subsequent heat treatment by artificial aging below the solvus temperature of $\beta$. In this solid-state reaction, the $\beta$ phase precipitates from Fcc in much finer microstructure compared to eutectic $\beta$. The hardness and strength of the alloy may increase due to this solid-state precipitation, as detailed in Section 5.6.3.

The Scheil simulation will give exactly the same results as the equilibrium solidification simulation if all crystallizing solid phases exhibit negligible solid solubility and only *eutectic-type* reactions occur in the phase diagram.

*Peritectic* or *transition-type* reactions, such as $L + \alpha \rightarrow \beta$ or $L + \alpha \rightarrow \beta + \gamma$, may follow after primary solidification of $\alpha$. That implies *consumption* of primary $\alpha$ while the growing $\beta-$ or $(\beta + \gamma)$ layer forms a solid-state barrier between $L$ and $\alpha$. Therefore, these reactions require solid-state diffusion to proceed, and they will be overrun under Scheil conditions. Solidification simply proceeds in the subsequent $L \rightarrow \beta$ or $L \rightarrow \beta + \gamma$ regions, leaving an unconsumed nonequilibrium rest of phase $\alpha$ behind. That is entirely different for the eutectic reaction type, $L \rightarrow \alpha + \beta$ or $L \rightarrow \alpha + \beta + \gamma$, which is a *decomposition-type* reaction. It does not require solid-state diffusion and proceeds also in real castings easily with finer microstructure at increased cooling rates (Schmid-Fetzer, 2014).

The Scheil and equilibrium paths are distinctly different for Al-rich ternary Al–Mg–Zn alloys just because of the significant Fcc solid solution range. For alloy Al–4Mg–6Zn, which was shown in Figure 5.14 and discussed partially in Section 5.5.2.1, the key distinctions of the two paths are analogous to the previous binary example. The equilibrium path of alloy Al–4Mg–6Zn produces only primary single-phase Fcc of that composition in the microstructure after a narrow freezing range of 81 K. The Scheil solidification path of alloy Al–4Mg–6Zn is different from the equilibrium path in three aspects: (1) After the primary Fcc, the secondary eutectic (Fcc $+ \tau$), and the tertiary eutectic (Fcc $+ \beta + \tau$) are predicted in the as-cast microstructure, (2) the freezing range of 183 K is much wider, and (3) the segregation profile in Fcc ranges from Al–1.2Mg–2.3Zn (in the core of an Fcc grain, wt.% throughout) to Al–11.9Mg–2.4Zn (at the border of the Fcc grain to the final eutectic microstructure). At this point, the total (average) Fcc composition is Al–2.5Mg–3.7Zn, and the balance to attain the overall Al–4Mg–6Zn composition is found in the eutectic phases $\tau$ and $\beta$.

For many alloys, the formation of the as-cast constitution is often better predicted approximately by the Scheil conditions than equilibrium condition, especially at higher cooling rates observed in thin-walled or die castings. While the main application of Scheil simulation is the prediction of solidification path and as-cast constitution, it is also important to predict the *incipient melting* temperature. Under the assumption of completely blocked solid-state diffusion during and after solidification, the solidification path may also be read in the reverse direction and on heating the nonequilibrium constitution of alloy Al–8Mg. The first droplet of liquid will form in the reverse eutectic reaction Fcc + $\beta$ → Liquid at 450°C, well below the equilibrium solidus of 535°C. That is very important for the heat treatment process. Heating the as-cast alloy rapidly above 450°C may result in formation of the initial liquid phase at the eutectic grain boundaries. This incipient melting must be avoided to maintain the integrity of the solid part during this process. On the other hand, the higher temperature will shorten the time required for dissolution of the $\beta$ phase, saving substantial cost. A smart solution is two-stage heat treatment, where in a first stage below the Scheil solidus (here $T < 450°C$), the main dissolution of the $\beta$ phase is achieved, and in a second stage below the equilibrium solidus (here $T < 535°C$) the final homogenization is achieved.

A better approximation of the *incipient melting* temperature is the nonequilibrium solidus temperature (NEST), where 2% of residual liquid exists, $f^{\text{Liquid}}(\text{Scheil}, T_{\text{NEST}}) = 0.02$, as suggested by Ohno et al. (2006b). The first reason is that a finite amount of liquid is required for some effect, arbitrarily set to 2%, as opposed to the Scheil solidus at $f^{\text{Liquid}}(\text{Scheil}) = 0$. The second reason is that in many multicomponent alloys, the $f^{\text{Liquid}}(\text{Scheil}, T)$ curve fades out down to very low temperature with a minute fraction of liquid. That is not the case for the example alloy Al–4Mg–6Zn, for which $T_{\text{NEST}} = 459°C$ with small difference to the true Scheil solidus at 447°C. For a general application, the concept of $T_{\text{NEST}}$ is recommended, and it was shown by comparison to experimental data on incipient melting in quaternary Mg–Al–Mn–Zn alloys that two groups of alloys can be identified: For the higher alloyed group, comprising also important Mg alloys AZ91 and AZ62, $T_{\text{NEST}}$ is a very good prediction of the experimental incipient melting. This prediction based on Scheil conditions is even valid for cooling rates as low as 1 K/min. For the low-alloyed composition range, the solidification proceeds closer to equilibrium conditions, with experimental incipient melting temperature significantly above $T_{\text{NEST}}$. Moreover, the predictions show a larger uncertainty defined by a range of 1–3% of residual liquid for the calculation of $T_{\text{NEST}}$ (Ohno et al., 2006b).

A recent example for the successful design of new die-cast Al alloys supported by Scheil simulations in the Al–Si–Mg–Mn–Fe system is given by Cai et al. (2020). The strategy involved prediction of the fraction of eutectic, aiming at promising mechanical properties reported for ultrafine eutectic and hypoeutectic multicomponent alloys. They started by calculation of the composition of completely eutectic ternary alloy as Al–13.9Si–5.55Mg (wt.%). The addition of 0.5 wt.% Mn was intended to prevent die soldering and to enable the formation of strengthening phases with Mn, and some 0.15 wt.% Fe is considered typical impurity in high-pressure die casting (HPDC). This is alloy C in their study, aiming at eutectic fraction of 1.0. From that base, two hypoeutectic alloys with calculated eutectic fractions of 0.5 (alloy B) and 0.3 (alloy

A) were also cast and characterized in detail. They have shown only a single graph presenting $f^{\text{Solid}}$ versus $T$ and the development of a total fraction solid, as a result of their Scheil calculations using a rather early version of Pandat 8, and do not mention the database (Cai et al., 2020).

Instead of the $f^{\text{Solid}}$ versus $T$ chart, it is recommended to use the phase fraction chart, such as in Figure 5.18, since it provides detailed information on the type and amount of primary, secondary, and subsequent phases in a single graph. This result is shown in Figure 5.19 for exactly the same alloys (A, B, C) studied by Cai et al. (2020) from Scheil simulations using Pandat and PanMg database (version 2021). Details of the alloys are listed in Table 5.5.

Figure 5.19 is given in logarithmic scale for the phase fractions to resolve the minor phases with a cutoff at $f^{\varphi-total} < 0.001$. Here $f^{\varphi-total}(T)$ denotes the accumulated or total fraction of any solid phase $\varphi$ during the Scheil simulation. For the leanest alloy A (see Figure 5.19a), the crystallizing phase sequence in the solidification path is given in (5.59):

$$L \rightarrow \text{Fcc}, \qquad\qquad\qquad \text{Start at } 618°\text{C, primary Al (Fcc)}$$
$$(5.59a)$$

$$L \rightarrow \text{Fcc} + \text{Al}_{18}\text{Mn}_4\text{Si}, \qquad\qquad \text{Start at } 608°\text{C, secondary eutectic}$$
$$(5.59b)$$

$$L \rightarrow \text{Fcc} + \text{Al}_{18}\text{Mn}_4\text{Si} + \text{Al}_{60}\text{Fe}_{15}\text{Si}_{25}, \qquad \text{Start at } 574°\text{C, tertiary eutectic}$$
$$(5.59c)$$

$$L \rightarrow \text{Fcc} + \text{Al}_{18}\text{Mn}_4\text{Si} + \text{Al}_{60}\text{Fe}_{15}\text{Si}_{25} + \text{Mg}_2\text{Si}, \quad \text{Start at } 566°\text{C, fourth eutectic}$$
$$(5.59d)$$

$$L \rightarrow \text{Fcc} + \text{Al}_{18}\text{Mn}_4\text{Si} + \text{Mg}_2\text{Si} + \text{Al}_8\text{FeMg}_3\text{Si}_6, \quad \text{Start at } 563°\text{C, fifth eutectic}$$
$$(5.59e)$$

In Figure 5.19a, the start temperatures of the various solidification stages at phase fraction zero of a new phase may appear lower than given in (5.59a–e) because of the cutoff in the logarithmic scale. Note that the minute fraction of $\text{Al}_{60}\text{Fe}_{15}\text{Si}_{25}$ is constant below 563°C where the fifth eutectic starts, as seen by the growth of the four solid phase fractions of (5.59e) and Figure 5.19a, while the liquid fraction decreases. That indicates a transition-type reaction at 563°C, where $\text{Al}_{60}\text{Fe}_{15}\text{Si}_{25}$ would be consumed under equilibrium conditions. It is overrun under Scheil conditions, leaving the rest of nonequilibrium $\text{Al}_{60}\text{Fe}_{15}\text{Si}_{25}$ behind. Finally, the only nonvariant eutectic is reached, which is correctly a six-phase equilibrium in the five-component system Al–Si–Mg–Mn–Fe at constant pressure:

$$L \rightarrow \text{Fcc} + \text{Si(diamond)} + \text{Mg}_2\text{Si} + \text{Al}_{18}\text{Mn}_4\text{Si} + \text{Al}_8\text{FeMg}_3\text{Si}_6, \text{ at } 559.6°\text{C}$$
$$(5.59f)$$

The fraction of this final eutectic in the microstructure (i.e., the sum of the five phases formed at 559.6°C) is simply given by the fraction of liquid at the start of the

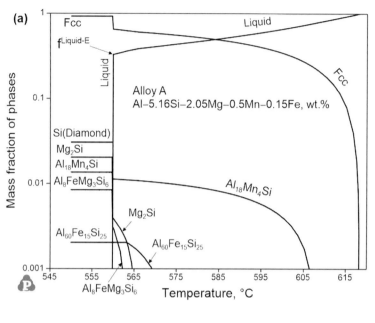

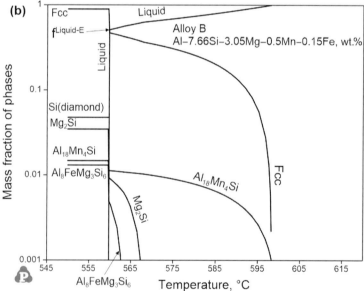

**Figure 5.19** Scheil solidification paths for the three Al–Si–Mg–Mn–Fe alloys studied by Cai et al. (2020) for the design of new die-cast Al alloys based on the CALPHAD method; the present calculations used Pandat the with current database PanMg2021 (www.computherm.com) for the alloys: (a) A, (b) B, and (c) C with compositions listed in Table 5.5.

eutectic reaction (5.59f), $f^{\text{Liquid}-\text{E}} = 0.3288 \text{ kg/kg} = 0.3289 \text{ mol/mol}$, for alloy A, as given in Table 5.5 and marked in Figure 5.19 for all three alloys. It is a coincidence that these mass and molar fractions are almost identical. For example, the fraction of

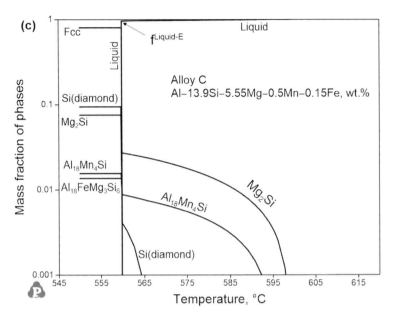

Figure 5.19 (*cont.*)

the monovariant secondary eutectic in the microstructure is given by the following difference:

$$f^{\text{Liquid}}(608°C) - f^{\text{Liquid}}(574°C) = 0.783 - 0.428 = 0.355 \text{ kg/kg.}$$

Even though this appears as a large value, it may be hard to discern in the experimental microstructure because this "secondary eutectic Fcc," crystallizing jointly with $Al_{18}Mn_4Si$, may grow mainly in continuation on the primary Fcc, and also because the phase fraction ratio of $Fcc/Al_{18}Mn_4Si$ crystallizing in this segment is large, 34:1. This effect may also apply to the tertiary, fourth, and fifth monovariant eutectic reaction. However, the final eutectic reaction, (5.59f), is expected to show up clearly in the microstructure of all three alloys. It is reported that two different morphologies of eutectic structure are found in all three alloys, and the major phases predicted in the eutectic by the present calculations are also seen in the experimental microstructure (Cai et al., 2020). This termination of Scheil solidification in a final eutectic at constant temperature distinguishes these alloys from some multicomponent alloys where the $f^{\text{Liquid}}$(Scheil, $T$) curve fades out, as discussed earlier. The highest yield and ultimate tensile strengths are provided in the alloy B with elongation of 2.3%, while alloy A provides elongation of 4.3% at somewhat lower strength, all in the as-cast state (Cai et al., 2020).

More examples of Scheil and equilibrium calculations applied as a tool for the design of Al alloys are given by Chen et al. (2018), for Al and Mg alloys by Schmid-Fetzer and Zhang (2018), and also in Chapter 8 and other chapters of this book. The key advantage and reason for these frequent applications are that only the thermodynamic database of the alloy system and major CALPHAD software are required without the need for kinetic and other material data of the alloy system.

**Table 5.5** Composition and experimental eutectic fractions of the three Al alloys in the design study of Cai et al. (2020) compared with the eutectic fractions from the present Scheil simulation.

| | Composition, wt.% | | | | | Fraction of (final) eutectic | |
|---|---|---|---|---|---|---|---|
| Alloy | Si | Mg | Mn | Fe | Al | Present calc. $f^{\text{Liquid-E}}$, kg/kg | Experimental (Cai et al., 2020), Area % |
| A | 5.16 | 2.05 | 0.5 | 0.15 | Bal. | 0.33 | $24 \pm 8$ |
| B | 7.66 | 3.05 | 0.5 | 0.15 | Bal. | 0.51 | $42 \pm 7$ |
| C | 13.9 | 5.55 | 0.5 | 0.15 | Bal. | 0.96 | $\approx 89$ |

## 5.6    Alloy Design Applications Using Extended CALPHAD-Type Databases

### 5.6.1    Overview

Extended CALPHAD-type databases are constructed on the thermodynamic basis of assessed compilation of $G^\varphi$ functions, generally starting with molar volume $V_m^\varphi(T)$ and mobility data, as briefly described in Section 5.4. Together with the proper software, these data are sufficient for diffusion simulation. Additional kinetic and thermophysical material parameters are necessary for dedicated simulation of solidification and solid-state precipitation. In alloy design, these tools are mainly applied for casting and heat treatment processes, summarized in Table 5.6. In these *extended simulations*, not only the alloy composition but also the quantitative *temperature–time profile* of the process are input parameters of a design study.

The simulation methods for solidification and precipitation in Table 5.6 are both focused on statistical description of microstructure characteristics as opposed to simulating in detail how a single dendrite, grain, or precipitate evolves in space and time. That is the subject of mesoscale methods, such as phase-field or cellular automaton methods discussed in Chapter 3. These mesoscale methods also benefit from the use of extended CALPHAD-type databases. The depth of information gained from the more elaborate mesoscale methods is often not required for industrial alloy design applications.

In the more common dendritic solidification, it may be sufficient to know how the secondary dendrite arm spacing evolves in time and how back-diffusion changes the solidification path depending on cooling rate. In the more common aging heat treatment, the most important information about the precipitation process is how the particle size and particle size distribution evolves in time rather than monitoring an individual particle. Established models to predict mechanical properties, such as hardness and yield strength, are also based on these statistical precipitation data and implemented in such simulation software. Some examples for this kind of simulation are given in the following two sections.

The solid-state diffusion simulation listed in Table 5.6 again focuses in more detail on the composition changes and phase transitions occurring in one-dimensional space

**Table 5.6** Results from extended CALPHAD-based computational thermodynamics simulations for multicomponent and multiphase cast and heat-treated alloys in addition to Table 5.4; Some kinetic and thermophysical data of the alloy system are required in addition to the thermodynamic database.

| Process | Method | Result | Software and comments |
|---|---|---|---|
| Casting | Microsegregation simulation in solidification | Microstructure constituents | • PanSolidification[a]<br>• Back-diffusion in solid, undercooling, secondary dendrite arm coarsening<br>• Solidification path: Phase sequence, phase fractions, and compositions<br>• Segregation in phase composition |
| Heat treatment | Solid-state precipitation simulation | Microstructure constituents | • PanPrecipitation[a]<br>• TC-PRISMA[b]<br>• Concurrent nucleation, growth, and coarsening<br>• Particle size and particle size distribution evolution<br>• Hardness and yield strength evolution |
| Heat treatment | Solid-state diffusion simulation | Composition gradient and phase transformations | • DICTRA[b]<br>• PanDiffusion[a]<br>• Multicomponent and multiphase simulation in 1D space<br>• Particle dissolution, carburization, decarburization, homogenization, phase transformation, diffusion couple |

[a] www.computherm.com.
[b] www.thermocalc.com.

by diffusion-controlled growth or dissolution, such as in a diffusion couple. That has been extensively described in Chapter 6, and only one example application is mentioned here, concerning solution heat treatment. In industrial applications, the quantitative information on the time required for substantial dissolution of nonequilibrium phases in as-cast microstructures is crucial. For binary Mg–Al, this dissolution process was studied using the CALPHAD-based simulation tool DICTRA and also by an analytical micromodel based on the Johnson–Mehl–Avrami–Kolmogorov (JMAK)-type equation and validated by experimental work (Yao et al., 2020).

## 5.6.2 Simulation of Solidification

The basic features of microsegregation simulation in alloy solidification applied here are given as keywords in Table 5.6. A more detailed background is found in Chapter 8 of this book and in the description of the PanSolidification module of the Pandat software package (Zhang et al., 2020). The popular Al alloy A356 is chosen as one example presented in Figure 5.20. It shows the secondary dendrite arm spacing (SDAS) as a function of the cooling rate (*CR*) for the alloy A356, and the composition is specified in the figure caption. The growth of SDAS is also calculated in each

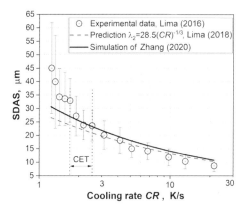

**Figure 5.20** Comparison between the calculated and experimental (Lima et al., 2018) secondary dendrite arm spacing (SDAS, $\lambda_2$) for alloy A356 (Al–7.235Si–0.336 Mg–0.148Fe wt.%) as a function of the cooling rate ($CR = dT/dt$), adapted from Zhang et al. (2020) with permission from Springer Nature.

simulation run at fixed $CR$, starting from an initial value taken as about twice the dendrite tip radius obtained also in the solidification model. Only the final values are shown in Figure 5.20. Instead of constant $CR$, a chosen temperature–time profile or measured cooling curve may be provided as one solidification condition, resulting in time-dependent values of $CR$. A second solidification condition must be specified, either the solidification rate ($V$, mm/s) or the temperature gradient ($G$, K/mm). These quantities are related by $CR = G \cdot V$. It is emphasized that the specified solidification conditions must reflect the real casting conditions as accurately as possible.

For the simulation performed by Zhang et al. (2020), as shown for alloy A356 in Figure 5.20, the values of $CR$ and $V$ are specified as solidification conditions for each simulation. This follows the experimental setup in the work of Lima et al. (2018), who used horizontal directional solidification with heat extraction only through the water-cooled lateral mold wall. Six thermocouples are located at distances from $z = 2$ to $z = 50$ mm from the mold wall monitored $T(t)$ during the casting experiment so that $CR$ and $V$ are determined as a function of position. The fitted function, $CR(z) = 35(z/\text{mm})^{-0.71} \text{K/s}$, is used to determine $CR$ at all positions of SDAS measurement. The variation of $CR$ in the experimental freezing range is not too strong, and the value at the separately measured liquidus temperature of 615°C was used by Lima et al. (2018) in their own evaluation of SDAS, which is given also in Figure 5.20 as a dashed line.

The same value of $CR$ and the related value of $V$ are taken as conditions for the PanSolidification simulation. Good agreement is obtained between the calculated and measured SDAS values at $CR \geq 1.9$ K/s. This is in the region of columnar to equiaxed transition (CET) of the microstructure. At lower values of $CR$, in the equiaxed part, the five measured values of SDAS are consistently above the simulated curve. However, the experimental uncertainties, which are shown by the reported error bar, are also significantly larger in this region. Exactly the same kinetic material parameters assessed for the Al–Si–Mg–Fe alloy system were also used to compare simulated and experimental SDAS for a slightly different alloy, Al–7Si–0.3 Mg–0.15Fe wt.% with a very good agreement (Zhang et al., 2020).

Additional example results obtained by this kind of microsegregation solidification simulation with PanSolidification are given in Section 5.7 and also in Chapter 8's Sections 8.2.1 and 8.3.1.2. Another microsegregation model based on CALPHAD-type thermodynamic data is also demonstrated to be successful for Al-rich alloys using the in-house software "Alstruc" applied in the Norwegian aluminum industry (Tang et al., 2018). Solidification microstructures and microsegregation of Al–Si–Mg alloys were also predicted using a two-dimensional cellular automaton model based on CALPHAD thermodynamic data by Fang et al. (2019).

## 5.6.3 Simulation of Heat Treatment

Both software packages listed in Table 5.6, PanPrecipitation and TC-PRISMA, can simulate solid-state precipitation involving concurrent nucleation, growth, and coarsening of precipitates in multicomponent and multiphase systems. The extended CALPHAD-type thermodynamic databases of the alloy system comprise molar volume and mobility data and additional material parameters. Some parameters are detailed in this section. They adopt the Kampmann–Wagner numerical (KWN) model (Kampmann and Wagner, 1984). Simulation results comprise particle size distributions (PSD) of various precipitate phases in addition to the temporal evolution of their average size and volume fractions as well as the compositions of matrix and the precipitate phase(s). More details are given in Cao et al. (2011, 2016) and Zhang et al. (2014, 2021).

One example result using TC-PRISMA for precipitation simulation of binary Mg alloys is given in Figure 5.21 (Zhang et al., 2021). The precipitation of $\gamma$ phase ($Mg_{17}Al_{12}$) at constant 170°C from an initially single-phase Mg (Hcp) matrix is simulated. This initial state is in practice created by a solution heat treatment, in this case for 24 hours at 410°C (Lee et al., 2013), generally followed by quenching and subsequent heating to the aging temperature. In this case, it is 170°C. The most important additional material parameter is the interfacial energy between precipitate and matrix, which was assumed as adjusted parameter, $\sigma(\gamma/Hcp) = 0.026 \text{ J/m}^2$ (Zhang et al., 2021), which is off the range 0.11–0.43 J/m$^2$ from the literature (Hutchinson et al., 2005), reported in the same work where the low value was suggested to be due to coherency (Zhang et al., 2021). Two additional parameters were adjusted: the "mobility enhancement factor," value $= 10$, and the "nucleation site," values $= 10^{21}$ for Mg–9Al and $10^{20}$ for Mg–6Al; the missing unit (Zhang et al., 2021) should be site/m$^3$. The agreement of the simulated volume fraction of $\gamma$ with the experimental data for the Mg–9Al alloy is better than that for the Mg–6Al alloy. A similar result is obtained for the two alloys Mg–8.8Al and Mg–5.9Al (wt.%) artificially aged at 200°C (Zhang et al., 2021).

The $\gamma$ phase is also the precipitate in the popular ternary Mg alloy AZ91, Mg–9Al–1Zn (wt.%), which was studied in the simulation work of Zhang et al. (2014) using

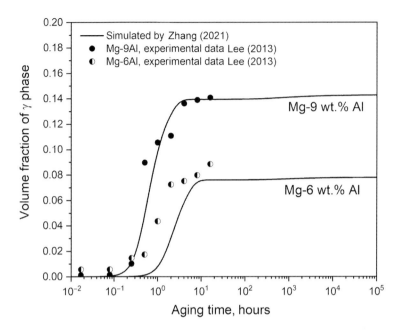

**Figure 5.21** Simulated volume fraction of $\gamma$-$Mg_{17}Al_{12}$ phase in Mg–9Al and Mg–6Al (wt.%) alloys during aging at 170°C with the experimental data (Lee et al., 2013), adapted from Zhang et al. (2021) with permission from Elsevier.

PanPrecipitation. In that work, a higher value of $\sigma(\gamma/Hcp) = 0.23\,J/m^2$ was used, according to the reported literature data (Hutchinson et al., 2005).

Also embedded in PanPrecipitation is a module to estimate the interfacial energy at the boundary between matrix and precipitate phase, using the generalized broken bond (GBB) method (Sonderegger and Kozeschnik, 2009a, b) for prediction. This may be used as an indication and guide for assessing this most critical kinetic parameter.

Based on the simulated evolution of PSD, both hardness and yield strength can be calculated, as shown in Figure 5.22 for the artificial aging of alloy Mg–3Nd–0.2Zn (wt.%) at three different heating temperatures using PanPrecipitation (Xia et al., 2018). The embedded mechanical property model comprises additional parameters, such as intrinsic strength, shear modulus, Burgers vector, and Taylor factor of the matrix phase, the critical radius of precipitate for shifting from shearing to looping mechanism, and other assessed parameters. That enables the prediction of yield strength and hardness quantitatively. As shown in Figure 5.22, the peak hardness shifts to shorter aging times at higher aging temperature.

The experimental data shown for comparison in Figure 5.22 are from alloy Mg–3Nd–0.2Zn–0.4Zr (wt.%) (Fu et al., 2008). For grain refinement of as-cast Mg alloys, the minor addition of some 0.5 wt.% Zr is common in Al-free and especially in Mg-RE (rare earth) alloys. This Zr is not involved in any phase transformation reactions, and its effect does not change during age treatment. To compare the simulated

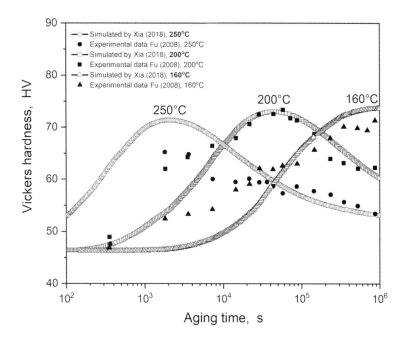

**Figure 5.22** Simulated age-hardening response of alloy Mg–3Nd–0.2Zn (wt.%) at 250, 200, and 160°C compared with the experimental data (Fu et al., 2008). Adapted from Xia et al. (2018) with permission from Elsevier.

mechanical properties with the experimental data, the intrinsic strength of the Hcp matrix was set to the yield strength of Mg–0.5Zr alloy in the work of Xia et al. (2018).

As the last example, the simulated time temperature–transformation (TTT) diagram of alloy Ni–14Al (at.%) is given in Figure 5.23. It follows the concept of solution heat treatment, quenching, and artificial aging at constant temperature. It is a truly isothermal transformation diagram instead of continuous cooling CCT diagram. Solutionizing is assumed to be performed in the single-phase Ni (Fcc) region above 821°C, the solvus point of the equilibrium precipitate phase L1$_2$ (Ni$_3$Al). Below 821°C, the Fcc phase is supersaturated and the driving force (*DF*) for precipitation increases (see Section 5.5.2). Therefore, the time to precipitate a given volume fraction of the L1$_2$ phase decreases up to the "nose" of the curves at some 750°C and then increases, mainly because the diffusion-controlled growth slows down with falling temperature. The times required to precipitate volume fractions of 2%, 10%, and 98% at 650°C are $10^{0.8}$ s, $10^{1.0}$ s, and $10^{5.5}$ s, respectively. These are the percentages relative to the equilibrium amount of the L1$_2$ (Ni$_3$Al) phase, often denoted as $\gamma'$.

The readers may perform the simulations for Figure 5.23 themselves using the databases and batch files provided as ancillary material and the PanPrecipitation module of the Pandat trial/Education software package that can be downloaded free of charge from CompuTherm's website (www.computherm.com). In the downloaded *Pandat Example Book*, the files and descriptions are included for a case with

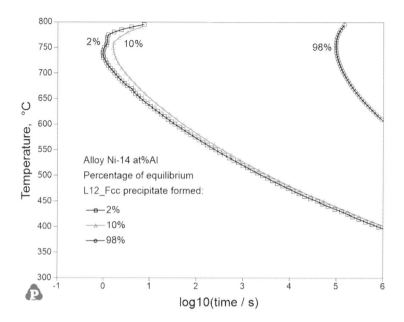

**Figure 5.23** The time–temperature–transformation (TTT) diagram of alloy Ni–14Al (at.%) showing the percentage of equilibrium phase L1$_2$ (Ni$_3$Al) formed after isothermal heat treatment starting from the supersaturated single phase Fcc. Calculations are performed with the PanPrecipitation module of the Pandat (version 2021) software package (www.computherm .com) and the database given in ancillary materials.

co-precipitation of different precipitates, $\gamma'$ and $\gamma''$, in the Ni–2.4Al–3.8Nb (at.%) alloy. This is a typical case producing a second "nose" in the TTT diagram due to the competing growth with two separate PSDs of the $\gamma'$ and $\gamma''$ precipitates.

More examples of the successful application of the simulation tools for such precipitation processes may be found in Section 6.9 on age hardening in AA6005 Al alloy, in Section 8.2.2 on the wrought Al alloy 7xxx series, in the work by Luo (2015); Schmid-Fetzer and Zhang (2018) on Mg and Al alloys; and Cao et al. (2011, 2016) and Zhang et al. (2018) on nickel-based superalloys.

## 5.7      A Case Study: CALPHAD Design of Al Alloys with High Resistance to Hot Tearing

The formation of shrinkage cracks during the solidification of alloys, often called hot tearing in casting and hot cracking in welding, is a serious defect that limits the suitability of the alloy. It causes costly repair measures or even scrapping the part. This issue may occur in the final stage of (local) solidification when obstructed solidification shrinkage and thermal contraction of mushy zones cannot be balanced by liquid feeding; a review is given by Li and Apelian (2011).

Experimental methods to determine the hot tearing susceptibility or crack suscepti-bility (CS) often use constrained rod casting and may be ex situ, where the CS is obtained from the length, location, and width of the crack formed after solidification (Cao et al., 2010). An alternative in situ method is to measure the temporal evolution of temperature together with load or displacement exerted by the shrinking sample during solidification (Li et al., 2011).

For Al alloy M206, it was found that at the time of crack initiation (load onset point) the measured temperatures at the center and surface of the rod are related to solid fraction ($f_S$) of 0.89 and 0.98, respectively, by Scheil simulations with Pandat (Li et al., 2011). In contrast to this alloy M206 (Al–4.55Cu–0.36Mn–0.25Mg, wt.%), no cracks were found in alloy A356 (Al–6.7Si–0.38 Mg, wt.%) under similar conditions. Trend correlations were also found between the CS determined ex situ and $f_S$ at the end of primary (dendritic) solidification obtained from Scheil simulations with Pandat for Mg alloys with Al and Sr (Cao et al., 2010). A study on Mg–Ca alloys using in situ constrained rod casting also confirmed a reasonably good correlation between experimental results and the CS obtained from the vulnerable time period and the $f_S$ from Scheil simulations (Song et al., 2015).

A new cracking criterion during alloy solidification has been derived by Kou (2015). Considering the $T$ versus $(f_S)^{1/2}$ profiles of adjacently growing grains near $(f_S)^{1/2} \approx 1$, it was shown that the steepness in the $T - (f_S)^{1/2}$ curves distinguishes shorter from longer feeding channels. The latter are more prone to cracking. This is suggested as a CS index that increases with the steepness, or absolute value, of slope $|d(T)/d(f_S)^{1/2}|$ near $(f_S)^{1/2} = 1$ by Kou (2015). This CS index, calculated from Scheil simulations with Pandat, was shown to compare well to the experimental data of some Al alloys (Kou, 2015).

In further advancement, the same definition of the CS index was applied. However, the $T - (f_S)^{1/2}$ curves were – in addition to the Scheil simulation – also calculated using kinetic solidification simulation for cooling rates of $CR = 100$ K/s and 20 K/s (Liu and Kou, 2017). The PanSolidification module of the Pandat software package with PanAl database (Zhang et al., 2020) was applied by Liu and Kou (2017). For each of the three ternary Al–Mg–Si, Al–Cu–Mg, and Al–Cu–Si alloy systems, a crack susceptibility map covering alloy compositions in the range from 0–5 wt.% was constructed for each value of $CR$.

It was shown that the regions of maximum values of the CS index for Scheil simulations do not correspond well with the experimental regions of high CS. With PanSolidification, however, even for the high $CR$ of 100 K/s and also for 20 K/s, this agreement is good, at least for the Al–Mg–Si and Al–Cu–Mg alloys. This major contribution (Liu and Kou, 2017) demonstrates that back-diffusion in the solid has a significant impact even at high $CR$, which was neglected in all previous studies that assumed Scheil conditions without any back-diffusion for the calculation of $f_S(T)$. Back-diffusion is one of the important aspects quantitatively considered in the micro-segregation solidification model implemented in PanSolidification, described in more detail in Section 5.6.2.

It was also suggested by Kou (2015); Liu and Kou (2017) that the maximum of $|dT/d(f_S)^{1/2}|$ before or at $(f_S)^{1/2} = 0.99$ should be used to define the CS index. That corresponds to a cutoff at $f_S = 0.98$, which is also well accepted in the

Rappaz–Drezet–Gremaud (RDG) criterion (Rappaz et al., 1999). With only 2% of residual liquid a sufficient interdendritic bridging is assumed and CS comes to an end (Kou, 2015; Rappaz et al., 1999). The calculations by Liu and Kou (2017) were done with a step size of 0.5 wt.%, which amounts to 121 calculations for a given ternary alloy system and $CR$. The CS index is in the order of some $10^3$ K; however, the scale was chosen differently for each of the CS index maps given by Liu and Kou (2017), which may impede the direct comparison. Further advancement of the method is suggested in the following to enable systematic applications to a wider range of alloy systems.

A more precise definition of the crack susceptibility index ($CSI$) of a given alloy at a fixed cooling rate $(CR = dT/dt)$ is given in (5.60):

$$CSI = \max \left\{ \left| \frac{dT}{df_S^{1/2}} \right| \right\} \quad \text{for } 0.7 < f_S^{1/2} < 0.99. \tag{5.60}$$

As an example, the solidification profiles, $T$ versus $(f_S)^{1/2}$, for five Al–Mg–Cu alloys are given in Figure 5.24 calculated with PanSolidification for $CR = 25$ K/s. Only the result in the region of interest, $0.95 < (f_S)^{1/2} < 1$, is shown. The short straight lines, shown as a guide for the eye, indicate the profile steepness, and the calculated values of $CSI$ are shown for the five alloys. It is seen that the maximum of

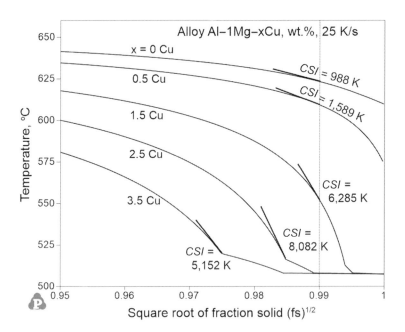

**Figure 5.24** Calculated solidification profiles, $T$ versus $(f_S)^{1/2}$, for Al–1Mg–$x$Cu (wt.%) alloys with five different Cu contents for fixed cooling rate $CR = dT/dt = 25$ K/s. The short straight lines show the profile steepness maximum values used for the definition of $CSI$ in (5.60). Calculation performed with PanSolidification module of Pandat (version 2021) software package (www.computherm.com) and the database is given in ancillary materials.

*CSI* occurs at the chosen limit of $(f_S)^{1/2} = 0.99$ for the three low-Cu alloys, whereas the maximum for the alloys with 2.5 and 3.5 wt.% Cu occurs at $(f_S)^{1/2} < 0.99$. As seen from the kinks in these profiles, this is where the primary solidification of Al(Fcc) ends and additional secondary phase crystallizes, $\theta$ (Al$_2$Cu) in this case. Solidification terminates at the ternary eutectic, Fcc $\rightarrow \theta +$ S(Al$_2$CuMg) at 508°C, for the three high-Cu alloys, whereas only a single Al(Fcc) phase formed during the solidification of the alloys with 0 and 0.5 wt.% Cu at this *CR*.

The lower limit in (5.60), not considered by Liu and Kou (2017), is crucial to exclude the early stage of solidification for alloys where high melting intermetallic phases crystallize as primary phase, before the main solidification of solid solution of the base metal, $\alpha$, starts. Examples are common Mg–Al–Mn alloys or common Al–Si–Ti alloys, where this effect must be also considered for calculation of the growth restriction factor (GRF) for multicomponent alloys (Schmid-Fetzer and Kozlov, 2011) which is the basis for grain refinement in the early stages of solidification (Easton and StJohn, 2001; StJohn et al., 2011). In general, the GRF is calculated from the initial slope of constitutional supercooling, $\mathrm{d}\left(T_{\text{liquidus}} - T\right)/\mathrm{d}f_S$ (Schmid-Fetzer and Kozlov, 2011), and artificially large values of GRF are obtained in case of primary intermetallic phase solidification that must be excluded.

In this range, it can be easily shown that the transformation of variable $f_S$ to $(f_S)^{1/2}$ always results in zero slope, thus, $|\mathrm{d}(T)/\mathrm{d}(f_S)^{1/2}| \rightarrow 0$ for $(f_S)^{1/2} \rightarrow 0$. However, in case of primary intermetallics, $\gamma$, the initial increase of $|\mathrm{d}(T)/\mathrm{d}(f_S)^{1/2}|$ (during Liquid $\rightarrow \gamma$) terminates at a first maximum due to the onset of $\alpha$ solidification (Liquid $\rightarrow \alpha + \gamma$). At this point, the value of $|\mathrm{d}(T)/\mathrm{d}(f_S)^{1/2}|$ drops back to almost zero, similar to the case of primary $\alpha$ solidification (Liquid $\rightarrow \alpha$). With further growing $(f_S)^{1/2}$, the value of $|\mathrm{d}(T)/\mathrm{d}(f_S)^{1/2}|$ rises to the relevant maximum at or before $(f_S)^{1/2} = 0.99$ defined in (5.60). The potential artifact is eliminated by the lower limit of $(f_S)^{1/2} = 0.7$ (or $f_S = 0.5$) for typical base-metal-rich alloys.

Systematic analyses of crack susceptibility for alloy optimization are efficiently done using the high-throughput calculation (HTC) function implemented in Pandat. It can perform thousands of PanSolidification simulations (Zhang et al., 2020), scanning a user-defined alloy compositional space and a range of cooling rates. The HTC function also includes the postprocessing data mining and analysis to extract the user-specified results as a function of the scanned variables. The exact definition of *CSI* in (5.60) is indispensable for the automatic data mining. A multidimensional variable space may be scanned in a single HTC job, and the chosen key result, such as *CSI*, can be displayed as a function of two chosen variables in a 3D or color map plot in Pandat. Example results for Al–Cu–Mg alloys are presented in Figure 5.25 using the same database as for Figure 5.24 and a higher resolution with step size of 0.25 or 0.125 wt.% and with 441 or 1,764 simulations for each graph, compared to the 121 manual calculations for the similar graphs presented by Liu and Kou (2017).

For a given alloy, the highest values of *CSI* are always obtained for the Scheil simulation. They decrease with the assumed values of *CR*, such as 100 or 25 K/s, down to the value calculated for equilibrium – or infinitely slow – solidification. Within a given alloy range, such as in Figure 5.25a, the maximum value of *CSI* is

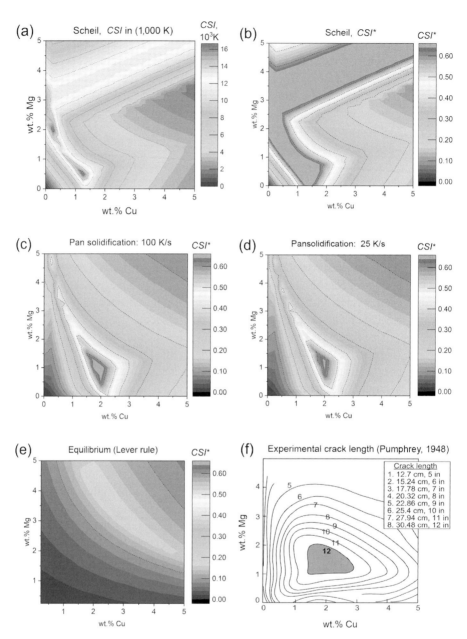

**Figure 5.25** Crack susceptibility maps for Al–Cu–Mg alloys: (a)–(e) calculated with PanSolidification and PanAl (www.computherm.com); (a) Scheil conditions with values of *CSI* from (5.60); (b) Scheil conditions with values of *CSI** from (5.61) normalized with $CSI_{max}$ = 16,672 K; (c) simulation with $CR$ = 100 K/s, (d) simulation with $CR$ = 25 K/s; (e) equilibrium (lever rule) solidification conditions; (f) experimentally determined map (Pumphrey and Moore, 1948) showing composition ranges in which cracking is greatest in ring castings. (f) is reproduced from Liu and Kou (2017) with permission from Elsevier. A black and white version of this figure will appear in some formats. For the colour version, please refer to the plate section.

found in the darkest red hot spot at Al–1.25Cu–0.5Mg; it is $CSI_{max} = 16,647$ K under Scheil conditions. This is also the maximum value of $CSI$ for the entire ensemble of alloys and $CR$s covered in Figure 5.25. The equilibrium case is shown for completeness in Figure 5.25e with the highest value of $CSI = 3,171$ K for alloy Al–2.75Cu–3.25Mg.

Scaling is an important issue for a realistic interpretation of the results for assumed "decreasing $CR$s" in Figures 5.25b–e. For direct comparability of these results, the following normalized, dimensionless quantity $CSI^*$ is defined using the constant value of $CSI_{max}$ obtained in Figure 5.25a:

$$CSI^* = CSI/CSI_{max}. \tag{5.61}$$

In order to improve the color resolution the full-color scale of $CSI^*$ is not used from 0–1, but the dark red level is set at $CSI^* = 0.65$ and color code gray is assigned to all regions of $CSI^* > 0.65$ in Figures 5.25b–d. This enables much better detection of the alloy regions with the highest values of $CSI^*$ in the more realistic maps of finite $CR$s in Figures 5.25c–d. For these $CR$s of 100 and 25 K/s, the peak of $CSI^*$ is found for alloy Al–2.0Cu–1.125Mg. This compares very well with the experimental crack susceptibility (crack length) in Figure 5.25f.

There are two key messages of the simulated Figures 5.25a–d: (i) For a substantial range of $CR$s from 100 to 25 K/s, the $CSI^*$ maps produce a well-defined peak region of $CSI^*$, which is quantitatively almost identical in the alloy composition range. This prediction of the narrow alloy range most vulnerable to hot cracking is in very good agreement with experimental data. (ii) The peak region of $CSI^*$ or $CSI$ in maps predicted by the simple Scheil calculations is significantly off the validated range. That demonstrates that solid-state back-diffusion cannot be neglected even at high $CR$ of 100 K/s.

The method demonstrated successfully in this case study is applicable for other alloys, ternary or multicomponent, provided that reliable thermodynamic and kinetic databases are available, such as PanAl for multicomponent Al alloys.

## References

Agarwal, R., Fries, S. G., Lukas, H. L., et al. (1992) Assessment of the Mg–Zn system. *Zeitschrift für Metallkunde*, 83(4), 216–223.

Agarwal, R., Fries, S. G., Lukas, H. L., et al. (1998) System Mg–Zn, in Ansara, I., Dinsdale, A. T., and Rand, M. H. (eds), *COST507 – Definition of Thermochemical and Thermophysical Properties to Provide a Database for the Development of New Light Alloys,* volume 2: *Thermochemical Database for Light Metal Alloys*. Luxembourg: European Commission, DG XII, 227–233.

Ågren, J., Cheynet, B., Clavaguera-Mora, M. T., et al. (1995) Workshop on thermodynamic models and data for pure elements and other endmembers of solutions: Schloß Ringberg 1995, Group 2. *CALPHAD*, 19(4), 449–480.

Ågren, J., and Hillert, M. (2019) Thermodynamic modelling of vacancies as a constituent. *CALPHAD*, 67, 101666.

Ågren, J., and Schmid-Fetzer, R. (2014) True phase diagrams. *Metallurgical and Materials Transactions A*, 45(11), 4766–4769.

Aldinger, F., Fernandez Guillermet, A., Iorich, V. S., et al. (1995) Workshop on thermodynamic models and data for pure elements and other endmembers of solutions: Schloß Ringberg 1995, Group 6. *CALPHAD*, 19(4), 555–571.

Andersson, J. O., Helander, T., Höglund, L., Shi, P., and Sundman, B. (2002) Thermo-Calc & DICTRA, computational tools for materials science. *CALPHAD*, 26(2), 273–312.

Ansara, I., Burton, B., Chen, Q., et al. (2000) Models for composition dependence. *CALPHAD*, 24(1), 19–40.

Ansara, I., Dinsdale, A., and Rand, M. (1998) *COST 507 Definition of Thermochemical and Thermophysical Properties to Provide a Database for the Development of New Light Alloys*, volume 2: *Thermochemical Database for Light Metal Alloys*. Luxembourg: European Commission. DG XII.

Ansara, I., Dupin, N., and Sundman, B. (1997) Reply to the paper: "When is a compound energy not a compound energy? A critique of the 2-sublattice order/disorder model": Of Nigel Saunders, Calphad 20 (1996) 491–499. *CALPHAD*, 21(4), 535–542.

Bale, C. W., Bélisle, E., Chartrand, P., et al. (2009) FactSage thermochemical software and databases – recent developments. *CALPHAD*, 33(2), 295–311.

Cacciamani, G., Dinsdale, A., Palumbo, M., and Pasturel, A. (2010) The Fe–Ni system: thermodynamic modelling assisted by atomistic calculations. *Intermetallics*, 18(6), 1148–1162.

Cai, Q., Mendis, C. L., Chang, I. T., and Fan, Z. (2020) Microstructure evolution and mechanical properties of new die-cast Al–Si–Mg–Mn alloys. *Materials and Design*, 187, 108394.

Campbell, C. E., Kattner, U. R., and Liu, Z.-K. (2014) The development of phase-based property data using the CALPHAD method and infrastructure needs. *Integrating Materials and Manufacturing Innovation*, 3(1), 12.

Cao, G., Zhang, C., Cao, H., Chang, Y. A., and Kou, S. (2010) Hot-tearing susceptibility of ternary Mg–Al–Sr alloy castings. *Metallurgy and Materials Transactions A*, 41(3), 706–716.

Cao, W., Chen, S. L., Zhang, F., Wu, K., Yang, Y., Chang, Y. A., Schmid-Fetzer, R., and Oates, W. A. (2009) PANDAT software with PanEngine, PanOptimizer and PanPrecipitation for multi-component phase diagram calculation and materials property simulation. *CALPHAD*, 33(2), 328–342.

Cao, W., Zhang, F., Chen, S.-L., et al. (2016) Precipitation modeling of multi-component nickel-based alloys. *Journal of Phase Equilibria and Diffusion*, 37(4), 491–502.

Cao, W., Zhang, F., Chen, S. L., Zhang, C., and Chang, Y. A. (2011) An integrated computational tool for precipitation simulation. *JOM*, 63(7), 29–34.

Chang, A., Colinet, C., Hillert, M., Moser, Z., Sanchez, J. M., Saunders, N., Watson, R. E., and Kussmaul, A. (1995) Workshop on thermodynamic models and data for pure elements and other endmembers of solutions: Schloß Ringberg 1995, Group 3. *CALPHAD*, 19(4), 481–498.

Chang, Y. A., Chen, S., Zhang, F., Yan, X., Xie, F., Schmid-Fetzer, R., and Oates, W. A. (2004) Phase diagram calculation: past, present and future. *Progress in Materials Science*, 49(3), 313–345.

Chang, Y. A., and Oates, W. A. (2010) *Materials Thermodynamics*. Hoboken: John Wiley & Sons.

Chartrand, P., and Pelton, A. D. (2001) The modified quasi-chemical model: Part III. Two sublattices. *Metallurgical and Materials Transactions A*, 32(6), 1397–1407.

Chase, M. W. (1998) *NIST-JANAF Thermochemical Tables*, fourth edition. New York: The American Chemical Society, and The American Institute of Physics for the National Institute of Standards and Technology.

Chase, M. W., Ansara, I., Dinsdale, A., et al. (1995) Workshop on thermodynamic models and data for pure elements and other endmembers of solutions: Schloß Ringberg 1995, Group 1. *CALPHAD*, 19(4), 437–447.

Chen, H.-L., Chen, Q., and Engström, A. (2018) Development and applications of the TCAL aluminum alloy database. *CALPHAD*, 62, 154–171.

Chen, L., Zhang, Z., Huang, Y., Cui, J., Deng, Z., Zou, H., and Chang, K. (2019) Thermodynamic description of the Fe–Cu–C system. *CALPHAD*, 64, 225–235.

Chen, Q., Hillert, M., Sundman, B., Oates, W. A., Fries, S. G., and Schmid-Fetzer, R. (1998) Phase equilibria, defect chemistry and semiconducting properties of CdTe(s) – thermodynamic modeling. *Journal of Electronics Materials*, 27(8), 961–971.

Chen, S. L., Daniel, S., Zhang, F., Chang, Y. A., Yan, X. Y., Xie, F. Y., Schmid-Fetzer, R., and Oates, W. A. (2002) The PANDAT software package and its applications. *CALPHAD*, 26(2), 175–188.

Chuang, Y.-Y., Chang, Y. A., Schmid, R., and Lin, J.-C. (1986) Magnetic contributions to the thermodynamic functions of alloys and the phase equilibria of Fe–Ni system below 1200 K. *Metallurgical Transactions A*, 17(8), 1361–1372.

Chuang, Y. Y., Schmid, R., and Chang, Y. A. (1985) Magnetic contributions to the thermodynamic functions of pure Ni, Co, and Fe. *Metallurgical Transactions A*, 16(2), 153–165.

Costa e Silva, A., Ågren, J., Clavaguera-Mora, M. T., et al. (2007) Applications of computational thermodynamics – the extension from phase equilibrium to phase transformations and other properties. *CALPHAD*, 31(1), 53–74.

de Fontaine, D. d., Fries, S. G., Inden, G., Miodownik, P., Schmid-Fetzer, R., and Chen, S.-L. (1995) Workshop on thermodynamic models and data for pure elements and other endmembers of solutions: Schloß Ringberg 1995, Group 4. *CALPHAD*, 19(4), 499–536.

Desai, P. (1987) Thermodynamic properties of nickel. *International Journal of Thermophysics*, 8(6), 763–780.

Dinsdale, A., Khvan, A., Smirnova, E. A., Ponomareva, A. V., and Abrikosov, I. A. (2021) Modelling the thermodynamic data for Hcp Zn and Cu–Zn alloys –an ab initio and CALPHAD approach. *CALPHAD*, 72, 102253.

Dinsdale, A. T. (1991) SGTE data for pure elements. *CALPHAD*, 15(4), 317–425.

Dupin, N., and Ansara, I. (1999) On the sublattice formalism applied to the B2 phase. *Zeitschrift für Metallkunde*, 90(1), 76–85.

Dupin, N., Ansara, I., and Sundman, B. (2001) Thermodynamic re-assessment of the ternary system Al–Cr–Ni. *CALPHAD*, 25(2), 279–298.

Easton, M. A., and StJohn, D. H. (2001) A model of grain refinement incorporating alloy constitution and potency of heterogeneous nucleant particles. *Acta Materialia*, 49(10), 1867–1878.

Eriksson, G. (1975) SOLGASMIX, a computer program for calculation of equilibrium compositions in multiphase systems. *Chemica Scripta*, 8, 100–103.

Fang, H., Tang, Q., Zhang, Q., Gu, T., and Zhu, M. (2019) Modeling of microstructure and microsegregation formation during solidification of Al–Si–Mg alloys. *International Journal of Heat Mass Transfer*, 133, 371–381.

Fernández Guillermet, A., Gustafson, P., and Hillert, M. (1985) The representation of thermo-dynamic properties at high pressures. *Journal of Physics and Chemistry of Solids*, 46(12), 1427–1429.

Fu, P., Peng, L., Jiang, H., Chang, J., and Zhai, C. (2008) Effects of heat treatments on the microstructures and mechanical properties of Mg–3Nd–0.2Zn–0.4Zr (wt.%) alloy. *Materials Science and Engineering A*, 486(1), 183–192.

Gibbs, J. W. (1874) On the equilibrium of heterogeneous substances, I. *Transactions of the Connecticut Academy of Arts and Sciences*, 3(96), 108–248.

Gibbs, J. W. (1878) On the equilibrium of heterogeneous substances, II. *Transactions of the Connecticut Academy of Arts and Sciences*, 3(96), 343–524.

Gröbner, J., and Schmid-Fetzer, R. (2013) Key issues in a thermodynamic Mg alloy database. *Metallurgical and Materials Transactions A*, 44(7), 2918–2934.

Gulliver, G. H. (1913) The quantitative effect of rapid cooling upon the constitution of binary alloys. *Journal of the Institute of Metals*, 9, 120–157.

Gustafson, P. (1985) A thermodynamic evaluation of the Fe–C system. *Scandinavian Journal of Metallurgy*, 14(5), 259–267.

Hallstedt, B., Dupin, N., Hillert, M., Höglund, L., Lukas, H. L., Schuster, J. C., and Solak, N. (2007) Thermodynamic models for crystalline phases. Composition dependent models for volume, bulk modulus and thermal expansion. *CALPHAD*, 31(1), 28–37.

Hertzman, S., and Sundman, B. (1982) A thermodynamic analysis of the Fe–Cr system. *CALPHAD*, 6(1), 67–80.

Hillert, M. (1981) Some viewpoints on the use of a computer for calculating phase diagrams. *Physica B*, 103(1), 31–40.

Hillert, M. (2001) The compound energy formalism. *Journal of Alloys and Compounds*, 320(2), 161–176.

Hillert, M. (2008) *Phase Equilibria, Phase Diagrams and Phase Transformations: Their Thermodynamic Basis*, second edition. New York and London: Cambridge University Press.

Hillert, M., and Jarl, M. (1978) A model for alloying in ferromagnetic metals. *CALPHAD*, 2(3), 227–238.

Hillert, M., Selleby, M., and Sundman, B. (2009) An attempt to correct the quasichemical model. *Acta Materialia*, 57(17), 5237–5244.

Hillert, M., and Staffansson, L. (1970) Regular-solution model for stoichiometric phases and ionic melts. *Acta Chemica Scandinavica*, 24(10), 3618–3626.

Huang, D., Liu, S., Du, Y., and Sundman, B. (2015) Modeling of the molar volume of the solution phases in the Al–Cu–Mg system. *CALPHAD*, 51, 261–271.

Hutchinson, C., Nie, J., and Gorsse, S. (2005) Modeling the precipitation processes and strengthening mechanisms in a Mg–Al–(Zn) AZ91 alloy. *Metallurgy and Materials Transactions A*, 36(8), 2093–2105.

Ilatovskaia, M., Savinykh, G., and Fabrichnaya, O. (2017) Thermodynamic description of the $ZrO_2$–$TiO_2$–$Al_2O_3$ system based on experimental data. *Journal of the European Ceramics Society*, 37(10), 3461–3469.

Inden, G. (1981) The role of magnetism in the calculation of phase diagrams. *Physica B*, 103(1), 82–100.

Janz, A., and Schmid-Fetzer, R. (2005) Impact of ternary parameters. *CALPHAD*, 29(1), 37–39.

Jung, J.-G., Cho, Y.-H., Lee, J.-M., Kim, H.-W., and Euh, K. (2019) Designing the composition and processing route of aluminum alloys using CALPHAD: case studies. *CALPHAD*, 64, 236–247.

Kampmann, R., and Wagner, R. (1984) Kinetics of precipitation in metastable binary alloys – theory and application to Cu-1.9 at % Ti and Ni-14 at % Al, in Haasen, P., Gerold, V., Wagner, R., and Ashby, M. F. (eds), *Decomposition of Alloys: The Early Stages: Proceedings of the 2nd Acta-Scripta Metallurgica Conference. Sonnenberg, Germany, 19-23/09/1983*. Oxford: Pergamon, 91–103.

Kattner, U. R., and Seifert, H. J. (2010) Integrated computational materials engineering, CALPHAD and Hans Leo Lukas. *CALPHAD*, 34, 385–386.

Kaufman, L., and Ågren, J. (2014) CALPHAD, first and second generation – birth of the materials genome. *Scripta Materialia*, 70(Supplement C), 3–6.

Kaufman, L., and Bernstein, H. (1970) *Computer Calculation of Phase Diagrams – with Special Reference to Refractory Metals*. New York: Academic Press.

Kevorkov, D., Schmid-Fetzer, R., and Zhang, F. (2004) Phase equilibria and thermodynamics of the Mg–Si–Li system and remodeling of the Mg–Si system. *Journal of Phase Equilibria and Diffusion*, 25(2), 140–151.

Kou, S. (2015) A criterion for cracking during solidification. *Acta Materialia*, 88, 366–374.

Lee, B. D., Kim, E. J., Baek, U. H., and Han, J. W. (2013) Precipitate prediction model of Mg–xAl(x=3,6,9) alloys. *Metals and Materials International*, 19(2), 135–145.

Li, S., and Apelian, D. (2011) Hot tearing of aluminum alloys. *International Journal of Metalcasting*, 5(1), 23–40.

Li, S., Sadayappan, K., and Apelian, D. (2011) Characterisation of hot tearing in Al cast alloys: methodology and procedures. *International Journal of Cast Metals Research*, 24(2), 88–95.

Liang, P., Su, H. L., Donnadieu, P., et al. (1998a) Experimental investigation and thermodynamic calculation of the central part of the Mg–Al phase diagram. *Zeitschrift für Metallkunde*, 89(8), 536–540.

Liang, P., Tarfa, T., Robinson, J. A., et al. (1998b) Experimental investigation and thermodynamic calculation of the Al–Mg–Zn system. *Thermochimica Acta*, 314(1), 87–110.

Liang, S.-M., Hsiao, H.-M., and Schmid-Fetzer, R. (2015) Thermodynamic assessment of the Al–Cu–Zn system, part I: Cu–Zn binary system. *CALPHAD*, (51), 224–232.

Liang, S.-M., and Schmid-Fetzer, R. (2013) Thermodynamic assessment of the Al–P system based on original experimental data. *CALPHAD*, 42(0), 76–85.

Liang, S.-M., and Schmid-Fetzer, R. (2014) Corrigendum to "thermodynamic assessment of the Al–P system based on original experimental data"[Calphad 42 (2013) 76–85]. *CALPHAD*, 100(45), 251–253.

Liang, S.-M., and Schmid-Fetzer, R. (2015) Thermodynamic assessment of the Al–Cu–Zn system, part II: Al–Cu binary system. *CALPHAD*, 51, 252–260.

Liang, S.-M., and Schmid-Fetzer, R. (2016) Thermodynamic assessment of the Al–Cu–Zn system, part III: Al–Cu–Zn ternary system. *CALPHAD*, 52, 21–37.

Liang, S.-M., Taubert, F., Kozlov, A., Seidel, J., Mertens, F., and Schmid-Fetzer, R. (2017) Thermodynamics of Li–Si and Li–Si–H phase diagrams applied to hydrogen absorption and Li–ion batteries. *Intermetallics*, 81, 32–46.

Lima, J., Barbosa, C., Magno, I., et al (2018) Microstructural evolution during unsteady-state horizontal solidification of Al–Si–Mg (356) alloy. *Transactions of the Nonferrous Metals Society*, 28(6), 1073–1083.

Liu, J., and Kou, S. (2017) Susceptibility of ternary aluminum alloys to cracking during solidification. *Acta Materialia*, 125, 513–523.

Liu, Y., Zhang, C., Du, C., et al. (2020) CALTPP: a general program to calculate thermophysical properties. *Journal of Materials Science and Technology*, 42, 229–240.

Liu, Z.-K. (2018) Ocean of data: integrating first-principles calculations and CALPHAD modeling with machine learning. *Journal of Phase Equilibria and Diffusion*, 39(5), 635–649.

Liu, Z.-K. (2020) Computational thermodynamics and its applications. *Acta Materialia*, 200, 745–792.

Liu, Z.-K., and Wang, Y. (2016) *Computational Thermodynamics of materials*. Cambridge: Cambridge University Press.

Löffler, A., Gröbner, J., Hampl, M., Engelhardt, H., Schmid-Fetzer, R., and Rettenmayr, M. (2012) Solidifying incongruently melting intermetallic phases as bulk single phases using the example of Al2Cu and Q-phase in the Al–Mg–Cu–Si system. *Journal of Alloys and Compounds*, 515(Supplement C), 123–127.

Löffler, A., Zendegani, A., Gröbner, J., Hampl, M., Schmid-Fetzer, R., Engelhardt, H., Rettenmayr, M., Körmann, F., Hickel, T., and Neugebauer, J. (2016) Quaternary Al–Cu–Mg–Si Q phase: sample preparation, heat capacity measurement and first-principles calculations. *Journal of Phase Equilibria and Diffusion*, 37(2), 119–126.

Lu, X.-G., Selleby, M., and Sundman, B. (2005a) Assessments of molar volume and thermal expansion for selected Bcc, Fcc and Hcp metallic elements. *CALPHAD*, 29(1), 68–89.

Lu, X.-G., Selleby, M., and Sundman, B. (2005b) Implementation of a new model for pressure dependence of condensed phases in Thermo-Calc. *CALPHAD*, 29(1), 49–55.

Lu, X.-G., Selleby, M., and Sundman, B. (2005c) Theoretical modeling of molar volume and thermal expansion. *Acta Materialia*, 53(8), 2259–2272.

Lukas, H., Henig, E. T., and Zimmermann, B. (1977) Optimization of phase diagrams by a least squares method using simultaneously different types of data. *CALPHAD*, 1(3), 225–236.

Lukas, H. L., Fries, S. G., and Sundman, B. (2007) *Computational Thermodynamics: The CALPHAD Method*. New York: Cambridge University Press.

Luo, A. A. (2015) Material design and development: from classical thermodynamics to CALPHAD and ICME approaches. *CALPHAD*, 50, 6–22.

Ma, X., Li, C., Zhang, W., and Schmid-Fetzer, R. (2005) The influence of nitrogen fugacity on the phase equilibria of Me–N systems. *CALPHAD*, 29(3), 247–253.

McDowell, M. T., Lee, S. W., Nix, W. D., and Cui, Y. (2013) 25th anniversary article: understanding the lithiation of silicon and other alloying anodes for lithium–ion batteries. *Advanced Materials*, 25(36), 4966–4985.

Mirković, D., Gröbner, J., Schmid-Fetzer, R., Fabrichnaya, O., and Lukas, H. L. (2004) Experimental study and thermodynamic re-assessment of the Al–B system. *Journal of Alloys and Compounds*, 384(1), 168–174.

Müller, M., Seebold, S., Wu, G., Yazhenskikh, E., Jantzen, T., and Hack, K. (2018) Experimental investigation and modeling of the viscosity of oxide slag systems. *Journal of Sustainable Metallurgy*, 4(1), 3–14.

Oates, W. A., Chen, S. L., Cao, W., Zhang, F., Chang, Y. A., Bencze, L., Doernberg, E., and Schmid-Fetzer, R. (2008) Vacancy thermodynamics for intermediate phases using the compound energy formalism. *Acta Materialia*, 56(18), 5255–5262.

Ohno, M., Kozlov, A., Arroyave, R., Liu, Z. K., and Schmid-Fetzer, R. (2006a) Thermodynamic modeling of the Ca–Sn system based on finite temperature quantities from first-principles and experiment. *Acta Materialia*, 54(18), 4939–4951.

Ohno, M., Mirkovic, D., and Schmid-Fetzer, R. (2006b) Liquidus and solidus temperatures of Mg-rich Mg–Al–Mn–Zn alloys. *Acta Materialia*, 54(15), 3883–3891.

Onderka, B., Unland, J., and Schmid-Fetzer, R. (2002) Thermodynamics and phase stability in the In–N system. *Journal of Materials Research*, 17(12), 3065–3083.

Otis, R., and Liu, Z.-K. (2017) Pycalphad: CALPHAD-based computational thermodynamics in Python. *JORS*, 5(1), 1–11.

Palumbo, M., Burton, B., Silva, A. C. E., et al. (2014) Thermodynamic modelling of crystalline unary phases. *Physica Status Solidi B*, 251(1), 14–32.

Pelton, A. D. (2014) Thermodynamics and phase diagrams, in Laughlin, D. E., and Hono, K. (eds), *Physical Metallurgy*, fifth edition. Amsterdam: Elsevier, 203–303.

Pelton, A. D., and Chartrand, P. (2001) The modified quasi-chemical model: part II. *Multicomponent Solutions. Metallurgical and Materials Transactions A*, 32(6), 1355–1360.

Pelton, A. D., Chartrand, P., and Eriksson, G. (2001) The modified quasi-chemical model: part IV. Two-sublattice quadruplet approximation. *Metallurgical and Materials Transactions A*, 32(6), 1409–1416.

Pelton, A. D., Degterov, S. A., Eriksson, G., Robelin, C., and Dessureault, Y. (2000) The modified quasichemical model I – binary solutions. *Metallurgical and. Materials Transactions B*, 31(4), 651–659.

Povoden-Karadeniz, E., Cirstea, D., Lang, P., Wojcik, T., and Kozeschnik, E. (2013) Thermodynamics of Ti–Ni shape memory alloys. *CALPHAD*, 41, 128–139.

Pumphrey, W., and Moore, D. (1948) Cracking during and after solidification in some aluminium copper magnesium alloys of high purity. *Journal of the Institute of Metals*, 74(9), 425–438.

Rappaz, M., Drezet, J.-M., and Gremaud, M. (1999) A new hot-tearing criterion. *Metallurgical and. Materials Transactions A*, 30(2), 449–455.

Redlich, O., and Kister, A. T. (1948) Algebraic representation of thermodynamic properties and the classification of solutions. *Industrial and Engineering Chemistry Research*, 40(2), 345–348.

Saal, J. E., Berglund, I. S., Sebastian, J. T., Liaw, P. K., and Olson, G. B. (2018) Equilibrium high entropy alloy phase stability from experiments and thermodynamic modeling. *Scripta Materialia.*, 146, 5–8.

Saunders, N. (1990) A review and thermodynamic assessment of the Al–Mg and Mg–Li systems. *CALPHAD*, 14(1), 61–70.

Saunders, N., Guo, U., Li, X., Miodownik, A., and Schillé, J.-P. (2003) Using JMatPro to model materials properties and behavior. *JOM*, 55(12), 60–65.

Scheil, E. (1942) Bemerkungen zur Schichtkristallbildung. *Zeitschrift für Metallkunde*, 34, 70–72.

Schmid-Fetzer, R. (2014) Phase diagrams: the beginning of wisdom. *Journal of Phase Equilibria and Diffusion*, 35(6), 735–760, Open Access: https://doi.org/10.1007/s11669 014-0343-5.

Schmid-Fetzer, R. (2022) Third generation of unary CALPHAD descriptions and the avoidance of re-stabilized solid phases and unexpected large heat capacity. *Journal of Phase Equilibria and Diffusion*, 43, 304–316, Open Access: https://doi.org/10.1007/s11669-022-00976-3.

Schmid-Fetzer, R., Andersson, D., Chevalier, P. Y., et al. (2007) Assessment techniques, database design and software facilities for thermodynamics and diffusion. *CALPHAD*, 31(1), 38–52.

Schmid-Fetzer, R., and Hallstedt, B. (2012) Is zinc HCP_ZN or HCP_A3? *CALPHAD*, 37(Supplement C), 34–36.

Schmid-Fetzer, R., and Kozlov, A. (2011) Thermodynamic aspects of grain growth restriction in multicomponent alloy solidification. *Acta Materialia*, 59(15), 6133–6144.

Schmid-Fetzer, R., and Zhang, F. (2018) The light alloy CALPHAD databases PanAl and PanMg. *CALPHAD*, 61, 246–263.

Schmid, R. (1983) A thermodynamic analysis of the Cu–O system with an associated solution model. *Metallurgical Transactions B*, 14(3), 473–481.

Schmid, R., and Chang, Y. A. (1985) A thermodynamic study on an associated solution model for liquid alloys. *CALPHAD*, 9(4), 363–382.

Schmitz, G. J., and Prahl, U. (eds) (2016) *Handbook of Software Solutions for ICME*, first edition. Weinheim: John Wiley & Sons.

Sommer, F. (1982) Association model for the description of the thermodynamic functions of liquid alloys, 1. Basic concepts, 2. Numerical treatment and results. *Z. Metallkd.*, 73(2), 72–86.

Sonderegger, B., and Kozeschnik, E. (2009a) Generalized nearest-neighbor broken-bond analysis of randomly oriented coherent interfaces in multicomponent Fcc and Bcc structures. *Metallurgical and Materials Transactions A*, 40(3), 499–510.

Sonderegger, B., and Kozeschnik, E. (2009b) Size dependence of the interfacial energy in the generalized nearest-neighbor broken-bond approach. *Scripta Materialia*, 60(8), 635–638.

Song, J., Wang, Z., Huang, Y., Srinivasan, A., Beckmann, F., Kainer, K. U., and Hort, N. (2015) Hot tearing susceptibility of Mg–Ca binary alloys. *Metallurgical and Materials Transactions A*, 46(12), 6003–6017.

Spencer, P. (2008) A brief history of CALPHAD. *CALPHAD*, 32(1), 1–8.

Spencer, P. J., Burton, B., Chart, T. G., et al. (1995) Workshop on thermodynamic models and data for pure elements and other endmembers of solutions: Schloß Ringberg 1995, Groups 5 and 6. *CALPHAD*, 19(4), 537–553.

StJohn, D. H., Qian, M., Easton, M. A., and Cao, P. (2011) The interdependence theory: the relationship between grain formation and nucleant selection. *Acta Materialia*, 59(12), 4907–4921.

Sundman, B., and Ågren, J. (1981) A regular solution model for phases with several components and sublattices, suitable for computer applications. *Journal of the Physics and Chemistry of Solids*, 42(4), 297–301.

Sundman, B., and Aldinger, F. (1995) The Ringberg workshop 1995 on unary data for elements and other end-members of solutions. *CALPHAD*, 19(4), 433–436.

Sundman, B., Chen, Q., and Du, Y. (2018) A review of CALPHAD modeling of ordered phases. *Journal of Phase Equilibria and Diffusion*, 39(5), 678–693.

Sundman, B., Kattner, U. R., Hillert, M., et al. (2020) A method for handling the extrapolation of solid crystalline phases to temperatures far above their melting point. *CALPHAD*, 68, 101737.

Sundman, B., Kattner, U. R., Palumbo, M., and Fries, S. G. (2015) OpenCalphad – a free thermodynamic software. *Integrating Materials Manufacturing Innovation*, 4(1), 1–15.

Tang, K., Du, Q., and Li, Y. (2018) Modelling microstructure evolution during casting, homogenization and ageing heat treatment of Al–Mg–Si–Cu–Fe–Mn alloys. *CALPHAD*, 63, 164–184.

Tsonopoulos, C. (1974) An empirical correlation of second virial coefficients. *AIChE Journal*, 20(2), 263–272.

Unland, J., Onderka, B., Davydov, A., and Schmid-Fetzer, R. (2003) Thermodynamics and phase stability in the Ga–N system. *Journal of Crystal Growth*, 256(1), 33–51.

Wang, P. S., Kozlov, A., Thomas, D., Mertens, F., and Schmid-Fetzer, R. (2013) Thermodynamic analysis of the Li–Si phase equilibria from 0 K to liquidus temperatures. *Intermetallics*, 42, 137–145.

Weiss, R. J., and Tauer, K. J. (1956) *Theory of Alloy Phases*. Cleveland: ASM.

Xia, X., Sanaty-Zadeh, A., Zhang, C., Luo, A. A., and Stone, D. S. (2018) Experimental investigation and simulation of precipitation evolution in Mg–3Nd-0.2Zn alloy. *CALPHAD*, 60, 58–67.

Xiong, W., and Olson, G. B. (2016) Cybermaterials: materials by design and accelerated insertion of materials. *NPJ Computational Materials*, 2(1), 15009.

Xiong, W., Zhang, H., Vitos, L., and Selleby, M. (2011) Magnetic phase diagram of the Fe–Ni system. *Acta Materialia*, 59(2), 521–530.

Yao, Z., Berman, T., and Allison, J. (2020) An ICME method for predicting phase dissolution during solution treatment in advanced super vacuum die cast magnesium alloys. *Integrating Materials Manufacturing Innovation*, 9(3), 301–313.

Zhang, C., Cao, W., Chen, S.-L., et al. (2014) Precipitation simulation of AZ91 alloy. *JOM*, 66(3), 389–396.

Zhang, F., Cao, W., Zhang, C., Chen, S., Zhu, J., and Lv, D. (2018) Simulation of co-precipitation kinetics of $\gamma'$ and $\gamma''$ in superalloy 718, in Ott, E. e. a. (ed), *Proceedings of the 9th International Symposium on Superalloy 718 and Derivatives: Energy, Aerospace, and Industrial Applications*. Cham: Springer TMS, 147–161.

Zhang, F., Zhang, C., Liang, S. M., Lv, D. C., Chen, S. L., and Cao, W. S. (2020) Simulation of the composition and cooling rate effects on the solidification path of casting aluminum alloys. *Journal of Phase Equilibria and Diffusion*, 41, 793–803. https://doi.org/10.1007/s11669-020-00834-0.

Zhang, Y., Liu, Y., Liu, S., et al. (2021) Assessment of atomic mobilities and simulation of precipitation evolution in Mg–X (X= Al, Zn, Sn) alloys. *Journal of Materials Science and Technology*, 62, 70–82.

Zhou, X., Zhang, F., Liu, S., Du, Y., and Jin, B. (2020) Phase equilibria and thermodynamic investigation of the In–Li system. *CALPHAD*, 70, 101779.

# 6 Fundamentals of Thermophysical Properties

## Contents

## 6.1        Definition of Thermophysical Properties

What are "thermophysical properties"? This question is hard to answer because there is a conflict between any strict definition and established terminology in the materials community. Attempting a strict approach, one may distinguish "thermophysical" and "thermochemical" properties. Thermophysical properties might be simply defined as material properties, which vary with temperature without altering the chemical identity of materials. This definition is close to Grimvall's book title, *Thermophysical Properties of Materials* (Grimvall, 1999), focusing on "how temperature enters various physical properties." That relates to lattice vibrations and electron transport of charge and heat. Examples for thermophysical properties in that book are entropy, heat capacity, thermal expansion, electrical conductivity, thermal conductivity, and elastic properties. On the other hand, "thermochemical" data published in recent books such as *NIST-JANAF Thermochemical Tables* (Chase, 1998) and many earlier books with similar titles compile for stoichiometric (or "pure") substances only the data in tabular form that are derived from (i) standard entropy, (ii) standard enthalpy of formation, (iii) heat capacity function, and (iv) enthalpy of fusion or transition. The overlap with the "strictly thermophysical" properties in the attempted futile definition is obvious.

Following Gibbs's thermodynamics and the CALPHAD approach, the key thermodynamic function (property) is the Gibbs energy ($G$), from which all other "thermodynamic properties" are mathematically derived, such as heat capacity, enthalpy, entropy, chemical potential, and so on. The prime example for a quantity denoted beyond that logic is the volume $V$. Even though $V$ can be calculated directly as derivative $V = \partial G / \partial p$ from the known pressure dependence of $G$, it is generally denoted as "thermophysical" property for solids and liquids. Going further, one may say that even bulk modulus / compressibility, thermal expansion, and such are all related to higher derivatives of $G(T, p)$. However, these are all clearly denoted as "thermophysical" properties despite their relation to $G(T, p)$, while their derivation from the Helmholtz energy is simpler (Jacobs et al., 2019), as also shown for the sound velocity (Jacobs et al., 2017; Nowick and Burton, 1975; Paul and Divinski, 2017; Philibert, 1991). As a conclusion, it is suggested to define *thermodynamic* properties as all those obtained from $G$ and its temperature and composition derivatives. By contrast, *thermophysical* properties comprise – but are not limited to – all those obtained from the pressure dependence of $G$ and higher derivatives of $\partial G(T, p) / \partial p$ with pressure or temperature.

In this book, we follow this widest definition of *thermophysical* properties to be "all phase-based materials properties other than *thermodynamic* properties." The aspect of "phase-based" properties is also a core concept of the CALPHAD method, recently successfully extended from the initial treatment of thermodynamics and phase diagrams to become a cornerstone of integrated computational materials engineering (ICME) and the materials genome (MG) (Du and Sundman, 2017; Olson and Kuehmann, 2014). The concept of "phase-based" properties is also important for the structure of comprehensive databases to be developed using the expanded CALPHAD method (Campbell et al., 2014). Many materials properties can be related to single phases, and some are inherently two-phase properties, such as the interfacial energy. Some thermophysical properties, such as viscosity, can be derived from models based on thermodynamic data. In the wide field of kinetics, the most important diffusion coefficients are preferably calculated from the atomic mobility and the thermodynamic factor. Building upon that knowledge of the thermodynamics is extremely useful since the atomic mobility is often a much less composition- or temperature-dependent thermophysical property compared to the diffusion coefficient. State-of-the-art kinetic materials databases of multicomponent systems are therefore assessed with mobilities in conjunction with thermodynamic (Gibbs energy) parameters. For practical reasons, these parameter sets are assembled in a "thermodynamic" database in the popular TDB file format, which is then sufficient for both thermodynamic and diffusion simulations. Additional "thermophysical" parameters, required, for example, for precipitation and aging simulation, are compiled in so-called "kinetic" databases. A practical example is given in Section 6.9.

Knowledge of thermophysical properties is of fundamental importance for many disciplines and key inputs for many numerical simulations associated with materials design. Short descriptions of the important aspects (i.e., definition, modeling, and its consequence for material design) of several thermophysical properties are given in this chapter. The chapter will not replace a textbook on thermophysical properties, but will help the reader to have a good command of the aforementioned important aspects. The brief introduction of these aspects will facilitate the understanding of materials design described in Chapters 7–12, which employ knowledge about thermophysical properties and thermodynamics.

In this chapter, firstly several important aspects for diffusion are described, including Fick's laws on diffusion, atomic mechanism of diffusion, interdiffusion in binary, ternary and multicomponent systems, diffusion of phases with narrow homogeneity ranges, and various computational methods (first-principles calculation, molecular dynamic simulation, empirical approaches, and the phenomenological DICTRA method) to calculate and/or estimate diffusivity. Many textbooks and professional books on diffusion are available (Heitjans and Kärger, 2005; Jost, 1969; Kirkaldy and Young, 1987; Mehrer, 1990, 2007), as listed in this chapter's references. The interested reader can refer to these books for more detail on diffusion. Secondly, a few other important thermophysical properties (interfacial energy, viscosity, volume, thermal conductivity, and so on) are briefly described. Thirdly, the procedure to establish the thermophysical databases is described from the materials design point of view. One case study for material design (i.e., precipitation and age hardening in AA6005 aluminum alloys is demonstrated mainly using thermophysical properties as input for simulation of microstructure evolution.

## 6.2     Diffusion Coefficient

### 6.2.1     Fick's Laws of Diffusion and Various Diffusion Coefficients

The equations governing diffusion processes and hence the redistribution of concentrations are Fick's laws (Fick, 1855). These laws are purely phenomenological descriptions. When considering the movement of particles (atoms, molecules, ions, etc.) caused by concentration gradient in one dimension for an isotropic medium, the Fick's first law defines that the diffusion flux is negatively proportional to the concentration gradient. This law is written as follows:

$$J = -D\frac{\partial C}{\partial x}, \tag{6.1}$$

where $J$ and $C$ are the diffusion flux and concentration of the particles, respectively, and $x$ is the position. $J$ is expressed in terms of number of particles (or moles) traversing a unit area per unit time, and $C$ in number of particles or moles per unit volume. The factor of proportionality $D$ is named as the diffusion coefficient or diffusivity, having the dimension of length$^2$ per time, for example the SI-unit of $[m^2s^{-1}]$. $D$ usually depends on temperature, pressure, and concentration.

It should be mentioned that the Fick's first law can be used only in a steady-state condition where the composition remains constant with time. In addition, for the utilization of the Fick's first law, there should be no external driving force other than the concentration gradient. Ionic diffusion under an external electric field can be classified into two cases: diffusion with or without chemical reaction. For the first case, the chemical reaction will consume the net charge in the system, thus the system can be considered to be electroneutral, and diffusion driving force is the gradient of chemical potential and external electric field. For the second case, ionic separation will generate net charge, and the internal electric field resulting from this net charge is also a driving force of ionic diffusion. Under most experimental conditions, the concentration at a particular position varies with time. In this situation, Fick's first law alone cannot be used to describe the diffusion components because of the absence of a time parameter. In such situations, a balance equation is necessary to describe the continuity of particles:

$$\frac{\partial C}{\partial t} + \nabla \cdot J = 0, \tag{6.2}$$

where $t$ is the diffusion time. Combining (6.1) and (6.2) yields in 1D space

$$\frac{\partial C}{\partial t} = \frac{\partial}{\partial x}\left(D\frac{\partial C}{\partial x}\right). \tag{6.3}$$

If $D$ is independent of concentration and hence of position $x$, (6.3) reduces to

$$\frac{\partial C}{\partial t} = D\frac{\partial^2 C}{\partial x^2}. \tag{6.4}$$

Equation (6.3) or (6.4) is Fick's second law. It is worth mentioning that although Fick's first law cannot be used to estimate the evolution of components in the interaction zone depending on time, it can be applied to compute diffusion flux at any time since it is proportional to the concentration gradient in a profile developed by a diffusion-controlled process.

From an experimental point of view, the diffusion coefficients can be divided into self-diffusion coefficient, impurity diffusion coefficient, tracer self-diffusion

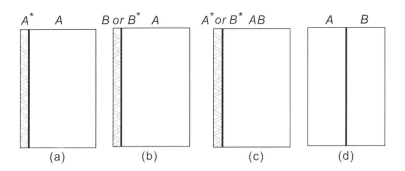

**Figure 6.1** Measurement of different diffusion coefficients: (a) self-diffusion of element $A$; (b) impurity diffusion of $B$ in pure $A$; (c) tracer self-diffusion of $A$ or $B$ in a homogeneous $AB$ alloy; (d) interdiffusion between $A$ and $B$ alloys. $A^*$ or $B^*$ are the tracer atoms of $A^*$ or $B^*$ of the same element $A$ or $B$.

coefficient, interdiffusion coefficient, and intrinsic diffusion coefficient. The experiments for the first four diffusion coefficients are schematically shown in Figure 6.1.

The diffusion of $A$ atom in a solid of element $A$ is called self-diffusion, which is denoted as $D_A^A$. Studies of self-diffusion usually need tracer atoms $A^*$ of the same element for the convenience of tracking. In most experiments, tracers are marked by their radioactivity, as shown in Figure 6.1a. The corresponding phenomenon is called tracer diffusion (or tracer self-diffusion), which is denoted as $D_A^*$.

If the diffusion rate of element $B$ using the radioisotope $(B^*)$ is measured in the matrix of element $A$ (solvent), it is called impurity diffusion of $B$ in $A$, and the corresponding diffusion coefficient is expressed as $D_A^{B*}$, as indicated in Figure 6.1b. It is noted that the amount of $B^*$ is so small that $B$ atoms may be regarded as diffusing quite independently of one another.

In a homogeneous binary $AB$ alloy, the tracer self-diffusion coefficients for $A^*$ and $B^*$ are denoted as $D_{AB}^{A*}$ and $D_{AB}^{B*}$, respectively. A schematic diagram for this kind of diffusion is illustrated in Figure 6.1c. The alloy composition is not changed during measurement since the concentration of radiotracer $A^*$ or $B^*$ is very small.

A diffusion coefficient measured under a concentration gradient can be regarded as a chemical diffusion coefficient or interdiffusion coefficient, which describes diffusion referred to sample-fixed axes. $\tilde{D}$ can be calculated from the composition profiles as functions of both concentration and temperature. An interdiffusion experiment between two alloys is illustrated in Figure 6.1d.

The intrinsic diffusion coefficient is related to the Kirkendall effect. This effect is the motion of the interface between two metals that happens due to different diffusion rates of the metal atoms. The effect can be observed, for example, by placing insoluble markers (fine insoluble wires, oxide particles, etc.) at the interface between a pure metal and its alloy. When heated to a temperature where atomic diffusion is activated, the interface can move relative to the markers.

The Kirkendall effect was discovered by Ernest Kirkendall and Alice Smigelskas in 1947, where they found through experiment that the diffusivities of the components in

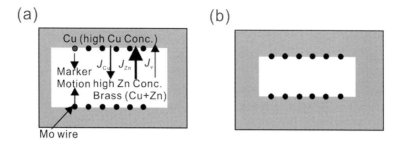

**Figure 6.2** Kirkendall experiment: (a) before annealing, (b) after annealing.

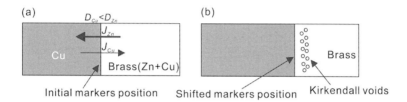

**Figure 6.3** Kirkendall voids: (a) before annealing, (b) after annealing.

brass are different (Smigelkas and Kirkendall, 1947). In their work, a bar of brass (70 at.% Cu, 30 at.% Zn) was used as a core with Mo wires stretched along its length, and the bar was then coated with a layer of pure Cu. The diffusion experiment was performed at 785°C for 56 days with six times of cross sections. Over time, the change of interface location can be observed during diffusion. It was noticed that the wire markers moved closer as the Zn diffused out of the brass and into the Cu layer. Their experimental schematic diagram is shown in Figure 6.2.

The observation of Kirkendall effect provides important evidence to the vacancy diffusion mechanism, which implies that the diffusion coefficients of the two elements are different. The intrinsic diffusion coefficients $D_A$ and $D_B$ describe diffusion of the components $A$ and $B$ in a binary alloy relative to the lattice planes. The diffusion rates of $A$ and $B$ are usually not equal. As a result, the net flux of atoms across any lattice plane can exist during an interdiffusion experiment. If the number of lattice sites is conserved, the lattice plane should move in order to compensate the unequal flux, since the creation of lattice sites on one side of diffusion couple cause annihilation on the other side. The shift of lattice planes with respect to sample-fixed axes is the microscopic mechanism of the Kirkendall effect (Smigelkas and Kirkendall, 1947). The Kirkendall effect can thus be described as the motion of the boundary layer between two diffusion couples that occurs due to the difference in diffusion rates of the metal atoms. As schematically shown in Figure 6.3, Kirkendall voids can generate due to the different diffusion rates of two different metals.

Different types of diffusivities can be transformed to each other via several quantities such as concentration, volume, and thermodynamic factor. In a binary system, for example, the relationship between interdiffusion coefficient and intrinsic diffusion coefficient is expressed as follows:

$$\tilde{D} = C_B V_B D_A + C_A V_A D_B, \tag{6.5}$$

in which $V_i (i = A, B)$ is the partial molar volume of element $i$. $\tilde{D}$ is the interdiffusion coefficient calculated from intrinsic diffusion coefficients $D_A$ and $D_B$, and $C_A$ and $C_B$ are the concentrations of species $A$ and $B$, respectively. When the variation of the molar volume with composition could be neglected, (6.5) is expressed as

$$\tilde{D} = x_B D_A + x_A D_B, \tag{6.6}$$

in which $x_A$ and $x_B$ are the molar fractions of species $A$ and $B$, respectively. The Kirkendall velocity $v$ is given by an equation of the form

$$v = (D_A - D_B)\partial x_A / \partial x, \tag{6.7}$$

where $\partial x_A / \partial x$ denotes the concentration gradient at the marker position. Using (6.6) and (6.7), $D_A$ and $D_B$ can be calculated when $\tilde{D}$ and $v$ are measured.

The intrinsic diffusion coefficients $D_A$ and $D_B$ as well as the tracer diffusion coefficients $D_{AB}^{A^*}$ and $D_{AB}^{B^*}$ in an $AB$ alloy differ fundamentally. The latter pertains to a homogeneous alloy whereas the former is measured under a composition gradient. The intrinsic diffusion coefficients and the tracer self-diffusion coefficients are related via the following equations:

$$D_A = D_{AB}^{A^*} \varphi_{jk} \tag{6.8}$$

$$D_B = D_{AB}^{B^*} \varphi_{jk}, \tag{6.9}$$

where $\varphi_{jk}$ is named as a thermodynamic factor. The relations (6.5), (6.6), (6.8), and (6.9) are denoted as Darken's equations (Darken, 1948, 1949). In his work, Darken (1949) investigated both experimentally and theoretically the uphill diffusion of $C$ in austenite Fe–C–Si. According to the Fick's first law, the driving force for diffusion is concentration gradient. When the diffusion direction is the same as the concentration gradient, this kind of diffusion is called downhill diffusion. However, uphill diffusion can happen (especially in multicomponent systems), where the direction of the diffusion is against the concentration gradient. It should be stressed that the chemical potential gradient is the real driving force for diffusion.

## 6.2.2    Atomic Mechanism of Diffusion

In this section, we will very briefly describe the diffusion mechanisms in metals from an atomistic point of view. Atom jumping in a crystal can occur by several basic mechanisms that depend on several factors, such as crystal structure, the bonding behavior in host matrix, the size and electrical charge of diffusion species, and the preferred diffusing path of different species (e.g., anion or cation, substitutional or

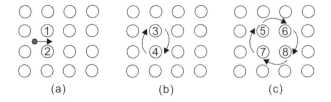

**Figure 6.4** Illustration of several diffusion mechanisms in crystals: (a) direct interstitial mechanism; (b) direct exchange mechanism of two neighboring atoms; (c) ring mechanism of 4 neighboring atoms.

interstitial). It is generally believed that mainly vacancies and interstitial atoms mediate the diffusion.

### 6.2.2.1 Interstitial Mechanism

Interstitial atoms like H, C, N, and O with small atomic radius may diffuse by jumping from one interstitial site to a nearby one, as indicated in Figure 6.4a. This is the so-called interstitial mechanism that involves no intrinsic point defects (vacancies, deviancies, self-interstitials, etc.). As a result, a direct interstitial diffusion is usually much faster than the diffusion of substitutional atoms.

### 6.2.2.2 Direct Exchange and Ring Mechanisms

Similar atom sizes of the solute and host species can usually result in the formation of substitutional solid solutions. Self- and substitutional solute diffusion in metals can also exhibit direct exchange of neighboring atoms (Figure 6.4b), in which two atoms move simultaneously. However, this process is quite unfavorable in a close-packed lattice due to the high activation barrier.

The ring mechanism (Figure 6.4c) corresponds to a rotation of three (or more) atoms as a group by one atom distance. Although this mechanism requires lower activation energy than direct exchange, the increase in atomic motion makes it more complex and unlikely for most crystalline substances.

### 6.2.2.3 Vacancy Mechanism

In the vacancy mechanism, vacant lattice sites act as diffusion vehicles. Species diffuse by jumping into a neighboring vacancy, as shown in Figure 6.5. Requiring a low activation energy, the vacancy mechanism exists in most metals and alloys.

### 6.2.2.4 Indirect Interstitial Mechanism

Diffusion may occur by the interstitial mechanism when the size of an interstitial atom is nearly equal to that of the lattice atom, which can also be stated as the indirect interstitial mechanism. A substitutional atom can migrate to a neighboring substitutional site by the two-step process, as illustrated in Figure 6.6. The first step is the exchange with an interstitial defect in which the migrating substitutional atom becomes an interstitial atom. The second step is to exchange the migrating atom with

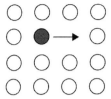

**Figure 6.5** Illustration of the vacancy mechanism: atom diffusion by jumping into the neighboring vacancy.

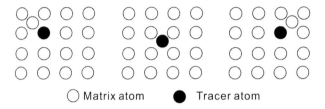

○ Matrix atom      ● Tracer atom

**Figure 6.6** Indirect interstitial mechanism of diffusion.

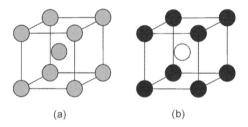

(a)                    (b)

**Figure 6.7** Atomic arrangements in binary $AB$ alloy with disordered $A2$ (a) and ordered $B2$ (b) structures. The $A$ and $B$ atoms are black and white spheres, respectively. The gray spheres in (a) represent sites at which the probabilities to find either an $A$ or $B$ atom are equal.

a neighboring substitutional atom. This mechanism is similar to the vacancy mechanism except the self-interstitial is the antidefect of the vacancy. This mechanism plays a dominate role in the self-diffusion of Si and possibly for some substitutional solutes.

### 6.2.2.5    Diffusion in Ordered Phase

In this subsection, the difference of atomic arrangements for disordered and ordered phase will be explained using a Bcc lattice as an example. As shown in Figure 6.7, $A$ and $B$ atoms are shown as black and white spheres, respectively. The gray spheres in Figure 6.7a represent sites at which the possibilities to find either an $A$ or $B$ atom are equal. The corresponding structure is called BCC_A2, in which different types of atoms are randomly distributed on the same crystalline lattice. In ordered Bcc $AB$ binary alloys, $A$ and $B$ atoms prefer for certain positions on the crystalline lattice, respectively, as shown in Figure 6.7b. The corresponding structure is called BCC_B2.

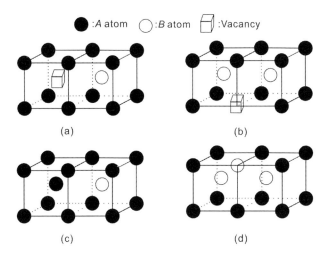

**Figure 6.8** Four types of substitutional defects in an ordered $AB$: (a) vacancy on the $B$ sublattice, (b) vacancy on the $A$ sublattice, (c) $A$ atom on the $B$ sublattice, and (d) $B$ atom on the $A$ sublattice.

In the latter case, two sublattices ($A$ and $B$ sublattices), which would preferably be occupied by $A$ and $B$ atoms, respectively, can be introduced.

In general, there are four types of basic substitutional defects in an ordered $AB$ compound: vacancies on an $A$ or $B$ sublattice, and antistructure atoms when atoms of $A$ occupy sites of $B$ or atoms of $B$ occupy sites of $A$, as shown in Figure 6.8a–d. In practice, these individual defects could occur simultaneously.

The intermetallic compounds with ordered structures geometrically limit the diffusion mechanism mediated by possible vacancy. Several diffusion mechanisms affect the diffusion in intermetallic compounds, including the six-jump cycle mechanism, sublattice diffusion mechanism, triple-defect diffusion mechanism, antistructure bridge mechanism, and interstitial diffusion mechanism. For more details about these mechanisms, see the book by Paul et al. (2014).

## 6.2.3 Interdiffusion in Binary, Ternary, and Multicomponent Systems

In this subsection, we will describe the widely used equations to calculate diffusion matrix. The diffusion matrix is one important input for many simulations for the sake of material design.

### 6.2.3.1 Interdiffusion in Binary Systems

In a binary system, Fick's first law is expressed as follows:

$$\tilde{J}_i = -\tilde{D}\frac{\partial C_i}{\partial x}(i = 1, 2), \tag{6.10}$$

in which $C_i$ is the concentration of component $i$, and $\tilde{J}_i$ the interdiffusion flux of the component $i$.

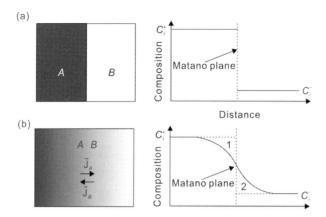

**Figure 6.9** Procedure of Boltzmann–Matano method to calculate the binary diffusion coefficient: (a) diffusion couples and composition profiles before annealing; (b) diffusion couples and composition profiles after annealing.

As shown in Figure 6.9, the Boltzmann–Matano method (Wagner, 1969) obtains the binary interdiffusivity via analyzing the concentration profile of one element in binary alloys. Supposing that the molar volume of the diffusion area is constant, the interdiffusion flux of the element $i$ at the concentration $C_i^*$ is as follows:

$$\tilde{J}_i = \frac{1}{2t} \int_{C_i^- \text{ or } C_i^+}^{C_i^*} (x - x_0)dC_i, \tag{6.11}$$

where the initial compositions of terminal alloys at left and right sides are $C_i^-$ and $C_i^+$, respectively, $x$ is the diffusion distance, and $x_0$ the Matano plane. $x_0$ is determined via the following condition:

$$\int_{C_i^-}^{C_i^+} (x - x_0)dC_i = 0. \tag{6.12}$$

Combining (6.11) and (6.12), one can obtain the binary interdiffusion coefficient:

$$\tilde{D}(C_i^*) = -\frac{1}{2t} \frac{\partial x}{\partial C_i} \left( \int_{C_i^-}^{C_i^*} (x - x_0)dC_i \right), \tag{6.13}$$

where $\tilde{D}(C_i^*)$ is the interdiffusion coefficient of component $i$ at the composition $C_i^*$. In principle, the Boltzmann–Matano method can be utilized only when the molar volume of the phase varies following Vergard's law. For most systems, the molar volume of the phase shows noticeable deviations from Vergard's law. In order to overcome this drawback associated with the Boltzmann–Matano method, a few methods for calculating interdiffusivities in binary alloys, such as the Sauer–Freise method (Sauer and Freise, 1962), the Wagner method (Wagner, 1969), and the Den Broeder method (Broeder, 1969), are proposed. Generally speaking, the calculated diffusion coefficients resulting from different methods are consistent with each other within estimated uncertainties. Refer to Wagner (1969) and Broeder (1969) for more detail on the Wagner and Den Broeder methods, respectively. In this section, the Sauer–Freise method is briefly described. This method introduced the concept of normalized concentration, which is expressed as follows:

$$Y_i = (C_i - C_i^-)/(C_i^+ - C_i^-). \tag{6.14}$$

Equation (6.15) can be obtained by Equations (6.11) and (6.14):

$$\tilde{J}_i = -\frac{1}{2t}\left(\frac{dC_i}{dY_i}\right)_{Y_i^*}\left[(1-Y_i)\int_{-\infty}^{x}Y_i dx + Y_i\int_{x}^{+\infty}(1-Y_i)dx\right]. \tag{6.15}$$

The interdiffusion coefficients in binary alloys can then be obtained as follows:

$$\tilde{D}(Y_i^*) = \frac{1}{2t}\left(\frac{dx}{dY_i}\right)_{Y_i^*}\left[(1-Y_i)\int_{-\infty}^{x}Y_i dx + Y_i\int_{x}^{+\infty}(1-Y_i)dx\right], \tag{6.16}$$

where $\tilde{D}(Y_i^*)$ is the interdiffusion coefficient at the normalized concentration $Y_i^*$. It is not necessary for the Sauer–Freise method to calculate the Matano plane, which can eliminate the error introduced by Matano plane calculation to a certain extent.

The present authors have written one code to implement the Sauer-Freise method for the sake of calculating the interdiffusion coefficients in binary alloys. The code is presented in Appendix A.

In a real system, it is rare to find the molar volume varying with the composition following Vegard's law. The deviation of the molar volume from the ideality will result in a change for the total volume of the diffusion couple. Here, we show the relation between the changes in total volume with the location of the initial contact planes estimated via a quantitative analysis.

In a hypothetical diffusion couple with $B$-lean alloy $C_B^-$ (or $C_A^-$ expressed concerning the element $A$) and $B$-rich alloy $C_B^+(C_A^+)$, the position of the initial contact plane can be obtained by the composition-normalized variable $Y_B/V_m$ versus $x$ and $Y_A/V_m$ versus $x$ plots, which is the derivation of Wagner's relation. Note the following:

$$Y_A = \frac{C_A - C_A^-}{C_A^+ - C_A^-} = 1 - Y_B.$$

The $Y_B/V_m$ versus $x$ and $Y_A/V_m$ versus $x$ plots are shown in Figure 6.10. The initial contact plane is determined by equalizing the areas $P$ and $Q$. In order for clear illustration, the initial contact plane by $Y_B/V_m$ versus $x$ is set to be $x_0^I$ and by $(1-Y_B)/V_m$ versus $x$ plot is $x_0^{II}$.

At a specified composition, suppose the uncertainty of molar volume by Vegard's law, or excess volume, is $\pm\Delta V_m$ $(V_m = V_m^- + Y_B(V_m^+ - V_m^-)\pm\Delta V_m)$. Then the following equation can be obtained:

$$x_0^{II} - x^{-\infty} + x^{+\infty} - x_0^I = \int_{x^{-\infty}}^{x^{+\infty}}\frac{V_m \pm \Delta V_m}{V_m}dx \tag{6.17}$$

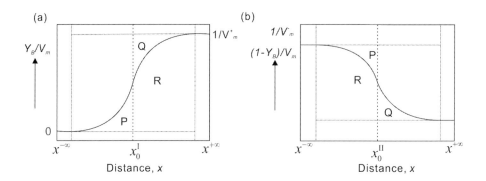

**Figure 6.10** Profiles utilized to derive the molar volume influence on the total volume of the diffusion couple. (a) $Y_B/V_m$ versus $x$ and (b) $(1 - Y_B)/V_m$ versus $x$ plots.

$$x_0^I - x_0^{II} = \pm \int_{x^{-\infty}}^{x^{+\infty}} \frac{\Delta V_m}{V_m} dx. \tag{6.18}$$

Equation (6.18) shows that the ideal case with no excess volume $\pm \Delta V_m = 0$ demonstrates the identical initial contact plane from $Y_B/V_m$ versus $x$ and $Y_A/V_m$ versus $x$ plots. When the molar volume changes ideally following Vegard's law with the composition, its excess volume will be zero. In (6.18),

$$\int_{x^{-\infty}}^{x^{+\infty}} \frac{\Delta V_m}{V_m} dx$$

measures the total volume change of the diffusion couple. In view of the linear profile in one diffusion couple, it can be supposed that the volume change is primarily due to the variation of the length of the diffusion couple. Therefore, the difference between the estimated positions of the initial contact plane should equal to the variation of the length of the diffusion couple.

### 6.2.3.2    Interdiffusion in Ternary Systems
*Matano–Kirkaldy Method*
For ternary phases, four interdiffusivities are needed to express the interdiffusion fluxes $\tilde{J}_1$ and $\tilde{J}_2$ of components 1 and 2. The relationship among $\tilde{J}_1$, $\tilde{J}_2$, and $\tilde{J}_3$ is $\tilde{J}_1 + \tilde{J}_2 + \tilde{J}_3 = 0$. The main interdiffusion coefficients $\tilde{D}_{11}^3$ and $\tilde{D}_{22}^3$ represent the effects of the composition gradients of elements 1 and 2 on their own fluxes, respectively. $\tilde{D}_{12}^3$ and $\tilde{D}_{21}^3$ are the cross-interdiffusivities, representing the effects of the composition gradients of element 2 on the fluxes of element 1 and of element 1 on the fluxes of element 2, respectively.

The differential equations for Fick's second law can be derived by substituting (6.3) into the continuity equation

$$\frac{\partial C_i}{\partial t} = \frac{\partial}{\partial x}\left(\tilde{D}_{i1}^3 \frac{\partial C_1}{\partial x}\right) + \frac{\partial}{\partial x}\left(\tilde{D}_{i2}^3 \frac{\partial C_2}{\partial x}\right) \quad (i = 1, 2). \tag{6.19}$$

Regarding the solutions of diffusion for the constant ternary interdiffusion coefficients, many textbooks on diffusion can be consulted. In this chapter, a general solution to obtain the concentration-dependent ternary interdiffusion coefficients is briefly described.

The Boltzmann–Matano analysis described earlier for binary systems can be extended into a ternary one (Kirkaldy et al., 1963):

$$
\int_{C_1^-}^{C_1^+} x \, dC_1 = -2t \left( \tilde{D}_{11}^3 \frac{\partial C_1}{\partial x} + \tilde{D}_{12}^3 \frac{\partial C_2}{\partial x} \right)
$$

$$
\int_{C_2^-}^{C_2^+} x \, dC_2 = -2t \left( \tilde{D}_{21}^3 \frac{\partial C_1}{\partial x} + \tilde{D}_{22}^3 \frac{\partial C_2}{\partial x} \right).
$$

(6.20)

Supposing that the volume variation can be neglected, the location of the Matano plane should be the same for composition profiles of elements 1 and 2 in theory. For avoiding the influence of the Matano plane on the diffusivities, the normalized concentrations $Y_i = \frac{\left(C_i - C_i^-\right)}{\left(C_i^+ - C_i^-\right)}$ suggested by Whittle and Green (1974) is introduced to avoid calculating the Matano plane. Consequently, (6.20) can be written as follows using the normalized concentrations $Y_i$:

$$
\tilde{D}_{11}^3 + \tilde{D}_{12}^3 \frac{dC_2}{dC_1} = \frac{1}{2t} \frac{dx}{dY_1} \left[ (1 - Y_1) \int_{-\infty}^{x} Y_1 \cdot dx + Y_1 \int_{x}^{+\infty} (1 - Y_1) \cdot dx \right]
$$

$$
\tilde{D}_{22}^3 + \tilde{D}_{21}^3 \frac{dC_1}{dC_2} = \frac{1}{2t} \frac{dx}{dY_2} \left[ (1 - Y_2) \int_{-\infty}^{x} Y_2 \cdot dx + Y_2 \int_{x}^{+\infty} (1 - Y_2) \cdot dx \right].
$$

(6.21)

Since there are four concentration-dependent ternary interdiffusion coefficients, the estimation of these four parameters is not straightforward as that in a binary system. Therefore, two couples are needed that are designed to have a common composition in their diffusion paths. As shown in a fictitious $A$–$B$–$C$ ternary system, Figure 6.11 presents two diffusion paths with a common intersection point $(C_1, C_2)$. The four equations from these two couples are solved by calculating the integrals and

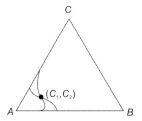

**Figure 6.11** Diffusion paths for two diffusion couples in the $A$–$B$–$C$ ternary system. $A$ is solvent, and solutes are $B$ and $C$. In the diffusion equations, $1 = B$ and $2 = C$.

derivatives in (6.21) at $(C_1, C_2)$ in order to obtain $\tilde{D}_{11}^3$, $\tilde{D}_{22}^3$, $\tilde{D}_{12}^3$, and $\tilde{D}_{21}^3$ at the intersection point. The Matano–Kirkaldy method is by far the most reliable method to calculate the ternary diffusivity from the concentration profiles.

Note that the obtained diffusion coefficients should satisfy the following conditions, which are obtained according to the thermodynamic stability of the solutions (Kirkaldy et al., 1963).

$$\tilde{D}_{11}^3 + \tilde{D}_{22}^3 > 0$$

$$\tilde{D}_{11}^3 \cdot \tilde{D}_{22}^3 - \tilde{D}_{12}^3 \cdot \tilde{D}_{21}^3 \geq 0 \tag{6.22}$$

$$\left(\tilde{D}_{11}^3 - \tilde{D}_{22}^3\right)^2 + 4 \cdot \tilde{D}_{12}^3 \cdot \tilde{D}_{21}^3 \geq 0.$$

The present authors have written a code to implement the Matano–Kirkaldy method for evaluating the interdiffusivities in ternary alloys. The code is presented in Appendix A.

### Numerical Methods for Evaluating a Diffusion Matrix along the Whole Diffusion Path of a Single Diffusion Couple

In order to conduct a high-throughput determination for obtaining the composition-dependent interdiffusivities along a single ternary diffusion couple, effective numerical methods are needed. In the following, we will describe such a method (Du et al., 2020), which concentrates on the aspects about the accuracy and stability of diffusivity calculation.

The method is based on the minimization of the difference between calculated and experimental concentrations, and the numerical process is implemented by the integration of both a finite element method (FEM) and a genetic algorithm (GA). FEM is applied to solve the forward problem. This means that (6.19) is employed to calculate the gradient of a metric function associated with the sum of the difference between the measured and calculated concentration profiles. GA is used to optimize model parameters. To obtain the interdiffusion coefficients, one could consider two cases for $\tilde{D}_{ij}^3$: constant and composition-dependent interdiffusivities. Inspired by Matano–Boltzmann analysis (Boltzmann, 1894; Matano, 1933), the concept of basis functions is utilized in the method (Du et al., 2020), in which some functions can be applied to describe the continuous change of interdiffusion coefficient. By introducing basis functions $\phi_{ij}^k(C_1, C_2, C_3)$, the interdiffusion coefficients can be written as follows:

$$\tilde{D}_{ij}^3 = \tilde{D}_{ij}^3(C_1, C_2, C_3) = \sum_{k=1}^{N} \alpha_{ij}^k \phi_{ij}^k(C_1, C_2, C_3), \tag{6.23}$$

where $\alpha_{ij}^k$ is the parameter, which is independent of the compositions $C_i$ ($i = 1, 2, 3$); $\phi_{ij}^k(C_1, C_2, C_3)$ are linearly independent basis functions; and the following types of basis functions are applied:

$$\left\{1, C_i, C_i C_j, \cdots\right\}_{i,j=1}^{3} \tag{6.24}$$

and

$$\left\{\exp\left(-C_i\right), C_i \exp\left(-C_j\right), C_i C_j \exp\left(-C_i C_j\right), \cdots\right\}_{i,j=1}^{3}, \tag{6.25}$$

$\alpha_{ij}^k (k = 1, 2, \ldots)$ are the coefficients to be optimized from the measured concentration profiles. For different alloy systems, the suitable basis functions can be selected based on the characteristics of the interdiffusivities for improving efficiency.

This method is performed by minimizing the deviation between evaluated compositions $C_i^{cal}$ and experimental compositions $C_i^{exp}$. Note that evaluating the diffusion coefficient $\tilde{D}_{ij}^3$ along the whole diffusion path is converted to evaluating coefficient $\alpha_{ij}^k$. Therefore, the inverse problem of parameter inversion is transformed into the optimization problem:

$$\min_{\alpha} \sum_{i=1}^{2} \left\| C_i^{exp}(z) - C_i^{cal}(z, \alpha) \right\|_2, \tag{6.26}$$

where $\|\cdot\|$ is a representation of the 2-norm in mathematical meaning; $z = (z_1, z_2, \ldots, z_s)$ is the position of experimental concentration data, $C_i^{cal}(z, \alpha)$ and $C_i^{exp}(z)$ are calculated and experimental concentrations, respectively, and $\alpha$ is the undetermined coefficient.

One can calculate coefficients $\alpha_{ij}^k$ using the genetic algorithm. The result can be guaranteed to be globally optimal. The stopping criterion for the genetic algorithm is as follows:

$$\sum_{i=1}^{2} \sqrt{\frac{1}{s} \sum_{k=1}^{s} \left[ C_i^{exp}(z_k) - C_i^{cal}(z_k, \alpha) \right]^2} < \varepsilon, \tag{6.27}$$

where $\varepsilon$ is a small parameter that can be selected according to the experimental feature. Meanwhile, the deterministic algorithms, such as the gradient descent method, can be also applied to the optimization.

A two-step procedure is proposed to compute the composition-dependent interdiffusivities along the whole diffusion path. In the first step, the constant diffusion coefficient is calculated, which serves as the initial value for the calculation of component-dependence diffusivities. In the second step, the composition-dependent diffusivities are calculated via a genetic algorithm and the gradient descent method. Figure 6.12 shows the diffusion paths and data measured in diffusion couples. Figure 6.13 shows the interdiffusion coefficients along the whole diffusion paths for diffusion couples A1, A2, A3, and A4 at 1,073 K calculated by the numerical method in comparison with the ones in symbols calculated by the Matano–Kirkaldy method, where a good agreement can be seen. A homemade code is made for implementing the novel algorithm. Such a code is available from the authors upon request.

*Average Diffusion Coefficients*
For practical applications in which diffusion coefficients do not change noticeably, using average diffusion coefficients within a composition range could be a reasonable treatment. The average effective interdiffusivity in a ternary system was developed by Dayananda and Sohn (1999), which can be obtained with a single diffusion couple.

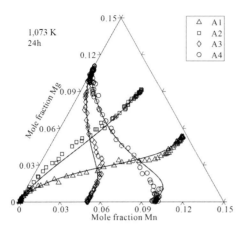

**Figure 6.12** Diffusion paths and composition profiles of Mn and Mg in the Fcc Ag–Mg–Mn alloys, in comparison with the ones measured in diffusion couples of A1: Ag/Ag–9.3Mn–5.3Mg, A2: Ag/Ag–4.4Mn–9.0Mg, A3: Ag–5.1Mn/Ag–10.9Mg and A4: Ag–10.0Mn/Ag–10.7Mg (at.%) at 1,073 K for 24 hours (Du et al., 2020). The figures are calculated by means of the calculation of thermophysical properties (CALTPP) program (Liu et al., 2020).

For a ternary system, there exists the following relation for the effective interdiffusion coefficient:

$$\tilde{J}_i = -\tilde{D}_i^{eff} \frac{\partial C_i}{\partial x}. \tag{6.28}$$

Integrating (6.28) from the left side of the Matano plane, we can obtain the following:

$$\int_{x^{-\infty}}^{x_0} \tilde{J}_i dx = -\int_{C_i^-}^{C_i^0} \tilde{D}_i^{eff} dC_i = \bar{\tilde{D}}_{i,L}^{eff} \left( C_i^- - C_i^0 \right). \tag{6.29}$$

$C_i^0$ is the concentration of element $i$ at the Matano plane. Then we can derive the expression of average effective interdiffusion coefficient on the left side of Matano plane, i.e., $\bar{\tilde{D}}_{i,L}^{eff}$:

$$\bar{\tilde{D}}_{i,L}^{eff} = \frac{\int_{x^{-\infty}}^{x_0} J_i dx}{C_i^- - C_i^0} = \frac{M_{Integ}}{C_i^- - C_i^0}. \tag{6.30}$$

$M_{Integ}$ is the integration of interdiffusion fluxes. Similarly, there exists the equation at the right side of the Matano plane:

$$\bar{\tilde{D}}_{i,R}^{eff} = \frac{\int_{x_0}^{x^{+\infty}} \tilde{J}_i dx}{C_i^0 - C_i^+} = \frac{M_{Integ}}{C_i^0 - C_i^+}. \tag{6.31}$$

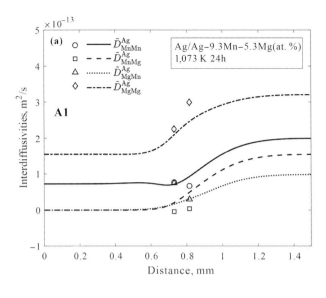

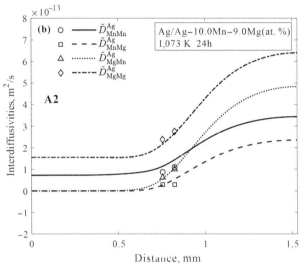

**Figure 6.13** Interdiffusivities along the whole diffusion paths for diffusion couples A1, A2, A3, and A4 at 1,073 K, shown in (a)–(d), calculated by the numerical method in comparison with those by the Matano–Kirkaldy method (denoted in symbols) (Du et al., 2020). The figures are reproduced from Du et al. (2020) with permission from Elsevier.

Instead of estimating one average effective interdiffusivity, Sohn and Dayananda (2002) developed another model to calculate the average values of the main and cross-interdiffusion coefficients. For a ternary system, there exists such an equation:

$$\tilde{J}_i = -\tilde{D}_{i1}^3 \frac{\partial C_1}{\partial x} - \tilde{D}_{i2}^3 \frac{\partial C_2}{\partial x}. \tag{6.32}$$

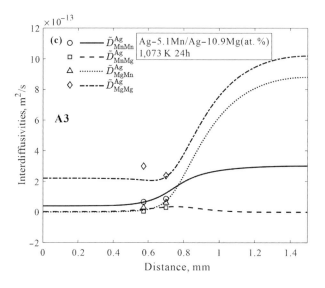

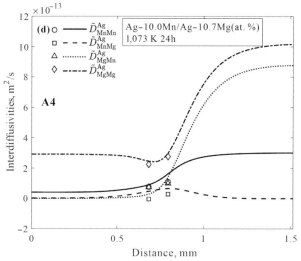

Figure 6.13 (*cont.*)

Integrating (6.32) from composition $x_1$ to $x_2$, one can obtain the following:

$$\int_{x_1}^{x_2} \tilde{J}_i dx = -\int_{C_1^{x_1}}^{C_1^{x_2}} \tilde{D}_{i1}^3 dC_1 - \int_{C_2^{x_1}}^{C_2^{x_2}} \tilde{D}_{i2}^3 dC_2 = -\bar{\tilde{D}}_{i1}^3 \left( C_1^{x_2} - C_1^{x_1} \right) - \bar{\tilde{D}}_{i2}^3 \left( C_2^{x_2} - C_2^{x_1} \right), \quad (6.33)$$

where $\bar{\tilde{D}}_{i1}^3$ and $\bar{\tilde{D}}_{i2}^3$ are average interdiffusion coefficients for element $i$ over the composition range $C_1^{x_1}$ and $C_2^{x_2}$, which means the composition at distance $x_1$ and $x_2$, respectively. Multiplying (6.33) by $(x - x_0)$ and then integrating the new equation from composition $x_1$ to $x_2$, the following equation can be obtained:

$$\int_{x_1}^{x_2} \tilde{J}_i(x - x_0)dx = 2t\left[\bar{\tilde{D}}_{i1}^3(\tilde{J}_i^{x_2} - \tilde{J}_i^{x_1}) - \bar{\tilde{D}}_{i2}^3(\tilde{J}_i^{x_2} - \tilde{J}_i^{x_1})\right],$$ (6.34)

where $\tilde{J}_i^{x_1}$ and $\tilde{J}_i^{x_2}$ are the interdiffusion fluxes of element $i$ at distance $x_1$ and $x_2$, respectively.

Based on Fick's first law, the integrated diffusion coefficients can be calculated for component $i$ by:

$$\tilde{D}_{int,i}^{\beta} = \int_{N_B^{\beta_1}}^{N_B^{\beta_2}} \bar{\tilde{D}}_i^{eff} dN_B = -v_m^{\beta}\int_{x^{\beta_1}}^{x^{\beta_2}} \tilde{J}_i dx.$$ (6.35)

Accordingly, the average effective interdiffusion coefficient can be also obtained by the integrated diffusion coefficient in the following equation:

$$\bar{\tilde{D}}_i^{eff} = \frac{\tilde{D}_{int}^{\beta}}{\Delta N_B^{\beta}}.$$ (6.36)

### 6.2.3.3    Interdiffusion in Multicomponent Systems

In multicomponent systems, the diffusion behavior is very complicated and the effect of cross-terms on interdiffusivity should be considered during diffusion. An extended form of Fick's law, which was proposed by Onsager (1931, 1945), is utilized for the treatment of the interdiffusion flux of component $k$ referred to a laboratory fixed frame of reference in a multicomponent system:

$$\tilde{J}_k = -\sum_{j=1}^{n-1} \tilde{D}_{kj}^n \frac{\partial C_j}{\partial x} \quad (k = 1, 2, \ldots, n-1),$$ (6.37)

in which $\tilde{J}_k$ is the interdiffusion flux of the component $k$, $\tilde{D}_{kj}^n$ the interdiffusion coefficient, $n$ the number of the components, and $C_j$ the concentration of component $j$. The relationship among these interdiffusion fluxes is as follows:

$$\sum_{k=1}^{n} \bar{v}_k \tilde{J}_k = \bar{v}_1 \tilde{J}_1 + \bar{v}_2 \tilde{J}_2 + \cdots + \bar{v}_n \tilde{J}_n = 0,$$ (6.38)

where $\tilde{v}_k$ is the partial molar volume of component $k$. Following (6.37) and (6.38), the interdiffusion fluxes of $(n-1)$ components are independent in a system with $n$ components. Thus, these $(n-1)$ independent interdiffusion fluxes should be described by the expressions of $(n-1)^2$ interdiffusion coefficients.

Similarly, the intrinsic diffusion flux $J_k$ of the component $k$ relative to a lattice-fixed frame in a multicomponent system is expressed by

$$J_k = -\sum_{j=1}^{n-1} D_{kj}^n \frac{\partial C_j}{\partial x} \quad (j = 1, 2, \ldots, n-1),$$ (6.39)

where $D_{kj}^n$ is the intrinsic diffusion coefficient.

Since the interdiffusion and intrinsic fluxes are related through the lattice or maker velocity $v$, the interdiffusion coefficients related to intrinsic diffusion coefficients under the condition of a constant molar volume can be expressed by the following equation:

$$\tilde{D}_{kj}^n = D_{kj}^n - x_k \sum_{i=1}^n D_{ij}^n \quad (k, j = 1, 2, \ldots, n-1). \tag{6.40}$$

Accordingly, the interdiffusion fluxes can also be described in terms of the intrinsic diffusion fluxes:

$$\tilde{J}_k = J_k - x_k \sum_{i=1}^n J_i \quad (k, j = 1, 2, \ldots, n-1). \tag{6.41}$$

In this chapter, the basic equations about diffusion in multicomponent systems are very briefly shown. The reader could refer to Bouchet and Mevrel (2003) for a detailed introduction about diffusion in multicomponent systems.

To demonstrate the application of basic diffusion equations in multicomponent systems, we will take the Cu–Mn–Ni–Zn quaternary alloys as the targeted system. In the Cu–Mn–Ni–Zn quaternary system, the interdiffusion fluxes in the volume-fixed frame of reference $J_i^v$ ($i$ = Cu, Mn, Ni, Zn) are given by the following equations:

$$
\begin{aligned}
J_{Cu}^V &= -\tilde{D}_{CuMn}^{Cu} \nabla C_{Mn} - \tilde{D}_{CuNi}^{Cu} \nabla C_{Ni} - \tilde{D}_{CuZn}^{Cu} \nabla C_{Zn} \\
J_{Mn}^V &= -\tilde{D}_{MnMn}^{Cu} \nabla C_{Mn} - \tilde{D}_{MnNi}^{Cu} \nabla C_{Ni} - \tilde{D}_{MnZn}^{Cu} \nabla C_{Zn} \\
J_{Ni}^V &= -\tilde{D}_{NiMn}^{Cu} \nabla C_{Mn} - \tilde{D}_{NiNi}^{Cu} \nabla C_{Ni} - \tilde{D}_{NiZn}^{Cu} \nabla C_{Zn} \\
J_{Zn}^V &= -\tilde{D}_{ZnMn}^{Cu} \nabla C_{Mn} - \tilde{D}_{ZnNi}^{Cu} \nabla C_{Ni} - \tilde{D}_{ZnZn}^{Cu} \nabla C_{Zn}.
\end{aligned} \tag{6.42}
$$

The cross-coefficients are significant in a multicomponent system, which may get a positive flux of a component in the direction of increasing concentration, leading to the so-called uphill diffusion, zero-flux, flux reversals planes, and some complicated diffusion behavior in multicomponent systems.

For such four interdiffusion fluxes defined in the volume-fixed frame of reference, it is valid that

$$J_{Cu}^V + J_{Mn}^V + J_{Ni}^V + J_{Zn}^V = 0. \tag{6.43}$$

Accordingly, 12 interdiffusion coefficients can be constrained by

$$
\begin{aligned}
\tilde{D}_{CuMn}^{Cu} + \tilde{D}_{MnMn}^{Cu} + \tilde{D}_{NiMn}^{Cu} + \tilde{D}_{ZnMn}^{Cu} &= 0 \\
\tilde{D}_{CuNi}^{Cu} + \tilde{D}_{MnNi}^{Cu} + \tilde{D}_{NiNi}^{Cu} + \tilde{D}_{ZnNi}^{Cu} &= 0 \\
\tilde{D}_{CuZn}^{Cu} + \tilde{D}_{MnZn}^{Cu} + \tilde{D}_{NiZn}^{Cu} + \tilde{D}_{ZnZn}^{Cu} &= 0.
\end{aligned} \tag{6.44}
$$

Therefore, nine interdiffusion coefficients must be solved in a quaternary system under the condition of three independent quaternary diffusion couples with a common composition. It is quite difficult to determine the accurate quaternary diffusivities based on the classic Matano–Kirkaldy method since the intersection of three diffusion paths in isothermal three-dimensional quaternary phase diagram space is required. Fortunately, on the basis of accurate atomic mobility parameters in ternary systems, it is possible to estimate the quaternary interdiffusion coefficients. For example, an effective approach to establish the atomic mobility parameters of quaternary Cu-rich Fcc Cu–Mn–Ni–Zn alloys (Zhang et al., 2014b) is presented by means of diffusion-controlled transformation (DICTRA) software. DICTRA modeling will be discussed in Section 6.2.6.4.

## 6.2.4  Diffusion in Phases with Narrow Homogeneity Ranges

In Section 6.2.3, we described the diffusion of a single-phase solid solution phase in binary, ternary, and multicomponent systems. The growth of the phases with a narrow homogeneity range is presented in this section. The phases with narrow homogeneity ranges are widely found in commercial alloys as intermetallic phases. Therefore, the description of their diffusion behavior is of both technological importance and theoretical interest.

### 6.2.4.1  Wagner's Approach

To describe the diffusion-controlled growth rate of compounds in a diffusion couple, the most common way is by parabolic growth constants, as shown in (6.45):

$$\Delta x^2 = kt, \tag{6.45}$$

where $\Delta x$ is the thickness of compound, and $k$ the parabolic growth constant. The assumption of diffusion-controlled growth may not be valid in the initial stage, where linear growth may be observed.

Based on the Arrhenius equation, the activation energy for the diffusion can be determined by implementing diffusion experiments at different temperatures:

$$k = k_0 \exp\left(-\frac{Q}{RT}\right) \tag{6.46}$$

$$\ln k = \ln k_0 - \frac{Q}{RT}, \tag{6.47}$$

where $R$ is the gas constant, $Q$ the activation energy, and $k_0$ the preexponential factor. The activation energy can be estimated from the slope $(-Q/R)$ of the $\ln k$ versus $(1/T)$ plot.

It should be pointed that the parabolic growth constant depends upon the composition of the endmembers of diffusion couple, and it is not a material constant. For this reason, we may draw the wrong conclusion if only the calculation of the parabolic growth constant is considered in the study of diffusion behavior. As the material constant, the diffusion coefficient is a safe choice when we discuss diffusion behavior.

The interdiffusion coefficient of a phase with a wide homogeneity range can be derived by Wagner's approach (Wagner, 1969), which is expressed as follows:

$$\tilde{D}(Y_B^*) = \frac{V_m^*}{2t(dY_B/dx)Y_B^*} \left[ (1 - Y_B^*) \int_{x^{-\infty}}^{x^*} \frac{Y_B}{V_m} dx + Y_B^* \int_{x^*}^{x^{+\infty}} \frac{(1 - Y_B)}{V_m} dx \right], \quad (6.48)$$

where $Y_B^* = \frac{C_B^* - C_B^-}{C_B^+ - C_B^-}$, the asterisk (*) means the concentration of interest, such as $C_B^*$. "+" and "−" represent the unreacted right- and left-hand side of the diffusion couple, respectively.

To determine the interdiffusion coefficient of a phase with a narrow homogeneity range, the calculation of the gradient $\frac{dY_B}{dx} = \frac{1}{x_B^+ - x_B^-} \frac{dC_B}{dx}$ is necessary. There is little prospect of obtaining the gradient in a narrow homogeneity range phase. To overcome this difficulty, the concept of integrated diffusion coefficient $\tilde{D}_{\text{int}}$ was proposed by Wagner (1969). $\tilde{D}_{\text{int}}$ is the interdiffusion coefficient $\tilde{D}$ integrated over the composition range $\left( \Delta C_B^\beta = C_B^{\beta_2} - C_B^{\beta_1} \right)$, which is expressed as follows:

$$\tilde{D}_{\text{int}}^\beta = \int_{C_B^{\beta_1}}^{C_B^{\beta_2}} \tilde{D} dC_B^\beta. \quad (6.49)$$

A fictitious phase diagram displayed in Figure 6.14a is taken as an example. The diffusion couple consists of an alloy in phase $\alpha(C_B^-)$ and an alloy in $\gamma$ phase $(C_B^+)$. $\alpha$, $\beta$, and $\gamma$ phases could grow during the diffusion process, and their composition profiles are shown in Figure 6.14b. The composition profile of the $\beta$ phase exhibits a negligible gradient due to the narrow composition range of the $\beta$ phase.

Using Wagner's approach (Wagner, 1969), the interdiffusion coefficient in the narrow composition range $\beta$ phase can be derived as (6.50):

$$\tilde{D}_{\text{int}}^\beta = \frac{\left(C_B^\beta - C_B^-\right)\left(C_B^+ - C_B^\beta\right)}{C_B^+ - C_B^-} \frac{(\Delta x_\beta)^2}{2t} + \frac{\Delta x_\beta}{2t} \frac{1}{C_B^+ - C_B^-}$$
$$\times \left[ \left(C_B^+ - C_B^\beta\right) \sum_{i=2}^{\beta-1} \frac{V_m^\beta}{V_m^i} \left(C_B^i - C_B^-\right) \Delta x_i + \left(C_B^\beta - C_B^-\right) \sum_{i=\beta+1}^{n-1} \frac{V_m^\beta}{V_m^i} \left(C_B^+ - C_B^i\right) \Delta x_i \right]. \quad (6.50)$$

### 6.2.4.2    Du and Schuster Approach

We will present another approach to describe the diffusion of the phase with a negligible composition range. To simulate the diffusion of the stoichiometric compounds, Du and Schuster (2001) proposed the concept of "average thermodynamic factor $\varphi_{\text{ave}}$." Based on this method, the diffusion growth in the Ni–Si system was simulated. The $\varphi_{\text{ave}}$ is defined by the following equation:

$$\varphi_{\text{ave}} = \left(c_\gamma^1 - c_\gamma^0\right)^{-1} \int_{c_\gamma^0}^{c_\gamma^1} \varphi dx, \quad (6.51)$$

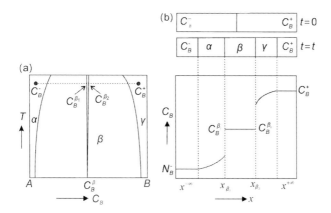

**Figure 6.14** (a) A fictitious phase diagram and (b) a composition profile of the diffusion couple.

where $c_\gamma^0$ and $c_\gamma^1$ are the homogeneity limits of $\gamma$ phase in the phase diagram. The relationship among the average interdiffusion coefficient $(D_\gamma^{ave})$ and tracer self-diffusion coefficient $(D_\gamma^{Ni*}$ and $D_\gamma^{Si*})$ is expressed as follows:

$$D_\gamma^{ave} = \left[ C_{Si} D_\gamma^{Ni*} + C_{Ni} D_\gamma^{Si*} \right] \varphi_{ave}, \tag{6.52}$$

where $C_{Si}$ and $C_{Ni}$ are the mole fractions of Si and Ni in the $\gamma$ phase, respectively.

Figure 6.15 schematically illustrates the growth of the $\gamma$ phase between the adjacent phases $Ni_3Si$ and $Ni_2Si$. The compositions at the interfaces and boundaries are obtained according to the thermodynamic calculation. The growth velocities of the two interface boundary positions of the $\gamma$ phase in such a diffusion couple are expressed in the following two equations:

$$\partial\xi_1/\partial t = \left(-D_{Ni_3Si}^{ave} \partial C_{Ni_3Si}/\partial x + D_\gamma^{ave} \partial C_\gamma/\partial x\right) / \left(C_{Ni_3Si}^1 - C_\gamma^0\right) \tag{6.53}$$

$$\partial\xi_2/\partial t = \left(-D_\gamma^{ave} \partial C_\gamma/\partial x + D_{Ni_2Si}^{ave} \partial C_{Ni_2Si}/\partial x\right) / \left(C_\gamma^1 - C_{Ni_2Si}^0\right). \tag{6.54}$$

And the thickness of the $\gamma$ phase $\Delta x_\gamma$ is as follows:

$$\Delta x_\gamma = \xi_2 - \xi_1. \tag{6.55}$$

The tracer diffusion coefficients of Ni and Si in the silicides can be expressed by the Arrhenius equation:

$$D_\gamma^{i*} = D_\gamma^{i0} \exp\left(-Q_\gamma^i/RT\right), (i = \text{Ni or Si}), \tag{6.56}$$

where $D_\gamma^{i0}$ is the preexponential factor and $Q_\gamma^i$ the activation enthalpy.

Figure 6.16 shows the calculated thicknesses of the reaction layer $\gamma$ ($Ni_5Si_2$) in the $Ni_3Si/Ni_2Si$ diffusion couple versus $\sqrt{t}$ (Du and Schuster, 2001). The unique feature

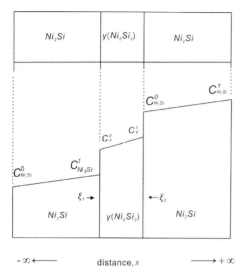

**Figure 6.15** Concentration profiles for the Ni₃Si/Ni₂Si diffusion couple. The concentrations denoted $C$ are obtained according to the thermodynamic calculation.

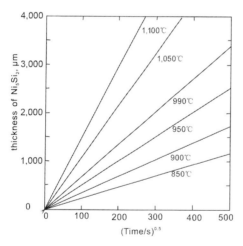

**Figure 6.16** Calculated growth rates of $\gamma$ (Ni₅Si₂) at various temperatures (Du and Schuster, 2001). The figures are reproduced from Du and Schuster (2001) with permission from Elsevier.

of Du and Schuster's approach is that the simulation can be performed by linking diffusion equations to thermodynamic calculations directly. Such a feature makes Du and Schuster's approach useful for simulations involving diffusion modeling for the phases with narrow homogeneity ranges.

Appendix A lists the thermodynamic and diffusion parameters in the Ni–Si systems, which are used to perform diffusion simulation in Figure 6.16.

## 6.2.5 Short-Circuit Diffusion

Compared with gases and liquids, crystals have several different structural paths for atoms to diffuse. In the preceding sections of this chapter, we have described lattice diffusion (also denoted as bulk diffusion), in which defects such as vacancies and interstitials can enhance diffusion in the crystal. Experiments demonstrate that along crystal imperfections such as dislocations, internal interfaces (such as grain boundaries in monophase materials and interphase boundaries in a multiphase material), and free surfaces, diffusion rates can be orders of magnitude faster than those in crystals containing only point defects (Darken, 1948). Because of the high diffusivity in these regions, the term *short-circuit diffusion* is utilized to describe such very fast diffusion phenomena, as illustrated in Figure 6.17. The effect of short-circuit diffusion can be illustrated by the deep penetrating frontiers around the grain boundary, the dislocation line, and also near the free surface.

The phenomena of short-circuit diffusion in metals and alloys have attracted extensive attention (Paul et al., 2014). These short-circuit paths, such as grain boundaries and dislocations, may greatly influence the process of interdiffusion, since the diffusion of atoms along these defects are significantly enhanced. Processes such as alloying, oxidation, and sintering are typical phenomena that demand defect-enhanced diffusion. Short-circuit diffusion is also important for various modern technologies. For example, short-circuit diffusion is often the dominating step for structural changes of magnetic storage devices, thin-film microelectronics, and optoelectronics. In addition, diffusion barriers are often used to avoid undesirable intermixing between individual layers of a thin-film device, and its efficiency is also affected by the short-circuit diffusion characteristics. Hence, it is important to understand and control these processes in order to improve the stability and ensure the integrity of the thin-film devices.

For grain-boundary diffusion, most mathematical treatments are based on the Fisher model (Fisher, 1951). In this model, the grain boundary, which is semi-infinite,

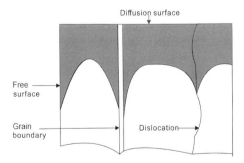

**Figure 6.17** Schematic illustration of short-circuit diffusion paths in a solid. Diffusion surface means the source for diffusion, and the gray part is the region in which atoms diffuse. Generally speaking, the diffusion along a free surface is fastest, followed by the diffusion along high-angle grain boundaries as well as slower diffusion along low-angle grain boundaries and dislocation in comparison with free surface and high-angle grain boundary.

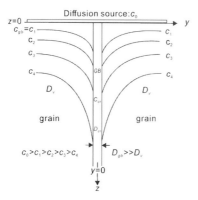

**Figure 6.18** Typical concentration contours according to Fisher model from a constant diffusion source. $D_v$: lattice diffusivity, $D_{gb}$: grain-boundary diffusivity, $\delta$: grain-boundary width, and GB: grain boundary.

uniform, and isotropic with high diffusivity, is located between two adjacent isotropic crystals with low diffusivity (Figure 6.18). When diffusion sources are dropped on the surface normal to the boundary, diffusion takes place both along the grain boundary and in the bulk. Since $D_{gb}$ (grain-boundary diffusivity) is much larger than $D_v$ (lattice or bulk diffusivity), the diffusion persists to much larger penetration depths along the grain boundary than in the bulk. Hence, after certain penetrating time and distance depending on $D$, materials in the grain boundary start diffusing laterally into the adjacent grains by volume diffusion. Different experimental conditions will result in different extents of lateral diffusion, and this extent determines the kinetics of diffusion and the required type of mathematical analysis for diffusion data evaluation. In this case, the diffusion kinetics will be classified by taking the grain sizes and annealing times into account. Conversely, the regime is important for the design of experiments and interpretation of the results. Also, from the various diffusion regimes, corresponding diffusion parameters that are essential to be obtained from the dependable studies can be extracted. Figure 6.19 presents the classification of diffusion kinetics by Harrison (1961), in which three regimes called A, B, and C types are introduced.

## 6.2.5.1    A-Type Kinetic Regime

This regime often happens at high annealing temperatures, and/or long annealing time, and/or when the grain size in materials is small, and/or $D_v$ not much smaller than $D_{gb}$. In this situation, the average volume diffusion depth exceeds the average grain size $d$, and the lateral diffusion fields of the grain boundaries can overlap extensively with each other, as shown in Figure 6.19a. The volume diffusion contribution and the GB diffusion contribution can hardly be distinguished. According to the parallel slab model of a polycrystal (Hart, 1957), an effective diffusivity $D_{eff}$, which represents a weighted average of the lattice diffusivity $D_v$ and grain-boundary diffusivity $D_{gb}$, is introduced:

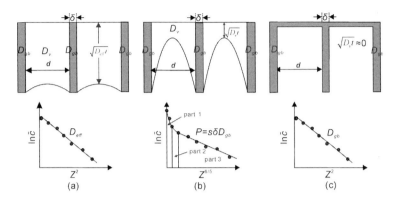

**Figure 6.19** Schematic classification of grain-boundary diffusion regimes after Harrison (1961): (a) A regime; (b) B regime; (c) C regime. The grain boundaries are considered isotropic and uniform slabs with width $\delta$ and diffusivity $D_{gb}$. $d$ is the grain size and $z$ the dimensionless diffusion depth. Part 1 is from volume diffusion, part 2 from both volume and grain-boundary diffusion, and part 3 only from grain-boundary diffusion in (b).

$$D_{eff} = t_g D_{gb} + \left(1 - t_g\right)D_v, \tag{6.57}$$

where $t_g$ is the time spent by diffusing atoms within a GB:

$$t_g = \frac{sf}{1 - f + sf} \tag{6.58}$$

in which $f$ is the fraction of GBs in the polycrystal, and $s$ the segregation factor, which is defined as $C_{gb}/C_v$, where $C_{gb}$ and $C_v$ are the concentration of component $C$ in grain boundary and in the bulk, respectively (Kaur et al., 1995).

The following solution is applicable for the case of instantaneous source. $M$ represents the total physical quantity of the traced atoms on the surface of the system at $t = 0$. $\delta$ is the grain-boundary width, as shown in Figure 6.18:

$$\bar{c} = \frac{\delta M}{d\sqrt{\pi D_{eff} t}} \exp\left(-\frac{z^2}{4 D_{eff} t}\right). \tag{6.59}$$

When plotting the logarithm of the average tracer concentration $\ln(\bar{c})$ against the square of penetration depth $z^2$, one can find that the penetration profile follows the Gaussian-type solution, and the value of effective diffusivity $D_{eff}$ can be determined by fitting of the slope of the linear region:

$$D_{eff} = -\frac{1}{4t}\left(\frac{\partial^2 \ln(\bar{c})}{\partial z^2}\right)^{-1}. \tag{6.60}$$

### 6.2.5.2 B-Type Kinetic Regime

This diffusion regime is often observed when the annealing temperatures are relatively low, or/and the annealing times are relatively short, or/and in materials that have sufficiently large grain size. Under these conditions, the volume diffusion length $\sqrt{D_v t}$ can be much shorter than the space $d$ between the boundaries. Meanwhile, the penetration width of grain-boundary diffusion, which is given by $\sqrt{D_{gb} t}$, can be significantly larger than the width $\delta$ of the grain boundary. If the solute diffusion is with segregation, an effective width $(s\delta)$ should be considered. In contrast to the A-type regime, the bulk diffusion fringes of neighboring grain boundaries do not overlap (Figure 6.19b). Therefore, B-type kinetics, annealing time $t$, and annealing temperature $T$ are classified as follows:

$$s\delta \ll \sqrt{D_v t} \ll d. \tag{6.61}$$

The parameters $\alpha$ and $\beta$ are used to identify whether the targeted grain boundary diffusion belongs to a B-type kinetic regime if the following conditions are satisfied:

$$\alpha = \frac{s\delta}{2\sqrt{D_v t}} < 0.1 \tag{6.62}$$

$$\beta = \frac{s\delta D_{gb}}{2D_v \sqrt{D_v t}} = \alpha \frac{D_{gb}}{D_v} > 10. \tag{6.63}$$

Although individually the bulk phase and the grain-boundary phase obey Fick's laws, the mathematical solutions for bulk diffusion cannot be applied under these conditions since on a macroscopic scale the system obviously does not obey these laws. It is necessary to mention that only the product $P$ $(P = s\delta D_{gb})$ can be obtained by experiments in a B-type regime.

The penetration profiles in the coordinates of the logarithm of the average tracer concentration $\ln(\bar{c})$ against diffusion depth to power 6/5 is plotted in Figure 6.19b. The near-surface part 1 of the penetration profiles is caused by volume diffusion from the surface, part 2 is caused by both volume diffusion and GB diffusion, and part 3 is only caused by GB diffusion. It can be clearly seen in Figure 6.19b that the GB diffusion-related parts of the profiles show a good linearity, and the value of the triple product $P$ is proportional to the slopes of the linear region.

### 6.2.5.3 C-Type Kinetic Regime

The diffusion in the C-type regime emerges only along grain boundaries, and no essential leakage into adjoining grains takes place. This case can be observed when the diffusion annealing is at very low temperature and/or the diffusion time is very short. This results in a rather small volume penetration depth compared to the diffusion width of grain boundaries. In this regime, the following equation is obtained:

$$\alpha = \frac{s\delta}{2\sqrt{D_v t}} > 1. \tag{6.64}$$

If the flux of volume diffusion can be neglected, the tracer sources localize in the GBs and the distribution of them obeys standard solutions for diffusion in uniform materials, which is characterized by grain-boundary diffusivity $D_{gb}$

Experimentally, after plotting the logarithm of the average tracer concentration $\ln(\bar{c})$ against the square of the penetration depth $z^2$, the GB diffusion coefficient $D_{gb}$ can be determined directly by estimating the slope of the linear fit:

$$D_{gb} = -\frac{1}{4t}\left(\frac{\partial^2 \ln(\bar{c})}{\partial z^2}\right)^{-1}. \tag{6.65}$$

Combining B- and C-type kinetics measurements, the GB width $\delta$ can be determined.

Usually, grain-boundary diffusion measurements cannot be performed under exact A-, B-, or C-type kinetic regime conditions. In consequence, they fall into transition kinetics from A to B regime (AB-type kinetic regime) or from B- to C-type regime (BC-type kinetic regime). For the detailed description of these two regimes, the interested reader can refer to Paul et al. (2014). The grain-boundary enhanced diffusion model has been implemented in DICTRA software. This implemented model will be briefly described in Section 6.2.6.4 since it has been used for materials design.

## 6.2.6    Computational Methods for Calculations of Diffusivity

### 6.2.6.1    Atomistic Description of Diffusion

Diffusion in crystals occurs by a series of atomic jumps. The basic quantities of diffusion are jump rate $\Gamma$ and jump distance $l$ of diffusing atoms. The link between the diffusion coefficient and the parameters of atomic jump in a cubic lattice is given by Mehrer (2005)

$$D = \frac{1}{6}fl^2 Z\Gamma, \tag{6.66}$$

in which $Z$ is the coordination number, and $f$ the correlation factor, which is related to crystal structure (0.7815 for Fcc_A1, 0.727 for Bcc_A2, and 0.653 for Hcp_A3). Equation (6.66) is the basic expression for atomistic diffusion in solid crystal.

*Self-Diffusion Coefficient*

In most cases, self-diffusion occurs through the vacancy-mediated mechanism. Thus, diffusion is usually assumed to be governed by a single-vacancy mechanism associated with the nearest-neighbor vacancy jumps.

For an Fcc lattice, such as Al and Cu, assume that the lattice constant is $a$, then the jump distance of the vacancy is $a\sqrt{2}/2$, and the coordination number is 12. Consequently, the self-diffusion coefficient is written as

$$D = \frac{1}{6}fl^2 Z\Gamma = \frac{1}{6}f\left(a\sqrt{2}/2\right)^2 12\Gamma = fa^2\Gamma, \tag{6.67}$$

Self-diffusion atoms migrate by jumping into a nearby vacant site, thus the diffusivity is linked with the availability of vacancies around the migrating atoms.

Under thermal equilibrium, the availability of vacancies is equal to the concentration of vacancy $C_0$, which is given by

$$C_0 = \exp\left(-\frac{\Delta_f G_{Va}}{k_B T}\right) = \exp\left(\frac{\Delta_f S_{Va}}{k_B}\right) \exp\left(-\frac{\Delta_f H_{Va}}{k_B T}\right), \tag{6.68}$$

where $k_B$ is Boltzmann constant, $\Delta_f S_{Va}$ and $\Delta_f H_{Va}$ denote entropy and enthalpy of formation for vacancy, respectively. $\Delta_f G_{Va}$ is the Gibbs energy for formation of vacancy $(\Delta_f G_{Va} = \Delta_f H_{Va} - T\Delta_f S_{Va})$. The jump rate of the diffusion species is given by

$$\Gamma = \omega C_0, \tag{6.69}$$

in which $\omega$ is the jump rate at which the jump of an atom into an empty neighboring site occurs. Atoms in a crystal oscillate around their equilibrium positions with frequency $v_0$, which is typically the value of phonon frequency with the order of $10^{12}$–$10^{13}$ Hz. Let us denote $G^{mig}$ as Gibbs energy of migration, which corresponds to the Gibbs energy difference between the configuration of the jumping atom at the saddle point and at its equilibrium position, then $\omega$ is obtained according to the transition state theory (TST) (Eyring, 1935):

$$\omega = v_0 \exp\left(-\frac{G^{mig}}{k_B T}\right) = v_0 \exp\left(\frac{S^{mig}}{k_B}\right) \exp\left(-\frac{H^{mig}}{k_B T}\right), \tag{6.70}$$

where $S^{mig}$ and $H^{mig}$ are entropy and enthalpy of migration, respectively. According to Vineyard's harmonic transition state theory (Vineyard, 1957), $S^{mig}$ is zero, and $v_0$ is defined as the ratio of the product of normal frequencies of the initial state $v_i$ and that of the transition state $v_i'$ with an imaginary frequency excluded:

$$v_0 = \frac{\prod_{i=1}^{3N-3} v_i}{\prod_{i=1}^{3N-4} v_i'}. \tag{6.71}$$

Combining (6.67–6.71), the self-diffusion coefficient for the Fcc lattice is written as

$$D = fa^2 \Gamma = fa^2 \omega C_0 = fa^2 v_0 \exp\left(-\frac{\Delta_f G_{Va}}{k_B T}\right) \exp\left(-\frac{G^{mig}}{k_B T}\right)$$

$$= fa^2 v_0 \exp\left(-\frac{\Delta_f G_{Va} + G^{mig}}{k_B T}\right). \tag{6.72}$$

Usually the diffusion equation (Arrhenius, 1899) is written in Arrhenius form in order to compare with measurements:

$$D = D_0 \exp\left(-\frac{Q}{kT}\right), \tag{6.73}$$

where $D_0$ is the diffusion prefactor, which is given by an equation of the following form:

$$D_0 = fa^2 \frac{\prod_{i=1}^{3N-3} v_i}{\prod_{i=1}^{3N-4} v_i'} \exp\left(\frac{\Delta_f S_{Va}}{k_B}\right). \tag{6.74}$$

The preceding equations are derived through the Fcc lattice. The diffusion equation (6.67) is also suitable for a Bcc lattice, being the starting equation for the derivation of self-diffusion coefficient for the Bcc phase.

## Impurity Diffusion Coefficient

Impurity diffusion in crystal can occur through interstitial mechanism or vacancy mechanism. For H, N, O, and other atoms with small atomic radius, they are prone to occupy the interstitial sites. For the interstitial mechanism, the diffusion coefficient is expressed as (6.66). For the interstitial diffusion in a very dilute interstitial alloy, $f$ is equal to 1. In the framework of harmonic transition state theory, the reaction or jump rate $\Gamma$ is also given by (6.69) and (6.70).

Larger atoms prefer to substitute lattice atoms. Thus, diffusion of such impurities is via vacancy mechanism. The diffusion ability is also related to jump distance, jump frequency, and the availability for finding vacancies nearby the impurity. The formulas for calculation of the impurity diffusion coefficient via the vacancy mechanism are similar to those of the self-diffusion coefficient. Let us assume A are host atoms and B are impurities, i.e., A is solvent and B is solute. Then the impurity diffusion coefficient in Fcc and Bcc lattices via the vacancy mechanism can be written as

$$D' = f'a^2\omega'p', \tag{6.75}$$

where $f'$ is the correlation factor, $\omega'$ the jump rate at which the impurity jumps into an empty neighboring lattice site occurs, and $p'$ the probability of formation of vacancy around the impurity. $p'$ and $\omega'$ are given by

$$p' = \exp\left(-\frac{\Delta_f G_{Va}'}{k_B T}\right) = \exp\left(\frac{\Delta_f S_{Va}'}{k_B}\right)\exp\left(-\frac{\Delta_f H_{Va}'}{k_B T}\right) \tag{6.76}$$

$$\omega' = v_0'\exp\left(-\frac{G'^{mig}}{k_B T}\right) = v_0'\exp\left(\frac{S'^{mig}}{k_B}\right)\exp\left(-\frac{H'^{mig}}{k_B T}\right). \tag{6.77}$$

$\Delta_f S_{Va}'$ and $\Delta_f H_{Va}'$ denote formation entropy and enthalpy of vacancy in the nearest neighboring of the impurity, respectively, and $\Delta_f G_{Va}'$ is the Gibbs energy for formation of such vacancy. $G'^{mig}$, $H'^{mig}$, and $S'^{mig}$ are the migration Gibbs energy, enthalpy, and entropy for the impurity jumping into a neighboring vacancy, respectively.

The attempt frequency of impurity $v_0'$ is also calculated through (6.71).

The correlation factor $f'$ in (6.75) is related to a series of atomic jumps of impurities and host atoms. To be specific, we consider the following Fcc solvents (for impurity

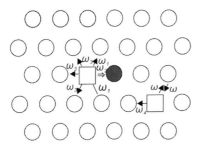

**Figure 6.20** Schematic illustration of the five-frequency model for vacancy jumps in the presence of an impurity in an Fcc lattice. The full circle, open circles, and squares denote impurity, solvent atoms, and vacancy, respectively.

diffusion in Bcc solvents, see Huang et al., 2010a). If only the nearest-neighboring interaction between impurity and vacancy occurs, four vacancy jumps, except for the jump in the pure solvent, must be considered. This is the so-called five-frequency model (LeClaire, 1978), as shown in Figure 6.20. $\omega$ is the host atom jump without an impurity, $\omega_1$ is the jump for a host atom (nearest neighbor to an impurity), $\omega_2$ is the jump for impurity atom, $\omega_3$ is also the host atom jump, and $\omega_4$ is the reverse of jump $\omega_3$. Note that $\omega_3$ makes the separation of the impurity and the vacancy, i.e., $\omega_3$ "dissociates" the impurity and vacancy, but $\omega_1$ does not "dissociate" the impurity from the vacancy.

The correlation factor of impurity diffusion can be written as

$$f' = \frac{2\omega_1 + 7F\omega_3}{2\omega_1 + 2\omega_2 + 7F\omega_3}. \tag{6.78}$$

$F$ was defined by Manning (1964) as

$$F(x) = 1 - \frac{10x^4 + 180.5x^3 + 927x^2 + 1{,}341}{7(2x^4 + 40.2x^3 + 254x^2 + 597x + 435)}, \tag{6.79}$$

where $x$ is $\omega_4/\omega$.

Every jump rate in the five-frequency model can be obtained via (6.70) and (6.71) based on the transition state theory.

The preceding expressions are suitable for cubic lattice. For a noncubic lattice, such as Hcp, diffusion is anisotropic due to the crystal anisotropy. The expression for diffusion coefficient in an Hcp lattice can be found in Huber et al. (2012).

### First-Principles Calculation of Self-Diffusion and Impurity Diffusivities

All the parameters in the preceding expressions for diffusion (6.67–6.79) can be obtained by first-principles calculation. In first-principles calculation for defects and diffusion events, the supercell approach, where the impurity or vacancy is surrounded by a few dozen to a few hundred atoms, is usually used to diminish the interactions of periodic atoms. The vacancy formation energy is easily obtained in DFT by just comparing the difference in total energy for the related systems involving creation of vacancy. The formation entropy of vacancy and the oscillate frequency of diffusion species can be computed through phonon calculation. For some less strict calculation of the diffusion coefficient, one can ignore the formation entropy and take the value of $10^{12}$–$10^{13}$ Hz for the oscillate frequency. The migration barriers can be directly obtained by the dimer method or the nudged elastic band (NEB) method. We recommend Huang et al. (2010), Mantina et al. (2009), and Marino and Carter (2008), which illustrate in detail the calculation of self- and impurity diffusivities by means of the first-principles method.

### Interdiffusion Coefficient

It remains challenging for the determination of the interdiffusion coefficient using first-principles calculations. Very limited work is reported in the literature about the prediction of interdiffusion coefficient by means of atomistic simulations. Ven and Ceder (2005) first attempted to use first-principles calculations for the evaluation of the interdiffusion coefficient for $Al_{1-x}Li_x$. Such a methodology that enables the prediction of the interdiffusion coefficient in binary alloys using first-principles calculations (Ven and Ceder, 2005) is based on the atomic migration mechanism mediated by vacancies. The calculation was parameterized via a consideration of the activation barrier and energy of the distinct configurational components in the solid, which was depicted by an accurate kinetic diffusion model.

It is widely recognized that in a binary alloy containing vacancies, the fluxes $J_1$ of the components in the alloy are described by concentration gradients, on the basis of $J_i = \sum_j - D_{ij} \nabla C_j$, wherein $\nabla C_j$ denotes the concentration gradient of the component $j$. The product of the matrix for kinetic factor $\mathbf{L}$ and thermodynamic factor $\varphi_{jk}$ can be utilized for calculating the diffusion coefficient matrix $D_{ij}$.

The thermodynamic factor matrix $\varphi_{jk}$ is generally defined by the partial derivative of the chemical potential for component $j$, $\mu_j$, with respect to the concentration of component $k$, $c_k$. It is very important in this definition that we take all the $n$ components in an $n$-component system as independent variables:

$$\psi_{jk} - \frac{\partial \mu_j}{\partial c_k} = V_m \frac{\partial \mu_j}{\partial x_k}. \tag{6.80}$$

The concentrations are related to molar fractions, $x_k$, by $x_k = c_k V_m$, where $V_m$ denotes the molar volume. Therefore, $\partial \mu_j / \partial x_k$ is the important core quantity and sometimes also called the thermodynamic factor.

If, however, we choose the component $n$ to be the solvent, i.e., $x_n$ becomes a dependent variable, we must calculate the thermodynamic factor $\varphi_{jk}$ as

$$\varphi_{jk} = V_m \left( \frac{\partial \mu_j}{\partial x_k} - \frac{\partial \mu_j}{\partial x_n} \right). \tag{6.81}$$

The thermodynamic factor $\varphi_{jk}$ is available as a generic function in the Pandat software package, and it is noted that, of course, exactly the same numerical results are obtained in any multicomponent system for both calculation paths in (6.80) or (6.81). When using the thermodynamic factor in diffusion equations, one must distinguish substitutional and interstitial elements as outlined in Andersson and Ågren (1992) and used in Section 6.2.6.4.

In an $n$-dimensional solid that exhibits volume $V$, the components of the matrix $\mathbf{L}$ and $\varphi$ can be calculated via a Kubo–Green formalism (Allnatt, 1982) when $\varphi = \partial \tilde{\mu}_i / \partial C_j$ and

$$L_{ij} = \frac{\left\langle \left( \sum_\eta \Delta \vec{R}^i_\eta(t) \right) \left( \sum_\xi \Delta \vec{R}^j_\xi(t) \right) \right\rangle}{(2n)t V k_B T}, \tag{6.82}$$

wherein the vector $\Delta \vec{R}^i_\eta$ denotes the connected endpoints of the trajectory for atom $\eta$ of type $i$ after time step $t$. $\tilde{\mu}_i$ represents the difference of chemical potential between component $i$ and vacancy (Cahn and Larche, 1983). The brackets in the equation for $L_{ij}$ mean the average value of the entire ensemble.

## 6.2.6.2    MD Simulation

It has been demonstrated in Chapter 2 that MD simulation is a feasible approach to predict diffusivity of liquid or even solid phases. The reliability of MD simulation strongly depends on the quality of interaction potentials. Currently, two types of MD simulations are widely used. One is the classic MD in which empirical potential functions are used, and the other is ab initio molecular dynamics using interaction potentials from density functional theory.

For the calculation of the self-diffusion coefficient in liquid or amorphous alloys, one can use the Einstein relation (Jang et al., 2010):

$$D = \lim_{t \to \infty} \frac{\langle R^2(t) \rangle}{6t} \tag{6.83}$$

$$\langle R^2(t) \rangle = \frac{1}{N} \sum_{i=1}^{N} |R_i(t) - R_i(0)|^2, \tag{6.84}$$

where $\langle R^2(t) \rangle$ is the mean square displacement (MSD), $N$ the number of atom $I$, and $R_i(t)$ the position of the $i$th atom at time $t$. If we have the MSD as a function of time, we simply need to perform a linear fitting to obtain the slope of MSD. Once the diffusion coefficient is known at several temperatures, the activation energy $Q$ and the prefactor $D_0$ are extracted by the least squares fit to the Arrhenius equation.

Figure 6.21 shows the calculated MSD for $Li^+$ in $Li_4BN_3H_{10}$, a promising hydrogen storage material, by AIMD simulation (Farrell et al., 2009). As shown in Figure 6.21b, the MSD for $Li^+$ reflects a diffusivity that exhibits an approximately Arrhenius behavior through the investigated temperature range. Note that at 300 K and 600 K, the MSD curve is flat, indicating very little diffusion or disordering of the $Li^+$ cations during the time studied. In contrast, it shows that the MSD for $Li^+$ and also the diffusivity of $Li^+$ of the 600 K "premelt" case are to be significantly larger and more liquid-like. Based on the 600 K premelt, as well as the data at 1,000 and 2,000 K, Farrell et al. (2009) estimated that the activation energy $Q$ and preexponential factor

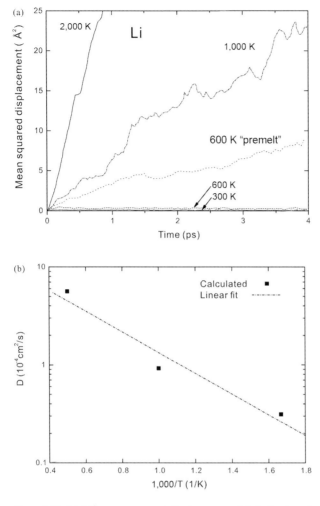

**Figure 6.21** (a) $Li^+$ mean squared displacement; (b) Arrhenius plot of the $Li^+$ diffusivity using the 600 K premelt case (the rightmost point).

$D_0$ for $Li^+$ diffusion are 20 kJ/mol-atoms and $15 \times 10^{-4}$ cm$^2$/s, respectively, which indicates a very high $Li^+$ diffusivity in $Li_4BN_3H_{10}$.

### 6.2.6.3   Semi-Empirical Methods

The semi-empirical methods are a class of computational methods for diffusivity prediction in addition to first-principles calculation and molecular dynamics simulation. The origin of various semi-empirical methods for diffusivity prediction began with Arrhenius equation of (6.46) (Arrhenius, 1899). This equation can accurately describe the temperature-dependent diffusivity in disordered materials. The following semi-empirical methods are devoted to predicting the activation energy and prefactor depending on temperature and concentration.

*Prediction of the Activation Energy*

There are several equations relating the activation energy to the melting point, heats of fusion and vaporization, compressibility, elastic module, and coefficients of linear expansion (Swalin, 1956). The correlation between the activation energy and the melting temperature is widely used:

$$Q = AT_m, \qquad (6.85)$$

where $A$ is a constant (32.5 for Bcc and 38 for Fcc), and $T_m$ the melting temperature in Kelvin. Considering the valence $V_{al}$ and crystal structure factor $K_0$, Sherby and Simnad (1961) proposed another correlation:

$$Q = (K_0 + V_{al})RT_m, \qquad (6.86)$$

in which $K_0$ is 17 for Fcc and Hcp structures, 14 for Bcc structures, and 21 for diamond structures, while $V_{al}$ is 1.5 for Group IVB (Ti, Zr, Hf), 3.0 for Group VB (V, Nb, Ta), 2.8 for Group VIB (Cr, Mo, W), 2.6 for Group VIIB (Mn, Re), and 2.5 for other transition elements. There is another similar equation (Su et al., 2010):

$$Q = RT_m(K_0 + 1.5V_{al}), \qquad (6.87)$$

with $K_0 = 15.5$ for Fcc and Hcp structures, 13 for Bcc structures, and 20 for diamond structures. Similarly, Cahoon (1997) proposed another equation:

$$Q = 0.17RT_m(16 + K_0), \qquad (6.88)$$

where $K_0$ is 1 for Bcc, 2 for Hcp, and 3 for Fcc.

There are also other correlations in literature in addition to the preceding correlations between the activation energy and the melting temperature. For instance, Gorecki (1990) gave simple relations between the activation energy of self-diffusion and impurity diffusion coefficients in metals on passing through the melting point. Recently, Du et al. (2003) proposed a semi-empirical correlation between the relative valence and the preexponential factor/activation energy.

The advantages of the semi-empirical methods are that the activation energy for diffusion can be easily estimated from available physical properties of the targeted alloys.

*Correlation between $D_0$ and $Q$*

Dushman and Langmuir (1922) derived the following correlation between the prefactor $D_0$ and the activation energy $Q$ for self-diffusion coefficient:

$$D_0 = Qa^2/N_A h = 2.50 \times 10^9 \left(J^{-1}s^{-1}\right)Qa^2, \qquad (6.89)$$

where $a$ is a lattice constant in Å, $N_A$ an Avogadro constant being equal to $6.02214076 \times 10^{23}$ mol$^{-1}$, and $h$ Planck's constant being equal to $6.626196 \times 10^{-34}$ J·s.

Zener (1951) proposed another correlation based on elastic and random walk theory:

$$D_0 = a^2 w \exp\left(\lambda_f \beta Q / R T_m\right), \tag{6.90}$$

in which $\lambda_f$ is the fraction of energy that goes into the straining lattice. $\beta$ is the dimensionless constant and equal to $-d(\mu/\mu_0)/(T/T_m)$, where $\mu$ is an appropriate elastic constant. The value of $\beta$ is from 0.25 to 0.45 for most metals, and $w$ is the vibrational frequency.

According to an electrostatic model using a Thomas–Fermi approximation, Swalin (1956) derived the following semi-empirical correlation:

$$\frac{\partial \log_{10} D_0}{\partial Q} = \frac{\alpha[(1 + qr) + 0.75(-q^3 r^3 + 6q^2 r^2 + 5qr + 5)]}{2.3R[1 - 0.25(q^2 r^2 - 5qr - 5)]}, \tag{6.91}$$

where $\alpha$ is the thermal expansion coefficients of the solvent, $r$ the interatomic distance of the solvent, and $q$ the screening constant.

## Correlation for Diffusivity

In this subsection, several methods estimating self- or impurity-diffusivity are presented.

Liu et al. (2006) developed an equation to predict impurity diffusivity in liquid metals by considering the atomic diameter and activation energy of solvent on diffusion of solute:

$$D_A^B = D_0^B (d_B / d_A) \exp\left[-Q_0^L / RT\right], \tag{6.92}$$

where $D_0^B$ is the prefactor for self-diffusivity, and $Q_0^L$ the activation energy for self-diffusivity of the solvent. $d_B$ and $d_A$ are the Goldschmidt atomic radius of solvent and solute, respectively.

Combining the Sutherland–Einstein formula with Kaptay's equation on the dynamic viscosity of liquid metals, Su et al. (2010) proposed another equation, which is given as follows:

$$D_A^B = \frac{k_B T^{1/2}}{4\pi} \frac{\left(\rho_A^m\right)^{1/3}}{0.644 \times 10^{-8} (M_A)^{1/3} \left(1 - 0.112\sqrt{T/T_A^m}\right)} \times \frac{V_B^{2/3}}{C_1 M_B^{1/2}} \exp\left(-C_2 \frac{T_B^m}{T}\right), \tag{6.93}$$

where $\rho_A^m$ is the mass density of element $A$, $M_i$ the atomic mass of $i$, and $V_B$ the absolute molar volume. $C_1$ and $C_2$ are two constants.

Brown and Ashby (1980) revisited the correlation between diffusivity and the melting point and suggested the following equation:

$$D = \frac{D_{T_m}}{\exp(-B)} \exp\left[-\frac{B T_m}{T}\left(1 + \frac{p}{T_m}\frac{dT_m}{dp}\right)\right], \tag{6.94}$$

where $D_{T_m}$ and $B$ are the diffusivity at melting point and the constant related to activation energy for a given crystal structure and bond type, respectively. $p$ is pressure. According to Brown and Ashby (1980), this relation remains valid for both bulk diffusion and grain-boundary diffusion.

Based on the detailed analysis of experimental interdiffusivities in some binary alloys, Vignes and Birchenall (1968) proposed an equation for the interdiffusivity that is characterized by the impurity diffusivity and thermodynamic factor, i.e.,

$$\tilde{D} = \left(D_A^B\right)^{N_B} \left(D_B^A\right)^{N_A} \exp\left(\frac{-16\Delta T_s^{AB}}{RT}\right)\varphi, \tag{6.95}$$

where $D_A^B$ and $D_B^A$ are the impurity diffusivities of $A$ in $B$ and that of $B$ in $A$, respectively. $\Delta T_s^{AB}$ is the difference between the solidus temperature that can be computed in a linearized A–B phase diagram. $\varphi$ is the thermodynamic factor.

Most recently, Xin et al. (2016) proposed a new relationship (the Xin–Du equation) among self- and impurity diffusion coefficients for binary solution phases and verified this relation via 30 solid solutions. This relation is described as follows:

$$Q_A^A + Q_B^B = Q_A^B + Q_B^A \tag{6.96}$$

$$\ln\left(D_A^{0A} \cdot D_B^{0B}\right) = \ln\left(D_A^{0B} \cdot D_B^{0A}\right), \tag{6.97}$$

in which $Q_i^j(i, j = A, B)$ is the activation energy of diffusion for element $i$ in pure $j$, and $D_i^{aj}$ is the corresponding prefactor. For one binary Fcc_A1 phase, there are two self-diffusivities and two impurity diffusivities. According to the proposed model (Xin et al., 2016), when the other three diffusion properties are known in a binary system, one can predict the activation energy and prefactor for the remaining diffusivity.

### 6.2.6.4   Diffusion Simulations Using DICTRA Software

For computational design of engineering materials, one often needs information about the diffusion as a function of temperature and composition for targeted alloys. With experimental diffusion coefficients and the estimated ones according to atomistic calculations and semi-empirical methods, DICTRA software can be used to assess atomic mobilities of various elements by parameter optimization (Cermak and Rothova, 2003; Jönsson, 1994). A simple introduction and general survey of DICTRA is given in Borgenstam et al. (2000), while a rigorous description of the mathematical basis is found in Andersson and Ågren (1992). In DICTRA, the matrix of diffusion coefficients in one multicomponent system can be established by using atomic mobilities of elements in phases and the thermodynamic factors to facilitate storage and calculations. The main advantage of the introduction of atomic mobility is that the number of independent parameters, which are critically needed for describing the matrix of diffusion coefficients for a multicomponent system, is considerably reduced. Based on the absolute reaction rate theory argument, the atomic mobility of element $B$, $M_B$, can be separated into two parts, i.e., a frequency factor $M_B^0$ and an activation enthalpy $Q_B$ (Jönsson, 1994; King and Woodruff, 1993). $M_B$ can be obtained by

$$M_B = \left(\frac{M_B^0}{RT}\right) \exp\left(\frac{-Q_B}{RT}\right) \left(^{mg}\Omega\right). \tag{6.98}$$

For convenience, $M_B$ is not stored directly in the mobility databases, but the terms of the following function are stored:

$$RT \ln \frac{RTM_B}{\text{m}^2\text{s}^{-1}} = RT \ln \frac{M_B^0}{\text{m}^2\text{s}^{-1}} - Q_B + RT \ln {}^{mg}\Omega. \tag{6.99}$$

The terms are stored in SI-units, and it is noted that the frequency factor $M_B^0$ has the same unit as a diffusivity. The first two terms are denoted as "+MF" and "+MQ" in the established syntax of TDB databases used by both DICTRA and the similar software PanDiffusion, which is embedded in the Pandat software package (Cao et al., 2009).

Both $RT\ln M_B^0$ and $Q_B$ generally depend on the composition, temperature, and pressure. ${}^{mg}\Omega$ is a parameter that includes the ferromagnetic contribution to the diffusivity. Meanwhile, for those phases without ordering, such as Fcc and liquid, the ferromagnetic effect on diffusion can be neglected (Jönsson, 1994), i.e., ${}^{mg}\Omega = 1$. Thus, $MF(B) = RT\ln M_B^0$ and $MQ(B) = -Q_B$ can be merged into one parameter, $\Phi_B = (RT\ln M_B^0 - Q_B)$, for these phases. In SI-units, we have the following:

$$RT \ln (RTM_B) = RT \ln M_B^0 - Q_B = MF(B) + MQ(B). \tag{6.100}$$

In the real TDB syntax for the example of element Ni diffusing in the Fcc phase of pure Ni, we may have MF(Fcc&Ni,Ni; 0) = $-69.8\,T$ and MQ(Fcc&Ni,Ni; 0) = $-287{,}000$. That is often merged to MQ(Fcc&Ni,Ni; 0) = $-287{,}000 - 69.8\,T$ with the MF parameter absent, i.e., equal to zero.

However, for the phases with magnetic ordering, such as Bcc–Fe, the effect of ferromagnetic contribution on atomic mobility should be considered, which can be represented by Jönssons's equation (Jönsson, 1994):

$$^{mg}\Omega = \exp\left(6\alpha\xi\right)\exp\left(\frac{-\alpha\xi Q_B}{RT}\right), \tag{6.101}$$

where $\alpha$ is regarded as a constant, which is approximately equal to 0.3 in Bcc alloys, while $\xi$ represents the state of the magnetic order $(0 < \xi < 1)$. Therefore, $RT\ln M_i^0$ and $Q_i$ are treated as different terms in those phases

On the basis of CALPHAD method, the composition dependence can be expressed as a linear superposition of the values at each endpoint of the composition space. For instance, $\Phi_B$ can be expressed for a ternary alloy by the Redlich–Kister polynomial:

$$\Phi_B = \sum_i x_i \Phi_B^i + \sum_i \sum_{j>i} x_i x_j \left[\sum_{r=0}^m {}^r\Phi_B^{i,j}(x_i - x_j)^r\right]$$
$$+ \sum_i \sum_{j>i} \sum_{k>j} x_i x_j x_k \left[\sum_s v_{ijk}^s \, {}^s\Phi_B^{i,j,k}\right]; \quad (s = i, j, k), \tag{6.102}$$

where $x_i$ is the mole fraction of components $i$, and $\Phi_B$ represents a temperature-dependent property, i.e., $-Q_B$ and/or $RT\ln M_B^0$. $\Phi_B^i$ is the value of $\Phi_B$ for $B$ in pure $i$, which is the value of the endmember. ${}^r\Phi_B^{i,j}$ and ${}^s\Phi_B^{i,j,k}$ are binary and ternary interaction parameters, respectively. Every $\Phi$ parameter, i.e., $\Phi_B^i$, $^r\Phi_B^{i,j}$ or $^s\Phi_B^{i,j,k}$, can be determined via a polynomial of temperature and pressure. The factor $v_{i,j,k}^s$ is expressed as

$$v_{ijk}^s = x_s + \left(1 - x_i - x_j - x_k\right)/3 \qquad (s = i, j, k), \qquad (6.103)$$

where $x_i$, $x_j$, $x_k$, and $x_s$ are mole fractions of elements $i$, $j$, $k$, and $s$, respectively.

Assuming a monovacancy atomic-exchange mechanism for diffusion and neglecting correlation factors (Jönsson, 1994; King and Woodruff, 1993), the tracer diffusion coefficient $D_i^*$ can be expressed by the atomic mobility $M_i$ via the Einstein relation:

$$D_i^* = RTM_i. \qquad (6.104)$$

For a substitutional solution phase, interdiffusivities in the volume-fixed reference frame are expressed by the following equation (Jönsson, 1994; King and Woodruff, 1993):

$$D_{kj}^n = \sum_i (\delta_{ik} - x_k) \cdot x_i \cdot M_i \cdot \left(\frac{\partial \mu_i}{\partial x_j} - \frac{\partial \mu_i}{\partial x_n}\right), \qquad (6.105)$$

in which $\delta_{ik}$ is the Kronecker delta ($\delta_{ik} = 1$ if $i = k$, otherwise $\delta_{ik} = 0$); $x_i$ and $M_i$ are the mole fraction and mobility of component $i$, respectively; and $n$th element is selected as the dependent component. Please refer to Section 6.2.6.1.4 for the proper definition of the thermodynamic factor $\varphi_{ik}$ in (6.81). The last term in (6.105) is just $\varphi_{ik}$. Equation (6.105) shows most clearly how the diffusivity matrix is calculated from the simpler mobility vector ($M_i$) and the purely thermodynamic factor matrix ($\varphi_{ik}$). The same relation as in (6.105) is also shown in equation 59 of Andersson and Ågren (1992); it just looks different because all $n$ components are treated as independent variables in that setting of the thermodynamic factor matrix ($\varphi_{ik}$), as explained in (6.80) in Section 6.2.6.1.

Supposing one system including order/disorder transitions, i.e., disordered phase Fcc_A1/ordered phase L1$_2$ and disordered phase Bcc-A2/ordered phase Bcc-B2, the chemical ordering influence on the atomic mobility should be taken into account. According to the research of Girifalco (1964), the increase of the activation energy resulted from chemical ordering is related to the long-range order parameter. Helander and Ågren (1999) developed one phenomenological model for describing diffusional phenomena for phases with the B2 ordering transition, in which the activation energy is determined by the following:

$$Q_B = Q_B^{dis} + Q_B^{ord}, \qquad (6.106)$$

where $Q_B^{dis}$ is the contribution of the disordered state and is determined by (6.100), and $Q_B^{ord}$ is the contribution of chemical ordering, which is expressed as follows:

$$Q_B^{ord} = \sum_i \sum_{i \neq j} Q_{Bij}^{ord} \left[y_i^\alpha y_j^\beta - x_i x_j\right], \qquad (6.107)$$

where $Q_{Bij}^{ord}$ is for determining the contribution of element $B$ resulted from the chemical ordering of the $i$-$j$ atoms on two sublattices. $y_i^\alpha$ is the site fraction of element $i$ on the $\alpha$ sublattice, and can be expressed as

$$y_i^\alpha = \frac{N_i^\alpha}{N_{total}^\alpha},$$

(6.108)

in which $N_i^\alpha$ is the number of sites on $\alpha$ sublattice occupied by the $i$ atom and $N_{total}^\alpha$ is the total number of sites on the $\alpha$ sublattice. The model proposed by Helander and Ågren (1999) is exactly applicable for the AB-type alloy (B2 structure), and it is also presented to be applicable for the $AB_3$-type alloy ($L1_2$ structure) by Campbell (2008).

The evaluation of grain-boundary-assisted diffusion was also included in DICTRA software according to the presumption that the grain boundary contributes to the diffusion, and the same preexponential factor and a modified bulk activation energy are utilized. The atomic mobility for diffusion in grain boundary $M^{gd}$ is expressed as follows:

$$M^{gd} = M_0^{bulk} \exp\left(\frac{F_{GB} Q^{bulk}}{RT}\right),$$

(6.109)

where $M_0^{bulk}$ and $Q^{bulk}$ are the frequency factor and the activation energy in the bulk phase, respectively. $F_{GB}$ represents the bulk diffusion activation energy multiplier, the value of which is suggested to be 0.5 in DICTRA.

As a case study for computation of the diffusion matrix in a multicomponent phase, combining with thermodynamic parameters for the Cu–Mn–Ni–Zn system, atomic mobilities for four ternary subsystems were evaluated based on the measured data from the literature by DICTRA software. Atomic mobilities for Fcc quaternary Cu–Mn–Ni–Zn alloys were directly extrapolated according to the four ternary subsystems. Meanwhile, the obtained mobility parameters, associated with the thermodynamic description, can be utilized for predicting composition profiles and diffusion paths in quaternary diffusion couples to validate the accuracy of obtained parameters. It is worth mentioning that quaternary interdiffusivities can be evaluated on the basis of the atomic mobility parameters. These computed diffusion matrixes can be used as input for many simulations associated with material design.

## 6.3 Interfacial Energy

In addition to diffusion coefficient, the other thermophysical properties (interfacial energy, viscosity, molar volume, thermal conductivity, and so on) are also of fundamental importance for both basic research and engineering. A very brief introduction to several important thermophysical properties is given here, with a special emphasis on the implementation for computer calculations for the sake of materials design. The thermophysical property diagrams presented in this chapter can be computed via the calculation of thermophysical properties (CALTPP) program (Liu et al., 2020).

Interfacial energy is an important thermophysical property that quantifies the excess free energy of the interface compared to the bulk phase. The solid–liquid/ solid–solid phase interfacial energy, solid (liquid)–vapor surface energy, and grain-boundary energy are typical types of interfacial energies. The interfacial energy is featured in various phenomena, for instance, crystal nucleation in supersaturated solution during aging (Sudbrack et al., 2008), grain growth of dendrites during directional solidification (Choi et al., 2015), and wetting between binder and ceramic phases during liquid-phase sintering (Long et al., 2017).

Various techniques, such as the supercooling nucleation method, grain growth method, sessile drop method, and boundary groove method, can be used to determine the interfacial energy experimentally. The boundary groove method is a quite common technique to measure the solid–liquid interfacial energy. During the measurement in a temperature gradient, the equilibria between the solid–liquid interface and the grain boundary is achieved, and the interfacial energy is determined based on the measured equilibrium shape of the groove profile. It is difficult to measure the accurate cusp shape and observe the grain-boundary groove shape. The measurements of interfacial energy involve the complex procedure and the immense number of valuable combinations to test, which inevitably leads to large uncertainties. Back-calculation is a common technique to measure the solid–solid interfacial energy, where the interfacial energy used in the coarsening model (e.g., Ostwald ripening) is adjusted with the aim to fit the observed evolution of precipitate size. The interfacial energies from back-calculation again are associated with serious uncertainties. One can refer to King and Woodruff (1993) for various experimental methods to measure interfacial energy.

In view of the difficulties to measure interfacial energies experimentally, models to obtain interfacial energies are of great value. When orientation and coherence are important for analyzing the interface region, atomistic calculation is needed to obtain interfacial energy. In general, one can obtain the interfacial energy of a targeted interface through first-principles calculations. Specifically, for a system with an interface area $A$, the interfacial energy $\gamma$ only contributed from chemical potentials can be defined as follows:

$$\gamma = \frac{1}{A}\left(E_{total} - \sum_i n_i \mu_i\right), \tag{6.110}$$

where $E_{total}$ is the total energy of the whole considered interfacial atomic model, $n_i$ represents the number of atoms in this interfacial atomic model of component $i$, and chemical potential $\mu_i$ is determined by the reservoir with which the system is in equilibrium. Using the unit cell of component $i$ to obtain the total energy $E_{total(i)}$ of component $i$ with $n_i$ atoms, $\mu_i$ can be calculated with (6.111) (Christensen and Wahnström, 2003; Johansson and Wahnström, 2010, 2012)

$$\mu_i = \frac{E_{total(i)}}{n_i}. \tag{6.111}$$

The interfacial energy $\gamma$ can finally be obtained with (6.110) after the total energy $E_{total}$ of the interface model is calculated from first-principles calculations.

This strategy has been extensively used to investigate various kinds of interface-related materials, and the quantitative analysis of interfacial energy and qualitative analysis of anisotropy are very important for both scientific research and engineering.

However, the establishment of an atomic model consistent with the actual situation for interface is quite time and vigor consuming. Frequently, theoretical simulation has to balance between an accurate atomic model and large computational cost. Normally, theoretical calculations deal with three types of interface models, i.e., coherent, semicoherent, and incoherent models, to mimic the complex actual interface. The most widely used and highly reliable coherent interfacial models have already been successfully used to investigate solid–solid two-phase boundary and grain boundary. After capturing the atomic structure details and interfacial energies, it is possible to simulate the scenarios for heterogeneous nucleation.

Currently, the DFT-based calculations are largely constrained to calculate the interfacial energy of the fully coherent interface. Only a few cases are focusing on the semicoherent interfaces with significantly large mismatches. Thousands of atoms are required to model the semicoherent interface with small to medium lattice misfit, including interface dislocations, which is usually beyond the capability of the present DFT calculations. By combing the ab initio data for the chemical interactions across the interface and a continuum description to illustrate the elastic distortions, a Peierls–Nabarro framework was developed to access the structure and energetics of semicoherent interface.

In addition to the preceding theoretical calculations, the CALPHAD approach is another method to estimate interfacial energy from thermodynamic properties of the phases. The solid–liquid interfacial energy is closely connected to the chemical potentials of the adjacent equilibrated bulk phases. It is possible to build the relationship between the solid–liquid interfacial energy and the Gibbs energies of bulk phases. Lippmann et al. (2016) conducted a comprehensive review of the models of solid–liquid interfacial energies based on thermodynamic considerations. Afterward, a thermodynamic model of solid–liquid interfacial energies was developed by Zhang and Du (2017), wherein the spherical geometry for interface was considered and the CALPHAD-type Gibbs energies were used. Warren (1980) regarded the interfacial energy $(\gamma_{SL})$ as the sum of chemical contribution $(\gamma_{SL(c)})$ and structure contribution $(\gamma_{SL(B)})$, which is adopted in the model of Zhang and Du (2017):

$$\gamma_{SL} = \gamma_{SL(c)} + \gamma_{SL(B)}. \tag{6.112}$$

The Gibbs energy curves of solid and liquid phases in a hypothetical binary $A$–$B$ system are displayed in Figure 6.22. The chemical contribution of solid–liquid interfacial energy can be obtained by calculating the difference between equilibrium molar Gibbs energy $\left({}^{E}G_{m}^{IF}\right)$ of the system and molar Gibbs energy of pure liquid phase $\left(G_{m}^{L}\right)$, in which the interfacial composition $x_{B}^{IF}$ is considered.

The $\gamma_{SL(c)}$ can be potentially expressed as follows according to the preceding assumption:

$$\gamma_{SL(c)} = \frac{G_{m}^{L} - {}^{E}G_{m}^{IF}}{A_{m}}, \tag{6.113}$$

where $A_{m}$ is the interfacial area per mole of atoms and ${}^{E}G_{m}^{IF}$ is defined as the molar Gibbs energy at composition $x_{B}^{IF}$ in the equilibrated two-phase region. Since solid and

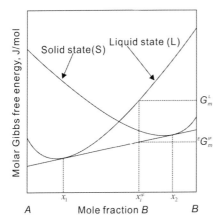

**Figure 6.22** Gibbs energy curves in a hypothetical binary $A$–$B$ system.

liquid layers comprise the interface, the effective interfacial composition $\left(x_i^{IF}\right)$ and molar interfacial area of solid and liquid phases are the premise to compute the $A_m$ and thus $\gamma_{SL(c)}$. The $\gamma_{SL(B)}$ is regarded to be equal to the interfacial energy between the solid phase and its melt. The expression of $\gamma_{SL(B)}$ proposed in Grimvall and Sjodin (1974) is adopted and displayed as follows:

$$\gamma_{SL(B)} = kT_m/A_S, \tag{6.114}$$

where $T_m$ is the melting temperature of solid phase, and $k$ an empirical constant. For detail, the reader can refer to Zhang and Du (2017).

The value of $\gamma_{SL}$ in binary system is of great interest since it provides a method to determine the solid–vapor surface energy $(\gamma_{SV})$ via simple wetting experiments (Warren and Waldron, 1972). Here, $\gamma_{SV}$ refers to the surface energy of the solid in the presence of the vapor associated with liquid. The following well-known Young's equation can depict the wetting behavior of a solid by a liquid sessile drop:

$$\gamma_{SV} = \gamma_{SL} + \gamma_{LV}\cos\theta, \tag{6.115}$$

where $\gamma_{LV}$ is the surface tension energy of the liquid and $\theta$ the angle of contact of the drop.

The coherent interfacial energy is a representative solid–solid interfacial energy, and it is often determined from the kinetics of coarsening. A thermodynamic model was developed by Kaptay (2012) to calculate coherent solid–solid interfacial energy using basic thermodynamic information, such as chemical potential and molar volumes. In addition, Brillo and Schmid-Fetzer (2014) deduced a model similar to that of the Butler equation (Butler, 1932) with the aim to calculate the immiscible liquid–liquid interfacial energy. Afterward, Liu et al. (2019) presented a detailed derivation for the general model of coherent solid–solid and immiscible liquid–liquid interfacial energies. According to the model of Liu et al. (2019), the interface is considered the thermodynamic phase with two-atomic layer, and its molar Gibbs

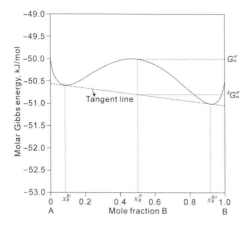

**Figure 6.23** Molar Gibbs energy curve in a hypothetical $A–B$ system, where the interface and two adjacent bulk phases have the same curve.

energy is assumed to be the average of the two adjacent bulk phases at any composition and temperature. In an immiscible $A–B$ system, the molar Gibbs energy curve of the interface and adjacent bulk phase are schematically presented in Figure 6.23.

The coherent solid–solid or immiscible liquid–liquid interfacial energy $(\gamma_{SS(LL)})$ can be captured by the following function:

$$\gamma_{SS(LL)} = 2 \cdot \frac{G_m^{IF} - {}^E G_m^{IF}}{A_A^0 (1 - x_B^{IF}) + A_B^0 x_B^{IF}}, \tag{6.116}$$

where $x_B^{IF}$ is the effective interfacial composition. $G_m^{IF}$ is defined as the molar Gibbs energy of the interface phase at the composition $x_B^{IF}$. $A_i^0$ $(i = A$ and $B)$ is the partial molar area of component $i$.

According to the assumption of a two-atomic layer of interface, the effective interfacial composition can be solved as follows:

$$x_B^{IF} = \frac{\left(1/A_m^{BI}\right)x_B^{BI} + \left(1/A_m^{BII}\right)x_B^{BII}}{\left(1/A_m^{BI}\right) + \left(1/A_m^{BII}\right)}, \tag{6.117}$$

in which $A_m^{BI}$ and $A_m^{BII}$ are the molar areas in bulk phases $BI$ and $BII$, respectively. $x_B^{BI}$ and $x_B^{BII}$ are the mole fraction of component $B$ in bulk phases $BI$ and $BII$, respectively, which can be calculated using the tangent construction at equilibrium state.

Clearly, the CALPHAD-based calculations of interfacial energy are sensitive to the choice of thermodynamic parameters, i.e., unreasonable thermodynamic descriptions might lead to the unreasonable predictions of interfacial energies.

## 6.4  Viscosity

Viscosity is the ratio of the shear stress to the velocity gradient and is used to define the fluid resistance to flow. In general, the temperature and composition of fluid affect the value of viscosity. There are a number of methods to measure the viscosity of liquids experimentally, such as rotational crucible, capillary, oscillating vessel, draining vessel, oscillating plate, and levitation using the damping of surface oscillations. One can refer to Yakymovych et al. (2014) for experimental measurements of viscosity in detail. For pure liquid, the Arrhenius equation (6.118) can be used to describe the temperature dependence of the viscosity by using the preexponential ($\eta_0$) and the activation energy ($Q$). The values of $\eta_0$ and $Q$ can be obtained through fitting experimental data.

$$\eta = \eta_0 \exp\left(\frac{Q}{RT}\right). \tag{6.118}$$

The AIMD approach has demonstrated its capability in obtaining the viscosity. In AIMD, the MSD and diffusion coefficient $D$ should be calculated firstly using the Einstein relation, which is expressed in (6.83).

The shear viscosity $\eta$ is assessed by means of Stokes–Einstein equation derived for the motion of a macroscopic particle in a viscous medium as follows (Brillo et al., 2008; Bueche, 1959; Das et al., 2008):

$$\eta = \frac{k_B T}{c_{bc} \pi r_{SE} D_A^A}, \tag{6.119}$$

where $c_{bc}$ is a constant relating to the specific boundary condition on the surface of the sphere ($c_{bc} = 4$ for slip boundary conditions and $c_{bc} = 6$ for nonslip boundary conditions), and $r_{SE}$ an effective hydrodynamic particle radius defined by the position of the first peak of the generalized pair coordination functions.

Several models were proposed to predict the viscosity for binary alloys, such as Moelwyn–Hughes (MH) (Moelwyn-Hughes, 1961), Kozlov–Romanov–Petrov (KRP) (Kozlov et al., 1983), Kucharski (1986), Hirai (H) (Hirai, 1993), Seetharaman–Du Sichen (SDS) (Seetharaman and Sichen, 1994), Kaptay (K) models (Kaptay, 2003), and so on. The predictions of all these equations depend on the thermodynamic parameters of liquid phase in binary systems. In this subsection, we will describe the Kaptay (K) equation briefly:

$$\eta = \frac{hN_A}{V} \exp\left(\frac{x_A \Delta G_A^* + x_B \Delta G_B^* - (0.155 \pm 0.015)\Delta_{\text{mix}} H_{\text{Liquid}}}{RT}\right), \tag{6.120}$$

where $\Delta_{\text{mix}} H_{\text{liquid}}$ is the heat of mixing of the liquid; $\Delta G_A^*$ and $\Delta G_B^*$ are the activation energies of pure $A$ and $B$ components, respectively; $V$ is the molar volume of the phase; and $x_A$ and $x_B$ are the mole fractions of the components $A$ and $B$, respectively.

The predictions based on the preceding equations are feasible for some systems but impracticable in some cases. The phenomenon is owing to the fact that the physical characteristics of the liquid fluid are not always regular and sometimes show associate phenomenon or short-range ordering (Li et al., 2014; Pilarek et al., 2014). Recently, the CALPHAD-type equation (Zhang et al., 2015) was proposed to calculate the viscosity, and this equation is expressed as follows:

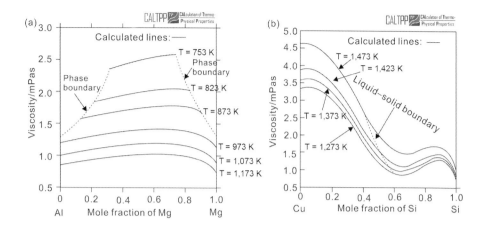

**Figure 6.24** Calculated viscosities of the Al–Mg and Cu–Si melts at different temperatures: (a) viscosity of the Al-Mg system; (b) viscosity of the Cu–Si system.

$$\eta = x_A\eta_A + x_B\eta_B + x_Ax_B\eta^E, \tag{6.121}$$

where $\eta_A$ and $\eta_B$ are the viscosities of $A$ and $B$, respectively. $\eta^E$ is the excess viscosity. The CALPHAD-type equation can extrapolate to binary and multicomponent liquids. Based on the CALPHAD-type equation, Zhang et al. (2015) calculated the viscosities of the Al–Mg and Cu–Si melts at different compositions and temperatures. The calculations are displayed in Figure 6.24.

However, none of the aforementioned models considered the effect of the associates on the viscosity. Recently, Zhang et al. (2019) developed a new model in order to estimate the viscosity of liquid alloys with the associates as shown in (6.122):

$$\eta \cong \frac{h \cdot N_A}{V} \cdot \exp\left(\frac{\sum_j x_j \cdot \Delta G_j^* - k \cdot \Delta H_{ass}}{RT}\right), \tag{6.122}$$

where $x_j$ is the mole fraction of model component $j$, and $j$ includes the unassociated atoms and molecules of the associates in the associated model; $\Delta G_j^*(\mathrm{J/mol})$ is the activation energy of viscous flow of a pure component $j$; and $\Delta H_{ass}(\mathrm{J/mol})$ is the enthalpy of mixing of the associated liquid alloy without regard to the heat of formation of the associates. It is noted that the predicted activation energies $\Delta G_j^*$ of associates in (6.122) exhibits a numerical interval due to some uncertainties. For the detailed information, one can refer to Zhang et al. (2019).

## 6.5     Volume

Several techniques could be utilized to measure the density of a liquid metal. Examples include the maximum-bubble-pressure, levitation, Archimedean, and capillary methods (Touloukian et al., 1975). Through many years of practices, density detecting means become more accurate than before. However, because of some shortcomings, such as the unavoidable chemical contamination on the melt surface and the surface tension, the accuracy for density measurement is still insufficient compared to that for lattice parameters. Density data could be converted to the form of molar volume by a simple formula:

$$V_m = \frac{W_{mol}}{\rho},$$

(6.123)

where $W_{mol}$ is the molar weight, and $\rho$ is the corresponding density.

The measurement of volume for the solid phase is a well-established procedure. The lattice parameters of solid phases measured by X-ray or neutron diffraction at ambient pressure could be utilized to calculate the corresponding molar volume by the following equation:

$$V_m = \frac{V_{cell}N_A}{N_{cell}},$$

(6.124)

where $V_{cell}$ is the molar volume of the unit cell, and $N_{cell}$ the number of atoms in the unit cell.

The coefficient of linear thermal expansion (CLE) can be used for calculating the molar volume of nonmagnetic materials at 1 bar, which is in terms of the following (Lu et al., 2005a, b):

$$V(T, P_0) = V_0 \exp(V_A)$$

$$= V_0 \exp\left(\int_{T_0}^{T} 3\alpha dT\right)$$

$$= V_0 \exp\left(\int_{T_0}^{T} (a + bT + cT^2 + dT^{-2})dT\right),$$

(6.125)

where $V_0$ and $V_A$ are molar volume at $T_0$ (reference temperature) and the integrated thermal expansion, respectively; adjustable parameters $a$–$d$ could be evaluated from experimental data for the volumetric thermal expansion coefficient at 1 bar, i.e., $3\alpha$.

For magnetic materials, the magnetic contribution is usually treated in a separate term. By taking the pressure dependence of the Curie temperature ($T_C$) into account, Guillermet (1987) proposed the following equation:

$$T_C(P) = T_C(P = 0) + P\omega,$$

(6.126)

in which $\omega$ is a constant, and

$$\Delta V_m^{magn}(T) = \Delta H_m^{magn} \frac{d \ln(T_C)}{dP},$$

(6.127)

where $\Delta H_m^{magn}$ is the magnetic enthalpy that can be calculated using Hillert and Jarl's model (Hillert and Jarl, 1978).

For a phase with a certain homogeneity range, the molar volume may vary with composition. A simple presumption could be given as the excess molar volume behaves just the same as other excess thermodynamic properties. Therefore, the composition dependence of the molar volume at 1 bar of a binary solution phase could be represented by a linear relationship of the molar volume of each endmember and given in the following Redlich–Kister form expression:

$$V = \sum_{i=1}^{2} x_i V_i + V^E,$$

(6.128)

where $V^E$ is the atmospheric excess volume and can be written as

$$V^E = x_i(1 - x_i)\left[A^0 + A^1(2x_i - 1) + A^2(2x_i - 1)^2 + \cdots\right],$$

(6.129)

in which $A^0$, $A^1$, and $A^2$ are interaction parameters to be assessed from experimental data.

Following the CALPHAD approach, the molar volumes of multicomponent solution phases are extrapolated from the binary and ternary parameters, and the excess molar volumes are defined by means of Redlich–Kister equation.

$$V = \sum_{i=1}^{m} x_i V_i + \sum_{i}\sum_{j>i} x_i x_j \sum_{k=0}^{n} A_{ij}^k (x_i - x_j)^k$$

(6.130)

where $A_{ij}^k$ are polynomial parameters related to the binary $i$–$j$ system.

Particularly, the molar volume of an intermetallic phase can be described by a sublattice model as $(A, B, \ldots)_x(C, D, \ldots)_y$, which is similar to the compound energy formalism (CEF) in the CALPHAD approach:

$$V_m^\varphi = \frac{1}{x+y}\left(\sum_i\sum_j y_i' y_j'' V_{i:j} + \sum_j\sum_i\sum_{k\neq i} y_i' y_k' y_j'' V_{i,k:j} + \sum_i\sum_j\sum_{l\neq j} y_i' y_j'' y_l'' V_{i:j,l}\right),$$

(6.131)

where $y_i'$ and $y_i''$ are the site fractions of $i$ on sublattices 1 and 2, $V_{i:j}$ is the molar volume of a hypothetical compound $i_x j_y$, and $V_{i,k:j}$ and $V_{i:j,l}$ are the interaction parameters to describe the interaction between elements on the first and second sublattices, respectively.

It is hard to assess all values for $V_{i:j}$ due to the existence of many metastable or unstable endmember compounds. An assumption is that the molar volume of those endmembers are in a form of a linear additive relationship between the constituent elements plus the determination of the actual value from the linear combinations. Figure 6.25 shows the calculated volumes of Mg and Al–Mg alloys by means of the CALPHAD approach (Huang et al., 2015).

## 6.6    Thermal Conductivity

The thermal conductivity describes the amount of heat transferred vertically to a surface of the unit area due to a temperature gradient of the unit in time units under stationary conditions. A high thermal conductivity ensures faster heat transfer that

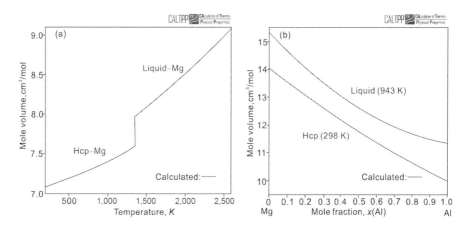

**Figure 6.25** Calculated molar volume of Mg and Al–Mg system at 1 bar: (a) molar volume of Mg; (b) molar volume of the Al–Mg system.

allows sufficient cooling and removes the formation of mechanically damaging hot spot defects. Further, higher thermal conductivity improves the uniform distribution of temperature, reducing the stresses induced by thermal gradient and thus improving the fatigue properties (Bauer et al., 2012). Thermal conductivity can be modeled as a function of temperature, composition, defects, and other similarly intense variables of the system (Pelleg, 2006). Therefore, the alloy composition, solid solubility, micro-structure, and heat treatment history have significant impact on thermal conductivity. Since the heat is mainly transported by electrons in metallic alloys, it is convenient to understand that the solute atoms, dislocations, and grain boundaries will lower the thermal conductivity.

The measurement of thermal diffusivity is normally performed by means of the laser-flash method, heat flow method, etc. The widely used method is the laser-flash technique. Prior to the measurement, the surfaces of the specimens were coated with graphite to enhance the absorption of the laser energy. During the measurement, the temperature was increased from room temperature to the specified temperature and then kept at the condition for about ten minutes. The equipment records the deter-mined data every three minutes, and each test point is an average of three measured values. One can refer to Bauer et al. (2012) for various experimental methods for obtaining thermal diffusivity, $\kappa(\mathrm{m}^2\,\mathrm{s}^{-1})$. Due to the effect of thermal expansion, at high temperatures the sample thickness increases. To compensate for such a thermal expansion, a correction should be considered for solid specimen during the thermal diffusivity measurement (Kaschnitz and Ebner, 2007):

$$\kappa_{corr} = \kappa_0 \left(1 + \frac{\Delta l}{l_0}\right)^2, \tag{6.132}$$

in which $\kappa_{corr}$ is the true (corrected) value of thermal diffusivity, $\kappa_0$ the measured one, and $\Delta l/l_0$ the thermal expansion.

Thermal conductivity $\lambda$ $\left(\mathrm{W\ m^{-1}K^{-1}}\right)$ can be obtained from thermal diffusivity by using the following equation:

$$\lambda = \kappa_{corr} C_p \rho, \tag{6.133}$$

where $C_p$ is the heat capacity at constant pressure (J kg$^{-1}$ K$^{-1}$) and $\rho$ is the mass density. The heat capacity can be accurately determined via differential scanning calorimetry (DSC) equipment or calculated using CALPHAD approach, while the Archimedes method is applied to obtain the density of materials. As the experimental determination of thermal conductivity is time consuming and expensive, the systematic investigations on the thermal conductivity of industrial alloys are still lacking. Therefore, an effective predictive method is needed to obtain the thermal conductivity of the alloys.

The contribution to thermal conductivity can be divided into two parts: One is the contribution from electrons, and the other is the contribution from phonons. The contribution from electrons can be estimated from electrical conductivity. As for the contribution from phonons, it can be calculated within lattice dynamics theory. Moreover, from atomistic point of view, the contribution from phonons can be calculated in at least two ways, by using equilibrium and nonequilibrium MD simulations. Generally speaking, the equilibrium MD method is based on the Green–Kubo formula, for which current fluctuations are used to calculate thermal conductivity by means of the fluctuation-dissipation theorem; while the nonequilibrium MD method is also called "direct method," which is based on the application of a temperature gradient on the simulation cell and is thus similar to that of experimental conditions.

For engineering materials, the CALPHAD approach is an effective tool to acquire information on thermal conductivity (Zhang et al., 2016). According to the CALPHAD approach, the model coefficients of thermal conductivity for pure elements, solid solutions, stoichiometric phases, and alloys in two-phase/multiphase range have been optimized to describe separate contributions. The thermal conductivities of elements and intermetallics are modeled as a temperature-dependent function as follows:

$$\lambda_p = \mathrm{A} + \mathrm{B}T + \mathrm{C}T^{-1}, \tag{6.134}$$

where $\lambda_p$ denote the thermal conductivity of pure elements or stoichiometric compounds, i.e., pure substances. The thermal conductivity of a binary solution phase $\varphi$, $\lambda^{\varphi}$, is described by the typical Redlich–Kister polynomials:

$$\lambda^{\varphi} = x_1 \lambda_1 + x_2 \lambda_2 + x_1 x_2 \sum_{v=0}^{n} L_{1,2}^{v,\varphi}(x_1 - x_2)^{v}, \tag{6.135}$$

where $x_1$ and $x_2$ are the mole fractions of elements 1 and 2 in the solution phase and $L_{1,2}^{v,\varphi}$ are the temperature-dependent interaction parameters describing the deviation from a linear mixing rule of thermal conductivity for this phase, which can be assessed

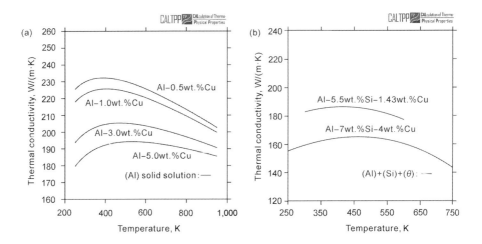

**Figure 6.26** Calculated thermal conductivities in (a) the (Al) solid solution of the Al–Cu system and (b) the Al–Cu–Si system in the $(Al) + (\theta) + (Si)$ three-phase region. The figures are reproduced from Zhang et al. (2016) with permission from Elsevier.

according to the experimental data. In Figure 6.26, the thermal conductivities for the Fcc_A1 solution phase in the Al–Cu and Al–Cu–Si systems are computed by the CALPHAD method (Zhang et al., 2016).

Zhang et al. (2016) also provide an empirical equation for two-phase mixtures. However, the analytical model of thermal conductivity for two-phase materials, applied to hard metals, was developed by combining the "parallel model" and "series model," which for the first time included all the main factors in the following equation (Wen et al., 2022):

$$
\lambda_{HM} = \frac{\left(1 - V_{WC}^{2/3}\right) \cdot \lambda_M}{1 + 0.5 \cdot V_{WC}}
$$
$$
+ \frac{V_{WC}^{2/3} \cdot \lambda_M \cdot \lambda_{WC} \cdot d_{WC}}{\left(1 - V_{WC}^{1/3}\right) \cdot d_{WC} \cdot \lambda_{WC} + V_{WC}^{1/3} \cdot d_{WC} \cdot \lambda_M + 2 \cdot V_{WC} \cdot R_{M/WC} \cdot \lambda_M \cdot \lambda_{WC}},
$$

$$(6.136)$$

where $\lambda_{HM}$ (W/mK) is the thermal conductivity of the two-phase hard metal; $V_{WC}$ (dimensionless), $d_{WC}$(m), and $\lambda_{WC}$ (W/mK) the volume fraction, grain size, and thermal conductivity of WC phase, respectively; $\lambda_M$ (W/mK) the thermal conductivity of the metal binder phase; and $R_{M/WC}$ (m²K/W) the interfacial thermal resistance (ITR) of the M/WC interfaces of a unit-specific interface area. This model was successfully applied in evaluating the thermal conductivity for WC–Co, WC–Ni, and WC–Ag systems (Wen et al., 2022). As shown in Figure 6.27 for the WC–Co system, the predicted 3D surface agrees well with the measured thermal conductivity (Wen et al., 2022).

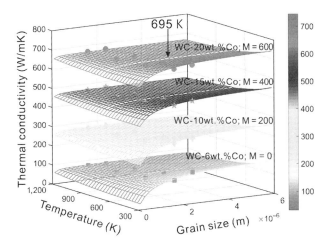

**Figure 6.27** Predicted thermal conductivities in comparison with the measured ones (Wen et al., 2022) for WC–Co. $M$ represents the value added to the corresponding thermal conductivity.

## 6.7    Some Other Thermophysical Properties

Besides interfacial energy, viscosity, molar volume, and thermal conductivity, there are some other thermophysical properties (such as thermal expansion coefficient, Seebeck coefficient, and thermal radiation) that describe the basic characteristics of materials.

Thermal expansion is a tendency of material to change its shape and size in response to a temperature change (Paul and Gene, 2008). Thermal expansion usually decreases with increase of the bond energy, which has an effect on the melting point of solids. Thus materials with high melting points are more likely to have low thermal expansions (Pelleg, 2006). The volumetric coefficient of thermal expansion is given by an equation of the following form:

$$\alpha_V = \frac{1}{V}\left(\frac{\partial V}{\partial T}\right)_p, \tag{6.137}$$

in which the subscript $p$ means that pressure is constant during expansion. The thermal expansion is related to heat capacity and modeled in a similar way as that for volume. This fact indicates that the thermal expansion varies linearly with composition unless there are experimental evidences to suggest a nonlinear feature. Therefore, the thermal expansion of a stoichiometric phase or each endmember compound in a nonstoichiometric phase can be written as follows:

$$\alpha = \sum_{i=1}^{m} x_i \alpha_i + \alpha^E, \tag{6.138}$$

where $\alpha^E$ is the deviation from the linearity relation between pure elements.

The Seebeck coefficient of a material is the magnitude of an induced thermoelectric voltage in response to a temperature difference across that material. The Seebeck coefficient is in the unit of volt per Kelvin ($V/K$). The high Seebeck coefficient is a critical factor for the efficiency of thermoelectric generators and thermoelectric coolers. One way to define the Seebeck coefficient is the voltage built up ($\Delta V$) in the case of a small temperature gradient applied to a material, where the material reaches a steady state at which the current density is zero everywhere. If the temperature difference $\Delta T$ between two ends of a material is small, the Seebeck coefficient is expressed as follows:

$$S = -\frac{\Delta V}{\Delta T}. \tag{6.139}$$

Thermal radiation is a kind of electromagnetic radiation generated by thermal motions of charged particles in material. All materials above 0 K emit thermal radiation. Typical thermal radiations include visible lights, infrared signals, and cosmic microwave background radiation.

## 6.8     Establishment of Thermophysical Property Databases

Scientific databases, such as thermodynamic and thermophysical databases, are of fundamental importance for materials design. Nowadays, thermodynamic databases for many engineering materials, such as iron and steel, light alloys, and superalloys, have been established mainly through the CALPHAD method. This method, which was initially developed for calculation of multicomponent phase diagrams, has now been expanded to describe other phase-based properties such as thermophysical properties (Campbell et al., 2014). The CALPHAD method can condense experimental thermodynamic, kinetic, and physical properties along with first-principles computed or empirically estimated properties into a consistent set of models and parameters for each phase and their relations (Du and Sundman, 2017). In comparison with thermodynamic database, currently the databases for thermophysical properties (diffusivity, thermal conductivity, and so on) are much less established, especially for multicomponent alloys.

Establishing thermophysical property databases for multicomponent alloys is extremely time consuming and costly. For some thermophysical properties, such as diffusivity, extensive experimental work should be performed to measure diffusivities for representative alloys. These experimental data are needed to establish the atomic mobility database for diffusion calculations. Even measuring a ternary diffusivity matrix is time consuming. As a result, a hybrid approach of key experiments (such as high-throughput methods, including diffusion couple multiples), the empirical approach, atomistic calculations, and the CALPHAD method is recommended to establish diffusivity databases in multicomponent alloys. For some thermophysical properties, such as thermal conductivity, which could be anisotropic, their modeling strongly depends on the microstructure including grain size and orientation, phase distribution, pores, grain boundary properties (e.g., thermal resistance across the grain boundaries), and so on (Du and Sundman, 2017). In consequence, a physically sound model that takes microstructural features and contributions from individual phases into account needs to be developed for the sake of describing thermal conductivity accurately.

Nowadays, CALPHAD-type programs, including ThermoCalc, Pandat, and OpenCalphad, are being used to establish databases for some thermophysical properties such as diffusivity and molar volume, which are not sensitive to microstructure. Most recently, some work has been done to develop phase-dependent databases within the frame of the CALPHAD approach, which describes thermophysical properties as functions of temperature, pressure, and composition of the phases (Liu et al., 2020). However, for anisotropic and strongly microstructure-dependent thermophysical properties, such as thermal conductivity and electrical resistivity, physically sound models and corresponding codes need to be developed (Du and Sundman, 2017). There is a long way to go for us to establish databases for technologically important thermophysical properties, which are as complete as the thermodynamic databases we have now for many commercial alloys. In spite of this, noticeable progress has been made for the establishment of thermophysical databases, in particular for a diffusion database.

Like a thermodynamic database, a thermophysical database is a merged set of thermophysical assessments of binary, ternary, and multicomponent systems. In order to create a database, one must first collect all the necessary assessments. There are a few hints that should be taken into account when establishing a database. First, the data for pure elements such as the self-diffusivity or impurity diffusivity must be the same in all assessments in which the element appears in order to make it possible to merge these into a database. Second, one should pay attention to model compatibility. When combining two assessments, it is essential that a phase with the same crystal structure is described with the same model in individual assessments. It is generally realized that a description for one binary system can be merged into a database when extrapolations from this binary system to several ternary systems are performed with a good agreement with experimental data. Thirdly, unassessed or missing parameters should be estimated by the database manager in order to extrapolate to metastable states or prevent a metastable phase in one binary system from being stable in other systems. Finally, a database needs continuous refinement, validation, and updating. It is necessary that there are experimental data for representative multicomponent systems, which can check if extrapolations from the database are reliable. Updating can include adding new assessments or replacing existing assessments by using the new and more reliable assessment.

## 6.9 A Case Study: Precipitation and Age Hardening in an AA6005 Al Alloy

In the following case study, by coupling the thermodynamic, diffusivity, and other thermophysical parameters through the precipitation module (PanPrecipitation) of

Pandat software (Cao et al., 2009, 2011), we demonstrate the simulation for precipitation and age hardening behavior of an AA6005 alloy during aging and reheating processes. The related thermodynamic (.tdb) and thermophysical parameters (.kdb) are provided so that readers can perform the simulations on their own by using PanPrecipitation. The Pandat demo version can be downloaded free of charge from CompuTherm's website: www.computherm.com/download/Pandat_2019_Demo_Setup.exe. The two example files "AA6005_yield_strength" and "AA6005_reheating" can subsequently also be found in the Pandat installation folder: "Pandat 2019 Examples\PanPrecipitation." The results obtained with the demo version, restricted to the ternary base system, differ somewhat from the full calculation for the six-component AA6005 alloy.

Aluminum alloys find wide applications in automotive and aerospace industries due to their excellent strength-to-density ratio. Al–Mg–Si (AA6xxx series) alloys are among the best candidates for such applications because they provide a medium-to-high strength and indicate good formability and weldability simultaneously. This series of alloys is usually precipitation hardened through the aging process. It is generally accepted that the precipitation sequence in AA6xxx alloys is as follows (Andersen et al., 1998; Miao and Laughlin, 1999):

$$\alpha(\text{SSS}) \rightarrow \text{GP Zones (clusters)} \rightarrow \beta''(\text{Mg}_5\text{Si}_6) \rightarrow \beta'(\text{Mg}_5\text{Si}_3) \rightarrow \beta(\text{Mg}_2\text{Si}),$$

in which $\alpha(\text{SSS})$ is the supersaturated solid solution matrix (Al). Among all precipitated phases, only $\beta$ is a thermodynamically stable phase. It was reported that the GP zones and $\beta''$ are the two important strengthening phases formed during the early stage of aging. During the simulation, the precipitate phases are approximately treated as one single phase with the composition $\text{Mg}_5\text{Si}_6$, and the particles are assumed to be spherical.

In this example, the precipitation behavior of an AA6005 alloy (Al–0.016Cu–0.55Mg–0.82Si–0.2Fe–0.5Mn in wt.%) was investigated by using PanPrecipitation. The alloy (Alloy IV in the original article) has been studied experimentally and theoretically by Myhr et al. (2001). Figure 6.28 shows the simulated temporal evolution of both number density and mean size for the particle as well as the hardness for Alloy IV aged at 185°C in comparison with the experimental data. The calculated maximum number density occurs at around $t = 0.6$ hour, while the peak alloy strength is not reached until a later time of 2.5 hours, which is long after the coarsening stage begins. This finding is consistent with the experimental observation, implying that the particle shearing mechanism outweighs the other one (particle bypassing) until $t = 2.5$ hours even after the coarsening has actually taken place since $t = 0.6$ hour. Both the hardness and yield strength are calculated during a PanPrecipitation simulation.

Figure 6.29 shows that the calculated response of the peak-aged Alloy IV (after the T6 heat treatment) to the reheating process at 350°C is in accord with the experimental data. Both calculation and experimentation reveal a sharp decrease in the particle number density and a considerable growth of the existing particles accompanying with a significant softening behavior during the reheating. This behavior has been a major

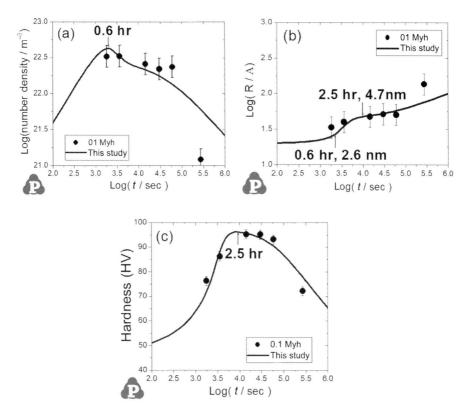

**Figure 6.28** Predicated temporal evolution of the particle (a) number density; (b) mean size, $R$; (c) hardness in an AA6005 alloy aged at 185°C in comparison with the experimental data (Myhr et al., 2001).

problem in the welding of the Al–Mg–Si alloys (Myhr et al., 2001). Similar phenomena can be found for the other three alloys with Alloy I (Al–0.54Mg–0.56Si–0.0062Mn–0.2Fe in wt.%), Alloy II (Al–0.35Mg–0.56Si–0.0064Mn–0.21Fe in wt.%) and Alloy III (Al–0.74Mg–0.58Si–0.0061Mn–0.21Fe in wt.%). The degree of agreement between simulation and experimentation is summarized in Figure 6.30. Indeed, a good agreement has been achieved for this class of Al alloys subjected to a variety of heat treatments.

To date, PanPrecipitation has been successfully applied to simulate the precipitation behaviors of many other series of Al alloys (AA2xxx, AA6xxx, AA7xxx, and A3xx) (Cao et al., 2011; Schmid-Fetzer and Zhang, 2018) as well as commercial Ni-based superalloys (Cao et al., 2016; Zhang et al., 2013, 2018) and Mg alloys (Luo et al., 2017; Schmid-Fetzer, 2019; Zhang et al., 2014a) in the cases of artificial aging, reheating, or continuous cooling processes. It is, therefore, of practical value to use such a modeling framework as an enabling tool for the development of new alloys and the optimization of the existing ones to meet the increasing demand for structural materials with a high strength.

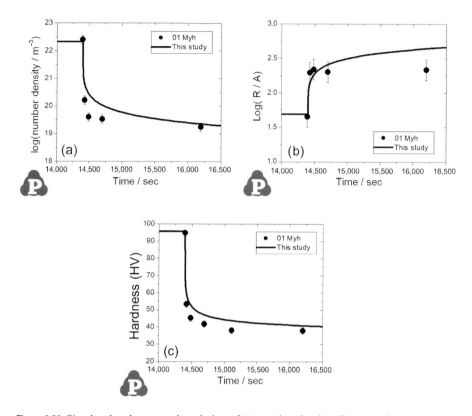

**Figure 6.29** Simulated and measured evolution of (a) number density; (b) mean size; (c) hardness for peak-aged Alloy IV during reheating at 350°C. The experimental data are from Myhr et al. (2001).

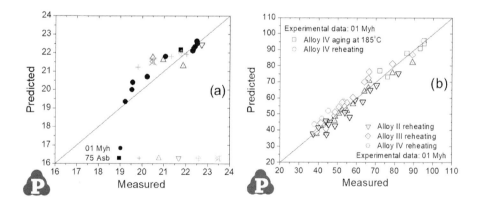

**Figure 6.30** Comparison between predicted and measured (a) number density (log10 of number/m³ value) and (b) hardness (Vickers pyramid number, HV) for a range of AA6xxx alloys under different heat treatment conditions. The experimental data are from Myhr et al. (2001).

The thermophysical parameter file (.kdb) is an XML file that can be viewed with any text editor (Notepad++ is recommended for XML color highlighting). This gives the reader a hands-on introductory insight on the "thermophysical" parameters to be used for the mechanical model of the matrix phase, for the nucleation, growth, and coarsening – such as the interfacial energy – and for the mechanical model of the precipitate phase(s). A comprehensive description of the models and parameters used in PanPrecipitation can be found in the Manual/Help-PDF on pages 137–185 in the free Pandat demo installation folder: "Pandat_Manual_2019.pdf."

# References

Allnatt, A. R. (1982) Einstein and linear response formulas for the phenomenological coefficients for isothermal matter transport in solids. *Journal of the Physics and Chemistry of Solids*, 15, 5605–5613.

Andersen, S., Zandbergen, H., Jansen, J., Traeholt, C., Tundal, U., and Reiso, O. (1998) The crystal structure of the $\beta''$ phase in Al–Mg–Si alloys. *Acta Materialia*, 46(9), 3283–3298.

Andersson, J.-O., and Agren, J. (1992) Models for numerical treatment of multicomponent diffusion in simple phases. *Journal of Applied Physics*, 72(4), 1350–1355.

Andriy, Y., Yuriy, P., Stepan, M., Jürgen, B., Hidekazu, K., and Herbert, I. (2014) Viscosity of liquid Co–Sn alloys: thermodynamic evaluation and experiment. *Physics and Chemistry of Liquids*, 52(4), 562–570.

Arrhenius, S. (1899) Chemical reaction velocities. *Zeitschrift für Physikalische Chemie*, 28, 317–335.

Bauer, A., Neumeier, S., Pyczak, F., Singer, R. F., and Göken, M. (2012) Creep properties of different $\gamma'$-strengthened Co-base superalloys. *Materials and Science Engineering A*, 550, 333–341.

Boltzmann, L. (1894) Zur integration der diffusionsgleichung bei variabeln diffusions coefficienten. *Annals of Physics*, 289, 959–964.

Borgenstam, A., Engstrom, A., Hoglund, L., and Agren, J. (2000) DICTRA, a tool for simulation of diffusional transformations in alloys. *Journal of Phase Equilibria and Diffusion*, 21(3), 269–280.

Bouchet, R., and Mevrcl, R. (2003) Calculating the composition-dependent diffusivity matrix along a diffusion path in ternary systems. *CALPHAD*, 27(3), 295–303.

Brillo, J., Chathoth, S. M., Koza, M. M., and Meyer, A. (2008) Liquid Al80Cu20: atomic diffusion and viscosity. *Applied Physics Letters*, 93(12), 121905.

Brillo, J., and Schmid-Fetzer, R. (2014) A model for the prediction of liquid–liquid interfacial energies. *Journal of Materials Science*, 49(10), 3674–3680.

Broeder, F. J. A. D. (1969) A general simplification and improvement of the Matano–Boltzmann method in the determination of the interdiffusion coefficients in binary systems. *Scripta Metallurgica*, 3, 321–325.

Brown, A. M., and Ashby, M. F. (1980) Correlations for diffusion constants. *Acta Metallurgica*, 28, 1085–1101.

Bueche, F. (1959) Mobility of molecules in liquids near the glass temperature. *Journal of Chemical Physics*, 30, 748–752.

Butler, J. A. V. (1932) Thermodynamics of the surfaces of solutions. *Proceedings of the Royal Society of London, Series A*, 135, 348–375.

Cahn, J. W., and Larche, F. C. (1983) An invariant formulation of multicomponent diffusion in crystals. *Scripta Metallurgica*, 27, 927–932.

Cahoon, J. R. (1997) A modified hole theory for solute impurity diffusion in liquid metals. *Metallurgical and Materials Transactions A*, 28, 583–593.

Campbell, C. E. (2008) Assessment of the diffusion mobilities in the gamma and B2 phases in the Ni–Al–Cr system. *Acta Materialia*, 56, 4277–4290.

Campbell, C. E., Kattner, U. R., and Liu, Z. K. (2014) The development of phase-based property data using the CALPHAD method and infrastructure needs. *International Journal of Materials and Manufacturing*, 3, 158–180.

Cao, W., Chen, S.-L., Zhang, F., et al. (2009) PANDAT software with PanEngine, PanOptimizer and PanPrecipitation for multi-component phase diagram calculation and materials property simulation. *CALPHAD*, 33(2), 328–342.

Cao, W., Zhang, F., Chen, S.-L., et al. (2016) Precipitation modeling of multi-component nickel-based alloys. *Journal of Phase Equilibria and Diffusion*, 37(4), 491–502.

Cao, W., Zhang, F., Chen, S.-L., Zhang, C., and Chang, Y. (2011) An integrated computational tool for precipitation simulation. *JOM*, 63(7), 29–34.

Cermak, J., and Rothova, V. (2003) Concentration dependence of ternary interdiffusion coefficients in Ni3Al/Ni3Al–X couples with X=Cr, Fe, Nb and Ti. *Acta Materialia*, 51(15), 4411–4421.

Chase, M. W. (1998) *NIST-JANAF Thermochemical Tables,* fourth edition. College Park: American Institute of Physics.

Choi, J., Park, S. K., Hwang, H. Y., and Huh, J. Y. (2015) A comparative study of dendritic growth by using the extended Cahn–Hilliard model and the conventional phase-field model. *Acta Materialia*, 84, 55–64.

Christensen, M., and Wahnström, G. (2003) Co-phase penetration of WC(10-10)/WC(101-0) grain boundaries from first principles. *Physical Review B*, 67(11), 045408.

Darken, L. S. (1948) Diffusion, mobility and their interrelation through free energy in binary metallic systems. *Transactions of AIME*, 175, 184–194.

Darken, L. S. (1949) Diffusion of carbon in austenite with a discontinuity in composition. *Transactions of AIME*, 180, 430–438.

Das, S. K., Horbach, J., and Voigtmann, T. (2008) Structural relaxation in a binary metallic melt: molecular dynamics computer simulation of undercooled $Al_{80}Ni_{20}$. *Physical Review B*, 78(6), 064208.

Dayananda, M. A., and Sohn, Y. H. (1999) A new analysis for the determination of ternary interdiffusion coefficients from a single diffusion couple. *Metallurgical and Materials Transactions A*, 30, 535–543.

Du, C. F., Zheng, Z. S., Min, Q. H., et al. (2020) A novel approach to calculate the diffusion matrix in ternary systems: application to Ag–Mg–Mn and Cu–Ni–Sn systems. *CALPHAD*, 68, 101708.

Du, Y., Chang, Y. A., and Huang, B. (2003) Diffusion coefficients of some solutes in Fcc and liquid Al: Critical evaluation and correlation. *Materials Science and Engineering A*, 363, 140–151.

Du, Y., and Schuster, J. C. (2001) An effective approach to describe growth of binary intermediate phases with narrow ranges of homogeneity. *Metallurgical and Materials Transactions A*, 32, 2396–2400.

Du, Y., and Sundman, B. (2017) Thermophysical properties: key input for ICME and MG. *Journal of Phase Equilibria and Diffusion*, 38, 601–602.

Dushman, S., and Langmuir, I. (1922) The diffusion coefficient in solids and its temperature coefficient. *Proceedings of the American Physical Society*, 113.

Eyring, H. (1935) The activated complex in chemical reactions. *Journal of Chemical Physics*, 3, 107–115.

Farrell, D. E., Shin, D., and Wolverton, C. (2009) First-principles molecular dynamics study of the structure and dynamic behavior of liquid Li4BN3H10. *Physical Review B*, 80, 224201.

Fick, A. (1855) Ueber diffusion. *Annals of Physics*, 94, 59–86.

Fisher, J. C. (1951) Calculation of diffusion penetration curves for surface and grain boundary diffusion. *Journal of Applied Physics*, 22, 74–77.

Girifalco, L. A. (1964) Vacancy concentration and diffusion in order-disorder alloys. *Physics and Chemistry of Solids*, 25, 323–333.

Gorecki, T. (1990) Changes in the activation energy for self- and impurity-diffusion in metals on passing through the melting point. *Journal of Materials Science Letters*, 9, 167–169.

Grimvall, G. (1999) *Thermophysical Properties of Materials*. Amsterdam: Elsevier Science B.V.

Grimvall, G., and Sjodin, S. (1974) Correlation of properties of materials to Debye and melting temperatures. *Physica Scripta*, 10, 340–352.

Guillermet, A. F. (1987) Critical evaluation of the thermodynamic properties of cobalt Int. *Journal of Thermophysics*, 8, 481–510.

Harrison, L. (1961) Influence of dislocations on diffusion kinetics in solids with particular reference to the alkali halides. *Transactions of the Faraday Society*, 57, 1191–1199.

Hart, E. W. (1957) On the role of dislocations in bulk diffusion. *Acta Materialia*, 5, 597–597.

Heitjans, P., and Kärger, J. (2005) *Diffusion in Condensed Matter*. Berlin and Heidelberg: Springer.

Helander, T., and Ågren, J. (1999) A phenomenological treatment of diffusion in Al–Fe and Al–Ni alloys having B2-Bcc. ordered structure. *Acta Materialia*, 47, 1141–1152.

Hillert, M., and Jarl, M. A. (1978) A model for alloying in ferromagnetic metals. *CALPHAD*, 2, 227–238.

Hirai, M. (1993) Estimation of viscosities of liquid alloys. *Journal of the Iron and Steel Institute of Japan*, 33, 251–258.

Huang, D., Liu, S., Du, Y., and Sundman, B. (2015) Modeling of the molar volume of the solution phases in the Al–Cu–Mg system. *CALPHAD*, 51, 261–271.

Huang, S., Worthington, D. L., Asta, M., Ozolins, V., Ghosh, G., and Peter, K. L. (2010) Calculation of impurity diffusivities in a-Fe using first-principles methods. *Acta Materialia*, 58, 1982–1993.

Huber, L., Elfimov, I., Rottler, J., and Militzer, M. (2012) Ab initio calculations of rare-earth diffusion in magnesium. *Physical Review B*, 85, 144301.

Jacobs, M. H., Schmid-Fetzer, R., and van den Berg, A. P. (2017) Phase diagrams, thermodynamic properties and sound velocities derived from a multiple Einstein method using vibrational densities of states: an application to MgO–SiO2. *Physics and Chemistry of Minerals*, 44(1), 43–62.

Jacobs, M. H., Schmid-Fetzer, R., and van den Berg, A. P. (2019) Thermophysical properties and phase diagrams in the system MgO–SiO2–FeO at upper mantle and transition zone conditions derived from a multiple-Einstein method. *Physics and Chemistry of Minerals*, 46(5), 513–534.

Jang, J., Kwon, J., and Lee, B. (2010) Effect of stress on self-diffusion in Bcc Fe: an atomistic simulation study. *Scripta Materialia*, 63, 39–42.

Johansson, S. A. E., and Wahnström, G. (2010) Theory of ultrathin films at metal-ceramic interfaces. *Philosophical Magazine Letters*, 90(8), 599–609.

Johansson, S. A. E., and Wahnström, G. (2012) First-principles study of an interfacial phase diagram in the V-doped WC–Co system. *Physical Review B*, 86(3), 035403.

Jönsson, B. (1994) Assessment of the mobility of carbon in Fcc C–Cr–Fe–Ni alloys. *Z. Metallkd.*, 85, 502–509.

Jost, W. (1969) *Diffusion in Solids, Liquids and Gases*, second edition. New York: Academic Press.

Kaptay, G. (2003) *Proceedings of MicroCAD 2003 International Section Metallurgy*. Hungary: University of Miskolc.

Kaptay, G. (2012) On the interfacial energy of coherent interfaces. *Acta Materialia*, 60, 6804–6813.

Kaschnitz, E., and Ebner, R. (2007) Thermal diffusivity of the aluminum alloy Al–17Si–4Cu (A390) in the solid and liquid states. *International Journal of Thermophysics*, 28, 711–722.

Kaur, I., Mishin, Y., and Gust, W. (1995) *Fundamentals of Grain and Interphase Boundary Diffusion*. Chichester: John Wiley.

King, D. A., and Woodruff, D. P. (1993) *The Chemical Physics of Solid Surfaces Distributors for the U.S. and Canada*. Amsterdam: Elsevier North-Holland.

Kirkaldy, J., Weichert, D., and Haq, Z. (1963) Diffusion in multicomponent metallic systems: VI. Some thermodynamic properties of the D matrix and the corresponding solutions of the diffusion equations. *Canadian Journal of Physics*, 41, 2166–2173.

Kirkaldy, J. S., and Young, D. J. (1987) *Diffusion in the Condensed State*. London: Institute of Metals.

Kozlov, L., Romanov, L. M., and Petrov, N. N. (1983) Prediction of multicomponent metallic melt viscosity. *Izv. Vuzov Chernaya Metall.*, 3, 7–11.

Kucharski, M. (1986) The viscosity of multicomponent systems. *Z. Metallkd.*, 77, 393–396.

LeClaire, A. D. (1978) Solute diffusion in dilute alloys. *Journal of Nuclear Materials*, 70, 70–96.

Li, D., Fürtauer, S., Flandorfer, H., and Cupid, D. M. (2014) Thermodynamic assessment and experimental investigation of the Li–Sn system. *CALPHAD*, 47, 181–195.

Lippmann, S., Jung, I.-H., Paliwal, M., and Rettenmayr, M. (2016) Modelling temperature and concentration dependent solid/liquid interfacial energies. *Philosophical Magazine*, 96, 1–14.

Liu, Y., Liu, S., Du, Y., Peng, Y., Zhang, C., and Yao, S. (2019) A general model to calculate coherent solid/solid and immiscible liquid/liquid interfacial energies. *CALPHAD*, 65, 225–231.

Liu, Y., Long, Z., and Wang, H. (2006) A predictive equation for solute diffusivity in liquid metals. *Scripta Materialia*, 55, 367–370.

Liu, Y. L., Zhang, C., Du, C. F., et al. (2020) CALTPP: a general program to calculate thermophysical properties. *Journal of Materials Science and Technology*, 42, 229–240.

Long, J., Zhang, W., Wang, Y., et al. (2017) A new type of WC–Co–Ni–Al cemented carbide: grain size and morphology of $\gamma'$-strengthened composite binder phase. *Scripta Materialia*, 126, 33–36.

Lu, X. G., Selleby, M., and Sundman, B. (2005a) Assessments of molar volume and thermal expansion for selected Bcc, Fcc and Hcp metallic elements. *CALPHAD*, 29(1), 68–89.

Lu, X. G., Selleby, M., and Sundman, B. (2005b) Implementation of a new model for pressure dependence of condensed phases in Thermo-Calc. *CALPHAD*, 29, 49–55.

Luo, A. A., Zhao, J.-C., Riggi, A., and Joost, W. (2017) High-throughput study of diffusion and phase transformation kinetics of magnesium-based systems for automotive cast magnesium alloys. Project final report. Ohio State University, CompuTherm LLC (Madison, WI). Contract No.: DE-EE0006450.

Manning, J. R. (1964) Correlation factors for impurity diffusion. Bcc, diamond, and Fcc structures. *Physical Review*, 136, A1758.

Mantina, M., Wang, Y., Chen, L. Q., Liu, Z. K., and Wolverton, C. (2009) First principles impurity diffusion coefficients. *Acta Materialia*, 57(14), 4102–4108.

Marino, K. A., and Carter, E. A. (2008) First-principles characterization of Ni diffusion kinetics in β-NiAl. *Physical Review B: Condensed Matter and Materials Physics*, 78(18), 184105/1–184105/11.

Matano, C. (1933) The relation between the diffusion coefficients and concentrations of solid metals (the nickel-copper system). *Japanese Journal of Physics*, 8, 109–113.

Mehrer, H. (1990) *Numerical Data and Functional Relationships in Science and Technology: Diffusion in Solid Metals and Alloys*. Germany: Landolt-Börnstein Springer-Verlag.

Mehrer, H. (2005) Diffusion: introduction and case studies in metals and binary alloys, in Heitjans, P., and Kärger, J. (eds), *Diffusion in Condensed Matter: Methods, Materials, Models*. Berlin, Heidelberg: Springer Berlin Heidelberg, 3–63.

Mehrer, H. (2007) *Diffusion in Solids: Fundamentals, Methods, Materials, Diffusion-Controlled Processes*. Berlin, Heidelberg: Springer Berlin Heidelberg.

Miao, W., and Laughlin, D. (1999) Precipitation hardening in aluminum alloy 6022. *Scripta Materialia*, 40(7), 873–878.

Moelwyn-Hughes, E. A. (1961) *Physical Chemistry*. Oxford: Pergamon Press.

Myhr, O., Grong, Ø., and Andersen, S. (2001) Modelling of the age hardening behaviour of Al–Mg–Si alloys. *Acta Materialia*, 49(1), 65–75.

Nowick, A. S., and Burton, J. J. (1975) *Diffusion in Solids*. London: Academic Press.

Olson, G. B., and Kuehmann, C. (2014) Materials genomics: from CALPHAD to flight. *Scripta Materialia*, 70, 25–30.

Onsager, L. (1931) Reciprocal relations in irreversible processes. *Physical Review*, 37, 405–426.

Onsager, L. (1945) Theories and problems of liquid diffusion. *Annals of the New York Academy of Sciences*, 46, 241–265.

Paul, A., and Divinski, S. V. (2017) *Diffusion Analysis in Material Applications, Handbook of Solid State Diffusion*. Oxford: Matthew Deans.

Paul, A., Laurila, T., Vuorinen, V., and Divinski, S. V. (2014) *Thermodynamics, Diffusion and the Kirkendall Effect in Solids*. Cham, Heidelberg, New York, Dordrecht, and London: Springer.

Paul, T. A., and Gene, M. (2008) *Physics for Scientists and Engineers*. New York: Worth Publishers.

Pelleg, J. (2006) Diffusion of 60Co in vanadium single crystals. *Philosophical Magazine*, 32(3), 593–598.

Philibert, J. (1991) *Atom Movements: Diffusion and Mass Transport in Solids*. Les Ulis: Les Éditions de Physique.

Pilarek, B., Salamon, B., and Kapała, J. (2014) Calculation and optimization of LaBr3–MBr (Li–Cs) phase diagrams by CALPHAD method. *CALPHAD*, 47, 211–218.

Sauer, F., and Freise, V. (1962) Diffusion in binary mixtures showing a volume change. *Z. Elektrochem. Angew. Phys. Chem.*, 66, 353–363.

Schmid-Fetzer, R. (2019) Recent progress in development and applications of Mg alloy thermodynamic database, in Joshi, V. V., Jordon, J. B., Orlov, D., and Neelameggham, N. R. (eds), *Magnesium Technology 2019*. Cham: Springer, 249–255.

Schmid-Fetzer, R., and Zhang, F. (2018) The light alloy CALPHAD databases PanAl and PanMg. *CALPHAD*, 61, 246–263.

Seetharaman, S., and Sichen, D. (1994) Estimation of the viscosities of binary metallic melts using Gibbs energies of mixing. *Metallurgical and Materials Transactions A*, 25B, 589–595.

Sherby, O. D., and Simnad, M. T. (1961) Prediction of atomic mobility in metallic systems. *ASM Transactions Quarterly*, 54, 227–240.

Smigelkas, A. D., and Kirkendall, E. O. (1947) Zinc diffusion in alpha brass. *Transactions of the AIME*, 171, 130–142.

Sohn, Y. H., and Dayananda, M. A. (2002) Diffusion studies in the $\beta$ (B2), $\beta'$ (Bcc), and $\gamma$ (Fcc) Fe–Ni–Al alloys at 1000 °C. *Metallurgical and Materials Transactions A*, 33, 3375–3392.

Su, X., Yang, S., Wang, J., et al. (2010) A new equation for temperature dependent solute impurity diffusivity in liquid metals. *Journal of Phase Equilibria and Diffusions*, 31, 333–340.

Sudbrack, C. K., Ziebell, T. D., Noebe, R. D., and Seidman, D. N. (2008) Effects of a tungsten addition on the morphological evolution, spatial correlations and temporal evolution of a model Ni–Al–Cr superalloy. *Acta Materialia*, 56, 448–463.

Swalin, R. A. (1956) Correlation between frequency factor and activation energy for solute diffusion. *Journal of Applied Physics*, 27, 554–555.

Touloukian, Y. S., Kirby, R. K., Taylor, R. E., and Desai, P. D. (1975) Thermophysical properties of matter – the TPRC data series. Volume 12. Thermal expansion metallic elements and alloys. (Reannouncement). Data book. *Thermal Conductivity*.

Ven, A. V. d., and Ceder, G. (2005) First principles calculation of the interdiffusion coefficient in binary alloys. *Physical Review Letters*, 94, 045901.

Vignes, A., and Birchenall, C. E. (1968) Concentration dependence of the interdiffusion coefficient in binary metallic solid solution. *Acta Metallurgica*, 16, 1117–1125.

Vineyard, G. H. (1957) Frequency factors and isotope effects in solid state rate processes. *Journal of the Physics and Chemistry of Solids*, 3, 121–127.

Wagner, C. (1969) Evaluation of data obtained with diffusion couples of binary single-phase and multiphase systems. *Acta Metallurgica*, 17, 99–107.

Warren, R. (1980) Research on the wettability of ceramics films by metal and their interfaces. *Journal of Materials Science*, 15, 2489–2496.

Warren, R., and Waldron, M. B. (1972) Surface and interfacial energies in systems of certain refractory-metal monocarbides with liquid cobalt. *Nature Physical Science*, 235, 73–74.

Wen, S., Du, Y., Tan, J., et al. (2022) A new model for thermal conductivity of "continuous matrix / dispersed and separated 3D-particles" type composite materials and its application to WC–M (M = Co, Ag) systems. *Journal of Materials Science and Technology*, 97, 123–133.

Whittle, D. P., and Green, A. (1974) The measurement of diffusion coefficients in ternary systems. *Scripta Materialia*, 8, 883–884.

Xin, J., Du, Y., Shang, S., et al. (2016) A new relationship among self- and impurity diffusion coefficients in binary solution phases. *Metallurgical and Materials Transactions A*, 47A, 3295–3299.

Yakymovych, A., Plevachuk, Y., Mudry, S., Brillo, J., Kobatake, H., and Ipser, H. (2014) Viscosity of liquid Co–Sn alloys: thermodynamic evaluation and experiment. *Physics and Chemistry of Liquids*, 52(4), 562–570.

Zener, C. (1951) Theory of $D_0$ for atomic diffusion in metals. *Journal of Applied Physics*, 22(4), 372–375.

Zhang, C., Cao, W., Chen, S.-L., et al. (2014a) Precipitation simulation of AZ91 alloy. *JOM*, 66(3), 389–396.

Zhang, C., and Du, Y. (2017) A novel thermodynamic model for obtaining solid–liquid interfacial energies. *Metallurgical and Materials Transactions A*, 48, 5766–5770.

Zhang, C., Du, Y., Liu, S., Liu, Y., and Sundman, B. (2016) Thermal conductivity of Al–Cu–Mg–Si alloys: experimental measurement and CALPHAD modeling. *Thermochimica Acta*, 635, 8–16.

Zhang, F., Cao, W., Chen, S., Zhang, C., and Zhu, J. (2013) The role of the CALPHAD approach in ICME. *Proceedings of the 2nd World Congress on Integrated Computational Materials Engineering (ICME)*. Cham: Springer, 195–200.

Zhang, F., Cao, W., Zhang, C., Chen, S., Zhu, J., and Lv, D. (2018) Simulation of co-precipitation kinetics of $\gamma'$ and $\gamma''$ in superalloy 718, in Ott, E., Liu, X., Andersson, J., et al. (eds), *Proceedings of the 9th International Symposium on Superalloy 718 and Derivatives: Energy, Aerospace, and Industrial Applications*. Cham: Springer, 147–161.

Zhang, F., Du, Y., Liu, S., and Jie, W. (2015) Modeling of the viscosity in the AL–Cu–Mg–Si system: database construction. *CALPHAD*, 49, 79–86.

Zhang, F., Wen, S., Liu, Y., Du, Y., and Kaptay, G. (2019) Modelling the viscosity of liquid alloys with associates. *Journal of Molecular Liquids*, 291, 111345.

Zhang, W., Zhang, L., Du, Y., Liu, S., and Tang, C. (2014b) Atomic mobilities in Fcc Cu–Mn–Ni–Zn alloys and their characterizations of uphill diffusion and zero-flux plane phenomena. *International Journal of Materials Research*, 105, 13–31.

# 7 Case Studies on Steel Design

## Chapter Contents

## 7.1      Brief Introduction about Steel

Steels are produced by combing iron with other elements. The most common element in steel is carbon, and its content is usually 0.002–2 wt.%, whereas higher additions, up to 4.3 wt.% C, are used in cast iron. Other commonly added elements are Si, Mn, Mo, Cr, B, Ti, V, and Nb. The main production processes for steel are as follows: (1) smelting iron ore, coke, and limestone into pig iron in a blast furnace; (2) smelting molten iron into steel in a basic oxygen furnace or melting of scrap in an electric arc

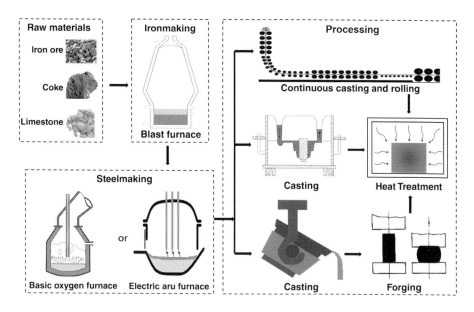

**Figure 7.1** Process illustrations for preparation of steel.

furnace; (3) casting molten steel into ingots, or continuous casting billet; and (4) using forging, rolling, or other plastic deformation methods to process steel for various purposes. Figure 7.1 presents simplified process illustrations for steel production. Steels are important structural materials with the largest output and widely used in a wide range of fields, including buildings, railways, shipbuilding, appliances, defense, aerospace, automotive industry, and so on.

Steel can be classified according to chemical composition, application field, manufacture method, metallographic structure, etc. According to the chemical composition, it can be divided into plain carbon steel (low, medium, high, or ultrahigh carbon steel) and alloy steel (stainless, chrome, manganese, chromium manganese, chrome nickel, chromium molybdenum, silicon manganese steel, etc., or low, medium, or high alloy steel). In view of the application fields, it can be divided into structural steel, tool steel, die steel, bearing steel, wear-resistant steel, bridge steel, marine steel, boiler steel, electrical steel, pressure vessel steel, and so on. The density of steel is one important physical property that should be considered in many applications. It is usually between 7.75 and 8.05 kg/m$^3$ according to the constituents.

Steels consist of a wide range of microstructures, which usually include austenite, ferrite, cementite, pearlite, bainite, martensite, Widmanstätten, ledeburite, etc. The austenite is also known as gamma phase ($\gamma$Fe) with Fcc structure, undergoing a eutectoid decomposition below a critical temperature. The ferrite is the alpha/delta phase with the Bcc structure, and this phase and gamma phase enable us to form the so-called gamma loop with the addition of ferrite stabilizing elements such as Cr, Al, and Mo. The cementite is a transition metal carbide with the formula of (Fe, M)$_3$C (M = alloying elements) and an orthorhombic structure, showing high hardness and brittleness. The pearlite, bainite, and martensite are the microstructures formed in austenite matrix through different undercooling/cooling rates. The pearlite microstructure consists of ferrite and cementite formed by a diffusion-controlled mechanism. In the cases of bainite and martensite, a deep undercooling is required to drive the diffusionless transformation. The Widmanstätten pattern is the structure of proeutectoid lamellar (needle) ferrite mixed with pearlite, and it is often formed in coarse austenite grain. The ledeburite is the eutectic mixture of austenite and cementite, while the austenite in ledeburite can decompose into ferrite, pearlite, and/or cementite during further cooling.

By optimizing the composition and controlling the casting and thermomechanical processes, the aforementioned microstructures can be observed simultaneously. The plain-carbon steel contains ferrite and pearlite, and its moderate strength and toughness properties are mainly due to the contribution from the ferrite matrix. For high-carbon steel used to fabricate bridge cables, the typical full-pearlite structure is required since it presents an extremely high strength after cold drawing (Anelli, 1992). Twinning-induced plasticity (TWIP) steel usually has a higher Mn content to stabilize the austenite at room temperature. The excellent tensile property and high work hardening rate in TWIP steel are traced to numerous twins induced by deformation, making TWIP steel candidate materials for the automotive industry (De Cooman et al., 2018). Ultrahigh-strength maraging steel normally possesses high hardenability to produce martensite matrix during quenching, while the dispersed nanosized particles can be precipitated during tempering. The martensite matrix along with fine precipitates enables maraging steel with ultrahigh strength without sacrificing much toughness (Ghosh and Olson, 2002).

Strength, toughness, corrosion resistance, and fatigue properties are the major concerns for steels. The interested reader can refer to the comprehensive review article by Xiong and Olson (2016) for the detail on the structure–property relation and the design model for these properties. These properties are determined by the microstructure, which are in turn affected by both chemical composition and processing during fabrication. A hierarchical formalism for the processing-structure–property-performance has been proposed by Olson (1997), and the properties are backward-related to the microstructure and corresponding manufacturing processes. For instance, the strength is mainly contributed by matrix, precipitates, and microsegregation, and the corrosion resistance is strongly affected by grain boundary chemistry and phase potentials. To facilitate the design of microstructure and properties, the theoretical basis of computational materials science is implemented into a variety of computational tools. Some representative computational tools employed for the design of steels are listed in Table 7.1. For more details about globally available computational tools for design of various materials, the interested reader could refer to a very comprehensive handbook by Schmitz and Prahl (2016).

Steels are typical multicomponent and multiphase materials, and their production process is rather complex. The change of one component or one individual process has a significant impact on the subsequent processes and even the performance of the final product. In the past, the development of steels largely depended on the knowledge and

**Table 7.1** Representative computational tools for microstructure and property design of steels.

| Computational tool (website) | Theoretical basis | Major simulation field |
|---|---|---|
| **VASP** (www.vasp.at) | Density functional theory | Atomic-scale phenomenon |
| **LAMMPS** (https://lammps.sandia.gov/) | Molecular dynamics | Large-scale atomic dynamics |
| **Thermo-Calc** (www.thermocalc.se) **Pandat** (www.computherm.com) **FactSage** （www.factsage.cn/） | Thermodynamics | Calculation of phase diagram and thermodynamic properties |
| **DICTRA** (https://thermocalc.com/products/add-on-modules/diffusion-module-dictra/) PanDiffusion (https://computherm.com/pandiffusion） | Transformation kinetics | Diffusion-controlled transformation |
| **TC-Prisma** (https://thermocalc.com/products/add-on-modules/precipitation-module-tc-prisma/) **PanPrecipitation** (https://computherm.com/panprecipitation) **MatCalc** (www.matcalc.at/index.php/documentation/examples) | Precipitation kinetics | Microstructure evolution of precipitates |
| **MICRESS** (www.micress.de/) **Openphase** (www.openphase.de/) | Cahn–Hilliard equation, Free energy functional | Microstructure evolution of solidification and solid-state transformation |
| **Abaqus** (www.abaqus.com/) | Finite element method | Prediction of mechanical properties |

experience of engineers. This kind of experimental trial-and-error procedure consumes lots of time and resources. In order to develop advanced new steels, computer simulations are becoming more and more important.

In this chapter, we are going to describe the computational design of an ultrahigh-strength and corrosion-resistant Ferrium S53 steel (Olson, 2013) as well as AISI H13 hot-work tool steel (Eser et al., 2016a, b). The Ferrium S53 was computationally designed and developed by QuesTek Innovations LLC of the USA using its proprietary Materials by Design® approach, which has brought the steel to market through licenses to the company. For the S53 steel, there are several design goals. The first goal is accurate control for the precipitation of $M_2C$ carbide, which is an efficient strengthener in steels because of its high modulus misfit relative to Bcc iron and coherent feature at the nanoscale. The second goal is to ensure a martensitic alloy by maintaining lath martensitic microstructure and keeping high strength. The key issue associated with these two goals is the development of a quantitative martensite kinetic model, which can design maximal strengthening effect and appropriate martensite start (Ms) temperature for the sake of maintaining an alloy with martensite as the dominant phase. The third goal is to obtain a good general corrosion resistance and resistance to stress corrosion cracking (SCC) as well as to avoid hydrogen embrittlement through maximizing Cr partitioning in the spinel oxide and enhancing grain boundary cohesion. It could be stressed that there is a trade-off for the three goals. We will demonstrate that a variety of calculations (including thermodynamic calculations, precipitation simulations, and atomistic calculations) and subsequent simulation-guided experiments, which were performed by QuesTek Innovations LLC, can reconcile the demands of different design aims and reduce the development cycle for S53 steel significantly in comparison with traditional expensive and time-consuming experimental efforts.

For the case study of AISI H13 tool steel, we will show that a thermomechanical-metallurgical model using multiscale commercial software can reasonably describe the microstructure evolution and the various mechanical properties at the macroscopic scale under different tempering conditions (Eser et al., 2016a, b). For that purpose, firstly the precipitations of carbides were simulated using a thermokinetic software with thermodynamic and thermophysical property databases. Secondly, simulated major microstructural parameters were coupled with structure–property models to predict yield stress, flow curve, and creep resistance. Subsequently, those simulated properties were coupled with a finite element method (FEM) to predict the relaxation of internal stresses and the deformation behavior at the macroscopic scale. This case study for AISI H13 tool steel will indicate the significant role of a variety of simulations for materials design at the macroscopic scale.

## 7.2     Ultrahigh-Strength and Corrosion-Resistant Ferrium S53 Steel

Landing gear is the undercarriage of an aircraft or spacecraft, and it is essential for takeoff and landing. In 1994, the *Commercial Jet Transport Safety Statistics* review of aircraft systems that have caused accidents showed that landing gear failure is the dominant factor leading to the accidents (Committee, 1994). An accident in aviation is defined as an occurrence "between the time any person boards the aircraft with the intention of flight until such time as all such persons have disembarked, if between those times any person suffers death or serious injury as a result of being in or upon the aircraft, or by direct contact with the aircraft or anything attached thereto, or the aircraft sustains substantial damage" (Committee, 1994, pp. 1–6). That review indicated that 456 accidents among the 1,395 accidents resulting from 19 aircraft parts, such as windows, hydraulic power, engine, and aeropower plant, are due to the failure of the landing gear. It is easy to imagine that landing gears require high strength in view of the heavy weight of the aircraft. Thus, ultrahigh-strength steels such as 4340 and 300M were used to manufacture the landing gears previously. However, these two steels show poor corrosion resistances. Corrosion is one of the most serious issues leading to the failure of the landing gear. According to a report in 2000, the cost of corrosion to the US Department of Defense is about $22.5 billion per year. Thus, landing gears made of 4340 or 300M steel were often plated with toxic Cd in order to improve the general corrosion resistance. Alternative alloys used as landing gears are

17-4PH and 15-5PH stainless steels, which provide a greater corrosion resistance but less strength in comparison with 4340 or 300M steel. In consequence, for a long period, the selection of alloys for the landing gears has often required a compromise between strength and corrosion resistance. It is of great interest to design new steels with ultrahigh strength and excellent corrosion resistance. Kuehmann et al. (2008) outlined the computational design for Ferrium S53. In the following, we will describe the systems design strategy and design of mechanical properties and corrosion resistance for Ferrium S53 in more detail.

## 7.2.1 Strategy for the Systems Design of Ferrium S53

For a long time, the discovery of new materials has been by a traditional trial-and-error method, which is laborious, costly, and time consuming. The definition of "materials design" given in Chapter 1 emphasizes the upmost importance to establish quantitative relationships among the four cornerstones (processing, structure, properties, and performance) of materials science and engineering in order to design and develop materials highly efficiently. The performance objectives can be mapped to specific property targets, which are determined by the structure of the materials, and the structure is, in turn, defined by processing. The Systems Design Chart developed by Olson and Kuehmann (2014) is used to capture the major processing–structure–properties–performance relationships for a targeted material. With this Systems Design Chart approach, a significant reduction for both time and cost associated with the design and development of Ferrium S53 can be achieved since the design goals can be accomplished through several prototypes instead of dozens.

Figure 7.2a illustrates the systems design chart for Ferrium S53. This figure shows how the processing determines the structure, which in turn determines the properties for Ferrium S53. The performance is mainly related to product design, which is beyond the scope of this book. For S53 steel, its strength is mainly controlled by matrix (lath martensite), strengthening dispersion [nanoscale $(Cr,Mo,V,Fe)_2C$ carbides], and microsegregation (Cr,Mo, and V). $(Cr,Mo,V,Fe)_2C$ carbide is named as $M_2C$ carbide. The aqueous corrosion is determined by passive film formation, and SCC is controlled by the formation of this passive film, grain-refining dispersion, and grain boundary chemistry. On the other hand, the fatigue resistance of S53 is mainly affected by strengthening dispersion and microsegregation, and its core toughness is controlled by the matrix, grain-refining dispersion, and grain boundary chemistry of B and Re. Figure 7.2b schematically shows the major structural features, the manipulation of which could achieve the desired design goals.

## 7.2.2 Design of Strength, Toughness, and Fatigue Resistance

As pointed out by Campbell and Olson (2000a), the design for Ferrium S53 steel was mainly completed by means of multicomponent thermodynamic and kinetic software packages to integrate a variety of mechanistic models, including the martensitic transformation behavior, coherent $M_2C$ carbide precipitation in a highly

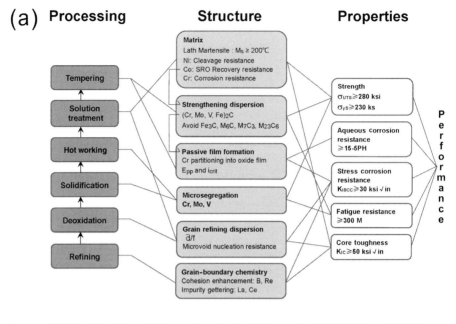

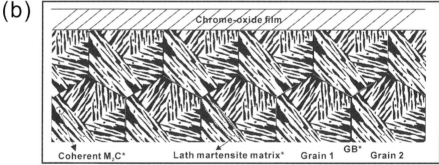

*1. Lath martensite: provide high strength
2. Nano-scale M₂C:coherent with the martensite / an efficient dispersion strengthener
3. Grain boundary: improve resistance to SCC and hydrogen embrittlement
4. Passive chrome-oxide film: provide good corrosion

**Figure 7.2** (a) Systems Design Chart for ferrium S53 (Olson and Kuehmann, 2014) and (b) schematic diagram indicating the major microstructure features achieving the design goals as indicated in the Systems Design Chart. Figure 7.2a is reproduced from Olson and Kuehmann (2014) with permission from Elsevier.

supersaturated Bcc solid solution, the formation and stability of passivating oxide films, and microsegregation during solidification. In this subsection, we will describe the major procedures to design the strength, toughness, and fatigue resistance, focusing on the thermodynamic and kinetic simulations integrating these mechanistic models. For details on the establishment of these models and corresponding calculations, the interested reader can refer to the PhD thesis by Campbell (1997).

*Strength* is one mechanical property that indicates the resistance of a material to fracture and excessive deformation under an applied load. Ultimate tensile strength is the maximum tensile stress that the material can withstand before breaking. As indicated in Figure 7.2a, the strength of S53 steel is mainly manipulated through matrix, strengthening dispersion, and microsegregation.

### 7.2.2.1 Martensitic Transformation Behavior

To reach the high strength, firstly it is essential to control the matrix structure of Ferrium S53 as the ductile lath martensite. The formation of the lath martensite during quenching is dependent on the Fcc→Bct (body-centered tetragonal) martensitic start temperature (Ms temperature). The information about the composition-dependent Ms temperature is critical for the design of an ultrahigh-strength steel. In the case of S53 steel, a high Ms temperature above 200°C must be maintained in order to form the lath martensite with the desired high strength. Using thermodynamic databases, the Ms temperature can be calculated by means of the Olson–Cohen model (Olson and Cohen, 1976a, b) for heterogeneous nucleation and the Ghosh–Olson solid solution strengthening model (Ghosh and Olson, 1994a, b) for the description of the frictional work associated with the interfacial motion. According to the Ghosh–Olson model, the critical chemical driving force per unit volume $\left(\Delta g^{ch}\right)$ needed for nucleation is given by an equation of the following form:

$$\Delta g^{ch} = -\left(e^{str} + \frac{2\gamma_s}{nd} + w_\mu + w_{th}\right), \tag{7.1}$$

where $e^{str}$ represents the elastic strain energy, $\gamma_s$ is the interfacial energy between austenite (Fcc) and martensite (Bct) nucleus under the coherent condition, $d$ the interplanar spacing between close-packed planes, and $n$ the number of close-packed planes describing the nucleus thickness. $w_\mu$ and $w_{th}$ represent athermal frictional work and thermal work, which are the two components of the interfacial frictional work, respectively. Athermal frictional work is the work produced during friction from the not thermally activated contribution, and thermal work is the additional work depending on thermal activation. The dislocations will dissociate when a critical driving force is provided through either cooling or an applied stress to overcome the athermal frictional work for the sake of pushing the martensite nucleus into the austenitic phase. In the design of S53 steel, $\Delta g^{ch}$ is estimated from the CALPHAD-type thermodynamic description (SGTE, 1994) with the revised thermodynamic description of the Fe–Co–Cr (Campbell, 1997). In view of the fact that the thermodynamic description for the Bct martensite is not reliable, it assumes that the Bcc and Bct phases have the similar Gibbs energy (i.e., the Gibbs energy of the Bct phase is above that of the Bcc phase in the unit of a few hundred J/mol-atoms). Currently, there is no clear knowledge on the relative energies of Bct and Bcc phases in steels.

By assuming a nucleating defect size and neglecting the weak temperature dependence of the shear modulus, the elastic strain and interfacial energy terms can be estimated (Campbell, 1997). Ghosh and Olson have modeled the compositional dependence of the athermal fraction work and thermal work, and they found that

$W_{th}$ can be neglected if Ms is above 450 K (Ghosh and Olson, 1994a, b). The athermal frictional work $W_\mu$ can be calculated by an equation of the following form (Ghosh and Olson, 1994a, b):

$$W_\mu = \sqrt{\sum_i \left(K_\mu^i X_i^{1/2}\right)^2} + \sqrt{\sum_j \left(K_\mu^j X_j^{1/2}\right)^2} + \sqrt{\sum_k \left(K_\mu^k X_k^{1/2}\right)^2} + K_\mu^{Co} X_{Co}^{1/2}, \quad (7.2)$$

in which $i = $ C, N; $j = $ Cr, Mn, Mo, Nb, Si, Ti, V; and $k = $ Al, Cu, Ni, W. The values for $K_\mu^i$, $K_\mu^j$, and $K_\mu^k$ are given elsewhere (Campbell, 1997). It is noted that $W_\mu$ [in (7.2)] equals $V_m \cdot w_\mu$ [in (7.1)], where $V_m$ is the molar volume. The same conversion to the unit J/mol is done for other molar quantities given in upper case, such as $\Delta G^{ch}$.

In order to predict Ms temperatures for the targeted alloy, it is necessary to solve (7.2) and use Thermo-Calc software to find out the temperature at which $\Delta G^{ch}$ satisfies (7.1). Figure 7.3 shows the calculated Ms temperatures along with the experimental ones, indicating a good agreement between calculation and experiment. Thus, the Ghosh–Olson model for solid solution strengthening model and the Olson–Cohen model for heterogeneous nucleation along with thermodynamic descriptions can be used to design Ms temperature for the S53 steel.

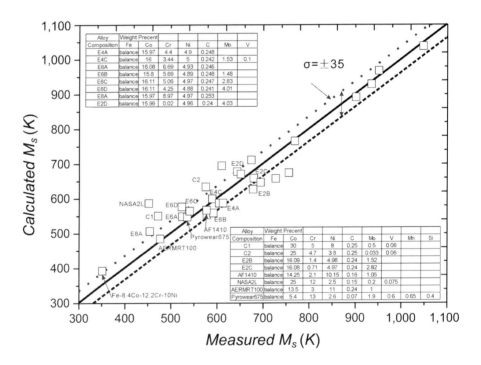

**Figure 7.3** Predicted $M_s$ temperatures along with the experimental results (Campbell and Olson, 2000a) for comparison. The standard deviation $\sigma$ for the deviation between the computed and measured $M_s$ is $\pm 35°C$. The compositions for the alloys are also given in the diagram. This figure is reproduced from Campbell and Olson (2000a) with permission from Springer.

### 7.2.2.2    Precipitation of Coherent $M_2C$ Carbides

While controlling the matrix structure as the lath martensite is the first step, another key point to achieve the designed high strength is the development of an efficient strengthening dispersion. According to the secondary hardening mechanism, the nanoscale precipitation of Hcp–$M_2C$ carbides further increases the strength due to its high modulus misfit relative to the Bcc matrix. The coherent $M_2C$ carbides were precipitated from the metastable Bcc + cementite paraequilibrium. The efficiency of the strengthening dispersion is dependent on the particle size and the spacing between particles. With increasing nucleation rate, the supersaturation also increases since the total volume fraction is constant when the temperature and concentration of materials are fixed. According to the Langer–Schwartz model (Langer and Schwartz, 1980), originally developed for demixing spinodal liquid, the precipitation in the highly supersaturated Bcc matrix causes the growth mechanism to be inhibited, which leads to precipitation through the restricted coarsening process. Thus, the mean carbide size remains near the critical particle size ($r^*$), which smoothly increases with the decrease of the supersaturation. The classical nucleation theory for homogeneous nucleation provides a well-known relation for $r^*$:

$$r^* = \frac{2\gamma}{\Delta G}, \tag{7.3}$$

where $\gamma$ is the specific interfacial energy ($J/m^2$) between the precipitate and matrix phase Bcc, which is associated with the critical particle size, and $\Delta G$ is the precipitation driving force ($J/m^3$). Technically, $\Delta G$ is the Gibbs energy difference per volume between the initial matrix phase Bcc and the $M_2C$ carbide phase precipitate.

To calculate the precipitation driving force, the initial state (Bcc + cementite paraequilibrium) from which precipitation starts must be defined first. The experiment (Ghosh et al., 1999) shows that the paraequilibrium between cementite and the Bcc matrix is the initial state. This paraequilibrium occurs within the temperature range of 400–550°C. At these temperatures, the mobility of C is much larger than that of substitutional elements. As a result, under the Bcc + cementite paraequilibrium condition, the chemical potentials of C are equal in Bcc and the cementite, and there is no redistribution for the substitutional elements. For the thermodynamic description of S53 steel (Campbell, 1997), the Bcc and cementite phases are described with the sublattice models $(Fe,Cr,Co,Ni,V...)_1(C,Va)_3$ and $(Fe,Cr,Co,Ni,V,...)_3C_1$, respectively. No redistribution for the substitutional elements in the first sublattice model means $y_i^{Bcc} = y_i^{Cem}$ ($i = Fe,Cr,Co,Ni,V...$). A thermodynamic database was developed to describe the Bcc + cementite paraequilibrium (Campbell, 1997).

It is required that a quantitative martensite kinetic model enables the prediction of both Ms and $M_2C$ precipitation driving force as a function of composition. After the Bcc + cementite paraequilibrium has been well defined, the total energy for the $M_2C$ carbides is needed to calculate its precipitation driving force from the Bcc matrix. The total Gibbs energy of the coherent $M_2C$ carbide (including the interface) is the summation of the chemical free energy ($G_{chemical}$), the elastic strain energy ($G_{elastic}$), and the interfacial energy ($G_{interfacial}$) between $M_2C$ and Bcc:

$$G_m^{M_2C} = G_{chemical} + G_{elastic} + G_{interfacial}. \qquad (7.4)$$

The interfacial energy $G_{interfacial}$ in (7.4) is different from $\gamma$ in (7.3); $\gamma$ is associated with the critical particle size for the nucleation, and $G_{interfacial}$ is the interfacial energy between precipitate (after nucleation and growth) and another phase, such as the matrix. The unit for $\gamma$ is the interfacial energy of particles per unit area, while that for $G_{interfacial}$ is the interfacial energy per mole (J/mol). $M_2C$ carbides were treated as the Hcp phase according to the crystal structure. An upper bound for the elastic strain energy can be estimated using King et al. (1991) and Liarng (1996), the linear-elastic model for an inhomogeneous inclusion. The elastic strain energy is a function of lattice parameters of both Bcc (matrix) and $M_2C$ phases as well as the elastic modulus and Poisson's ratio of Bcc. To calculate the interfacial energy, it is assumed that this energy is composition and size independent for a given state of coherency. The Johnson–Cahn model (Johnson and Cahn, 1984) was used to calculate the interfacial energy. These calculated values of elastic strain energy and interfacial energy were calibrated using experimental data of small angle neutron scattering (SANS) and transmission electron microscopy (TEM) (Allen et al., 1993; Campbell, 1997). The SANS and TEM analysis can provide data on the particle size distribution, volume fraction, and aspect ratio for the precipitates that are needed for the calibration.

The strengthening effect of the nanoscale $M_2C$ carbide is optimized by maximizing the driving force for its precipitating from the Bcc matrix in paraequilibrium with cementite. Maximizing the driving force promotes the smaller particle size and is thus beneficial to the strength. Under the assumption that the molar volumes for matrix and precipitates are identical, Figure 7.4 plots the calculated driving force for $M_2C$ carbides. Note that the molar driving force in Figure 7.4 is not the same as $\Delta G$ in (7.3). In Figure 7.4, the driving force $\Delta G$ from (7.3) is multiplied by molar volume

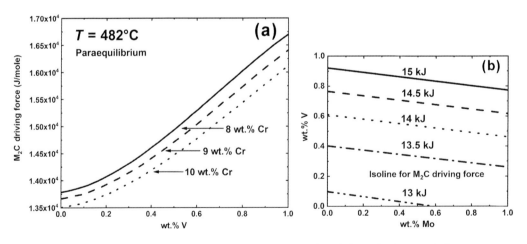

**Figure 7.4** Molar driving force for the precipitation of $M_2C$ from the Bcc matrix in paraequilibrium with cementite versus wt.% of Cr and V (a) and the contour plot of $M_2C$ driving force against wt.% of V and Mo (b). This figure is reproduced from Campbell and Olson (2000a) with permission from Elsevier.

($m^3$/mol). This figure indicates that the driving force of the $M_2C$ carbide is sensitive to the V content but not to the Cr content.

### 7.2.2.3 Solidification Microsegregation and Castability

Microsegregation during solidification is another factor that contributes to the strength because of its impact to solution strengthening. The reduction of microsegregation is beneficial to the increase of the strength. Increasing the amount of carbide forming elements Cr, Mo, and V will increase the tendency of solidification microsegregation. The DICTRA module in the Thermo-Calc software can be used to predict the degree of microsegregation. The simulation assumes that the solid–liquid interface maintains equilibrium during solidification (i.e., local equilibrium) and that diffusion occurs in both the solid and liquid phases. The simulation also assumes that the initial liquid composition is uniform and the liquid–solid interface is planar. Figure 7.5 shows the comparison between the calculated and measured compositional profiles for the AerMet100 alloy, indicating that the thermodynamic and atomic mobility parameters are reliable (Campbell, 1997). Thus, these thermodynamic and atomic mobility

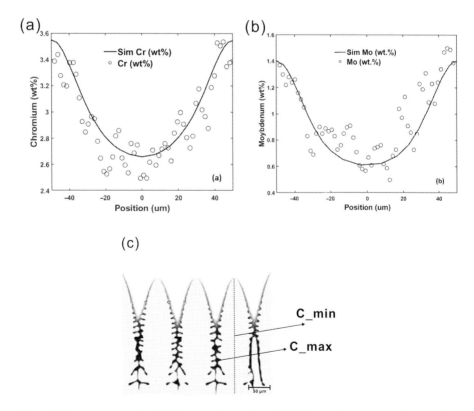

**Figure 7.5** For the as-cast AerMet100, experimental and DICTRA simulated composition profiles across a 100 µm secondary dendrite arm for (a) Cr and (b) Mo; (c) shows a schematic for the segregation ratio. Figures 7.5a,b are reproduced from Campbell and Olson (2000a) with permission from Elsevier.

databases can be used to design new alloys in a similar composition range, including S53 steel, for the sake of minimizing solidification microsegregation. The DICTRA simulations performed by Campbell (1997) indicate that Mo has the largest segregation ratio among the added elements, which is defined as $\delta = C\_max / C\_min$, where $C\_max$ is the concentration between the secondary dendrite arm, and $C\_min$ the concentration of the primary dendrite arm, as indicated in Figure 7.5c.

The castability of the alloy is another factor that needs to be considered for solidification simulation. Decreasing the microsegregation and improving the fluidity of the liquid (i.e., the ability of the liquid to fill a mold) are beneficial to the castability. Reducing the freezing temperature range enables the improvement to the fluidity. The fluidity is best when freezing occurs at a constant temperature, which permits the slow inward growth of the solid phase(s). Even for multicomponent and multiphase systems, thermodynamic calculations can predict the freezing temperature ranges and illustrate the effect of composition on these temperature ranges accurately.

*Toughness* is the ability of a material to absorb energy by plastic deformation without fracturing. A good toughness requires both high strength and high ductility. There is a high demand of substantially improving the toughness of the high-strength stainless steels and overcoming the strength/ductility trade-off. For example, to improve the efficiency of an airplane engine, its operating temperature needs to be increased. Low-toughness stainless steels render bearings vulnerable to rapid fracture mechanisms, which can cause potentially serious secondary damage to the engine. Achieving a high core toughness is one of the design objectives for S53 steel. The core toughness and strength levels are critical to preventing catastrophic failure. The fracture toughness is mainly controlled by the prior austenite grain size, the martensitic microstructure, and the grain boundary cohesion. Sufficiently decreasing the prior austenitic grain size can increase the toughness, which is achieved by adding Ti carbide $(Ti,Mo)_1(C,N)_1$ grain refiners and optimizing the solution temperature to achieve high hardness with the minimum solution treatment time. The lath martensitic is an essential microstructural attribute for high strength and toughness. In addition, the $M_2C$ carbide precipitation further enhances the strength. In corrosive environments where embrittlement could occur, grain boundary cohesion becomes an important factor in maintaining toughness. To optimize grain boundary cohesion, B and Re are added during the vacuum-melt processing. Impurities will reduce the interfacial cohesion, which may cause segregation of the impurities to the grain boundaries and lead to a brittle fracture. The harmful impurities, which were identified by thermodynamic calculations, are P, Sn, Sb, and S (Anderson et al., 1987). According to first-principle calculations, B is a superior cohesive enhancer (Olson, 1997). Subsequent first-principles calculations (Kantner, 2002) demonstrated that Re is also a highly effective cohesion enhancer in view of its cohesion enhancement potency and boundary segregation energy. La and Ce can be used to getter both P and S impurities effectively according to Watton et al. (1987). Thus, in Ferrium S53 steel, trace amounts of La and Ce were used to getter impurities such as P, Sn, Sb, and S.

*Fatigue* is the weakening of a material caused by cyclic loading that leads to progressive and localized structural damage and the growth of cracks. As shown in

Figure 7.2a, the fatigue resistance was mainly determined by the microsegregation and strengthening dispersion. In addition, to achieve the desired fatigue strength, it is crucial to avoid the formation of the primary carbides such as $Fe_3C$ and $M_6C$. Primary carbides can form at the interface and facilitate crack initiation. Furthermore, the formation of primary carbides removes Cr from the matrix and thus decreases the aqueous corrosion resistance.

## 7.2.3 Design of Resistance to General Corrosion and Stress Corrosion Cracking

*General corrosion resistance* refers to how a substance (especially a metal) can withstand damage caused by oxidization or other chemical reactions. The formation of a stable passive film on the surface of a substance provides the corrosion resistance for S53 steel. This passivating oxide film is mainly composed of $M_2O_3$ (M = Cr, Fe, Ni) and $M_3O_4$ oxides. The amount of Cr partitioning to the film controls the formation and stability of the film. Thermodynamic calculations are performed to maximize Cr partitioning in the oxides, which is primarily driven by Co content. Sufficient Co is increasing the activity of Cr above that for the 12 wt.% Cr steel. The corrosion resistance for Ferrium S53 steel with a fixed Cr content can be improved by increasing the activity of Cr in the matrix, thus promoting the partition of Cr to the passive oxide film. However, the increase of Cr usually leads to the reduction of the Ms temperature and promotes the formation of harmful carbides. Thus, there is a trade-off between good corrosion resistance and high strength.

The Pourbaix diagram (Pourbaix, 1973) can be applied to identify the compositional region with a good corrosion resistance. This diagram is a graphical method in which potential versus pH plots (the electrode potential is drawn against the pH value in an aqueous solution) are established for chemical and electrochemical reactions associated with corrosion. In aqueous chemistry, pH is defined as the negative value of the decimal logarithm of the hydrogen ion activity $(a_{H^+})$ in a solution, i.e., $pH = -\log_{10}(a_{H^+})$, and an electrode potential is the potential difference referring to the standard hydrogen electrode when an electrode of an element is placed in a water-based aqueous solution containing ions of that element. Three areas (corrosion, immunity, and passivation) are identified in the Pourbaix diagram. In the area of *immunity* or cathodic protection, corrosion is impossible. In the case of the area of *passivation*, insoluble products resulting from electrochemical and/or chemical reactions are so adherent and impermeable that corrosion of the underlying metal is essentially suspended. That is to say that the metal is "passive." In his original publication, Pourbaix (1973) assumed that the dilute solution concentration is within the range of $10^{-4}$–$10^{-6}$ molar (molar concentration unit, 1 mol solute per 1 liter solution). The multicomponent pH-potential diagram calculations suppose that a total ionic concentration of the solution is below $10^{-4}$ molar, and this ionic concentration scales with the base alloy composition. The latter assumption is supported by Kirchheim et al. (1989), who found that the Cr content partitioning to the oxide film increases almost linearly with decadic logarithm of the matrix Cr content when investigating the passivation of Fe–Cr alloy at long times (i.e., steady-state conditions).

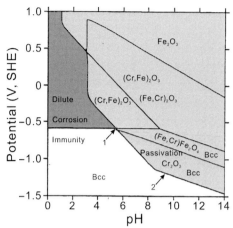

1. Intersection of the passive, immunity, and corrosion
regions
2. This line represents the point at which a passive film
just begins to form

**Figure 7.6**  Potential-pH diagram for Fe–12Cr (wt.%) considering nonstoichiometric oxides. This figure is reproduced from Campbell and Olson (2000a) with permission from Elsevier.

In the case of Ferrium S53 steel, the Pourbaix diagram is a thermodynamic representation for the stability of oxide films formed at specified pH and potential for the targeted solid phases and aqueous solution. Figure 7.6 is the computed potential-pH diagram for a Fe–12Cr (wt.%) alloy in a water-based aqueous solution (Campbell and Olson, 2000a). This diagram indicates that the passivating oxides $M_3O_4$ and $M_2O_3$ are thermodynamically stable in the neutral and alkaline aqueous solution environment. As shown in Figure 7.6, the addition of Cr in the Fe matrix shifts the passive boundaries to lower pH levels, decreasing the corrosive region and thus increasing the region over which the passivation occurs. In addition, a miscibility gap for the $M_2O_3$ phase [$(Cr,Fe)_2O_3$ and $(Fe,Cr)_2O_3$] is observed in this potential-pH diagram. The potential-pH diagrams of multicomponent alloys can be utilized to indicate whether the desired oxide passivation is favored in terms of thermodynamics.

Thermodynamic modeling of the corrosion processes such as the potential-pH diagrams is useful in finding an appropriate alloy with a good corrosion resistance, but it cannot describe the time scale of the reactions associated with corrosion. In order to understand the mechanism for the formation of passive films and simulate the corrosion process such as the composition profiles of the oxide films, a large amount of diffusion-based data (such as the mobility parameters of vacancy, oxygen, cation, and anion) and a detailed description of the electronic transport (i.e., electronic field) are needed. The very limited data for diffusion under corrosion places a significant restriction on the dynamic modeling possible.

To further increase the corrosion resistance of Ferrium S53 steel, $M_2C$ carbides are designed to be sufficiently smaller than the scale of the passive oxide film to enable the oxidation of carbides during passivation. This frees the Cr to be incorporated into the passive film.

*Stress corrosion cracking* (SCC) is the growth of crack formation under the combination of stress and a corrosive environment. As indicated in the Systems Design Chart (see Figure 7.2a), the SCC can be influenced by passivation of oxide film, grain-refining dispersion, and grain boundary chemistry. In fact, corrosion modeling is rather complex. An indirect simulation of the microstructure feature of the steel itself could be considered as a possible pathway to mitigate the sophisticated corrosion simulation and thus provide a useful design guideline. The grain boundary chemistry of S53 was designed through first-principles calculations of impurity and alloying effects on the interfacial cohesion of high-angle grain boundaries in the Bcc phase. Through these calculations, it was found that the enrichment of B and Re along the grain boundary is beneficial to the resistance of S53–SCC. In addition, the grain-refining dispersion design is according to results from the interfacial adhesion calculations. High interfacial adhesion is beneficial to increase the resistance to microvoid formation and thus reduces the SCC. Analogous to the general corrosion, the formation of a stable passive film can also improve the resistance of SCC. By manipulating the microstructure along the interfaces and oxide films, S53 provides very good SCC resistance.

## 7.2.4   Hydrogen Embrittlement

As previously mentioned, the traditional high-strength steel landing gear (300M and SAE 4340) has poor corrosion resistance, and improving its resistance requires cadmium plating (Campbell and Olson, 2000a). However, the cadmium plating may cause hydrogen embrittlement in addition to being toxic. Hydrogen embrittlement is the embrittling of metal after being exposed to hydrogen, and hydrogen embrittlement failure is caused by the presence of H segregating to lath martensite or grain boundaries, which reduces ductility and leads to brittle, fast-growing cracks that are usually intergranular or transgranular. Cyclic fatigue enhanced by hydrogen embrittlement and SCC can lead to unpredictable and unacceptable failures. The strategy to improve or even eliminate SCC has been described in the preceding paragraph. To avoid hydrogen embrittlement failure, hydrogen bakeout operations are necessary, which increases the cost. S53 uses nanoscale $M_2C$ particles to trap a substantial amount of H due to its large surface area to volume ratio. The S53 is thus less sensitive to hydrogen embrittlement than the landing gears of 300M and SAE 4340 steels.

## 7.2.5   Prototype and Applications of Ferrium S53

As just described, for the design of S53 steel, conflicting objectives exist. The Ms temperature is mainly controlled by the contents of C, Co, Cr, and Ni. The addition of Co increases the Ms temperature, while the addition of C, Cr, and Ni leads to a different effect on Ms. Maintaining a constant Ms temperature by limiting C conflicts with the desired strength objective, which requires a high level for C. A good aqueous corrosion resistance needs the high level of Cr that partitions to the oxide film. However, increasing Cr content decreases the Ms temperature and promotes the formation of undesirable carbides. Increasing the amounts of carbide formers Cr, Mo, and V increases the tendency for solidification microsegregation. The complexity of the design objectives reveals that the composition window meeting all the objectives is very narrow. Thus, it is almost impossible to find an alloy that meets the requirements using the experimental trial-and-error method. Consequently, performing computational design of S53 steel is essential. The most remarkable part for the successful design of this steel is the high-quality scientific databases (i.e., thermodynamic and thermophysical databases) and simulation codes.

As described by Campbell and Olson (2000a), the stages of the systems design applied to materials design include four primary steps. Step 1 is systems analysis, which defines application and material property objectives. Step 2 is the design and synthesis. During this step, the property objectives and their interactions are prioritized, quantitative and mechanistic models are developed, and then these models are integrated to design the targeted material for a special application. Once the optimal alloy composition is determined and models are developed, the characterization of prototype alloy(s) is performed in step 3, focusing on whether the material property objectives are obtained. The fourth (last) step is the operation of the targeted material in a (prototype) device. Generally speaking, step 1 defined by Campbell and Olson (2000a) is similar to the user demand stage indicated in Figure 1.2 of Chapter 1 in this book, and steps 2 and 3 from Campbell and Olson (2000a) almost overlap with alloy design and preparation stage in Figure 1.2. Step 4 from Campbell and Olson, (2000a) is identical to the start of industrial production stage in Figure 1.2. The Subsections 7.2.2–7.2.4 detail the main contents in steps 1 and 2 of the development of S53 steel. Step 4 is mainly associated with industrial production, and it is not described in this subsection. In the following, only the main aspects of the prototype characterization (step 3) will be briefly mentioned. For the details on the prototype characterization, the interested reader can refer to the publication Campbell and Olson (2000b).

In step 3, firstly the mechanistic models are integrated to design an optimal alloy composition that could meet the previously defined property objectives. The designed composition should ensure that an actual composition with small deviations from the designed one will not result in noticeably different properties. After the optimal composition is determined, the prototype alloy is made and characterized. The characterization for prototype alloy of S53 includes the preparation of the alloy, measurement of martensitic transformation temperature, obtainment of the desired microstructure through heat treatments, measurement of mechanical properties (strength, toughness, and fatigue), and corrosion resistance. These measured properties are compared with the model-predicted ones. If some derivations exist, either alloy composition should be modified, or further modeling of the system is needed. As a matter of fact, there could be a few cycles between modeling and prototype characterization. Using the systems design approach, Olson and

**Table 7.2** Chemical composition (wt.%) of Ferrium S53®. [a]

| Fe | C | Co | Cr | Ni | Mo | W | V |
|---|---|---|---|---|---|---|---|
| 66.99 | 0.21 | 14 | 10 | 5.5 | 2 | 1 | 0.3 |

[a] Chris and Rich Kooy (2010).

**Table 7.3** Typical mechanical property data of Ferrium S53®.[a]

| | U.T.S. | | 0.2% Yield | | El,% | | Impact energy | | Fracture toughness | | |
|---|---|---|---|---|---|---|---|---|---|---|---|
| Orientation | ksi | Mpa | ksi | Mpa | in 4D | %RA | ft-lbs | J | ksi·√in | MPa·√m | HRC |
| Long. | 288 | 1986 | 225 | 1511 | 15 | 57 | 18 | 24 | 65 | 71 | 54 |
| Trans. | 288 | 1986 | 225 | 1511 | 15 | 55 | 18 | 24 | 65 | 71 | 54 |

[a] Chris and Rich Kooy (2010).

coworkers developed S53 highly efficiently. Table 7.2 lists the chemical composition of S53 steel, and Table 7.3 shows its typical mechanical properties (Chris and Rich Kooy, 2010).

The Ferrium S53 designed and developed by Olson and his colleagues (Olson, 2013) is an ultrahigh-strength and corrosion-resistant steel. The typical application fields for S53 are flight-critical aerospace components (such as landing gear), critical flight components on orbital class rocket programs, rotorcraft power transmission components (such as rotor shafts), and structural components for the energy industry. In the military, it has been used for landing gears on US Air Force (USAF) aircraft A-10, T-38, C-5, and KC-135 (Kuehmann et al., 2008). Additional applications of S53 include the Boeing 787 Ram Air Turbine (RAT) pins and CH-53K helicopter components (Taskin, 2017). In addition, most recently S53 steel has been used on many SpaceX's Falcon rocket flight-critical components (Lancaster, 2019).

## 7.3    AISI H13 Hot-Work Tool Steel

Hot-work tool steel AISI H13 (DIN 1.2344, X40CrMoV5–1), the composition of which is given in Table 7.4, possesses excellent mechanical properties, such as high hardness, good wear resistance, and thermal-cycle stability as well as high strength at elevated temperatures. Consequently, AISI H13 steel is widely used as a tool for extruding and forging molds, particularly as die-casting molds for light metal processing. A good combination of these properties for this steel can only be obtained by a controlled heat treatment schedule, which includes austenitization, quenching, and multiple tempering processes. During the whole heat treatment procedure, both diffusion-induced phase transformation (such as ferrite/pearlite to austenite transformation during austenitization, austenite to bainite transformation upon quenching,

**Table 7.4** Composition of hot-work tool steel AISI H13 (wt.%).[a]

| C | Cr | Mo | V | Mn | Si | Fe |
|------|------|------|------|------|------|------|
| 0.40 | 5.37 | 1.34 | 1.22 | 0.30 | 0.97 | 90.4 |

[a] Eser et al. (2016a, b).

first-step tempered martensite to double-tempered martensite) and diffusionless transformation (austenite to martensite transformation upon quenching) can occur. The excellent mechanical properties are mainly attributed to the martensite formed during quenching as well as high number density and homogeneously distributed nanosize carbides precipitated during tempering.

In this case study, we will demonstrate the multiscale modeling by Eser et al. (2016a, b). In their work, the microstructure evolution of various carbides and cementite was simulated by means of the MatCalc software (Kozeschnik et al., 2004a, b; Svoboda et al., 2004). The computer-simulated microstructure parameters (i.e., the particle size and volume fraction of carbides) and the concentrations of alloying elements in the matrix are used as input parameters for the physics-based models to predict the mechanical properties, followed by the calculations of stress relaxation and deformation during tempering process through employing the finite element method in conjunction with predicted mechanical properties. Figure 7.7a presents schematically the simulation of microstructure evolution and mechanical properties as well as the modeling of stress relaxation during the heat treatment procedure of AISI H13 steel, and Figure 7.7b outlines the corresponding phase transformations, which lead to the desired microstructure.

## 7.3.1    Simulations of Microstructure Evolution, Yield Stress, Flow Curve, and Creep

### 7.3.1.1    Simulation of Microstructure

The simulation for the microstructure evolution of precipitates was performed by means of the MatCalc software package, which is based on the Kampmann–Wagner numerical (KWN) model (Kampmann and Wagner, 1984). The CALPHAD-type thermodynamic and thermophysical databases were used as input parameters for MatCalc software. Based on the calculations from MatCalc, detailed information about the evolution of size, phase fraction (or volume fraction), number density, and compositions of the precipitates, as well as matrix composition, can be obtained. The initial precipitation state should be known as one input condition for the simulation of the precipitation process during heat treatment. In this case, the initial microstructure (carbide size and content) after austenitizing and quenching was determined by SEM (Eser, 2014). The undissolved carbides after austenitizing and quenching are MC carbides with an average radius of 57 nm and a volume fraction of 0.65%. The simulation results for the precipitation of carbides at 600°C are shown in Figure 7.8.

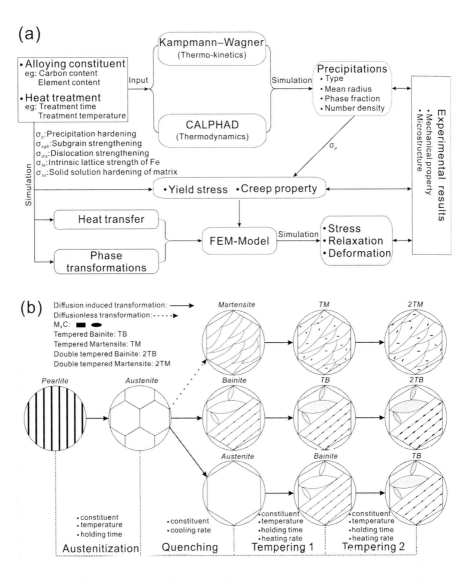

**Figure 7.7** (a) Schematic diagram for the simulation of microstructure evolution and mechanical properties during the heat treatment procedure of AISI H13 steel and (b) the phase transformations under different heat treatment conditions of AISI H13 steel.

This temperature is the common tempering temperature for this kind of steel. Eser et al. (2016a, b) also performed the simulations at 520°C and 650. The tempering temperature at which the maximum hardness is achieved is 520°C, while 650°C is the highest temperature at which the hot-work tool steel can be tempered.

As can be seen from Figure 7.8, cementite is formed during the first tempering step, and it dissolves rapidly. When the tempering temperature is reached, $M_7C_3$ carbide is formed, and its phase fraction is the highest among all the carbides. However, its fraction decreases with further soaking time. On the contrary, during tempering, the

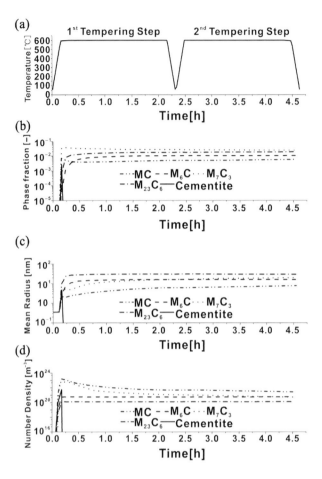

**Figure 7.8** Simulation of the microstructure evolution of steel AISI H13 during tempering at 600°C: (a) temperature–time profile; (b) phase fractions of the five carbides; (c) mean particle radius; and (d) number density. This figure is reproduced from Eser et al. (2016a) with permission from Elsevier.

second Cr-rich carbide $M_{23}C_6$ gradually increased. The calculated phase fraction of $M_6C$ carbide has a large value during tempering at 600°C. The simulation suggests the existence of a small amount of cementite. These simulated microstructure parameters are key inputs for prediction of mechanical properties using structure–property models, which will be described later.

### 7.3.1.2    Simulation of Yield Stress

Several strengthening mechanisms contribute to the overall yield strength $(\sigma_{yield})$ of AISI H13 steel, which can be considered as a sum of individual strengthening mechanisms (Li, 2003; Wang et al., 2006; Young and Bhadeshia, 1994):

$$\sigma_{yield} = \sigma_{Fe} + \sigma_{ss} + \sigma_P + \sigma_{dis} + \sigma_{sgb}, \tag{7.5}$$

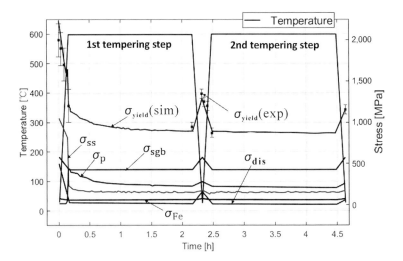

**Figure 7.9** Simulated individual contributions to yield stress and the total yield stress value according to (7.5) at 600°C compared with the experimental data. This figure is reproduced from Eser et al. (2016a) with permission from Elsevier.

in which $\sigma_{Fe}$ is the intrinsic lattice strength of pure Fe as the major component in steel, $\sigma_{ss}$ the solid solution hardening of matrix, $\sigma_P$ the precipitation hardening due to carbides, $\sigma_{dis}$ the dislocation strengthening, and $\sigma_{sgb}$ the subgrain strengthening. A detailed description of the individual strengthening terms is given in the review article by Xiong and Olson (2016). In Eser et al. (2016a), the parameters in all of the strengthening formulas are presented.

Figure 7.9 summarizes the simulated overall yield stress and individual strengthening terms along with the experimental data at the tempering temperature of 600°C. The experimental yield stress was measured by compression test (Eser, 2014). By comparing the calculated values with the experimental ones, a good agreement can be observed in view of experimental errors. It should be noted that the microstructure is unstable during tempering, and the experiment has a large error bar, especially for the first tempering step (Eser et al., 2016a).

### 7.3.1.3 Simulation of the Flow Curve

During quenching, significant residual stress may develop, and the major mechanism of stress relaxation during the tempering process is due to the flow of the material. In order to simulate this process, the temperature-dependent flow curve is needed, which is simulated using the work-hardening model based on the Kocks–Mecking equation (Kocks, 1976). Assuming that the moving dislocation moves an average free path $L$ before it is annihilated or immobilized, the change of dislocation density $(\rho_{dis})$ is described by the following equation (Lindgren et al., 2008):

$$\frac{d\rho_{dis}}{dt} = \frac{M\dot{\varepsilon}^{pl}}{bL}, \tag{7.6}$$

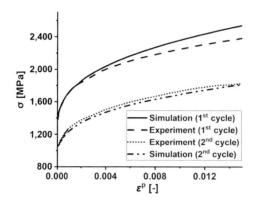

**Figure 7.10** Simulated flow curves along with the experimental ones at 600°C. This figure is reproduced from Eser et al. (2016a) with permission from Elsevier.

in which $\dot{\varepsilon}^{pl}$ is the plastic strain rate, and $M$ the Taylor factor. All of the parameters are listed in Eser et al. (2016a) for the calculation of the flow curve.

The critical process point of this stress relaxation is assumed where the tempering temperature is just reached. At this point, the time-dependent relaxation caused by creep (or flow) will be the most important mechanism. Therefore, the flow curve is simulated immediately after the tempering temperature is reached.

The flow curves calculated and measured at 600°C in the first and second tempering steps are shown in Figure 7.10. The measured and simulated flow curves are in good agreement with each other. It should be mentioned again, especially for the first tempering step, the measured value has a large uncertainty due to the high instability of microstructure at high temperatures (Eser, 2014). With this in mind, it can be expected from Figure 7.10 that there is a better consistency between the measured and calculated flow curves for the second tempering cycle compared to the first cycle.

### 7.3.1.4    Simulation of Creep

The creep (time-dependent/viscous deformation) is one important material property for deformation, in particular at high temperatures. For the tool steel uniformly tempered for a long time at temperatures above $0.4\ T_M$ (where $T_M$ is the melting or liquidus temperature of an alloy), creep plays a crucial role during the stress relaxation. The physics-based model used to describe the creep of AISI H13 steel is the composite model developed by Mughrabi (1979, 1983). This model has been utilized to describe the high-temperature creep behavior of martensitic steel (Blum et al., 1989a, b; Straub et al., 1995). The model assumes that the heterogeneous dislocation distribution leads to the heterogeneous stress distribution. Under this condition, the material can be considered a composite of a hard phase with a high dislocation density and a soft phase with a low dislocation density. The total strain

($\varepsilon_{tot}$) is assumed to be identical in both phase regions, and it is thus described by the following equation:

$$\varepsilon_{tot} = \frac{\sigma_S}{E} + \varepsilon_S = \frac{\sigma_H}{E} + \varepsilon_H, \tag{7.7}$$

in which $\varepsilon_{tot}$ is the sum of elastic and plastic strains, $E$ the elastic modulus, and $\sigma_S$ and $\sigma_H$ are the local stresses of the soft and hard phases, respectively. The elastic strain of the soft phase is $\sigma_S/E$, and that of the hard phase $\sigma_H/E$. The plastic strains in the soft and hard regions are given by $\varepsilon_S$ and $\varepsilon_H$, respectively.

For the sake of defining the creep strain, the Norton–Bailey model in Abaqus® is utilized where the creep strain rate $\dot{\varepsilon}^{cr}$ can be calculated by an equation of the following form:

$$\dot{\varepsilon}^{cr} = A\sigma^n t^m, \tag{7.8}$$

where $t$ is time, and the parameters $A$, $n$, and $m$ depend on the material and temperature. In the simulation, only the secondary creep is considered. Consequently, $m$ has been set to 0, and $A$ and $n$ are measured experimentally.

Based on (7.7) and (7.8), modeling for the creep of AISI H13 steel can be performed. More details about the material and microstructure parameters used for the modeling of creep are given by Eser et al. (2016a).

## 7.3.2 Simulation of Heat Transfer, Phase Transformation, and Stress Relaxation

In Section 7.3.1, the mechanical properties including yield stress, flow curve, and creep are simulated by means of physics-based structure–property models with the microstructure parameters from precipitation simulations. These properties are needed for the prediction of stress relaxation and deformation during the tempering of AISI H13 hot-work tool steel. In addition to these mechanical properties, the thermal effect resulting from the latent heat of phase transformation should be considered for the modeling of tempering of AISI H13 hot-work tool steel.

### 7.3.2.1 Simulation of Heat Transfer

Considering that the heat field changes due to the latent heat of a phase transformation, the transient heat transfer during the heat treatment of AISI H13 steel can be described by Fourier's heat conduction equation:

$$\rho C_p \dot{T} = \nabla \cdot (\nabla(\lambda T)) + \dot{Q}, \tag{7.9}$$

where $\rho$ is density, $C_p$ the heat capacity, $\lambda$ the thermal conductivity, and $\dot{Q}$ the power of the internal heat source resulting from the latent heat of a phase transformation (Eser et al., 2016b). For the heat treatment simulation of steel, the necessary modification in this equation is to describe the thermophysical properties of material related to the phase and temperature, and the heat generated due to the phase transformation $(\dot{Q})$. Abaqus software is used to simulate the heat transfer of this steel during the heat treatment.

### 7.3.2.2    Simulation of Phase Transformations

As mentioned before, both the diffusional and diffusionless phase transformations can occur during the heat treatment of AISI H13 steel. The diffusionless transformation from austenite to martensite during quenching is approximated using the Koistinen–Marburger phenomenological model (Eser et al., 2014; Koistinen and Marburger, 1959):

$$\xi_m = \overline{\xi}_m(\lambda_{ct})\left[1 - \exp\left\{-\left(\frac{M_s(\lambda_{ct}) - T}{b(\lambda_{ct})}\right)^{n(\lambda_{ct}, T)}\right\}\right], \qquad (7.10)$$

in which $\overline{\xi}_m(\lambda_{ct})$ is the maximum volume fraction of martensite, $\lambda_{ct}$ the cooling time from 800 to 500°C, $M_s(\lambda_{ct})$ the martensite starting temperature, and $n(\lambda_{ct}, T)$ and $b(\lambda_{ct})$ are specific parameters of the material. The other parameters in (7.10) are listed elsewhere (Eser et al., 2014).

For the modeling of diffusion-controlled phase transformation, the Johnson–Mehl–Avrami–Kolmogorov (JMAK) equation (Avrami, 1941; Johnson and Mehl, 1939; Kolmogorov, 1937) is utilized. For the present case, the transformation of phase $i$ to $k$ can be modeled as follows:

$$\dot{\xi}_k = n(T)\left(\frac{K_{ij}\xi_i - \tilde{K}_{ij}\xi_k}{TR}\right)\left(\ln\left(\frac{K_{ij}\cdot(\xi_i + \xi_k)}{K_{ij}\xi_i - \tilde{K}_{ij}\xi_k}\right)\right)^{\left(\frac{n(T)-1}{n(T)}\right)} \qquad (7.11)$$

$$K_{ij} = \frac{\overline{\xi}_k(T)}{\tau(T)}\cdot f(\dot{T}) \qquad (7.12)$$

$$\tilde{K}_{ij} = \frac{1 - \overline{\xi}_k(T)}{\tau(T)}\cdot \tilde{f}(\dot{T}) \qquad (7.13)$$

$$TR = \frac{\tau(T)}{\overline{\xi}_k(T)f(\dot{T}) + (1 - \overline{\xi}_k(T))\tilde{f}(\dot{T})} \qquad (7.14)$$

where $\dot{\xi}_k$ is the formation rate for the product phase $k$, $\overline{\xi}_k(T)$ the maximum phase fraction of phase $k$ at a specified temperature, $\tau(T)$ the time delay of a phase transformation, and $n(T)$ the index of transformation rate. $f(\dot{T})$ and $\tilde{f}(\dot{T})$ are temperature-rate-dependent functions that can be extracted from the continuous cooling transformation (CCT) diagrams. To calculate the phase fraction $\xi_k$, (7.11) is integrated by using an implicit Newton–Raphson integration scheme. Detail on the integration scheme and the corresponding phase transformation parameters can be found elsewhere (Eser et al., 2014). In addition, the thermal strains and volume change during phase transformations are also modeled by Eser et al. (2016b).

### 7.3.2.3 Simulation of Stress Relaxation

The total strain rate $(\dot{\varepsilon}^t)$ is the sum of several strain rate terms:

$$\dot{\varepsilon}^t = \dot{\varepsilon}^{el} + \dot{\varepsilon}^{pl} + \dot{\varepsilon}^{thr} + \dot{\varepsilon}^{tp} + \dot{\varepsilon}^{cr}, \qquad (7.15)$$

in which $\dot{\varepsilon}^{el}$ and $\dot{\varepsilon}^{pl}$ are the elastic and plastic strain rates, respectively. $\dot{\varepsilon}^{thr}$ is the contribution from both thermal strain and transformation strain due to the volume change during a phase transformation. The transformation-induced plasticity strain rate and creep-induced strain rate are $\dot{\varepsilon}^{tp}$ and $\dot{\varepsilon}^{cr}$, respectively. The detailed expressions for these individual strain terms are given elsewhere (Eser et al., 2016b).

Using (7.12), the heat treatment cycle (austenitizing, quenching together with secondary tempering) of a cylindrical workpiece ($\phi = 250$ mm, $h = 250$ mm) is simulated (Eser et al., 2016b). The discretization is finer when a higher temperature gradient (at the corner) is observed and higher stress (in the middle) is simulated. The impact of gravity is also considered, which has an important effect on deformation, especially at higher austenitizing temperatures, since the mass of the workpiece itself is about 100 kg. The simulated workpiece was heat-treated in a vacuum quenching furnace at 1,040°C. Quenching was carried out at 12 bars in a pressurized $N_2$ atmosphere. Figure 7.11 shows the axial-symmetric FEM model with a second-order quadratic element.

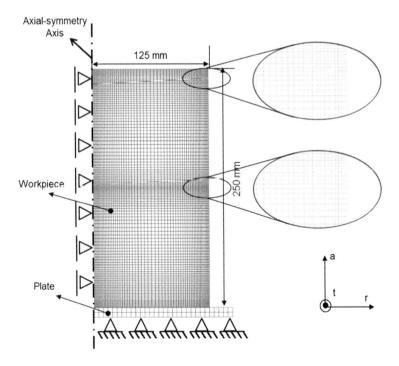

**Figure 7.11** FEM-model of the cylindrical workpiece. This figure is reproduced from Eser et al. (2016b) with permission from Elsevier.

The simulation uses the user-defined material (UMAT) subroutine in Abaqus. To avoid the problem of accurately defining the Jacobian matrix when using UMAT, several other user-defined subroutines (UHARD, UEXPAN, UFIELD, HETVAL, etc.) are coupled in order to implement the microstructure–temperature–stress–strain interactions during heat treatment (Eser et al., 2016b). In Abaqus, UHARD is used to define the isotropic yield behavior and the size of the yield surface in the combined hardening model, and UEXPAN is utilized to simulate the thermal expansion behavior of a material. The subroutine UFIELD allows a user to prescribe predefined field variables at the nodes of a model, and the subroutine HETVAL can be employed to define a heat flux due to internal heat generation in a material.

The simulations for stress relaxation were carried out at 600 and 650°C, and the simulated results at 600°C were compared with the experimental results. Figure 7.12 shows the development of the axial stress in the middle of the side surface point for two different simulations, where the material parameters are taken either from direct measurements or from the prior simulation. In addition, the measured axial stress after the quenching and tempering steps is also shown in Figure 7.12. During the process of heating the cylinder to the tempering temperature, purely compressive thermal stress ($<0$) is observed up to one hour tempering due to the hotter outer surface and the colder core. After the homogenization of the workpiece temperature, this effect decreases gradually, and the stress state becomes tensile again ($>0$), just like the original residual stress after quenching. Above 450°C, the stress relaxation can occur due to creep.

For the residual stress after quenching, the maximum tensile stress appears in the middle of the outer surface of the cylinder, as indicated in Figure 7.13 in the left-outer simulation result. Compared with the yield stress of tempered martensite up to about

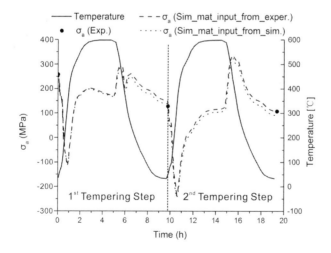

**Figure 7.12** Evolution of the axial stress and temperature at the middle of lateral surface during tempering at 600°C. This figure is reproduced from Eser et al. (2016b) with permission from Elsevier.

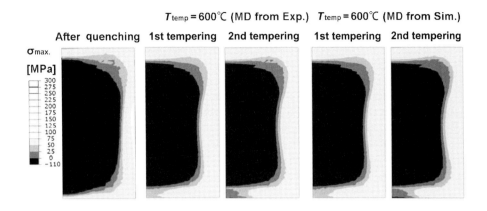

**Figure 7.13** Maximum principal stresses after heat treatment steps of hot-work tool steel AISI H13. In this figure, "MD from Exp." means that material data (parameters) in the simulation are from the experiment, while "MD from Sim." means that material parameters in the simulation are from the calculations such as the microstructure parameters from the MatCalc program. This figure is reproduced from Eser et al. (2016b) with permission from Elsevier.

850 MPa, as shown in Figure 7.9, the magnitude of residual stress is quite low (about 333 MPa). Therefore, no relaxation of residual stress due to the reduction of yield stress was observed during tempering. The relaxation of residual stress results from the creep effect.

In summary, for the case study of AISI H13 tool steel, the important mechanical properties (yield stress, flow curve, and creep), which are necessary for modeling the stress relaxation of tool steel during tempering, are calculated by physics-based structure–property models. In addition, the key microstructural parameters, which are needed for mechanical property prediction, are provided by means of a precipitation simulation code using thermodynamic and thermophysical parameters. Considering that the simulation results agree reasonably with the experimental measurements, it can be concluded that the multiscale simulations due to Eser et al. (2016b) is helpful to replace costly and time-consuming experiment work. The present case study for AISI H13 tool steel demonstrates that the multiscale modeling performed by Eser et al. (2016a, b) can be used to simulate the complex macroscopic stress relaxation and deformation under different soaking time and tempering temperature.

# References

Allen, A. J., Gavillet, D., and Weertman, J. R. (1993) SANS and TEM studies of isothermal $M_2C$ carbide precipitation in ultrahigh strength AF1410 steels. *Acta Metallurgica et Materialia*, 41(6), 1869–1884.

Anderson, P. M., Wang, J. S., and Rice, J. R. (1987) Thermodynamic and mechanical models of interfacial embrittlement, *34th Sagamore Army Materials Research Conference*. Washington: U.S. Government Printing Office, 619–649.

Anelli, E. (1992) Application of mathematical modelling to hot rolling and controlled cooling of wire rods and bars. *ISIJ International*, 32(3), 440–449.

Avrami, M. (1941) Granulation, phase change, and microstructure kinetics of phase change. III. *Journal of Chemical Physics*, 9, 177–184.

Blum, W., Rosen, A., Cegielska, A., and Martin, J. L. (1989a) Two mechanisms of dislocation. motion during creep. *Acta Metallurgica*, 37(9), 2439–2453.

Blum, W., Vogler, S., and Biberger, M. (1989b) Stress dependence of the creep rate at constant dislocation structure. *Materials Science and Engineering A*, 112, 93–106.

Campbell, C. E. (1997) *Systems Design of High-Performance Stainless Steels*. PhD thesis, Northwestern University.

Campbell, C. E., and Olson, G. B. (2000a) Systems design of high performance stainless steels I. Conceptual and computational design. *Journal of Computer-Aided Materials Design*, 7(3), 145–170.

Campbell, C. E., and Olson, G. B. (2000b) Systems design of high performance stainless steels II. Prototype characterization. *Journal of Computer-Aided Materials Design*, 7(3), 171–194.

Chris, K., and Rich Kooy, P. E. (2010) New corrosion-resistant, ultra-high-strength Steel. *Fastener Technology International*, 33, 68–69.

Committee, I. (1994) The dollars and sense of risk management and airline safety. *Flight Safety Digest*, 13(12), 1–6.

De Cooman, B. C., Estrin, Y., and Kim, S. K. (2018) Twinning-induced plasticity (TWIP) steels. *Acta Materialia*, 142, 283–362.

Eser, A. (2014) *Skalenübergreifende Simulation des Anlassens von Werkzeugstählen*. PhD thesis, Faculty of Mechanical Engineering, RWTH Aachen University.

Eser, A., Bezold, A., Broeckmann, C., Schruff, I., and Greeb, T. (2014) Simulation des Anlassens eines dickwandigen Bauteils aus dem Stahl X40CrMoV5–1. *HTM Journal of Heat Treatment Materials*, 69(3), 127–137.

Eser, A., Broeckmann, C., and Simsir, C. (2016a) Multiscale modeling of tempering of AISI H13 hot-work tool steel – part 1: prediction of microstructure evolution and coupling with mechanical properties. *Computational Materials Science*, 113, 280–291.

Eser, A., Broeckmann, C., and Simsir, C. (2016b) Multiscale modeling of tempering of AISI H13 hot-work tool steel – part 2: coupling predicted mechanical properties with FEM simulations. *Computational Materials Science*, 113, 292–300.

Ghosh, G., Campbell, C. E., and Olson, G. B. (1999) An analytical electron microscopy study of paraequilibrium cementite precipitation in ultra-high-strength steel. *Metallurgical and Materials Transactions A*, 30(3), 501–512.

Ghosh, G., and Olson, G. B. (1994a) Kinetics of F.C.C→ B.C.C heterogeneous martensitic nucleation – I. the critical driving force for athermal nucleation. *Acta Metallurgica et Materialia.*, 42(10), 3361–3370.

Ghosh, G., and Olson, G. B. (1994b) Kinetics of F.C.C→ B.C.C heterogeneous martensitic nucleation – II. thermal activation. *Acta Metallurgica et Materialia*, 42(10), 3371–3379.

Ghosh, G., and Olson, G. B. (2002) Precipitation of paraequilibrium cementite: experiments, and thermodynamic and kinetic modeling. *Acta Materialia*, 50(8), 2099–2119.

Johnson, W. A., and Mehl, R. F. (1939) Reaction kinetics in processes of nucleation and growth. *Transactions of the American Institute of Mining Metallurgical Engineers*, 135, 416–458.

Johnson, W. C., and Cahn, J. W. (1984) Elastically induced shape bifurcations of inclusions. *Acta Metallurgica*, 32(11), 1925–1933.

Kampmann, R., and Wagner, R. (1984) Kinetics of precipitation in metastable binary alloys – theory and application to Cu –1.9 at% Ti and Ni–14 at% Al, in Haasen, P., Gerold, V., Wagner, R., and Ashby, M. F. (eds), *Decomposition of Alloys: The Early Stages: Proceedings of the 2nd Acta-Scripta Metallurgica Conference. Sonnenberg, Germany, 19-23/09/1983*. Oxford: Pergamon Press, 91–103.

Kantner, C. D. (2002) *Designing Strength, Toughness, and Hydrogen Resistance: Quantum Steel*. PhD thesis, Northwestern University.

King, K. C., Voorhees, P. W., Olson, G. B., and Mura, T. (1991) Solute distribution around a coherent precipitate in a multicomponent alloy. *Metallurgical Transactions A*, 22, 2199–2210.

Kirchheim, R., Heine, B., Fischmeister, H., Hofmann, S., Knote, H., and Stolz, U. (1989) The passivity of iron–chromium alloys. *Corrosion Science*, 29(7), 899–917.

Kocks, U. F. (1976) Laws for work-hardening and low-temperature creep. *Journal of Engineering Materials and Technology*, 98(1), 76–85.

Koistinen, D. P., and Marburger, R. E. (1959) A general equation prescribing the extent of the austenite–martensite transformation in pure iron–carbon alloys and plain carbon steels. *Acta Metallurgica*, 7(1), 59–60.

Kolmogorov, A. N. (1937) On the statistical theory of metal crystallization. *Izvestiya Akademii Nauk SSSR, Seriya Matematicheskie*, 3, 355–360.

Kozeschnik, E., Svoboda, J., and Fischer, F. D. (2004a) Modified evolution equations for the precipitation kinetics of complex phases in multi-component systems. *CALPHAD*, 28(4), 379–382.

Kozeschnik, E., Svoboda, J., Fratzl, P., and Fischer, F. D. (2004b) Modelling of kinetics in multi-component multi-phase systems with spherical precipitates: II: numerical solution and application. *Materials Science and. Engineering A*, 385(1), 157–165.

Kuehmann, C., Tufts, B., and Trester, P. (2008) Computational design for ultra high-strength alloy. *Advanced Materials and Processes*, 166(1), 37–40.

Lancaster, J. (2019) *Frontiers of Materials Research: A Decadal Survey (2019)*. Washington: Press, T. N. A.

Langer, J. S., and Schwartz, A. J. (1980) Kinetics of nucleation in near-critical fluids. *Physical Review A*, 21(3), 948–958.

Li, Q. (2003) Modeling the microstructure–mechanical property relationship for a 12Cr–2W–V–Mo–Ni power plant steel. *Materials Science and Engineering A*, 361(1), 385–391.

Liarng, R.-H. (1996) *Applications of the Eigenstrain Method in Inclusion Problems and Micromechanics of Coherent $M_2C$ Carbide Precipitation in Steel*. PhD thesis, Northwestern University.

Lindgren, L.-E., Domkin, K., and Hansson, S. (2008) Dislocations, vacancies and solute diffusion in physical based plasticity model for AISI 316L. *Mechanics of Materials*, 40 (11), 907–919.

Mughrabi, H. (1979) Microscopic mechanisms of metal fatigue, in Haasen, P., Gerold, V., and Kostorz, G. (eds), *Strength of Metals and Alloys*. Aachen: Pergamon, 1615–1638.

Mughrabi, H. (1983) Dislocation wall and cell structures and long-range internal stresses in deformed metal crystals. *Acta Metallurgica*, 31(9), 1367–1379.

Olson, G. B. (1997) Computational design of hierarchically structured materials. *Science*, 277 (5330), 1237–1242.

Olson, G. B. (2013) Genomic materials design: the ferrous frontier. *Acta Materialia*, 61(3), 771–781.

Olson, G. B., and Cohen, M. (1976a) A general mechanism of martensitic nucleation: part II. Fcc → Bcc and other martensitic transformations. *Metallurgical Transactions A*, 7(12), 1905–1914.

Olson, G. B., and Cohen, M. (1976b) A general mechanism of martensitic nucleation: part III. kinetics of martensitic nucleation. *Metallurgical Transactions A*, 7(12), 1915–1923.

Olson, G. B., and Kuehmann, C. J. (2014) Materials genomics: from CALPHAD to flight. *Scripta Materialia*, 70, 25–30.

Pourbaix, M. (1973) *Lectures on Electrochemical Corrosion*. New York and London: Plenum Press.

Schmitz, G. J., and Prahl, U. (2016) *Handbook of Software Solutions for ICME*. Carmel: John Wiley and Sons.

SGTE (Scientific Group Thermochemical Europe) (1994) *Solution database*. July 27.

Straub, S., Polcik, P., Besigk, W., Blum, W., König, H., and Mayer, K. H. (1995) Microstructural evolution of the martensitic cast steel GX12CrMoVNbN9–1 during long-term annealing and creep. *Steel Research*, 66(9), 402–408.

Svoboda, J., Fischer, F. D., Fratzl, P., and Kozeschnik, E. (2004) Modelling of kinetics in multi-component multi-phase systems with spherical precipitates: I: theory. *Materials Science and Engineering A*, 385(1), 166–174.

Taskin, K. (2017) *Innovations on New Metals for Aerospace*. Available online: https://rockfordil.com/wp-content/uploads/2017/03/Innovations-on-New-Metals-for-Aerospace-QuesTek.pdf [Accessed November 16, 2021].

Wang, Y., Appolaire, B., Denis, S., Archambault, P., and Dussoubs, B. (2006) Study and modelling of microstructural evolutions and thermomechanical behaviour during the tempering of steel. *International Journal of Microstructure and Materials Properties*, 1(2), 197–207.

Watton, J. F., Olson, G. B., and Cohen, M. (1987) *A Novel Hydrogen-Resistant UHS Steel, the 34th Sagamore Army Materials Research*. Lake George: New York, U.S. Government Printing Office.

Xiong, W., and Olson, G. B. (2016) Cybermaterials: materials by design and accelerated insertion of materials. *NPJ Computational Materials*, 2(1), 15009.

Young, C. H., and Bhadeshia, H. K. D. H. (1994) Strength of mixtures of bainite and martensite. *Materials Science and Technology*, 10(3), 209–214.

# 8 Case Studies on Light Alloy Design

## Contents

## 8.1 Introduction

By definition, light alloys are those based on low-density metals, and the most important ones are based on aluminum and magnesium, and to some extent also titanium. They are widely used to reduce the weight of structural components, but also offer other specific processing and property advantages. A recent comprehensive overview on such aspects of light alloys and their applications in automotive, aviation, and a few other industries is given by Polmear et al. (2017).

In this chapter, selected examples demonstrate how we can use the CALPHAD simulation tool to understand and predict the effect of alloying elements and processing conditions on the alloy properties and how this is used in the design of Al and Mg alloys. While only a few general comments on Al alloys are made at the beginning of Section 8.2, a brief overview of the less well-established Mg alloys, also in comparison with Al alloys, is given in Section 8.3.

For the case studies on Al alloys design, the use of extended CALPHAD-type databases is demonstrated. This involves more elaborate mobility and kinetic databases in addition to the thermodynamic database. In the first example on cast alloy A356, the solidification simulation involving dedicated microsegregation modeling is presented. The second example on wrought alloy 7xxx involves elaborate heat treatment simulation and precipitation kinetics.

The case studies on Mg alloy design are first given for three examples of structural Mg components and detail the selection of cast Mg alloy composition and optimized heat treatment conditions. The first two examples summarize selected case studies for new creep-resistant Mg alloys and present the concept of solidification path and T6 heat treatment of AZ series alloys. These are alloy design cases using solely thermo-dynamic CALPHAD databases. The third example on computational design and development of new Mg–Al–Sn based (AT) cast alloys presents an industrial automotive application highlight: the award-winning die-cast side-door inner panel. In addition, the dedicated microsegregation modeling is also applied to this AT72 alloy. The last Mg alloy design example is on biomedical Mg alloy implants. Compared to structural Mg components, this application field is less known but rapidly growing, including economically. One state-of-the-art bioresorbable Mg alloy stent to cure coronary artery disease is presented together with the CALPHAD-based design strategy that was part of the development process.

Ancillary material for this chapter is available online at Cambridge University Press. These are related data files for the PanSolidification (Zhang et al., 2020) and PanPrecipitation (Cao et al., 2011) module of the Pandat software package (www .computherm.com) (Cao et al., 2009; Chen et al., 2002). They are designed to run not only with the licensed Pandat software but also with the Pandat trial version, which can be downloaded free of charge from CompuTherm's website (https://computherm .com/). This is intended to provide the reader with hands-on experience in solidification simulation (with microsegregation modeling) and heat treatment simulation (with precipitation modeling) as applied in the present case studies. Step-by-step instructions to perform the simulations related to Sections 8.2.2 and 8.3.1.3 are given in a readme file. Additional ancillary material is given in the case studies of Chapter 5 for PanSolidification and in Chapter 6 for PanPrecipitation.

## 8.2    Aluminum Alloys

There are literally hundreds of specified Al alloys in the two classes of cast and wrought alloys. Cast alloys are classified (www.aluminum.org) by their main alloying elements: For example, the "3xx.x series" is based on Al–Si, plus some Cu and/or Mg. In this series, the 356.0 and A356.0 alloys are based on the Al–Si–Mg system. Alloy design/optimization of these high-strength casting alloys will thus focus on both composition design and on the cooling conditions. This is highlighted by the solidification simulation in the first example.

The huge number of wrought Al alloys are classified in eight series, from 1xxx to 8xxx, again indicating the major alloying element. For example, the 7xxx series is based on Al–Zn, plus some Mg and Cu. Of course, all wrought alloys are cast before any extrusion, heat treatment, or thermomechanical processing is applied. Therefore, the initial as-cast microstructure is also relevant for through-process modeling. However, the main focus here is on heat treatment for microstructure and property optimization. This is highlighted by the simulation of precipitation hardening in the second example.

### 8.2.1     Cast Al Alloy A356: Solidification Simulation and Microsegregation

The A356 alloy with nominal composition Al–7Si–0.3Mg (wt.%) is a widely used hypoeutectic Al–Si alloy for sand and permanent mold castings with excellent castability (fluidity), high mechanical strength and hardness, good ductility, corrosion resistance, and machinability. Hardness/strength is mainly due to precipitation of the intermetallic $Mg_2Si$ phase, especially after a T6 heat treatment, involving solutionizing, quenching, and artificial aging (Polmear et al., 2017). The presence of iron impurity often results in the formation of Fe-containing intermetallics, mainly $\beta$-AlFeSi or $\alpha$-AlFeSi. The $\beta$-AlFeSi phase is most detrimental because of its needle or plate-like appearance in the microstructure. Its reduction is possible by lowering the Fe content, such as in the A356 alloy (maximum 0.2 wt.% Fe), making it more costly compared to the cheaper 356 alloy (maximum 0.6 wt.% Fe). Thus, controlling the actual Fe content is important for cost and properties. In the higher Mg content range also, the phase $\pi$-$Al_8FeMg_3Si_6$ may form. Controlling the interplay of all these tertiary intermetallics, typically forming in the interdendritic region of the primary (Al) matrix and the secondary eutectic silicon, is crucial for designing the cast microstructure and properties.

Given the fact that at least the compositions in the quaternary Al–Si–Mg–Fe system need to be controlled, any quantitative solidification simulation must start from a reliable CALPHAD-type thermodynamic database and software. As a first step, it is always recommended to select one alloy composition and perform an equilibrium (lever rule) and Scheil solidification simulation as explained in detail in Section 5.5. That was done for the Al–7.235Si–0.336 Mg–0.148Fe (wt.%) alloy (Zhang et al., 2020) using the PanAl database and Pandat software package (www.computherm .com) (Cao et al., 2009). Under equilibrium conditions, only the $\beta$-AlFeSi forms. Under Scheil conditions, involving essentially blocked diffusion and transitions in the solid phases, local solid/liquid equilibrium, and a homogeneous liquid phase, the $\beta$-AlFeSi, additional $\pi$-$Al_8FeMg_3Si_6$ and finally $Mg_2Si$ form. All tertiary phase fractions are well below 0.01. These results provide an important overview and useful bounds between extreme cases of very slow and much more rapid solidification, with the Scheil case being often closer to real casting conditions. These simulations are readily obtained because the only material parameters of the alloy required are given by the thermodynamic database. However, time and space are out of the scope of these calculations. Thus, quantitative cooling rates or microstructural details, such as the important secondary dendrite arm spacing, are inaccessible.

These limitations are resolved by employing a proper microsegregation model. For such a typical model, both effort and result details are in a medium range between the Scheil simulation and elaborate phase-field or cellular automaton methods described in Chapter 3. Most microsegregation models allow, first of all, back-diffusion into the already solidified part in one dimension, which has a significant impact on the microsegregation (Clyne and Kurz, 1981; Hillert et al., 1999; Kraft et al., 1996a).

The importance of reliable thermodynamic descriptions and phase equilibria of the alloys for any microsegregation modeling was highlighted earlier (Du et al., 2005; Liang et al., 2000). Such a proper CALPHAD-type thermodynamic database is indispensable for industrial multicomponent alloys and also ensures the consistent progress from the results of the Scheil approximation to the quantitative kinetics.

In addition to thermodynamic parameters, kinetic material parameters are required that are also assessed with the CALPHAD method. On one hand, a mobility database is needed, which – combined with the thermodynamic database – enables the software to generate consistent thermodynamic factors and diffusion coefficients. On the other hand, an alloy-specific kinetic parameter database (*.sdb) is needed with parameters such as the assumed dendrite geometry (plate, cylinder, or sphere), interfacial energy, latent heat, sound velocity, solute trapping parameter, coarsening geometric factor, dendrite tip factor, boundary-layer factor, and solid diffusivity factor. The latter three are by default 1 but introduced to give the user some flexibility to better fit with the experimental data. For example, the solid diffusivity factor allows to scale the diffusivity in the solid phase obtained from the mobility database up or down by a common factor.

For the present case study, the PanSolidification module of the Pandat software package is applied (Zhang et al., 2020), incorporating the effects of back-diffusion, undercooling, and secondary dendrite arm coarsening (Kraft and Chang, 1997; Kraft et al., 1996b; Liang et al., 2000; Yan et al., 2002). A simulation run typically takes only seconds on an ordinary PC and provides these valuable statistical microstructure features as compared to the detailed morphological development of an individual dendrite obtained in the elaborate phase-field method with much higher computational cost.

For each simulation run, the alloy composition and solidification conditions are specified. These comprise the cooling rate ($CR$, K/s), which is given by a temperature–time profile or measured cooling curve, and in addition either the solidification rate ($V$, µm/s) or temperature gradient ($G$, K/µm) is specified. These quantities are related by $CR = G \cdot V$.

Example results for the A356 alloy are given in Figure 8.1 for the effect of cooling rate on the fractions of intermetallic phases. Note that volume fractions can be given because molar volumes are also assessed in addition to the thermodynamic description, whereas ordinary equilibrium and Scheil simulations can only provide

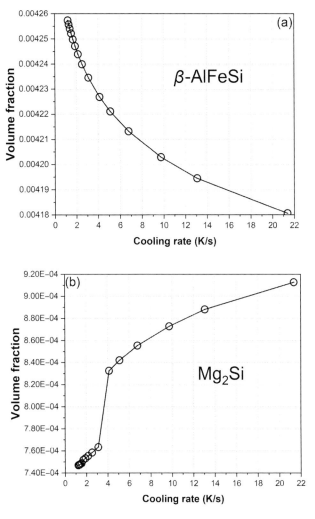

**Figure 8.1** Effect of cooling rate on the volume fractions of intermetallic phases in the A356 alloy Al–7.235Si–0.336Mg–0.148Fe (wt.%): (a) $\beta$-AlFeSi, (b) $Mg_2Si$, (c) $\pi$-$Al_8FeMg_3Si_6$. Adapted from Zhang et al. (2020) with permission from Springer Nature. The symbols in the figures denote the simulated results for each CR.

atomic or mass fractions of phases. In agreement with experimental observation, the formation of detrimental $\beta$-AlFeSi is suppressed by higher cooling rates (Zhang et al., 2020). Interestingly, there appears to be a threshold value at a cooling rate of 3–4 K/s. At such a slow cooling rate, the fraction of $Mg_2Si$ drops to virtually nil, as also seen in the equilibrium calculation. Faster cooling tends to increase the fractions of both $Mg_2Si$ and $\pi$-$Al_8FeMg_3Si_6$. More detailed analysis of the simulation runs shows that the cooling rate impacts the fraction of the intermetallics but not the phase sequence, which is the same as that for the Scheil simulation. Furthermore, the total freezing range is almost unchanged for this alloy under different cooling rates (Zhang et al., 2020).

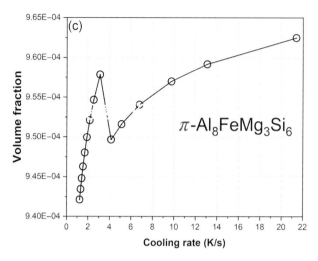

Figure 8.1 (*cont.*)

More detailed results are generated for each simulation run over time or temperature of the freezing range, such as the growth of secondary dendrite arm spacing (SDAS) or composition distribution in each phase due to microsegregation. However, instead of the complete information on the solidification path of a single alloy under a fixed solidification condition, it is often more relevant for design applications to extract key information from a very large number of simulation runs covering the area of interest for the crucial variables that may actually be changed by the engineer controlling the alloy composition and/or the process variables. This is efficiently done using the high-throughput calculation (HTC) function implemented in Pandat also for the PanSolidification module. It can perform thousands of simulations, scanning a user-defined alloy compositional space and a range of cooling rates. The HTC function also includes the postprocessing data mining and analysis to extract the user-specified results, such as SDAS or intermetallic fractions as functions of the scanned variables. A multidimensional variable space may be scanned in a single HTC job. However, it is most enlightening to display a chosen key result/property as function of two chosen variables in a 3D or color map in Pandat, where the HTC analysis function automatically includes chosen property contour lines.

An example is given in Figure 8.2, where the color map with contour lines of the $\beta$-AlFeSi phase fraction is presented in a scanned composition range of Al–8Si–0.35 Mg–$x$Fe–$y$Mn (wt.%) alloys at a fixed cooling rate of 30 K/s (Zhang et al., 2020). The Fe content range is higher compared to the A356 alloy, and the Mn addition can be tailored to suppress the detrimental $\beta$-AlFeSi phase fraction to a tolerable level. A series of these figures may be produced for different cooling rates that relate to the location and wall thickness in the casting. Alternatively, the HTC data mining feature may extract the $\beta$-AlFeSi phase fraction in the scanned space of Mn content and cooling rate for a fixed Fe content of 0.3 wt.% Fe, as shown in Figure 8.3. These

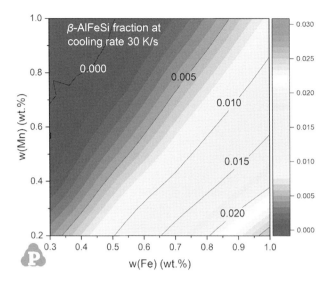

**Figure 8.2** Color map of β-AlFeSi phase fraction as function of the Fe and Mn compositions for Al–8Si–0.35Mg–xFe–yMn (wt.%) alloys at a fixed cooling rate of 30 K/s. Adapted from Zhang et al. (2020) with permission from Springer Nature. A black and white version of this figure will appear in some formats. For the colour version, please refer to the plate section.

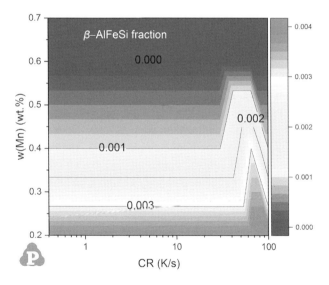

**Figure 8.3** Color map of the β-AlFeSi phase fraction as function of the Mn composition and the cooling rate (CR) for Al–8Si–0.35Mg–0.3Fe–yMn (wt.%) alloys. Adapted from Zhang et al. (2020) with permission from Springer Nature. A black and white version of this figure will appear in some formats. For the colour version, please refer to the plate section.

maps provide useful design guidance on the necessary Mn addition in the alloy for a target β-AlFeSi phase fraction when the composition of Fe impurity and the cooling rate are known (Zhang et al., 2020).

The prediction of SDAS in these simulation runs is also important because smaller SDAS is known to correlate with less porosity, more finely dispersed secondary phases, and ultimately improved tensile properties. For the example of quaternary Al–7Si–$x$Fe–$y$Mg and Al–8Si–$x$Fe–$y$Mg, this was studied with contour maps of SDAS at a fixed cooling rate of 5 K/s. The studies showed that within the composition range (Mg: 0.2–0.5wt.%, Fe: 0.1–1.0 wt.%), the largest SDAS variation is only about 2 µm. The spread in SDAS for the Al–7Si–$x$Fe–$y$Mg alloy is $18.2 - 19.4$ µm and for the Al–8Si–$x$Fe–$y$Mg alloy it is $17.2 - 18.6$ µm at this $CR$. This is in agreement with the experimental experience that the cooling rate has much larger impact compared to the limited composition variation. Therefore, it is more useful to scan cooling rates in solidification simulation for a prediction of SDAS. It has been demonstrated that the predicted SDAS in the A356 alloy discussed earlier and one other alloy agrees well with experimental data over a wide range of cooling rates. For the example of the A356 alloy solidified for $CR = 1.2 - 21.4$ K/s, the predicted SDAS ranges from $30.7 - 10.8$ µm (Zhang et al., 2020). However, in the range of $CR < 1.9$ K/s, the predictions tend to be low but within the experimental error bar. Here, at transition from columnar to equiaxed grain growth (CET), the experimental error bar significantly increases.

## 8.2.2    Wrought Al Alloy 7xxx: Heat Treatment Simulation and Precipitation Kinetics

The 7xxx series alloys are based on Al–Zn (0.8–12 wt.% Zn) with minor Mg/Cu additions and comprise some of the highest-strength aluminum alloys, often used in high-performance applications such as aircraft, aerospace, and competitive sporting equipment. The high strength is due to the great age hardening potential in the Al–Zn–Mg ternary system. While the highest strength alloys in 7xxx series alloys always contain additional Cu, the important medium-strength alloys with little or no Cu have the advantage of a better weldability (Polmear et al., 2017). The following case study presents simulations for 7xxx alloys based on the Al–Zn–Mg ternary system. However, it is emphasized that this methodology can be applied to other 7xxx alloys classified to contain additional Cu and other minor additions, such as Cr or Zr. It is obvious that there are many variables for the design of alloy composition and the optimization of the heat treatment process with customized temperature–time profiles. Thus, a powerful simulation tool is of invaluable assistance.

This is demonstrated by using the PanPrecipitation module (Cao et al., 2011) of the Pandat software package. This tool is developed for the simulation of precipitation kinetics of multicomponent alloys under arbitrary heat treatment conditions. It is based on the Kampmann–Wagner numerical (KWN) model (Kampmann and Wagner, 1984) and extended to handle both homogeneous and heterogeneous nucleation, dealing with co-precipitation of phases with various morphologies (sphere and lens), always with concurrent processes of nucleation, growth, and coarsening. Simulation results comprise particle size distributions (PSD) of various precipitate phases in addition to the temporal evolution of their average size and volume fractions as well as the compositions of matrix and the precipitate phase(s). Based on that, both hardness and yield strength can be calculated. Also included in PanPrecipitation is a module to estimate the interfacial energy at the boundary between the matrix and precipitate phase. This is the most critical kinetic parameter in the precipitation simulation, and the module uses the generalized broken bond (GBB) method (Sonderegger and Kozeschnik, 2009a, b) for prediction. The kinetic database (*.kdb) of the alloy required by PanPrecipitation is also developed by the CALPHAD method and comprises all the parameters assessed for the precipitation and the strength simulation. The simulation also builds on the same CALPHAD-type thermodynamic and mobility databases for aluminum alloys, PanAl, which is also used in Section 8.2.1 for the A356 alloy. Running one precipitation simulation typically takes only seconds on an ordinary PC, while the simulation of a complete TTT curve with PanPrecipitation takes some tens of seconds for a Ni–14Al alloy. It provides these valuable statistical microstructure and property features as compared to the detailed morphological development of an individually precipitated particle obtained in the elaborate phase-field method, which has a much higher computational cost.

The first simulation is performed for the Al–6.1Zn–2.3Mg (wt.%) alloy aged at 160°C for 1,000 hours. Results are based on the precipitate $\eta'$ studied experimentally by Deschamps et al. (2001) and compared in Figure 8.4. The simulated particle size shows that they grow and coarsen with aging time, as obtained by the simulated particle size distributions not shown here. The simulated yield strength reaches peak between 1–10 hours and decreases quickly after 10 hours of aging at 160°C in agreement with the experimental data.

The effect of four different ageing temperatures on the yield strength evolution for the same alloy is shown in Figure 8.5. In this temperature range of 140–200°C, the yield strength reaches peak faster for higher aging temperature. However, a trade-off is important for industrial practice between a rapid process to increase the level of capacity utilization of heat treatment furnaces and the achieved maximum yield strength. For this Al–6.1Zn–2.3Mg alloy, the best compromise appears to be aging around 180°C, assuming prior complete solutionizing.

One other important parameter in the complete temperature–time profile of such a T6 heat treatment is the cooling rate between the solution temperature and the artificial aging temperature. This effect is simulated for three different cooling rates between 480°C and 160°C. As shown in Figure 8.6, the slower cooling rate tends to increase the peak yield strength slightly.

Obviously, such simulations can support an optimization of alloy chemistry as well, interdependent with the heat treatment process. For example, the variation of Mg and Zn contents indicates that the total yield strength increases with the increase of Mg composition in the Al–6.1Zn–2.3Mg alloy, as shown in Schmid-Fetzer and Zhang (2018).

These quantitative simulation results provide not only key data for the design of the alloy but also the optimization of the heat treatment process. Depending on the furnace/plant design and wall thickness of the treated component, one needs to take a range of heating and cooling rates into consideration in the practical application.

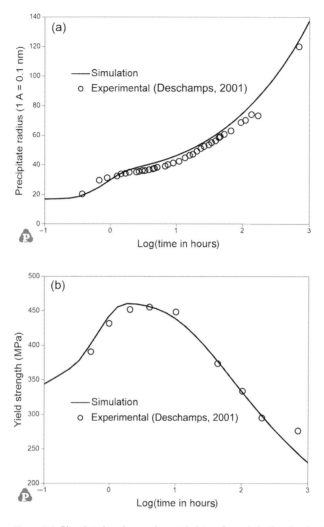

**Figure 8.4** Simulated and experimental data of precipitation hardening in the Al–6.1Zn–2.3Mg (wt.%) alloy aged at 160°C for up to 1,000 hours: (a) particle size; (b) yield strength. Adapted from Schmid-Fetzer and Zhang (2018) with permission from Elsevier.

Therefore, such simulation results are also relevant to optimize the very costly investment in a new heat treatment facility with a custom-fit level of capacity. One more case study can be found in Chapter 6 for the precipitation hardening of the 6xxx series Al–Si–Mg-based alloy with six components, the AA6005 alloy (Al–0.82Si–0.55Mg–0.5Mn–0.2Fe–0.016Cu, wt.%).

## 8.3     Magnesium Alloys

The three most commonly used structural metals are steel, Al alloys, and Mg alloys, in that sequence. The decreasing density in this series makes magnesium the lightest

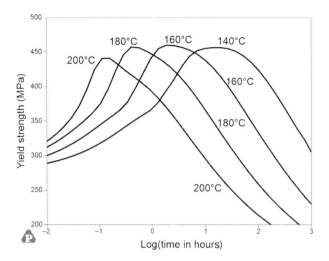

**Figure 8.5** Simulated effect of aging temperature on the yield strength evolution in the Al–6.1Zn–2.3Mg (wt.%) alloy. Adapted from Schmid-Fetzer and Zhang (2018) with permission from Elsevier.

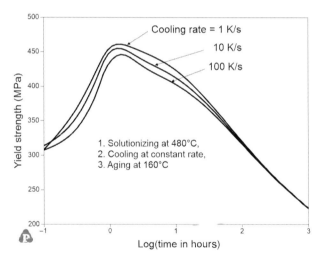

**Figure 8.6** Simulated effect of cooling rate on the yield strength evolution in the Al–6.1Zn–2.3Mg (wt.%) alloy. Adapted from Schmid-Fetzer and Zhang (2018) with permission from Elsevier.

structural metal, offering a significant potential for lightweight industrial applications in automotive, aerospace, power tools, and 3C (computer, communication and consumer/sporting products) (Pekguleryuz et al., 2013). The lightweight saving potential of Mg alloys is typically between 10–33% at the automotive component level compared to Al alloys (Klaumünzer et al., 2019). Special highlights for fabrication of components are the excellent machinability of Mg alloys, the best among all the structural metals, and the excellent castability for thin-wall components. The provision of electromagnetic shielding is an additional advantage for mobile computers and telephones (Polmear et al., 2017).

Compared to Al alloys, much fewer Mg alloys have already been classified. The classification was developed by the American Society for Testing and Materials (ASTM B951, www.astm.org). Two (or more) letters designate the main alloying elements followed by their rounded weight percentage. Some important Mg cast alloys are AZ91, AM50, AS41, AE44, AXJ530, AT72, ZK51, QE22, and WE54. The letters mean A (Al), Z (Zn), M (Mn), S (Si), X (Ca), J (Sr), T (Sn), K (Zr), Q (Ag), W (Y), and E for rare earths (RE, mischmetal). For example, AZ91 indicates the nominal alloy Mg–9Al–1Zn, while the actual composition ranges are $8.3 - 9.7$ wt.% Al and $0.4 - 1.0$ wt.% Zn (Polmear et al., 2017). As of 2019, the vast majority of the automotive applications is limited to die castings, such as gearbox houses, interior support brackets, instrument-panel beam, or steering wheels, also due to the better castability and less energy consumption on melting compared to Al alloys. Examples for the design of cast Mg alloys along with a potential heat treatment are given in the next subsection.

Applications of wrought Mg alloys, such as sheets or extruded profiles, are still very limited. A prime example of the outer-skin application of magnesium sheets is the roof of the Porsche 911 GT3RS, where the high lightweight construction potential of magnesium surpasses even that of carbon fiber reinforced plastics (Klaumünzer et al., 2019). For this application, the new process of twin-roll casting was used, demonstrating the close interdependence of alloy composition and process design (Park et al., 2009). Forming of Mg is more difficult compared to Al. Because of the hexagonal (Hcp) crystal structure, Mg possesses fewer slip systems than cubic (Fcc) Al, which restricts its ability for plastic deformation, particularly at room temperature. Therefore, fabrication of wrought Mg alloy components is normally carried out by hot working above about 250°C. Some conventional wrought magnesium alloys, which are applied as sheet, extrusions, or forgings, are AZ31, AM30, or ZK30 (Polmear et al., 2017).

Mg is more reactive and does not build a native protecting oxide layer, as is the case for Al. Therefore, Mg components generally suffer from a relatively high corrosion rate and require a designed protection for exterior use and/or in contact with aqueous media (Pekguleryuz et al., 2013). However, this property can be turned into an advantage, as shown by the biomedical application in the second case study.

## 8.3.1    Selection of Cast Mg Alloy Composition and Optimized Heat Treatment

### 8.3.1.1    Selected Case Studies for New Creep-Resistant Mg Alloys

The majority of case studies for Mg alloys are based on key results from CALPHAD-type software calculations for the design of cast and heat-treated alloys where only the thermodynamic database of the alloy system is required. The principles of this method for alloy design are explained in Section 5.5. Documented applications for design of new Mg alloys and related processing are reviewed in Schmid-Fetzer (2019) and

Schmid-Fetzer and Zhang (2018). They cover a wide range of topics, such as new creep-resistant cast alloys; castability and hot-tearing susceptibility; semisolid metal processing; ductility design and extrusion; corrosion and recycling alloys; low-density, high-entropy alloys; and oxidation and surface design.

Major efforts on the development of new creep-resistant Mg alloys were spurred by extending the application range for lightweight automotive components to higher temperature. Examples are die-casting alloys for powertrain applications, such as engine blocks for temperatures up to 190°C or gearbox housings operating under high loads at temperatures up to 150°C. Strontium (code J) is one of the new components tried early to improve creep resistance (Pekguleryuz and Baril, 2001) with castings of ternary Mg–Al–Sr (AJ) alloys. The AJ62 alloy was actually used by BMW in the development of a composite six-cylinder engine block with AJ62 for the external block and an aluminum alloy A390 for the inner parts. That saved 24% weight compared to a conventional aluminum block (Schmid-Fetzer and Zhang, 2018). The thermodynamic modeling of this Mg–Al–Sr ternary system in combination with detailed experiments was performed by Janz et al. (2007). Other examples of successful commercial development for these demands by adding the additional element calcium (code X) are the so-called MRI alloys by Dead Sea Magnesium Ltd. (www.dsmag.co.il) and applied by Volkswagen AG. The alloys MRI230 and MRI153 are based on the Mg–Al–Ca–Sr–Mn (AXJM) system. The first thermo-dynamic description with key experimental validation of this system was given by Janz et al. (2008, 2009) and provided the basis for computational design.

General Motors has developed the alloy AXJ530 for powertrain applications with a good balance of tensile properties, creep strength, castability, and low cost (Luo, 2004). The commercial specification range of nominal composition of that alloy is wide for Ca, ranging between 1.7–3.3 wt.% Ca. Scheil solidification simulation of AXJ530 alloys can be successfully used to predict the impact of the Ca content variation between these low- and high-ends to solidification microstructure. It is noteworthy that not only the as-cast fractions of precipitated phases differ significantly, but also the Scheil-freezing range of "AXJ530-*low*" is significantly wider (176°C) compared to "AXJ530-*high*" (100°C) (Gröbner et al., 2008; Janz et al., 2008)

The availability of kinetic data for more elaborate simulation of Mg alloys is constantly growing but has not reached the level attained for Al alloys or steel. Despite that, six examples demonstrating kinetic and precipitation simulations are also summarized in Schmid-Fetzer and Zhang (2018).

### 8.3.1.2 Solidification Path and T6 Heat Treatment of AZ Series Alloys

The most prominent Mg alloy in application is the cast alloy AZ91. The design of alloys in the AZ series with optimized corrosion properties involves the variation of Mn and Fe, which are always present, making the AZ91 effectively a quinary alloy in the Mg–Al–Zn–Mn–Fe system (Chen et al., 2021). In order to understand such computational design in detail, it is advisable to consider first the solidification path of AZ91 in the ternary system Mg–Al–Zn. This path is defined by the liquid composition variation of the alloy during solidification, which implies the sequence of solid phases crystallizing from the melt. That is best visualized by superimposing the solidification path on the liquidus projection and separately by a graph showing the calculated phase fractions as function of temperature. These graphs can be found for the ternary AZ91 with a detailed explanation in figures 20–22 of Schmid-Fetzer (2014), not repeated here. The path for the equilibrium solidification predicts single-phase solidification of only the (Mg) solid solution phase. Under Scheil conditions, in addition to this primary phase (Mg), the secondary phases $\gamma$ ($Al_{12}Mg_{17}$), $\Phi$ (AlMgZn), $\tau$ $[Mg_{32}(Al, Zn)_{49}]$, and MgZn are predicted to crystallize in that sequence. It is noteworthy that even for cooling rates as slow as 1 K/min, the Scheil conditions are a better approximation of reality than the equilibrium conditions for AZ91. That is demonstrated by the experimental microstructure. The basic of the solution heat treatment window is also explained in Schmid-Fetzer (2014). The freezing range of AZ91 is shown to be much wider under Scheil conditions compared to equilibrium, which is generally detrimental for the cast quality. Especially the main secondary phase $\gamma$, if crystallized coarse from the melt, is detrimental for ductility and strength of the cast AZ91. These properties may be significantly improved by a T6 heat treatment comprising a solutionizing step to dissolve the coarse $\gamma$ followed by artificial aging.

This T6 treatment has been applied as a multistage heat treatment to the alloy AZ64 by Ma et al. (2007). The computational design basis for an optimization of that process is given in Figure 8.7, presenting the Scheil and equilibrium solidification simulations of alloy AZ64. The actual composition of AZ64 alloy (Mg–5.87Al–3.75Zn–0.52Mn,

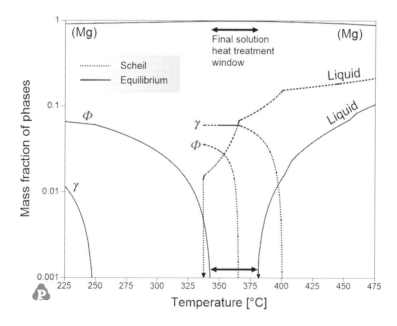

**Figure 8.7** Solidification simulation of AZ64 alloy (Mg–5.87Al–3.75Zn–0.52Mn, wt.%) under Scheil condition (dotted lines, termination at 337°C) and equilibrium condition (solid lines), phase fractions in logarithmic scale; $\gamma$ is $Al_{12}Mg_7$, and $\Phi$ is AlMgZn.

wt.%) (Ma et al., 2007) and software Pandat with database PanMg2020 (Schmid-Fetzer and Zhang, 2018) were used for these calculations. The small fraction of additional $Al_8Mn_5$ (or $Al_xMn$) phase crystallizing even before the (Mg) matrix is not relevant for this discussion and omitted in Figure 8.7 for clarity. The dotted lines in Figure 8.7 for the Scheil simulation reveal that the liquid phase fraction decreases with the fall of temperature, and the first two kinks at 401°C and 365°C occur at the crystallization start of phases $\gamma$ and $\Phi$, respectively, until the solidus at 337°C, according to Scheil simulation (the "Scheil solidus"), terminates solidification. The final phase fractions are $f(\gamma) = 0.060$ and $f(\Phi) = 0.035$. The simulation is verified by the coarse occurrence of $\gamma$ phase along the grain boundaries embedded in the final eutectic in the as-cast microstructure (Ma et al., 2007). This detrimental $\gamma$ phase, as well as $\Phi$, can be completely dissolved by heating in the "final solution heat treatment window" indicated by the equilibrium simulation and double-arrows in Figure 8.7. The window ranges between the equilibrium solidus at 379°C and the start of solid-state precipitation of $\Phi$ at 344°C. This window is also seen where the fraction of single-phase (Mg) reaches unity. However, heating the casting immediately into this range may result in most detrimental incipient melting at the nonequilibrium constitution present at the grain boundaries. This first temperature limit is given by the Scheil solidus at 337°C.

The concept of predicting the "incipient melting temperature" by Scheil solidus of as-cast Mg alloys was verified experimentally and by simulations more generally for the higher alloyed Mg–Al–Zn–Mn alloys with 0.2 wt.% Mn (Ohno et al., 2006). Only for the lower alloyed as-cast Mg alloys, with the approximate limit (wt.% Al + 2 wt.% Zn < 8), does the solidification proceed closer to equilibrium conditions and the predictions show a larger uncertainty (Ohno et al., 2006). An experimental method to determine Scheil solidus and solidification curves for various AZ and AM alloys is given in Mirković and Schmid-Fetzer (2007). The concept is also confirmed by considering the terminal freezing range (TFR) for the example of Mg–Al–Zn–Ca alloys (Djurdjevic and Schmid-Fetzer, 2006).

In the present example, the conclusion is that a useful multistage heat treatment would start safely below the Scheil solidus (337°C), and in subsequent stages the temperature can be increased after sufficient dissolution of the low-melting multiphase assembly had been accomplished. The maximum temperature to shorten the necessary duration of completing the solutionizing heat treatment as much as possible is the equilibrium solidus (379°C). In agreement with this simulation, the actual solutionizing schedule reported by Ma et al. (2007) was first 300°C (2 hours), next 345°C (12 hours), finally 370°C (4 hours), followed by artificial aging at 180°C for 16 hours.

Further refinement and property improvement were obtained by addition of 0.5 wt.% Sb even in the as-cast state compared to AZ64 (Ma et al., 2007). The present calculations with the thermodynamic database PanMg2020 were also performed for this AZ64–Sb alloy (Mg–5.87Al–3.75Zn–0.52Mn–0.5Sb, wt.%). The simulation results, not shown here, are virtually identical to those of Figure 8.7, with the significant exception that the additional phase $Mg_3Sb_2$ starts crystallizing at 545°C (Scheil) and almost the final phase fraction of $f(Mg_3Sb_2) = 0.006$ is already reached at 401°C, where the phase $\gamma$ again starts crystallizing. This is suggested to explain the reported significant refinement of the as-cast microstructure of the AZ64–Sb alloy compared to AZ64 (Ma et al., 2007) because $Mg_3Sb_2$ is present in this case as a potential nucleant for the eutectic reaction Liquid $\rightarrow$ (Mg) $+ \gamma + Mg_3Sb_2$ predicted in the range from 401–365°C.

### 8.3.1.3  Computational Design and Development of New Mg–Al–Sn-Based (AT) Cast Alloys

Figure 8.8 shows the outstanding automotive application of the Mg–Al–Sn alloy AT72 for the inner panel of a side door. This Mg alloy provides significant weight savings over the conventional steel inner panel design. In addition, the excellent castability of AT72 with a high fluidity for large thin-walled components enables significant part consolidation in a single die-cast piece, eliminating the assembly cost of the steel design. Moreover, casting enables the innovative joint and section designs for stiffness and strength, also implemented in this case. This is a prime example for the computational design of a material integrated with the redesign of the engineering component and its production. The die-casting process is favorable for mass production, and maximum performance is often achieved in the as-cast state, also for AT72, without heat treatment.

The computational design of this AT72 alloy is based on the work of Luo et al. (2012). Following the concept of solidification path described previously, they applied such calculations with Pandat and the PanMg database for precipitate prediction, as shown in Figure 8.9. This superposition of the solidification paths associated with nine different candidate AT alloys on the liquidus projection of the Mg-rich corner of the Mg–Al–Sn ternary system guided the selection of the promising alloys for the more costly and detailed studies. This figure shows that the secondary phases $\gamma$ and $Mg_2Sn$ must occur in the final microstructure, but their sequence on the path is controlled by the alloy composition. One key point is to avoid the occurrence of $\gamma$ along the monovariant secondary eutectic line Liquid $\rightarrow$ (Mg) $+ \gamma$ because that is known to result in the coarse

**Figure 8.8** Magnesium alloy AT72 die-cast side-door inner panel, which received an International Magnesium Award Association (IMA) award and was presented in Carter (2017), developed at General Motors and Dongguan EONTEC Co., Ltd., by courtesy of Alan Luo.

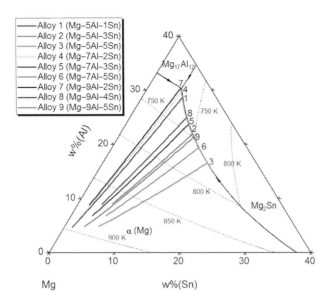

**Figure 8.9** Calculated Mg-rich corner of the Mg–Al–Sn liquidus projection and the solidification paths of the experimental AT alloys. Reprinted from Luo et al. (2012) with permission from Springer Nature. A black and white version of this figure will appear in some formats. For the colour version, please refer to the plate section.

and detrimental formation of $\gamma$ described in the previous section. All of the nine alloys shown in Figure 8.9 meet that requirement with secondary $Mg_2Sn$ along this complete path: (1) Liquid $\rightarrow$ (Mg), (2) Liquid $\rightarrow$ (Mg) + $Mg_2Sn$, and (3) Liquid -› (Mg) + $Mg_2Sn + \gamma$ at the final eutectic (430°C) with typically finely dispersed phases.

The next requirement is the precipitation of an appropriate total fraction of secondary and tertiary $Mg_2Sn$ in relation with the ternary eutectic $\gamma$ phase to balance the essentially increasing strength and ductility. That was studied by detailed calculations of phase fraction charts, such as in Figure 8.7, under Scheil conditions for all alloys. The mechanical testing is in agreement with the trends indicated by these calculations. The two most promising alloys are found to be Mg–7Al–2Sn (AT72) and Mg–7Al–5Sn (AT75) compared with AZ91, all in the as-cast state. The AT72 alloy has the most balanced tensile properties with similar yield strength (YS), 92 MPa, but significantly higher ultimate tensile strength (UTS), 176 MPa, and elongation of $A = 5.3\%$ compared with the AZ91, YS $= 89$ MPa, UTS $= 151$ MPa, and $A = 2.4\%$. The AT75 alloy has significantly increased YS $= 110$ MPa, with still increased UTS $= 161$ MPa and $A = 3.2\%$ compared to AZ91 (Luo et al., 2012).

While the key results from the relatively simple Scheil simulation were instrumental for the original development of alloy AT72, more comprehensive data for further optimization are obtained by the more recent application of a microsegregation solidification model involving the kinetics by Zhang et al. (2019). The same PanSolidification module of the Pandat software package is applied (Zhang et al., 2019), which was described in Section 8.2.1 in detail and also summarized in Zhang

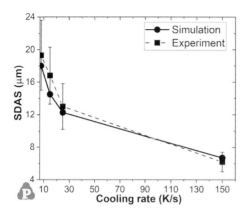

**Figure 8.10** Comparison between simulated and measured secondary dendrite arm spacing of wedge casting AT72 alloy. Reprinted from Zhang et al. (2019) with permission from Springer Nature.

et al. (2019). The development of the mobility and kinetic data for the Mg–Al-based alloys using the CALPHAD method is experimentally validated in Zhang et al. (2019) and applied to three series of Mg alloys, Mg–Al, Mg–Al–Ca, and Mg–Al–Sn. The microsegregation composition profiles and SDAS were determined experimentally and compare well with the simulation results at a wide range of cooling rates from 0.12–0.2 K/s (directional solidification) and from 8 to 150 K/s (wedge and mold casting, AT72). These higher cooling rates are more relevant to industrial die castings of Mg alloys with 50–200 K/s or even higher. The comparison of simulated and experimental SDAS data in Figure 8.10 is reproduced from Zhang et al. (2019). The very good agreement verifies that simulation approach. The simulated and experimental composition profiles of Sn and Al also agree to some extent. Problems in the EPMA measurements arising from the very fine microstructure at high cooling rates are discussed in detail in Zhang et al. (2019). This validated simulation tool is expected to be most helpful for further computational design of cast Mg alloys and process optimization.

Another new class of Mg alloys was developed in Luo's group using the CALPHAD approach by adding Si (code S) to AT alloys (Klarner et al., 2016, 2017). These Mg–Al–Sn–Si alloys (ATS alloys) contain minor additions of Si and some 0.2–0.4 wt.% Mn. The Scheil and equilibrium calculations were used to predict the microstructure and design the T6/T5 heat treatment schedules. Some improvement of mechanical properties is achieved in these preliminary studies (Klarner et al., 2016, 2017).

## 8.3.2   Biomedical Mg Alloy Implants

Corrosion of structural components exposed to water or humidity and subsequent loss of mechanical integrity is often a major limitation of Mg alloy application. This is due to the very low electrochemical standard potential of Mg, also dependent on the formation of MgO or Mg(OH)$_2$. The corrosion rate also depends on the local pH, and it dramatically increases if impurities such as Ni, Fe, or Cu are present beyond a threshold concentration. Because of their low solid solubility in (Mg), they form precipitates, and these are more noble, acting as local cathodes and favoring the evolution of $H_2$ gas. For Fe, the tolerance limits are in the order of 170 ppm in Mg or 50 ppm in alloy AZ91 in salt water (Pekguleryuz et al., 2013). Iron impurity is hard to avoid in the process chain, and its control by Mn additions is also supported by the CALPHAD method (Chen et al., 2021).

For biomedical applications, however, the complete but controlled "dissolution" of a Mg-based implant is the actual purpose if the implant is needed only temporarily. That also solves problems that may occur with permanent implants, such as long-term irritation or their fixed size in a growing patient, eventually requiring a follow-up surgery for implant removal. The use of biodegradable magnesium implants is nothing new, and the history dating back two centuries ago was reviewed by Witte (2010). The overall corrosion reaction of Mg in aqueous environments, given as $(Mg) + 2[H_2O]_{aq} \leftrightarrow Mg(OH)_2 + H_2(gas)$, also governs the so-called biodegradation, a very complex process (Witte et al., 2008). The generally high corrosion rate of Mg alloys must be limited to allow the harmless absorption of intermediate corrosion products over time without forming local gas cavities. These products can either be absorbed or metabolized in the form of $Mg^{2+}$ by reacting with $Cl^-$ ions in the body fluid or digested by macrophages (Zhao et al., 2017). Major advances in magnesium alloys as temporary biomaterials have been reported in the years 2004–2008. Advanced clinical applications at that time were biodegradable cardiovascular magnesium stents but also bone implants, such as screws, plates or other fixture devices (Witte et al., 2008). However, investigated magnesium alloys in that comprehensive review were obtained as off-the-shelf, purchasable standard alloys or alloys that can be easily cast (Witte ct al., 2008). One example is the alloy WE43, originally developed by Magnesium Elektron Ltd. for improving high-temperature strength, creep, and corrosion resistance for lightweight aeroengine applications (MEL data sheet 467, 1989, in Lyon et al., 1991). Further development is highly multidisciplinary, which involves material design and preparation, medical device design and fabrication, biological evaluation, clinical assessment, and product registration. Recent reviews arc focusing on orthopaedic implants (Zhao et al., 2017) and cardiovascular stents (Li et al., 2018). Research on Mg alloys for stents is ongoing, and one example is the alloy Mg–2.25Nd–0.11Zn–0.43Zr (wt.%), showing advantages if compared to traditional materials, such as medical stainless steel and polymer (Mao et al., 2017).

In the following, just one example will be presented, highlighting the metallurgical challenges in designing a new Mg alloy for such stents, which goes back to the CALPHAD-based design work by Hänzi et al. (2009a). While successful clinical trials by the company BIOTRONIK AG with a Mg-7% RE alloy were reported as early as 2005 with a first generation of absorbable magnesium stent (AMS-1), further development to AMS-2 with a modified alloy and improved stent design provided prolonged mechanical stability and enhanced surface passivity. The third generation was a drug-eluting absorbable metal scaffold (AMS-3, DREAMS), as summarized by Witte in chapter 10 of the book (Pekguleryuz et al., 2013) and by Luthringer et al.

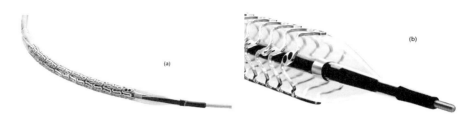

**Figure 8.11** A commercially available bioresorbable magnesium scaffold (Magmaris®), shown at two stages: (a) The Mg alloy is produced as a narrow and meshed tube placed on a delivery system; (b) after insertion in a narrowed coronary artery, the inflated balloon expands the scaffold. Courtesy of BIOTRONIK AG, Bülach, Switzerland.

(2014). In this context, the implant is envisaged to perform more as an assisting healing device and is called a "scaffold" instead of a "stent." The latest generation is shown in Figure 8.11, the commercially available resorbable magnesium scaffold (Magmaris®, www.biotronik.com/en-de/products/coronary/magmaris) displayed at two stages. First, as indicated in Figure 8.11a, the Mg alloy is produced as a narrow and meshed tube called a scaffold and placed on a delivery system of typically 1.5 mm diameter crossing profile. Second, as shown in Figure 8.11b, after insertion in a narrowed coronary artery during a minimal invasive surgery, the inflated balloon expands the scaffold, depending on the scaffold size, typically to 3 mm and to a maximum diameter of 4.1 mm as per its instruction for use. There it remains and resorbs at 95% in 12 months. The expansion step demonstrates the challenging metallurgical demand on high ductility of the Mg alloy at 37°C.

The CALPHAD-based design strategy by Hänzi et al. (2009a) succeeded in developing ultraductile Mg alloys based on a comprehensive concept, aiming to restrict grain growth significantly during alloy casting and forming. Conventional processing steps could be used, such as direct chill casting (DCC) and direct extrusion, as opposed to the more complex severe plastic deformation (SPD). Achieving a small grain size, which is known to enhance low-temperature formability of Mg alloys, was instrumental at each step. During casting and solidification, the growth restriction concept developed by StJohn et al. (2005) for efficient grain refinement was used. This concept is based on the combined action of nucleants and solutes, and the growth restriction factor (GRF) can be obtained from thermodynamic calculations using Pandat to assess the usefulness of certain alloying components, as demonstrated in more detail for multicomponent systems (Schmid-Fetzer and Kozlov, 2011). To avoid grain growth due to dynamic recrystallization during subsequent hot extrusion, a microalloying concept was employed for the desired pinning of grain boundaries by second-phase particles and also to prevent the formation of coarse intermetallic phase particles. The choice of alloying elements was restricted to those that are harmless to the human body since the application field of medical implant technology was the target. All these aspects led to the choice of alloying components Zn, Ag, Ca, and Zr. Manganese was added because of its known efficiency in cleaning the melt from Fe impurities. Finally, two alloys of L1 and L2 were developed, which are denoted here as ZQX3 (measured composition Mg–2.9Zn–0.5Ag–0.25Ca–0.15Mn, wt.%) and ZKQX3 (Mg–2.7Zn–0.74Zr–0.44Ag–0.17Ca–0.07Mn) (Hänzi et al., 2009a).

Figure 8.12 shows the calculated equilibrium phase fractions for alloy ZQX3 and ZKQX3 by means of the thermodynamic data given in Schmid-Fetzer and Zhang

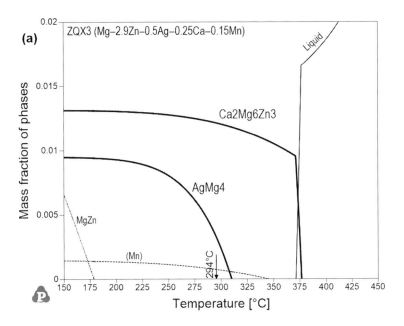

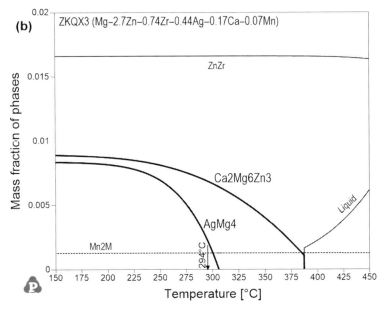

**Figure 8.12** Equilibrium phase fractions of alloy (a) ZQX3 and (b) ZKQX3, calculated from the thermodynamic data in Schmid-Fetzer and Zhang (2018). The Scheil solidus at 294°C is indicated by an arrow.

(2018). The first two important intermetallic phases highlighted by thick solid lines are $Ca_2Mg_6Zn_3$ in both alloys and additionally $Zn_2Zr$ in ZKQX3. These are identified to further promote grain refinement during hot extrusion as pinning particles at the grain boundaries (GBs), and a large number are desired. It is noted that the phase fraction of $Ca_2Mg_6Zn_3$ decreases significantly beyond 300°C. The next important intermetallic phase is $AgMg_4$, which, as a stable Ag-phase, would generate localized electrochemical potential differences toward the matrix, thereby decreasing the corrosion resistance. Thus, a minimum amount for the $AgMg_4$ phase is desired. The calculations predict that $f(AgMg_4) = 0$ is reached at 310°C for ZQX3, as shown in Figure 8.12a, and at 306°C for ZKQX3, as indicated in Figure 8.12b. The experimental microstructure observations verify these Pandat calculations (Hänzi et al., 2009a).

The alternative but mutually valid approach to a minimum $AgMg_4$ phase fraction is given in Figure 8.13. Here the goal is to maximize the equilibrium content of Ag dissolved harmlessly in the (Mg) solid solution. These data result from the same Pandat equilibrium calculations for both Figures 8.12 and 8.13. This dissolved content of Ag increases in both alloys with rising temperature in Figure 8.13, parallel to the decreasing phase fraction of the compound $AgMg_4$ shown in Figure 8.12, until reaching the kinks marked for the zero phase fractions. The following maximum plateaus of Ag in the (Mg) solid solution correspond to the measured total Ag content in the alloys. The content of Ag dissolved in (Mg) drops again with the appearance of equilibrium liquid at 371°C (ZQX3) and 387°C (ZKQX3).

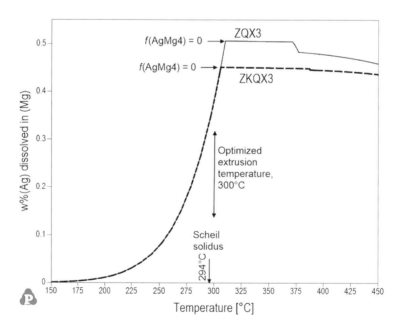

**Figure 8.13** Equilibrium composition of Ag in the (Mg) solid solution in alloys ZQX3 and ZKQX3, calculated from the thermodynamic data given in Schmid-Fetzer and Zhang (2018) with ZQX3 (Mg–2.9Zn–0.5Ag–0.25Ca–0.15Mn, wt.%) and ZKQX3 (Mg–2.7Zn–0.74Zr–0.44Ag–0.17Ca–0.07Mn).

Another process limit is to avoid incipient melting of the cast alloys at the beginning of hot forming, as detailed for the example in Section 8.3.1.2, and this is determined by separate Scheil simulations. The same calculated Scheil solidus at 294°C is obtained for both alloys, and this value is marked as a guide for the eye in Figures 8.12 and 8.13. Finally, as a best compromise for all the preceding temperature limits, the optimized extrusion temperature of 300°C was selected (Hänzi et al., 2009a), and this value is marked by the double arrow in Figure 8.13.

Impressive results for small grain size and room temperature mechanical properties were obtained following this design strategy guided by computational thermodynamics. The as-extruded mean grain size was 6 μm in ZQX3 and 2 μm in ZKQX3. In accord with the theoretical expectations the as-extruded alloys exhibit high ductility at moderate strength (YS = 150 MPa; UTS = 250 MPa, elongation to fracture 27%) in ZQX3 and high strength at reasonable ductility (YS = 320 MPa; UTS = 350 MPa, elongation 21%) in ZKQX3 (Hänzi et al., 2009a). Moreover, these data for the new alloys compare favorably with those for the as-extruded commercial alloy ZK31 (YS = 245 MPa; UTS = 300 MPa, elongation 15%), lacking the designed Ag and Ca additions. It was observed that as-extruded ZK31 displays occasionally very coarse grains (>20 μm), and thus it is not surprising that the mechanical properties of as-extruded ZK31 are inferior to those observed for ZKQX3 (Hänzi et al., 2009a).

As a final remark, surface design is an important tool to tailor the biodegradation rate of a Mg implant to meet specific target values of temporary service time. One example process is highlighted, the selective high-temperature oxidation of Mg alloys, for which quantitative CALPHAD-type thermodynamic calculations have been demonstrated.

The selective high-temperature oxidation as a cheap and efficient process for surface design by growing a thin oxide layer on the Mg–Y–RE (WE43) alloy has been demonstrated by Hänzi et al. (2009b). The heat treatment of the alloy at 500°C in air enhanced the biodegradation performance of the alloy in simulated body fluid (SBF). Experimentally, the formation of a thin layer of $Y_2O_3$ was found to grow during oxidation, and its thickness can be tailored by the oxidation time (Hänzi et al., 2009b). Thermodynamic calculations using Pandat with the PanMg database have been demonstrated to enable a deeper understanding of the complex interactions during oxidation (Schmid-Fetzer, 2019). According to the reported WE43 composition (Hänzi et al., 2009b), Nd is considered the leading RE element in the CALPHAD-type study, summarizing the other RE, which simplifies the WE43 composition to the ternary Mg–4Y–3Nd alloy (Schmid-Fetzer, 2019). The principles of applying the calculated phase diagrams are first explained by considering the oxidation of binary Mg–Nd alloys, which can be best understood by reading the ternary Mg–Nd–O phase diagram. Next, the oxidation of Mg–Y–Nd alloys is explained analogously through calculated Mg–Y–Nd–O phase diagrams. It is demonstrated that a "classical" approach by interpretation of the "chemical reaction equations" is more cumbersome. The advantage of the proposed approach to understand alloy oxidation through calculated phase diagrams is that the impact of process variables, such as temperature and alloy composition, is directly readable. The importance to distinguish the overall alloy composition from the composition at the interface to the oxide was demonstrated. Initially, only $Y_2O_3$ may form at the surface of this alloy as shown in the calculated Mg–Y–Nd–O phase diagrams. However, the consumption of Y by the growing $Y_2O_3$ layer results in depletion of Y below the initial 4 wt.% Y, and a concentration gradient curve forms with decreasing value at the interface. A critical limit was obtained by the Pandat calculations to be about 0.85 wt.% Y. Below this critical limit, the additional formation of MgO together with $Y_2O_3$ at the inner metal–oxide interface is possible (Schmid-Fetzer, 2019). This method may be suggested to be helpful to guide the optimization of such oxidation processes for surface design of Mg alloys.

## 8.4    Summary

The case studies on Al and Mg alloys in this chapter demonstrate the successful applications of a CALPHAD-based approach at two levels.

### 8.4.1    Alloy Design Applications Using Solely Thermodynamic CALPHAD Databases

Next to phase diagram calculations of the alloy system, the equilibrium and Scheil simulations for selected alloys are applied. Mostly Mg alloys are studied here, and examples summarize selected case studies for new creep-resistant Mg alloys and the concept of solidification path and T6 heat treatment of AZ series alloys. As one industrial automotive application, the side-door inner panel that is die-cast from alloy AT72 is highlighted. The computational design was followed for development of these new Mg–Al–Sn-based (AT) cast alloys. In addition to these Mg alloys designed for structural components, a concrete design study of a biomedical Mg alloy implant is given to emphasize this rapidly growing application field. The CALPHAD-based design strategy is detailed that was part of the development process that culminated in a current bioresorbable Mg alloy stent to cure coronary artery disease.

### 8.4.2    Alloy Design Applications Using Extended CALPHAD-Type Databases and Kinetic Simulations

Mostly Al alloys are studied here, and one reason for focusing on Al alloys is the better availability of kinetic data for Al alloys as compared to Mg alloys. It must be stressed that any of such kinetic simulations reasonably starts with the calculation methods of the previous subsection. That is because they provide a good overview of the phases involved and an approximated range of results. Also, any of the kinetic simulations in this next level requires the same high-quality thermodynamic CALPHAD-database. In addition, the mobility data for diffusion and a set of kinetic alloy data are required for the elaborate solidification simulation and for the phase formation and transformation modeling during heat treatment with precipitation kinetics. Accordingly, two case studies for Al alloys are presented. Solidification simulation with microsegregation modeling for the cast Al alloy A356 shows how to control the desired and detrimental secondary phases by alloy composition and cooling rate. Also, the HTC function is demonstrated that can perform thousands of simulations and include the postprocessing data mining and analysis to extract the results of interest, such as SDAS in a user-defined alloy compositional space and a range of cooling rates.

The HTC function is also implemented in other modules of the Pandat software package and enables this powerful search also for the first-level thermodynamic calculations or the PanPrecipitation module applied in the second case study on wrought 7xxx Al alloy for heat treatment simulation and precipitation kinetics. In this example, on one hand, data for a better understanding of the precipitation process, such as temporal particle size development or distributions (PSD), are obtained. On the other hand, the prediction of mechanical properties (yield strength and hardness) as a function of the selected heat treatment temperature–time profile provides immediate design guidance.

# References

Cao, W., Chen, S. L., Zhang, F., Wu, K., Yang, Y., Chang, Y. A., Schmid-Fetzer, R., and Oates, W. A. (2009) PANDAT software with PanEngine, PanOptimizer and PanPrecipitation for multi-component phase diagram calculation and materials property simulation. *CALPHAD*, 33(2), 328–342.

Cao, W., Zhang, F., Chen, S. L., Zhang, C., and Chang, Y. A. (2011) An integrated computational tool for precipitation simulation. *JOM*, 63(7), 29–34.

Carter, J. (2017) Magnesium Die-Cast Side-Door Inner Panels, award presentation at 2017 World Magnesium Conference, Singapore, May 21–23.

Chen, S. L., Daniel, S., Zhang, F., Chang, Y. A., Yan, X. Y., Xie, F. Y., Schmid-Fetzer, R., and Oates, W. A. (2002) The PANDAT software package and its applications. *CALPHAD*, 26(2), 175–188.

Chen, T., Yuan, Y., Liu, T., Li, D., Tang, A., Chen, X., Schmid-Fetzer, R., and Pan, F. (2021) Effect of Mn addition on melt purification and Fe tolerance in Mg alloys. *JOM*, 73, 892–902.

Clyne, T. W., and Kurz, W. (1981) Solute redistribution during solidification with rapid solid-state diffusion. *Metallurgical Transactions A*, 12(6), 965–971.

Deschamps, A., Dumont, D., Brechet, Y., Sigli, C., and Dubost, B. (2001) Process modeling of age-hardening aluminum alloys: from microstructure evolution to mechanical and fracture properties, *Proceedings of the James T. Staley Honorary Symposium on Aluminum Alloys*. Russell Township: ASM International, 298–305.

Djurdjevic, M. B., and Schmid-Fetzer, R. (2006) Thermodynamic calculation as a tool for thixoforming alloy and process development. *Materials Science and Engineering A*, 417(1), 24–33.

Du, Y., Chang, Y. A., Liu, S. H., Huang, B. Y., Xie, F. Y., Yang, Y., and Chen, S. L. (2005) Thermodynamic description of the Al–Fe–Mg–Mn–Si system and investigation of microstructure and microsegregation during directional solidification of an Al–Fe–Mg–Mn–Si alloy. *Zeitschrift für Metallkunde*, 96(12), 1351–1362.

Gröbner, J., Janz, A., Kozlov, A., Mirkovic, D. and Schmid-Fetzer, R. (2008) Phase diagrams of advanced magnesium alloys containing Al, Ca, Sn, Sr, and Mn. *JOM*, 60(12), 32–38.

Hänzi, A. C., Dalla Torre, F. H., Sologubenko, A. S., Gunde, P., Schmid-Fetzer, R., Kuehlein, M., Loffler, J. F., and Uggowitzer, P. J. (2009a) Design strategy for microalloyed ultra-ductile magnesium alloys. *Philosophy Magazine Letters*, 89(6), 377–390.

Hänzi, A. C., Gunde, P., Schinhammer, M., and Uggowitzer, P. J. (2009b) On the biodegradation performance of an Mg–Y–RE alloy with various surface conditions in simulated body fluid. *Acta Biomaterialia*, 5(1), 162–171.

Hillert, M., Höglund, M. H. L., and Schalin, M. (1999) Role of back-diffusion studied by computer simulation. *Metallurgical and Materials Transactions A*, 30(6), 1635–1641.

Janz, A., Gröbner, J., Mirković, D., Medraj, M., Zhu, J., Chang, Y. A., and Schmid-Fetzer, R. (2007) Experimental study and thermodynamic calculation of Al–Mg–Sr phase equilibria. *Intermetallics*, 15(4), 506–519.

Janz, A., Gröbner, J., and Schmid-Fetzer, R. (2009) Thermodynamics and constitution of Mg–Al–Ca–Sr–Mn alloys: Part II. Procedure for multicomponent key sample selection and application to the Mg–Al–Ca–Sr and Mg–Al–Ca–Sr–Mn systems. *Journal of Phase Equilibria and Diffusion*, 30(2), 157–175.

Janz, A., Groebner, J., and Schmid-Fetzer, R. (2008) The Mg–Al–Zn–Mn–Ca–Sr alloy system: backbone of understanding phase formation in AXJ alloys and modifications of AZ and AM alloys with Ca or Sr, in Pekguleryuz, M. O., Neelameggham, N. R., Beals, R. S., and Nyberg, E. A. (eds), *Magnesium Technology*. Pittsburgh: Minerals, Metals and Materials Society, 427–429.

Kampmann, R., and Wagner, R. (1984) Kinetics of precipitation in metastable binary alloys – theory and application to Cu-1.9 at % Ti and Ni-14 at % Al, in Haasen, P., Wagner, V. G. R., and Ashby, M. F. (ed), *Decomposition of Alloys: The Early Stages*. Oxford: Pergamon, 91–103.

Klarner, A., Sun, W., Meier, J., and Luo, A. (2016) Development of Mg–Al–Sn–Si alloys using a CALPHAD approach, in Singh, A., Solanki, K., Manuel, M. V., and Neelameggham, N. R. (eds), *Magnesium Technology 2016*. Carmel: John Wiley & Sons, Inc., 79–82.

Klarner, A. D., Sun, W., Miao, J., and Luo, A. A. (2017) Microstructure and mechanical properties of high pressure die cast Mg–Al–Sn–Si alloys, in Solanki, K. N., Orlov, D., Singh, A., and Neelameggham, N. R. (eds), *Magnesium Technology 2017*. Cham: Springer, 289–295.

Klaumünzer, D., Hernandez, J. V., Yi, S., Letzig, D., Kim, S.-h., Kim, J. J., Seo, M. H., and Ahn, K. (2019) Magnesium process and alloy development for applications in the automotive industry, in Joshi, V. V., Jordon, J. B., Orlov, D., and Neelameggham, N. R. (eds), *Magnesium Technology 2019*. Cham: Springer, 15–20.

Kraft, T., and Chang, Y. A. (1997) Predicting microstructure and microsegregation in multicomponent alloys. *JOM*, 49(12), 20–28.

Kraft, T., Rettenmayr, M., and Exner, H. E. (1996a) An extended numerical procedure for predicting microstructure and microsegregation of multicomponent alloys. *Modelling and Simulation in Materials Science and Engineering*, 4(2), 161–177.

Kraft, T., Roósz, A., and Rettenmayr, M. (1996b) Undercooling effects in microsegregation modelling. *Scripta Materialia*, 35(1), 77–82.

Li, P., Zhou, N., Qiu, H., Maitz, M. F., Wang, J., and Huang, N. (2018) In vitro and in vivo cytocompatibility evaluation of biodegradable magnesium-based stents: a review. *Science China Materials*, 61(4), 501–515.

Liang, H., Kraft, T., and Chang, Y. A. (2000) Importance of reliable phase equilibria in studying microsegregation in alloys: Al–Cu–Mg. *Materials Science and Engineering A*, 292(1), 96–103.

Luo, A. A. (2004) Recent magnesium alloy development for elevated temperature applications. *International Materials Reviews*, 49(1), 13–30.

Luo, A. A., Fu, P., Peng, L., Kang, X., Li, Z., and Zhu, T. (2012) Solidification microstructure and mechanical properties of cast magnesium–aluminum–tin alloys. *Metallurgical and Materials Transactions A*, 43(1), 360–368.

Luthringer, B. J. C., Feyerabend, F., and Willumeit-Romer, R. (2014) Magnesium-based implants: a mini-review. *Magnesium Research*, 27(4), 142–154.

Lyon, P., King, J. F., and Fowler, G. A. (1991) Developments in magnesium based materials and processes, in *Magnesium Based Materials and Processes: Proceedings of the ASME 1991 International Gas Turbine and Aeroengine Congress and Exposition*, volume 5. New York: American Society of Mechanical Engineers, V005T12A002.

Ma, Y. Q., Chen, R. S., and Han, E.-H. (2007) Keys to improving the strength and ductility of the AZ64 magnesium alloy. *Materials Letters*, 61(11), 2527–2530.

Mao, L., Shen, L., Chen, J., et al. (2017) A promising biodegradable magnesium alloy suitable for clinical vascular stent application. *Scientific Reports*, 7(1), 46343.

Mirković, D., and Schmid-Fetzer, R. (2007) Solidification curves for commercial Mg alloys determined from differential scanning calorimetry with improved heat-transfer modeling. *Metallurgical and Materials Transactions A*, 38(10), 2575–2592.

Ohno, M., Mirkovic, D., and Schmid-Fetzer, R. (2006) Liquidus and solidus temperatures of Mg-rich Mg–Al–Mn–Zn alloys. *Acta Materialia*, 54(15), 3883–3891.

Park, S. S., Park, W. J., Kim, C. H., You, B. S., and Kim, N. J. (2009) The twin-roll casting of magnesium alloys. *JOM*, 61(8), 14–18.

Pekguleryuz, M. O., and Baril, E. (2001) Development of creep resistant Mg–Al–Sr alloys. In Mathaudhu, S. N., Luo, A. A., Neelameggham, N. R., Nyberg, E. A., and Sillekens, W. H. (eds), *Essential Readings in Magnesium Technology 2001*. Pittsburgh: Minerals, Metals, and Materials Society, 119–125.

Pekguleryuz, M. O., Kainer, K., and Kaya, A. A. (2013) *Fundamentals of Magnesium Alloy Metallurgy*. Cambridge: Woodhead.

Polmear, I., StJohn, D., Nie, J.-F., and Qian, M. (2017) *Light Alloys: Metallurgy of the Light Metals*, fifth edition. Woburn: Butterworth-Heinemann.

Schmid-Fetzer, R. (2014) Phase diagrams: the beginning of wisdom. *Journal of Phase Equilibria and Diffusion*, 35(6), 735–760.

Schmid-Fetzer, R. (2019) Recent progress in development and applications of Mg alloy thermodynamic database, in Joshi, V. V., Jordon, J. B., Orlov, D., and Neelameggham, N. R. (eds), *Magnesium Technology 2019*. Cham: Springer, 249–255.

Schmid-Fetzer, R., and Kozlov, A. (2011) Thermodynamic aspects of grain growth restriction in multicomponent alloy solidification. *Acta Materialia*, 59(15), 6133–6144.

Schmid-Fetzer, R., and Zhang, F. (2018) The light alloy CALPHAD databases PanAl and PanMg. *CALPHAD*, 61, 246–263.

Sonderegger, B., and Kozeschnik, E. (2009a) Generalized nearest-neighbor broken-bond analysis of randomly oriented coherent interfaces in multicomponent Fcc and Bcc structures. *Metallurgical and Materials Transactions A*, 40(3), 499–510.

Sonderegger, B., and Kozeschnik, E. (2009b) Size dependence of the interfacial energy in the generalized nearest-neighbor broken-bond approach. *Scripta Materialia*, 60(8), 635–638.

StJohn, D. H., Qian, M., Easton, M. A., Cao, P., and Hildebrand, Z. (2005) Grain refinement of magnesium alloys. *Metallurgical and Materials Transactions A*, 36(7), 1669–1679.

Witte, F. (2010) The history of biodegradable magnesium implants: a review. *Acta Biomaterialia*, 6(5), 1680–1692.

Witte, F., Hort, N., Vogt, C., et al. (2008) Degradable biomaterials based on magnesium corrosion. *Current Opinion on Solid State Materials Science*, 12(5), 63–72.

Yan, X., Chen, S., Xie, F., and Chang, Y. A. (2002) Computational and experimental investigation of microsegregation in an Al-rich Al–Cu–Mg–Si quaternary alloy. *Acta Materialia*, 50(9), 2199–2207.

Zhang, C., Miao, J., Chen, S., Zhang, F., and Luo, A. A. (2019) CALPHAD-based modeling and experimental validation of microstructural evolution and microsegregation in magnesium alloys during solidification. *Journal of Phase Equilibria and Diffusion*, 40(4), 495–507.

Zhang, F., Zhang, C., Liang, S. M., Lv, D. C., Chen, S. L., and Cao, W. S. (2020) Simulation of the composition and cooling rate effects on the solidification path of casting aluminum alloys. *Journal of Phase Equilibria and Diffusion*, 41, 793–803.

Zhao, D., Witte, F., Lu, F., Wang, J., Li, J., and Qin, L. (2017) Current status on clinical applications of magnesium-based orthopaedic implants: a review from clinical translational perspective. *Biomaterials*, 112, 287–302.

# 9 Case Studies on Superalloy Design

## Content

## 9.1 Introduction

Superalloys are metallic alloys that can be used at high temperatures because of their high mechanical strength at elevated temperatures, exceptional resistance to creep deformation and rupture, good fatigue behavior and surface stability, as well as excellent resistance to corrosion or oxidation. Superalloys can improve operating efficiency, reduce fuel emissions to the environment, and most importantly solve the urgent demands for the durability and strength of materials in aeroengines, gas turbines, advanced ultrasupercritical units, and other important components (Darolia, 2019; Reed, 2006).

Based on the primary metal, superalloys can be broadly classified into three categories: Ni-, Fe-, and Co-based superalloys. Ni-based superalloys, the most important superalloys, are most widely applied in parts that experience very high temperatures. Fe-based superalloys are of interest because they are similar to Ni-based superalloys but have good creep resistance and oxidation resistance at considerably lower costs. Co-based superalloys, which are typically applied in low-stress/high-temperature stationary vanes, have higher strength at elevated temperatures than Ni-based alloys, owing to their higher melting points, and also have good resistance to fatigue, oxidation, and corrosion.

For these materials, there is a need to balance and compromise between numerous property requirements and also the competitive requirements for material properties and compositions. Therefore, superalloys are ideal case studies for real-life materials design that not only can serve as a stringent test of alloy design methods but also lead to potential commercially viable alloys (Conduit et al., 2017). Traditionally, superalloys have been designed empirically. However, the traditional trial-and-error alloy design approach for superalloys is quite challenging. Superalloys usually contain more than 10 alloying elements and need to meet many performance requirements. It is very difficult to precisely describe the complex nonlinear relationship among so many elements and performances empirically (Montakhab and Balikci, 2019; Reed et al., 2016). For example, even if only three elements are chosen for the alloy with a composition interval gap of 10% and their concentration is measured with an accuracy of 0.1%, the number of allowable candidate alloys may exceed $10^6$. In addition, changing the concentration of a single element usually affects more than one alloy property. Consequently, it is difficult to determine the individual effects. If a possible change of microstructures is considered, it is obvious that the design of a new superalloy is not a task that can be solved empirically. The risk is quite high of missing the unknown "optimal" alloy by unguided costly testing of a limited number of alloys.

The development of a new superalloy is an expensive and time-consuming task, resulting in a typical complete development cycle that may exceed 10 years. This delay is not only because of the need for extensive mechanical performance testing and validation to ensure the safety of key components, but also because it is difficult to find a material composition and related heat treatment schedules that can correctly balance the mechanical properties and long-term stability. Computational methods can shorten the material development cycle to keep up with the rapid pace of product design and development and to develop new alloys that meet the cost/performance requirements. With the significant advances in understanding the correlation among composition, processing, microstructures, and properties, these methods provide new possibilities to reconcile the intrinsic properties of materials with external requirements.

With the development of integrated computational materials engineering (ICME), there have been numerous efforts to develop computational methods for designing materials and predicting material properties. The advent of the Materials Genome Initiative (MGI) and the emergence of artificial intelligence technology have provided new opportunities for the research and development of new strategies for alloy design. The latest developments in computing and analysis capabilities have enabled machine learning to be applied to materials design and physics-based material models at various length scales. In recent years, with the development of MGI, ICME, and materials informatics, the design of superalloys has become an active area of research in materials science, and significant progress has been made (Hu et al., 2018; Jiang et al., 2018; Long et al., 2018; Markl et al., 2018; Menou et al., 2019; Rettig et al., 2015; Suzuki et al., 2019; Tancret, 2012; Tin et al., 2018).

In this chapter, we take Ni-based single-crystal superalloys and Ni–Fe-based superalloys for advanced ultrasupercritical (A-USC) units as examples to demonstrate how alloy design is accomplished in these alloy systems. The Ni-based single-crystal superalloy was designed using a multistart optimization algorithm. For the Ni–Fe-based superalloy, the artificial neural network (ANN) was employed to design the alloy based on an experimental dataset.

## 9.2  Ni-Based Single-Crystal Superalloys

Ni-based single-crystal (SX) superalloys are widely used as turbine blades of aero-engines and gas turbines owing to the remarkable creep and fatigue resistance at elevated temperatures. Ni-based SX superalloys are "super" because of their two-phase structure. Figure 9.1 shows a typical microstructure of an SX superalloy, i.e., a precipitate-hardened microstructure comprising the $L1_2$ phase ($\gamma'$) with a cuboidal morphology distributed in the solid-solution matrix with the Fcc crystal structure ($\gamma$-Ni). The $\gamma'$ phase is largely responsible for the good high-temperature strength and the incredible creep resistance because $\gamma'$ precipitates are coherent with the matrix $\gamma$ phase, and their strength increases with temperature. Ni-based SX superalloys are based on the Ni–Cr–Al ternary system but have many other alloy additions, including

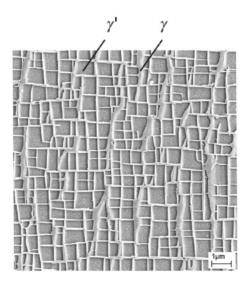

**Figure 9.1** Microstructure of a typical single-crystal superalloy, showing presence of cuboidal $\gamma'$ precipitates within a matrix of $\gamma$. Reprinted from Darolia (2019) by permission from Taylor & Francis.

**Table 9.1** Representative properties for consideration in designing SX superalloys.

| Category | Properties |
| --- | --- |
| Physical properties | Density, thermal expansion, thermal conductivity, modulus |
| Mechanical properties | Yield strength, ultimate tensile strength, ductility, creep life, fatigue life (high cycle, low cycle, hold time), crack growth rate, long-term thermal stability |
| Environmental resistance | Oxidation, corrosion, coating life |
| Processing, manufacturing | Casting, heat treatment, welding, brazing, machining, repair |

Adapted from Suzuki et al. (2019) by permission from Springer Nature.

light elements and heavy refractory elements. Typical Ni-based SX superalloys contain more than 10 elements, including Al, Ti, Co, Cr, Mo, W, Re, Ta, Hf, C, and B.

The performance of a material is determined by both its inherent properties (such as composition and element interactions) and its processing and service conditions. To design a new Ni-based SX superalloy, there are a variety of performance requirements, such as physical, mechanical, and environmental properties, as well as processing ability, manufacturability, and cost requirements. Table 9.1 shows representative properties for the design of SX superalloys. The mechanical performance of SX superalloys is significant for the successful operation of turbine blades. Among these properties, high-temperature creep resistance, alloy density, microstructural stability, and cyclic oxidation are the most critical properties to be investigated. Depending on the combination and balance of the major alloying additions in Ni-based SX superalloys, the microstructural parameters, such as the volume factions of $\gamma$ and $\gamma'$ phases and the lattice misfit, can change remarkably, resulting in dramatic variations in creep strength. Therefore, one of the most important strategies in alloy design is to determine the dependence of mechanical properties and environmental performance on compositions.

Figure 9.2 schematically illustrates a typical alloy design procedure for superalloys proposed by Reed et al. (2009). The "alloy design system" describes a feasible method and a path from the design constraints to the selection of alloy composition. The alloy design procedure is as follows: (1) define a list of elements and the upper and lower limits of the composition, (2) calculate the volume fraction and composition of the strengthening phase at equilibrium, (3) define the microstructural architecture, (4) determine the merit indices, (5) apply the design constraints, and (6) optimize the compositions. In practice, the alloy design system typically consists of both the quick evaluation of new alloys and the efficient exploration in a search space containing billions of unknown alloys.

In the present case, we focus only on the composition design and not on the additional effort to design/optimize the processing conditions for heat treatment. Significant progress has been made on the heat treatment optimization for the important Ni-based superalloy 718, for example on the simulation of co-precipitation kinetics of $\gamma'$ and $\gamma''$ (Zhang et al., 2018) and on simulation of phase transformation, plastic deformation, and microstructural evolution during precipitation (Zhou et al., 2014).

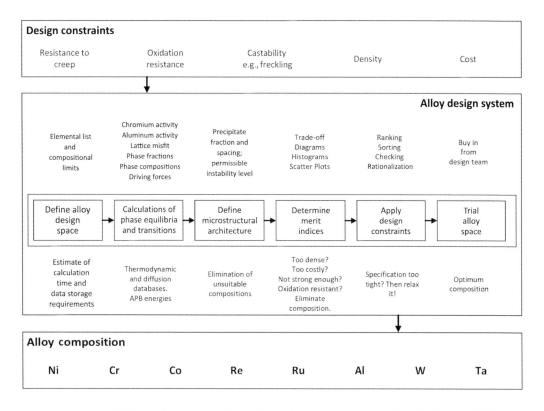

**Figure 9.2** Typical procedures of an alloy design system. Adapted from Reed et al. (2016) by permission from Elsevier.

Alloy design for Ni-based SX superalloys should consider numerous factors that are interlinked in a complex manner, including the size, distribution, and volume fraction of $\gamma'$; the lattice misfit between $\gamma'$ and $\gamma$; and the temperature, stress, and environmental conditions. Thus, in alloy design, many properties (multiple criteria) must be considered simultaneously. However, multicriteria optimization is not an easy task because these multiple criteria are often in conflict with one another. To optimize multiple criteria, a variety of computational methods have been proposed. Among them, data mining tools, such as artificial neural networks and Gaussian processes, have been successfully employed to search for the intricate linkages among composition, microstructure, and properties.

Considerable investigation effort on designing SX superalloys has indicated potential composition spaces for optimization. Harada and Murakami (1999) developed an alloy design program for Ni-based superalloys by establishing a mathematical model using multiregression methods based on a composition–microstructure–property database. This program has been successfully applied to design several Ni-based superalloys. Using an ANN model, Fink et al. (2010) developed two new Ni-based superalloys (René N500 and René N515) that provide current second-generation properties (creep resistance close to René N5) but with a lower Re concentration.

With the aid of the CALPHAD approach, the group of Reed et al. proposed an alloys-by-design method, which allows the design of compositions for SX superalloys by numerically estimating the creep resistance, microstructural stability, density, cost, and castability. Using these procedures, they also successfully designed three new alloys: ABD-1/2/3 (Reed et al., 2009, 2016; Zhu et al., 2015). By applying a numerical multicriteria optimization algorithm, Rettig et al. (2015) developed Re-free Ni-based SX superalloys. Based on the QuesTek's proprietary ICME design platform, Gong et al. (2017) designed the low-Re alloy QTSX. Applying to thermodynamic calculations and regression analysis, Zhang et al., 2019) developed computer-aided modeling procedures to explore the design space of Re-free Ni-based SX superalloys.

In the following, we take the work of Rettig et al. (2015, 2016) as an example to show how Ni-based SX superalloys are developed by numerical multicriteria optimization techniques. Because of the high cost and limited availability of Re, the market for improved SX superalloys with low or no Re is attractive. Therefore, the target was to design a new Re-free SX superalloy, which has similar performance compared with existing commercial alloys, such as CMSX-7 and René N500. They applied a multistart optimization algorithm to find the optimum alloy compositions considering many property requirements and constraints, and successfully designed a Re-free superalloy with a creep strength that was only slightly below CMSX-4 under high-temperature/low-stress conditions.

## 9.2.1    Model Description

In alloy design, it is necessary to calculate certain properties of the alloy based on its composition. Rettig et al. (2015) employed the CALPHAD method to calculate phase equilibrium and utilized a qualitative numerical model to predict the creep strength.

### 9.2.1.1    Thermodynamic Properties

The phase transition temperatures, equilibrium compositions, phase fractions, driving forces, and so on, can be calculated by the CALPHAD method, which is particularly powerful for multicomponent alloy systems. For Ni-based SX superalloys, the amount of $\gamma'$ depends on the chemical composition and temperature, and related alloy properties can be predicted in terms of composition and temperature. With the CALPHAD method, the dependence of the phase compositions and phase fractions on alloy composition and temperature can be calculated. If the chemical composition and phase fraction can be further linked with mechanical properties, an alloy can be designed to meet specific requirements.

### 9.2.1.2    Density

When optimizing the composition of a superalloy to obtain a high creep resistance, it is necessary to lower the density of the alloy, especially for applications in aerospace. The density of the alloy depends on the lattice parameters of all phases and all elements that form the phase. To predict the density of SX superalloys at room temperature, Caron derived a semi-empirical equation based on multiple linear regression analysis, and the predicted results were in good agreement with experimental results when the formula was tested for lots of first-to-third-generation superalloys (Caron, 2000).

### 9.2.1.3 Misfit

The lattice misfit between two phases is determined by the lattice parameters of the phases, which are governed by the molar volume and the thermal expansion coefficients of both phases. Caron (2000) proposed a semi-empirical equation to calculate the lattice misfit between $\gamma$ and $\gamma'$ phases at room temperature via Vegard's law from chemical composition. The variation of the $\gamma/\gamma'$ misfit with temperature can be obtained from the thermal expansion effect and thermodynamic equilibrium.

### 9.2.1.4 Creep-Rupture Lifetime

Creep is a major deformation mode of superalloys at operating temperature. It is a complex long-term plastic deformation process based on the movements of thermally activated dislocations (Rettig et al., 2015). The creep resistance of SX superalloys strongly depends on the coarsening resistance of the $\gamma'$ phase and the diffusivity of vacancy in the alloy. A reliable model for creep should require information such as the morphology and distribution of precipitates, dislocation density, and lattice misfit. Taking the lattice misfit parameter as an example, when the lattice mismatch is too large, the $\gamma'$ phase will become incoherent with the $\gamma$ phase during heat treatment. On the other hand, when it is too small, only spherical $\gamma'$ phase will precipitate. These two cases are detrimental to the creep resistance. As early as 1987, Nathal (1987) suggested that alloys with a misfit ranging from –0.5% to –0.1% and an initial size of the $\gamma'$ phase between 300 and 700 nm have the highest creep-rupture lifetimes. Rettig et al. (2015) defined a solid-solution strengthening index ($I_{SSS}$) to weight the relative effectiveness of the elements for reducing creep:

$$I_{SSS} = 2.44x^{\gamma}_{Re} + 1.22x^{\gamma}_{W} + x^{\gamma}_{Mo}, \tag{9.1}$$

where $x^{\gamma}_i$ is the equilibrium concentration of element $i$ in the $\gamma$ phase (in at.%), which can be calculated via CALPHAD from the alloy composition. However, in the low-temperature/high-stress regime, the $\gamma'$ phase is frequently cut by dislocations. In this case, other creep deformation mechanisms may occur. Accordingly, if the entire temperature range is considered, more sophisticated models for predicting creep strength are needed.

### 9.2.1.5 Design Criteria

For Ni-based SX superalloys, creep and oxidation resistance are the primary design criteria. Four guidelines for optimizing the chemistry of Ni-based SX superalloys were summarized by Reed (2006): (1) To ensure that the $\gamma'$ fraction is $\sim 70\%$, the contents of $\gamma'$-forming elements (such as Al, Ti, and Ta) should be high. (2) To restrict $\gamma'$ coarsening, the $\gamma/\gamma'$ interfacial energy should be minimized; thus, the composition must be selected to ensure that the $\gamma/\gamma'$ lattice misfit is small. (3) To avoid the precipitation of topologically close-packed (TCP) phases, the addition of strengthening elements (particularly Re, W, Ta, Mo, and Ru) should not be too high. (4) To avoid surface degradation, the composition must be chosen carefully. These guidelines have been widely accepted in the design of Ni-based SX superalloys. According to these guidelines, to design a new Re-free SX superalloy comparable to CMSX-7 and René N500, Rettig et al. (2015) assumed the following requirements as the design criteria: (1) The optimal molar fraction of $\gamma'$ phase at 1,100°C is about 46%; (2) the optimal misfit between $\gamma$ and $\gamma'$ phases at 1,100°C is negative and in the range between within the range of –0.5~–0.1%, and (3) the solid-solution strengthening index $I_{SSS}$ should be enhanced as high as possible.

## 9.2.2　Alloy Design Procedure

In alloy design, it is almost impossible to improve one property without degrading the others. Thus, many properties must be balanced and optimized, including mechanical, physical, and thermodynamic requirements. Figure 9.3 shows an overview of this type of optimization procedure. This multicriterion constrained multistart optimization algorithm was divided into three primary tasks (Rettig et al., 2015): (1) choosing suitable alloy components to represent the entire alloy design space; (2) calculating the properties of the sample alloys and subsequently generating a surrogate model from these datasets; and (3) performing multicriteria optimizations. These three tasks are iteratively repeated to increase the quality of the surrogate models in the vicinity of the optimum solution. To refine the surrogate model, the current global optima are also added into the samples after each iteration. The iteration is not stopped until the

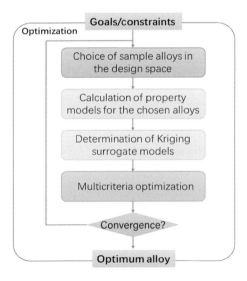

**Figure 9.3** General outline of the multicriteria optimization algorithm. Adapted from Rettig et al. (2016) by permission from John Wiley and Sons.

improvement in the global optimum drops below a threshold of 0.1 at.% compared to the previous cycle.

To formulate a precise surrogate model with fewer property calculations, the sample alloys should be chosen to ensure that the entire multidimensional alloy design space can be well represented. Typically, computer experiments use the Latin Hypercube sampling method to ensure the representativeness of the dataset to the entire design space.

### 9.2.2.1    Surrogate Models

During optimization, multiple property function evaluations are required. Surrogate models, which are essentially based on nonlinear interpolation techniques, can provide fast access to property models and considerably increase the optimization speed. As shown in Figure 9.3, the construction of a surrogate model comprises four steps: the choice of an appropriate number of sample compositions in the design space, the property calculation based on the selected compositions, and the determination of the surrogate model and multicriteria optimization. First, Rettig et al. (2015) evaluated the property models for all 300 initial sample alloys. Subsequently, they refined the generated surrogate model using 200 additional and independent test alloys. Finally, 15 test alloys with the largest errors were added to the sample alloys to further refine the surrogate model. The procedure was iterated until the model contained at least 700 sample alloys. For a design space containing eight alloying elements, it is sufficient to evaluate a few hundred members, but they must be well distributed.

### 9.2.2.2    Optimization Algorithm

To find the best possible solution that fulfils all constraints, Rettig et al. (2015) used a multistart algorithm based on the sequential quadratic programming (SQP) method. By searching different starting points, this deterministic solver combines high performance with convergence to the global optimum and can evaluate up to 2,000 property functions for each call of the SQP optimizer. Initially, the starting points are selected from the entire design space. The Pareto front is defined as a set of nondominated solutions, being chosen as optimal, if no objective can be improved without sacrificing at least one other objective (Manne, 2019). Then, along the Pareto front, seven starting points are selected within $\pm 5$ at.% around the composition of the previous step, and this composition itself is also chosen as a starting point. Such a procedure is intended to avoid having the algorithm getting stuck into the current local optimum (Rettig et al., 2015).

### 9.2.3    Alloy Design and Experimental Validation

To demonstrate the multicriteria alloy design procedure, Rettig et al. (2015) considered the design of a Re-free low-density SX superalloy with a desired creep strength close to that of the commercial SX superalloy CSMX-4. The design criteria for the optimization were chosen to find a new alloy with the highest possible creep strength at 1,100°C but without Re. Additionally, the alloy should exhibit the lowest possible density. Therefore, they assumed an identical solid-solution strengthening index as CMSX-4 owing to the identical creep strength. Based on metallurgical experience, the composition space was chosen to be as wide as possible and with the fewest artificial constraints to not miss potentially good alloys. Minimum Al and Ti concentrations were set to ensure good oxidation and high-temperature corrosion resistance. The Cr concentration was limited to 7.6 at.% to ensure sufficient hot corrosion resistance. The Mo concentration was fixed to the typical value of most commercial SX superalloys. The solidus temperature ($T_S$) must be high enough for rapid solution heat treatments.

Finally, with the multicriteria alloy design approach, Rettig et al. (2015) successfully developed a Re-free SX superalloy: ERBO/13. Further experimental validation indicated that the primary properties of the newly designed alloy were comparable to those of the commercial René N500 and CMSX-7 alloys. Table 9.2 shows the compositions of the designed Re-free SX superalloy ERBO/13 and the reference alloy CMSX-4. Figure 9.4 shows the microstructure of directionally solidified test rods of the newly designed superalloy. Similar to CMSX-4, this new superalloy has typical dendritic structures with an arm spacing about 260 μm, and the $\gamma'$ phase has a distinct cuboidal shape with sharp corners. The physical properties were also investigated experimentally. As summarized in Table 9.3, the predicted alloy properties agree well with the measured results.

A series of creep tests under various conditions for the newly designed alloy were also performed (Rettig et al., 2015). The comparison between the creep strength of the

**Table 9.2** Chemical compositions of the newly designed SX superalloy ERBO/13 and the reference alloy CMSX-4 (in wt.%).

| Alloy | Al | Co | Cr | Mo | Re | Ta | Ti | W |
|---|---|---|---|---|---|---|---|---|
| **ERBO/13** | 4.8 | 8.6 | 5.0 | 1.4 | – | 10.1 | 1.3 | 8.8 |
| **CMSX-4** | 5.6 | 9.0 | 6.5 | 0.6 | 3.0 | 6.5 | 1.0 | 6.0 |

Adaption from Rettig et al. (2015) by permission from IOP Publishing.

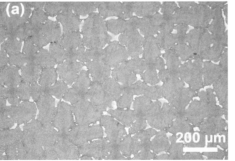 

**Figure 9.4** (a) Optical microscopy images of the dendritic microstructure and (b) SEM images of the $\gamma/\gamma'$-microstructure of the ERBO/13 alloy. Adapted from Rettig et al. (2015) by permission from IOP Publishing.

**Table 9.3** Comparison of the predicted and measured properties of the optimized alloy along with a commercial alloy CMSX-4.

| Property | ERBO/13 calc. | ERBO/13 meas. | CMSX-4 calc. |
|---|---|---|---|
| Density (g cm$^{-3}$) | 8.95 | 9.0 | 8.7 |
| Cost (€/kg) | 103 | — | 207 |
| Liquidus temperature (°C) | 1,368 | 1,371 | 1,381 |
| Solidus temperature (°C) | 1,322 | 1,316[*] | 1,338 |
| $\gamma'$ solvus temperature (°C) | 1,262 | 1,242 | 1,257 |
| W-$\gamma/\gamma'$ part. coeff. (1,100°C) (at.%/at.%) | 2.63 | 2.4 ± 0.4 | 1.92 |
| Mo-$\gamma/\gamma'$ part. coeff. (1,100°C) (at.%/at.%) | 5.88 | 3.6 ± 1.2 | 4.76 |
| W-matrix content (1,100°C) (at.%) | 4.20 | 4.4 ± 0.3 | 2.52 |
| Mo-matrix content (1,100°C) (at.%) | 1.45 | 1.7 ± 0.3 | 0.59 |
| SSS-index (1,100°C) (at.%) | 6.6 | 6.1 ± 0.4 | 7.7 |
| Misfit (1,100°C) (%) | −0.2 | −0.02[**] | −0.2 |
| $\gamma'$ fraction (1,100°C) (mol%) | 46.0 | 56[***] | 46 |

Notes: *As-cast state, **at room temperature, ***determined with the lever rule using the measured W concentrations in $\gamma$ matrix and $\gamma'$ phase as well as the nominal concentration of W. Adapted from Rettig et al. (2015) by permission from IOP Publishing.

newly designed alloy and that of the reference alloy CMSX-4 in the Larson–Miller plot shows that the newly designed alloy has the same creep strength as that of CMSX-4 if the service temperature is chosen only 14 K lower. The service temperature of CMSX-4 is approximately 15–20 K below that of another commercial SX superalloy, CMSX-7. Therefore, the creep resistance of ERBO/13 is comparable to that of the best commercial Re-free or low-Re superalloys available currently, such as CMSX-7 and René N500, which is quite remarkable because the alloy development was based solely on computations without any prior assumptions about the compositions. However, in the low-temperature/high-stress regime, the creep strength of ERBO/13 is notably lower than that of CMSX-4. This is because the solid-solution strengthening mechanism favored in this alloy design only dominates at high temperatures.

In summary, using a numerical multicriteria global optimization algorithm, Rettig et al. (2015) successfully proposed a method to optimize alloy compositions to meet different requirements and constraints for Ni-based SX superalloys. The capability of the developed multicriteria optimization method was also experimentally validated by the newly designed Re-free alloy ERBO/13.

## 9.3    Ni–Fe-Based Superalloys for Advanced Ultrasupercritical Units

A-USC power plants can remarkably increase the efficiency of power plants and thus reduce pollutant emissions if steam parameters are pushed to the limits of temperature and pressure. Developing stronger high-temperature materials is always the primary task enabling the improved steam parameters. Future A-USC power plants are intended to operate at temperatures between 700°C and 760 with steam pressures of up to 35–37.5 MPa, which requires that the creep-rupture strength corresponding to the $10^5$ hour endurance life is not less than 100 MPa, which is a challenge at 750°C (Zhong et al., 2013). Conventional materials, such as ferritic steels and austenitic steels, do not meet these requirements because their maximum operating temperature is limited to 650°C. At present, only a few Ni-based or Ni–Co-based superalloys can meet these creep-rupture property requirements. However, these alloys are difficult and/or expensive to manufacture. Ni–Fe-based superalloys, which have a high creep-rupture strength, excellent corrosion resistance, good workability, and low cost, have become a potential candidate for A-USC units at 700–760°C. Some examples include the GH2984 (Ni–33Fe–19Cr–2.2Mo–1Ti–1Nb–0.4Al) (Wang et al., 2013), HR6W (Ni–24Fe–23Cr–7W–Ti–Nb) (Tokairin et al., 2013), and HT700 (Ni–30Fe–17.5Cr–0.5Mo–1.8Ti–1.2Nb–1.6Al–0.03C–0.005B) alloys (Zhong et al., 2013).

To accelerate the development and qualification of new alloys for the next generation of A-USC steam turbine systems, it will be necessary to combine materials informatics and physics-based models to design new Ni–Fe-based superalloys. In the following, we will take the work of Hu et al. as an example to show how the Ni–Fe-based superalloy was designed using the machine learning method (Hu et al., 2018).

## 9.3.1     Model Description

To realize the global optimization and two-way design of alloys, Hu et al. employed an ANN model combined with a genetic algorithm (GA) during the alloy design process. For alloy design using machine learning approaches, first a related dataset should be established. Hu et al. (2018) built a dataset in which the data are taken from the literature and measurements on a few key alloys. In this dataset, there are 580 instances and five attributes related to microstructure and mechanical properties, in which 150 instances were employed as testing data. As shown in Table 9.4, the five attributes are phase fractions of $\gamma$ and $\gamma'$, lattice mismatch of $\gamma/\gamma'$, yield strength, and creep-rupture lifetime at operating temperatures. The alloy composition includes the matrix element (Ni), precipitation strengthening elements (Al and Ti), solid-solution strengthening elements (Co, W, Nb, and Mo), corrosion resistance elements (Cr and Si), structural stability strengthening elements (B and C), and a cost-reducing element (Fe). The microstructural information that has significant effects on the performance at

**Table 9.4** List of the first five instances with five features used.

| $\gamma$ content (vol%) | $\gamma'$ content (vol%) | Mismatch of $\gamma/\gamma'$ | Yield strength (MPa) | LOG (creep-rupture life) (h) |
|---|---|---|---|---|
| 52.93 | 29.24 | 0.0123 | 952.74 | 5.16 |
| 60.56 | 25.60 | 0.0343 | 909.43 | 5.06 |
| 56.69 | 26.07 | 0.0474 | 860.89 | 5.13 |
| 53.06 | 29.19 | 0.0642 | 876.76 | 5.15 |
| 50.53 | 30.39 | 0.0307 | 975.52 | 5.16 |

Reprinted from Hu et al. (2018) by permission from Elsevier.

operating temperatures, such as the average size of grain and the $\gamma'$ phase, was also included in the dataset. Ultimately, the input to the network consisted of the composition variables of the 12 elements and the two microstructural parameters, phase fractions of $\gamma$ and $\gamma'$. The five attributes of the first five samples in the dataset are shown in Table 9.4, while these target properties were considered the output.

During training of the back propagation (BP) neural network, the number of nodes in a hidden layer was adaptively adjusted according to the training error until the actual error was lower than a predetermined one. The optimization algorithm GA simulates biological evolution, which optimizes the target as an individual chromosome and determines its survival probability using the fitness of the problem as the criterion. After each iteration in the search space, the variable generated in the previous step is updated by the operation, including duplicate, cross, and mutation.

The combination of these two algorithms is the key to establish an accurate prediction model. The prediction results of the BP network were combined with the GA by the fitness function:

$$f(A, B, C) = \frac{1}{1 + 2e^{|A_E - A|}} + \frac{1}{1 + 2e^{|B_E - B|}} + \frac{1}{1 + 2e^{|C_E - C|}}, \tag{9.2}$$

where $A$, $B$, and $C$ are the predicted properties, and $A_E$, $B_E$, and $C_E$ are the expected ones. When the predictions are equal to the expected properties, the function reaches a maximum value of 1, and the corresponding alloy composition is considered optimal. The fitness optimization function was used to transform the optimization problem of the alloy into the extreme value problem of the function. A flowchart showing the combination of these two algorithms is given in Figure 9.5.

The parameters in the BP network model, such as the hidden-layer number and the hidden-node number, were determined according to the training error. Repetitive network training was performed using the control variable method, and the corresponding parameters were selected when the error reached the minimum value. After debugging, it was determined that the hidden-layer number was 1, the hidden-node number 12, the learning rate 0.06, the training frequency 100, and the remaining parameters were algorithm default values. The parameters for the GA were determined based on the principle of minimum error, and the control variable method was run and debugged many times.

After training the network and achieving an acceptable accuracy, the next step is to verify the feasibility of the model calculations. Six unseen alloys were selected in the dataset by Hu et al. (2018) to complete the verifications. Figure 9.6 shows the comparison of results from the literature and ANN prediction of five properties for the six targeted alloys. This shows that the predicted results are quite consistent with the results from the literature.

## 9.3.2 Alloy Design Procedure

To design a Ni–Fe-based superalloy for A-USC units operated at $700 \sim 750°C$, the reference alloy In740H was chosen, also included in Figure 9.6. In740H has a yield

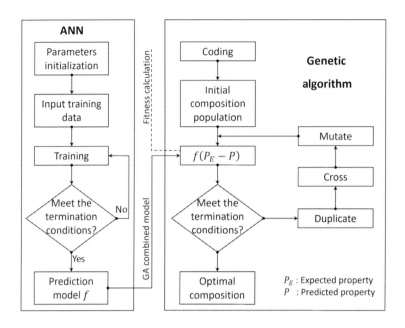

**Figure 9.5** Combination of BP artificial neural network and genetic algorithm, the back propagation (BP)–genetic algorithm (GA) model.

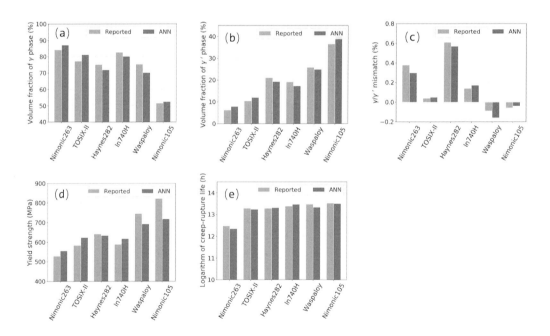

**Figure 9.6** Validation of the model ANN prediction. The results from two different methods were compared in six different alloys. The data of five features, defined in Table 9.4, for the alloys are from the literature and ANN prediction. Adapted from Hu et al. (2018) by permission from Elsevier.

strength of approximately 580 MPa at 750°C and a creep-rupture life of approximately 10,000∼15,000 hours at 750°C/150 MPa. Therefore, the expected properties of the designed alloy are a yield strength of 580 MPa and a 10,000-hour creep-rupture life at 750°C/150 MPa. During alloy design, it is noted that some prior knowledge was embedded, i.e., the Co content was constrained to be as little as possible to reduce the cost, while the Fe content was increased to improve the processing performance and reduce the cost. The average sizes of the $\gamma$ grains and $\gamma'$ phase were controlled to be 70 μm and 50 nm by heat treatments, respectively.

Each optimization can only provide the best composition of a group of alloys. However, only two target properties, i.e., yield strength and creep-rupture life, are considered in the alloy design. The optimized compositions cannot fully guarantee that the other properties can also meet the requirements in subsequent experiments. Therefore, more calculations are needed to optimize the compositions through the distribution of elements in the interval. As a result, the alloy compositions can be determined from the final optimal alloy composition range. The alloy design procedure is shown in Figure 9.7. According to the statistical results, the interval with a concentrated element distribution was chosen as the optimal interval. If the interval is small enough, the iteration is stopped, and then a set of alloy compositions for experimental validation can be selected. Otherwise, the next iteration will be performed.

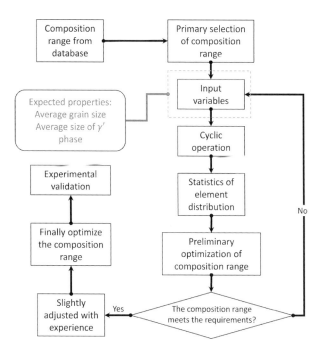

**Figure 9.7** Design process of alloy based on the BP-GA model.

### 9.3.3    Alloy Design and Experimental Validation

As shown in Table 9.5, based on the composition range of the dataset and the composition features of In740H, a preliminary alloy composition range was selected first. Then, within this range, the trained model was used to predict the distribution of elements. For example, Al is mainly distributed within 1–2.5 wt.%, while Ti is slightly more concentrated in 1–2.4 wt.%. Thus, the preliminary optimized composition intervals of Al and Ti are 1–2.5 wt.% and 1–2.4 wt.%, respectively, as shown in Table 9.5. However, this composition range obtained from the initial optimization was still too large to determine a suitable alloy for experimental verification. Thus, this procedure was repeated to reduce the composition interval further. After the second round of optimization, a smaller composition range was obtained. When the composition range was sufficiently small, the finally optimized composition could be determined, as shown in Table 9.5.

To validate the alloy design approach, a set of alloy compositions based on the final optimization range was selected, as shown in Table 9.6. An alloy with this composition was prepared for experimental validation. To be consistent with the input parameters of the model, the samples were carefully prepared to ensure that the average sizes of the $\gamma$ grains and $\gamma'$ phase were approximately 70 μm and 50 nm, respectively. The microstructure of the designed Ni–Fe-based superalloy is shown in Figure 9.8, revealing that the microstructure of the new alloy is very close to that of In740H. Some carbides precipitated discontinuously in the grain as well as at twin boundaries and grain boundaries, and fine spherical $\gamma$ particles precipitated within the interior of the matrix grains.

Tensile tests and uniaxial tensile creep tests were also conducted at various temperatures and stress conditions, and the results are shown in Figure 9.9. The yield strength at 750°C is 597.8 MPa, which is slightly higher than the target value (580 MPa). Figure 9.9b further suggests that, even after a 3,691-hour creep test at 750°C and 150 MPa, no creep rupture occurred. Further calculations by the Larson–Miller method

**Table 9.5** Alloy composition range for three optimization levels, in wt.%.

|     | Ni   | Al      | Co  | Cr     | Fe     | Mo    | Nb    | Ti      | W     | Si    | B       | C      |
|-----|------|---------|-----|--------|--------|-------|-------|---------|-------|-------|---------|--------|
| I   | Bal. | <5      | <5  | 15~28  | 15~30  | <4    | <4    | <5      | <4    | <0.5  | <0.5    | <0.8   |
| II  | Bal. | 1~2.5   | 1~4 | 20~26  | 20~27  | <1.5  | <2.2  | 1~2.4   | <1.5  | <0.5  | <0.1    | <0.1   |
| III | Bal. | 1.6~2.2 | <2  | 20~25  | 23~27  | ≤0.5  | <0.5  | 1.8~2.4 | <0.5  | <0.3  | ≤0.003  | ≤0.05  |

Notes: I: Preliminary selection; II: preliminary optimization; III: final optimization.
Reprinted from Hu et al. (2018) by permission from Elsevier.

**Table 9.6** Experimental alloy composition, in wt.%.

| Ni   | Al + Ti | Co | Cr | Fe | Mo + W + Nb | Si + B + C |
|------|---------|----|----|----|-------------|------------|
| Bal. | 4~4.2   | 2  | 20 | 26 | 0.5         | <0.1       |

Reprinted from Hu et al. (2018) by permission from Elsevier.

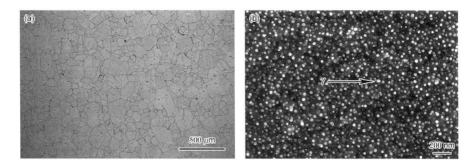

**Figure 9.8** Microstructures of the designed Ni–Fe-based superalloy: (a) grain; (b) $\gamma'$ phase. Reprinted from Hu et al. (2018) by permission from Elsevier.

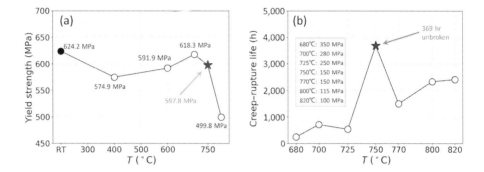

**Figure 9.9** Experimental results of alloy properties, data at 750°C, are highlighted by the asterisks: (a)Yield strength at room temperature is marked by solid symbol; (b) creep-rupture life. Adapted from Hu et al. (2018) by permission from Elsevier.

show that the rupture life under this condition (750°C/150 MPa) can be 5,800 hours, which is in line with the service requirements. Therefore, the newly designed alloy can meet the microstructure and property requirements and thus can be a new candidate material for A-USC units.

In summary, to design a Ni–Fe-based superalloy for A-USC units at 750°C, first the effects of alloying elements, grain size, phase composition, and precipitate size on the properties of the alloy were analyzed based on a collection of a large amount of experimental data. Then a database was established containing the composition, microstructure, and properties of Ni–Fe-based superalloys with a reasonable composition of variables, logical structure, and unified standard. After that, the BP-ANN model combined with GA was performed on this database to develop the corresponding alloy design program to predict the volume fractions of $\gamma$ and $\gamma'$, $\gamma/\gamma'$ mismatch, creep-rupture life, and yield strength at 750°C of the Ni–Fe-based superalloys used in 750°C A-USC units. Finally, a new Ni–Fe-based superalloy for A-USC units was successfully designed after experimental validation.

This case study shows that the ANN model is a powerful alloy design method that can accurately predict alloy properties and help to find a good composition window and optimized microstructure features of the target alloy, thereby shortening the material development cycle significantly.

# References

Caron, P. (2000) High γ′ solvus new generation nickel-based superalloys for single crystal turbine blade applications, in Pollock, T. M., et al. (eds), *Superalloys 2000*. Pittsburgh: TMS (The Minerals, Metals and Materials Society), 737–746.

Conduit, B. D., Jones, N. G., Stone, H. J., and Conduit, G. J. (2017) Design of a nickel-base superalloy using a neural network. *Materials and Design*, 131, 358–365.

Darolia, R. (2019) Development of strong, oxidation and corrosion resistant nickel-based superalloys: critical review of challenges, progress and prospects. *International Materials Reviews*, 64(6), 355–380.

Fink, P. J., Miller, J. L., and Konitzer, D. G. (2010) Rhenium reduction – alloy design using an economically strategic element. *JOM*, 62(1), 55–57.

Gong, J., Snyder, D., Kozmel, T., et al. (2017) ICME design of a castable, creep-resistant, single-crystal turbine alloy. *JOM*, 69(5), 880–885.

Harada, H., and Murakami, H. (1999) Design of Ni-base superalloys, in Saito, T. (ed), *Computational Materials Design*. Berlin, Heidelberg: Springer, 39–70.

Hu, X. B., Wang, J. C., Wang, Y. Y., et al. (2018) Two-way design of alloys for advanced ultra supercritical plants based on machine learning. *Computational Materials Science*, 155, 331–339.

Jiang, X., Yin, H. Q., Zhang, C., et al. (2018) A materials informatics approach to Ni-based single crystal superalloys lattice misfit prediction. *Computational Materials Science*, 143, 295–300.

Long, H. B., Mao, S. C., Liu, Y. N., Zhang, Z., and Han, X. D. (2018) Microstructural and compositional design of Ni-based single crystalline superalloys – a review. *Journal of Alloys and Compounds*, 743, 203–220.

Manne, J. R. (2019) Swarm intelligence for multi-objective optimization in engineering design, in Mehdi Khosrow-Pour, D. B. A. (ed), *Advanced Methodologies and Technologies in Artificial Intelligence, Computer Simulation, and Human-Computer Interaction*, fourth edition. Hershey: IGI Global, 180–194.

Markl, M., Müller, A., Ritter, N., et al. (2018) Development of single-crystal Ni-base superalloys based on multi-criteria numerical optimization and efficient use of refractory elements. *Metallurgical and Materials Transactions A*, 49(9), 4134–4145.

Menou, E., Rame, J., Desgranges, C., Ramstein, G., and Tancret, F. (2019) Computational design of a single crystal nickel-based superalloy with improved specific creep endurance at high temperature. *Computational Materials Science*, 170, 109194.

Montakhab, M., and Balikci, E. (2019) Integrated computational alloy design of nickel-base superalloys. *Metallurgical and Materials Transactions A*, 50(7), 3330–3342.

Nathal, M. V. (1987) Effect of initial gamma prime size on the elevated temperature creep properties of single crystal nickel base superalloys. *Metallurgical Transactions A*, 18(11), 1961–1970.

Reed, R. C. (2006) *The Superalloys: Fundamentals and Applications*. New York: Cambridge University Press.

Reed, R. C., Tao, T., and Warnken, N. (2009) Alloys-by-design: application to nickel-based single crystal superalloys. *Acta Materialia*, 57(19), 5898–5913.

Reed, R. C., Zhu, Z., Sato, A., and Crudden, D. J. (2016) Isolation and testing of new single crystal superalloys using alloys-by-design method. *Materials Science and Engineering A*, 667, 261278.

Rettig, R., Matuszewski, K., Müller, A., Helmer, H. E., Ritter, N. C., and Singer, R. F. (2016) Development of a low-density rhenium free single crystal nickel-based superalloy by application of numerical multi-criteria optimization using thermodynamic calculations, in Hardy, M., Huron, E., Glatzel, U., et al. (eds), *Superalloys 2016: Proceedings of the 13th International Symposium on Superalloys*. Hoboken: John Wiley and Sons, 35–44.

Rettig, R., Ritter, N. C., Helmer, H. E., Neumeier, S., and Singer, R. F. (2015) Single-crystal nickel-based superalloys developed by numerical multi-criteria optimization techniques: design based on thermodynamic calculations and experimental validation. *Modelling and Simulation in Materials Science and Engineering*, 23(3), 035004.

Suzuki, A., Shen, C., and Kumar, N. C. (2019) Application of computational tools in alloy design. *MRS Bulletin*, 44(4), 247–251.

Tancret, F. (2012) Computational thermodynamics and genetic algorithms to design affordable $\gamma'$-strengthened nickel–iron based superalloys. *Modelling and Simulation in Materials Science and Engineering*, 20(4), 045012.

Tin, S., Detrois, M., Rotella, J., and Sangid, M. D. (2018) Application of ICME to engineer fatigue-resistant Ni-base superalloys microstructures. *JOM*, 70(11), 2485–2492.

Tokairin, T., Dahl, K. V., Danielsen, H. K., Grumsen, F. B., Sato, T., and Hald, J. (2013) Investigation on long-term creep rupture properties and microstructure stability of Fe–Ni based alloy Ni–23Cr–7W at 700°C. *Materials Science and Engineering A*, 565, 285–291.

Wang, T. T., Wang, C. S., Guo, J. T., and Zhou, L. Z. (2013) Stability of microstructure and mechanical properties of GH984G alloy during long-term thermal exposure. *Materials Science Forum*, 747–748, 647–653.

Zhang, F., Cao, W. S., Zhang, C., Chen, S. L., Zhu, J., and Lv, D. (2018) Simulation of co-precipitation kinetics of $\gamma'$ and $\gamma''$ in superalloy 718, in Ott, E., et al. (eds), *Proceedings of the 9th International Symposium on Superalloy 718 and Derivatives: Energy, Aerospace, and Industrial Applications*. The Minerals, Metals and Materials Series. Cham: Springer, 147–161.

Zhang, L. F., Huang, Z. W., Pan, Y. M., and Jiang, L. (2019) Design of Re-free nickel-base single crystal superalloys using modelling and experimental validations. *Modelling and Simulation in Materials Science Engineering*, 27(6), 065002.

Zhong, Z. H., Gu, Y. F., Yuan, Y., and Shi, Z. (2013) A new wrought Ni–Fe-base superalloy for advanced ultra-supercritical power plant applications beyond 700°C. *Materials Letters*, 109, 38–41.

Zhou, N., Lv, D. C., Zhang, H. L., et al. (2014) Computer simulation of phase transformation and plastic deformation in IN718 superalloy: microstructural evolution during precipitation. *Acta Materialia*, 65, 270–286.

Zhu, Z., Höglund, L., Larsson, H., and Reed, R. C. (2015) Isolation of optimal compositions of single crystal superalloys by mapping of a material's genome. *Acta Materialia*, 90, 330–343.

# 10 Case Studies on Cemented Carbide Design

## Contents

## 10.1 Brief Introduction to Cemented Carbides

Cemented carbides (also called hard metals) are usually composed of soft and ductile Co-based binder phase and one or more refractory metal carbides with a high hardness (such as WC, TiC, and NbC) as the hard phase. Cemented carbides are usually prepared by the powder metallurgy technique, which includes powder batching, ball milling, spray drying, sieving, blending, pressing, dewaxing, and sintering, as presented in Figure 10.1. The refractory metal carbide in the cemented carbides imparts high hardness (i.e., excellent wear resistance) to the material, while the binder phase provides the necessary toughness for the deformation of a material. In view of microstructure and chemistry, cemented carbides are systematically categorized into four groups (García et al., 2019): *group 1* – WC morphology and chemistry (WC-Co, fine-grained WC, bimodal WC, doped hex-WC, and platelet WC); *group 2* – cubic carbide or carbonitride containing cemented carbide or cermets (WC–Co–$\gamma$, WC–Co–$\eta$, particle reinforcement, NbC-based, and cermets) in which $\gamma$ is a NaCl-type cubic carbide or carbonitride and $\eta$ the M$_6$C; *group 3* – functionally graded cemented carbides ($\gamma$-phase free, $\gamma$-enriched, dual properties, double-layer structure, and

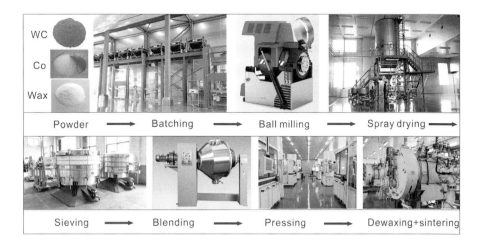

**Figure 10.1** Process illustrations for preparation of cemented carbide. A black and white version of this figure will appear in some formats. For the colour version, please refer to the plate section.

Co-migration); and *group 4* – binder design of cemented carbides (Co–Fe–Ni, precipitate reinforcement, high-entropy alloys, Ni–Al/Fe–Al–B/Ru, and binder-free). As indicated, each group contains five categories according to the categorization by García et al. (2019). In particular, group 2 includes cermets, which are Ti(C,N)-based refractory materials embedded in (Co,Ni)-based binder phase and usually contain other carbides, such as $Mo_2C$, WC, and TaC. The hard cubic phase, i.e., (Ti,M)(C,N) solid solution (M = W/Mo/Ta/Nb), normally exhibits a core-rim structure. The core is formed during solid-state sintering, while the rim is formed during liquid phase sintering by means of the dissolution-precipitation mechanism with the same crystal structure for the rim as that of the core. Cermets can replace some WC–Co cemented carbides due to their excellent wear resistance, high-temperature hardness, low-friction coefficient to steels, and superior thermal deformation resistance. For the microstructures associated with the four groups of cemented carbides, the reader may refer to García et al. (2019).

The properties of cemented carbides are closely dependent on the microstructure resulting from different constituents of the structure mainly including binder and wear-resistant carbides. Due to various appropriate combinations of hardness and toughness, cemented carbides find wide applications in industry, i.e., as cutting tools for machining metal components in automotive and aerospace industry, as wear parts in wire drawing dies, and as components of drill bits in mining and so on. Although the first patent (German patent DRP 420689) on cemented carbides was issued to the German company Osram Studiengesellschaft as early as 1923, cemented carbides are still very important materials and enjoy the reputation of "industrial teeth."

Cemented carbide is a multicomponent composite material usually with several phases. Its production process involves complex thermodynamic and kinetic phenomena. Alloy compositions and several processing parameters, including sintering temperature, time, and atmosphere, affect the microstructure and properties of the cemented carbides in a complex way. The traditional trial-and-error method or the empirical development of cemented carbide consumes a lot of manpower and material resources, but also does not accurately describe the relationship among composition/process, structure, properties, and performance. On the contrary, the advantages of calculation and simulation methods in the design of cemented carbide materials have become more apparent.

In this chapter, we will demonstrate the computational designs of three types of cemented carbides, ultrafine cemented carbides, cemented carbides with composite binder phases of Co and $\gamma'$-Ni$_3$Al, and gradient cemented carbides. For the ultrafine cemented carbides, we will show how to avoid or minimize the segregation of the (Ta,W)C cubic phase and obtain a good combination of Rockwell hardness (HRA), transverse rupture strength (TRS), and fracture toughness by means of thermodynamic calculations, Weibull distribution calculation, and calculation-guided decisive experiments. In the case of cemented carbides with a composite binder, we will demonstrate that thermodynamic calculations in conjunction with interfacial energy calculation can find an optimal composition of the composite binder, which will lead to the desired phase assemblage. Subsequently, the phase-field simulation can capture the major structural characteristics of microstructural evolution for composite binder phases of Co and $\gamma'$-Ni$_3$Al during heat treatment. Based on the knowledge of microstructure evolution, it is possible to design optimal processing parameters efficiently instead of time-consuming and costly experiments. Key experiments based on thermodynamic calculations and phase-field simulations can then be performed efficiently to manipulate the microstructure with the desired mechanical properties. For the case study of gradient cemented carbides, we will demonstrate that diffusion modeling using reliable thermodynamic and atomic mobility databases can accurately reproduce the experimentally observed concentration profiles and volume fractions of the phases under the given alloy composition, sintering temperature, time, and atmosphere. Based on these diffusion simulations, key experiments can be designed efficiently in order to obtain the required gradient structure. Subsequently, an established microstructure-based hardness model can be used to predict the hardness profile from the surface to the core of the gradient cemented carbides with key inputs from thermodynamic calculation and diffusion simulation.

Ancillary material for this chapter is available online at Cambridge University Press. The related data files with thermodynamic and atomic mobility parameters are used in diffusion simulations of Subsection 10.4.

## 10.2    Ultrafine Cemented Carbides

The hardness, toughness, and wear resistance of WC–Co cemented carbides can be improved by controlling the grain size of WC to the submicron or nanometer level. Due to their excellent comprehensive performance, the ultrafine cemented carbides with grain size below 0.5 μm have been widely used in microelectronics industry, precision mold processing industry, wood processing, medicine, etc. The key steps for the preparation of ultrafine cemented carbides are powder preparation, selection of suitable grain growth inhibitors, and sintering processes. The grain size of WC first depends on the size of the starting powders. The powder used to make the ultrafine cemented carbide is much finer than the conventional powder. This powder has a very high activity, and the grain grows abnormally during the sintering process. Therefore, it is necessary to add a suitable grain growth inhibitor to control the grain growth of WC. The type and amount of the added inhibitor as well as the sintering temperature have an important impact on the microstructure, including the grain size of WC and mechanical properties of the ultrafine cemented carbide. The increase of the additive inhibitor can effectively inhibit the growth of WC grains, but it increases the porosity and thus decreases the densification of the alloy. When the amount of the added inhibitor is small, it completely dissolves into the binder phase. When the added amount is higher than the saturated solubility of the binder phase, the dissolved inhibitor mainly precipitates in the form of a brittle honeycomb cubic phase during cooling, forming an unevenly distributed microstructure, which reduces the mechanical properties of the alloy. On the other hand, the amount of liquid in the sintering process increases with sintering temperature, accelerating the densification of the alloy and reducing the porosity of the alloy. However, the higher the sintering temperature, the more obvious the grain growth. Consequently, the reasonable selection for the amount of added inhibitor and the sintering temperature is required to inhibit grain growth of WC effectively. In this subsection, we will describe an effective method to control the segregation of the cubic (Ta,W)C phase based on the thermodynamic calculations and calculation-designed experiment (Li et al., 2015, 2016).

## 10.2.1 Segregation of the (Ta,W)C Cubic Phase in Ultrafine Cemented Carbides

It is well established that doping transition metal carbide is one of a few techniques for controlling grain growth of WC. The inhibition mechanism is to wrap the WC grains through a continuous and coherent segregation layer that acts as a diffusion barrier to prevent the grain growth of WC. Among a variety of grain growth inhibitors, the addition of TaC to cemented carbides can improve their plastic deformation resistance and high-temperature properties. The cubic TaC phase in the Ta–C binary system shows a wide carbon homogeneity range at high temperatures. In the presence of hexagonal WC, the TaC phase dissolves W atoms and is thus denoted as the (Ta,W)C cubic phase, while WC remains stoichiometric in equilibrium. Figure 10.2a shows the calculated phase diagram by using CSUTDCC1 thermodynamic database for cemented carbides (Peng et al., 2014), which shows the detail for the sintering region of the alloy WC–10Co–0.5Ta (wt.%). As shown in the diagram, the carbon content in the cemented carbide should be carefully selected in order to avoid the formation of the two harmful phases [graphite (C) and $M_6C$ ($\eta$)] among the phases. In the preferable phase region (WC + $\gamma$ + $\delta$), the composition range for C content is very narrow (less than 0.2% by weight). In consequence, it is extremely difficult to prepare the targeted alloys in this favorable phase region experimentally without the guidance from thermodynamic calculations. Figure 10.2b shows the calculated atomic phase fraction of the (Ta,W)C cubic phase, indicating that the fraction of undissolved

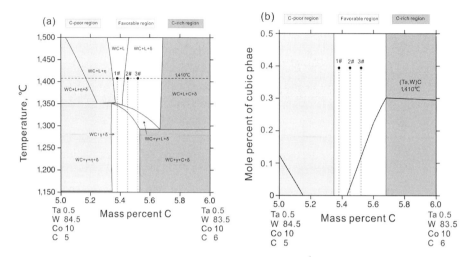

**Figure 10.2** Thermodynamically calculated (a) phase diagram and (b) mole % of the (Ta,W)C cubic phase in WC–10Co–0.5Ta cemented carbide at 1,410°C. Overall composition range covers 0.5Ta–84.5W–10Co–5C to 0.5Ta–83.5W–10Co–6C (wt.%). Samples 1–3 are prepared according to the computed phase diagram and sintered at 1,410 °C for 1 hour under a vacuum. L = liquid, $\eta$ = M$_6$C, $\gamma$ = Co binder, $\delta$ = (Ta, W)C cubic phase, C = graphite phase. WC is the hexagonal stoichiometric binary compound. The figures are reproduced from Li et al. (2015) with permission from Elsevier.

(Ta,W)C increases with C content at 1,410°C. Figure 10.2 shows that the sintering temperature and C content are important to obtain the desired phases and their amounts, which may result in different microstructures and performance of the cemented carbide.

Based on the calculated phase diagram, Li et al. (2015) prepared three WC–10Co–0.5Ta cemented carbides with 5.38 (1#), 5.45 (2#), and 5.52 (3#) wt.% C using the standard powder metallurgy method. The electron backscattered diffraction (EBSD) maps for the samples are shown in Figure 10.3. Figure 10.3a shows that in sample 1 the segregation of the (Ta,W)C cubic phase has the same orientation, representing a large honeycomb Fcc particle. In comparison with sample 1, the size of the honeycomb (Ta,W)C cubic particles in sample 2 is reduced. Some small cubic phase particles in different orientations can also be detected in Figure 10.3b. Compared to samples 1 and 2, only a small amount and uniform dispersed cubic phase particles having multiple Fcc orientations are observed in sample 3, as indicated in Figure 10.3c. The (Ta,W)C cubic phase has almost no segregation in sample 3. Therefore, the increase in carbon content can significantly impede the segregation of the (Ta,W)C cubic phase and improve the microstructure.

The microstructure of samples 1–3 can be explained qualitatively by thermo-dynamic calculation. At 1,410°C, different phases are observed from three sintered samples, as shown in Figure 10.2. In sample 1, only the Co-rich liquid phase and WC phase are found, and the TaC inhibitor is completely dissolved in the Co-rich liquid phase. During the cooling stage after the sintering, some Ta atoms precipitate and

**Figure 10.3** EBSD maps of (a) sample 1, (b) sample 2, and (c) sample 3, where the WC phase is shown in the band comparison chart and the orientation of the Fcc phase is shown in the color inverse pole figure (IPF). The figures are reproduced from Li et al. (2015) with permission from Elsevier. A black and white version of this figure will appear in some formats. For the colour version, please refer to the plate section.

form the (Ta,W)C cubic phase with W and C atoms. The first precipitated (Ta,W)C particle will be the nucleation center of the cubic phase. The surrounding Ta atoms migrate rapidly in the liquid phase and then precipitate into the nucleated cubic phase. Due to the poor wettability between the liquid Co phase and cubic phase, the rapid growth of the preferentially oriented (Ta,W)C cubic phase can replace the liquid phase in the cooling process and form a honeycomb shape with the WC phase. Sample 2 shows three phases of WC, liquid, and cubic phases. In sample 2, most of the added TaC inhibitors are dissolved in liquid Co at 1,410°C, while the rest are not distributed. During the cooling process, some dissolved Ta atoms precipitate on the undissolved TaC and become the nucleation center for the cubic phase. However, in the cooling process, the existing (Ta,W)C phase has only a small increase for the preferred orientation. Therefore, the separated cubic phase particles and small size honeycomb cubic particles can be observed in the microstructure. In sample 3, a large amount of inhibitor added at sintering temperature did not dissolve in liquid Co. It is evenly dispersed in the whole microstructure, providing the nucleus for the cubic phase remaining in the liquid phase. At the beginning of cooling from sintering temperature, the dissolved Ta atoms precipitate from the supersaturated binder phase to the undissolved cubic phase. Therefore, fine isolated cubic particles are formed in the microstructure. The results show that the segregation degree of the (Ta,W)C cubic phase is mainly determined by the molar percentage of the cubic phase in the liquid phase at the sintering temperature.

As we have described, the elimination for the segregation of the (Ta,W)C cubic phase in ultrafine cemented carbides is achieved by adjusting the carbon content at the sintering temperature. This elimination is beneficial to improve the mechanical properties of cemented carbides. In fact, sample 3 shows the best TRS and HRA properties among the three samples due to the elimination of the cubic phase segregation and a uniform microstructure in this sample (Li et al., 2015).

### 10.2.2 Optimization of Composition, Sintering Temperature, and Inhibitors

In Section 10.2.1, we described a method of controlling (Ta,W)C cubic phase segregation by thermodynamic calculation and key experiments for TaC inhibitor addition into cemented carbides. Next, we will demonstrate the influence of alloy composition, sintering temperature, and the use of two kinds of grain growth inhibitors on the mechanical properties and reliability of WC–10 wt.% Co with different Cr and V additions through the combined use of thermodynamic calculation, Weibull distribution calculation, and experiment (Tian et al., 2019).

First, we describe the effect of C content on the mechanical properties and performance of WC–10Co–0.5Cr. Prior to experiment, thermodynamic calculations were conducted by means of Thermo-Calc software using a thermodynamic database CSUTDCC1 (Peng et al., 2014). The thermodynamic calculations were performed for the sake of avoiding the formation of the two harmful phases [$M_6C$ ($\eta$) and graphite (C)]. Figure 10.4a shows the calculated phase equilibria of WC–10Co–0.5Cr near the favorable sintering region. In Figure 10.4b, the weight fraction of the liquid phase in WC–10Co–0.5Cr with different C contents at 1,360°C is calculated. Based on the thermodynamic calculations, three samples were prepared and sintered at 1,360°C for 1 hour. The mechanical properties of these samples are listed in Table 10.1. As indicated in the table, sample 2 with 5.46 wt.% C has the best mechanical properties among the prepared samples. This can be explained qualitatively according to thermodynamic calculations and the microstructure feature. SEM images of three samples (not shown here) show homogeneous microstructures without the formation of the harmful phases and abnormal grain growth. The images also indicate that the grain size of WC gradually increases with C content. On the other hand, as shown in Figure 10.4b, the weight fraction of liquid phase in the samples increases with C content, which is beneficial to inhibit the grain growth of WC and enhances the

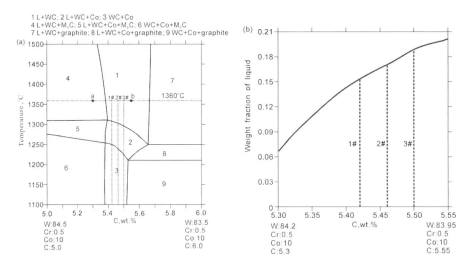

**Figure 10.4** (a) Calculated phase equilibria of WC–10 wt.% Co–0.5 wt.% Cr cemented carbides close to the sintering region. Overall composition range covers 84.5W–0.5Cr–10Co–5C to 83.5W–0.5Cr–10Co–6C (wt.%). (b) Calculated liquid phase mass fraction of the cemented carbide at 1,360°C from point a to point b. Overall composition range covers 84.2W–0.5Cr–10Co–5.3C to 83.95W–0.5Cr–10Co–5.55C (wt.%). The figures are reproduced from Tian et al. (2019) with permission from Elsevier.

**Table 10.1** Rockwell hardness (HRA), transverse rupture strength (TRS), fracture toughness ($K_{IC}$), and Weibull parameter (m) of WC–10 wt.% Co–0.5 wt.% Cr cemented carbides with different C contents.

| C (wt.%) | HRA | TRS (MPa) | $K_{IC}$ (MPa•m$^{1/2}$) | m |
|---|---|---|---|---|
| 5.42 | 93.2 | 4,240 | 14.5 | 39 |
| 5.46 | 92.8 | 4,380 | 15.5 | 52 |
| 5.5 | 92.3 | 4,292 | 15.4 | 51 |

Tian et al. (2019).

densification. In view of grain size and the amount of liquid phase, sample 2 is expected to show the overall best mechanical properties, as indicated experimentally (Tian et al., 2019).

The performance of cemented carbide cannot be judged by mechanical properties alone, because the influence of alloy defects or special properties of the binder phase on mechanical properties should also be considered. For cemented carbides, a uniform crack size frequency density function $g(a)$ of the form $g \propto a^{-P}$ (Hunt and McCartney, 1979; Weibull, 1951) with $P$ as the material constant can be assumed. This function further derives the well-known probability function (cumulative Weibull distribution) (Danzer, 1992):

$$P_f(\sigma, V) = 1 - \exp\left[-V/V_0 (\sigma/\sigma_0)^m\right], \tag{10.1}$$

in which $P_f$ is fracture probability at stress $\sigma$, $\sigma_0$ a scaling parameter, and $V$ and $V_0$ are the volume and average volume of the sample, respectively. $m$ is the Weibull parameter, the value of which is defined as the slope of fracture possibility against fracture strength for the targeted sample. In fact, the steep slope at the inflection point of the Weibull curve with a large value of $m$ (for example, $m = 5$) is approximated by the linear fit to the data points. The large value of $m$ means the small disparity for the strength of materials, and vice versa. Thus, a large $m$ indicates a good performance for the cemented carbides. One can use the Weibull distribution to reveal the strength reliability of brittle materials. The results show that the Weibull parameters of WC–Co cemented carbide are obviously affected by the characteristics of alloy defects or binder phases (Bolognini et al., 2001; Lambrigger, 1999), which are in turn influenced by grain growth inhibitor and C content. Figure 10.5 shows the Weibull plot of WC–10Co–0.5Cr with different C contents. Twenty data points representing individual samples are shown in this figure. The straight line is a linear fitting of these data points. As described previously, the slope of this fitting is the Weibull parameter $m$. As indicated in Table 10.1, for the alloy (1#) with carbon content of 5.42 wt.%, the strength data are highly dispersive and the Weibull parameter is small. When the C content is 5.46 (2#) and 5.5 (3#) wt.%, the strength data are concentrated, and the Weibull parameters are large. The Weibull parameter for sample 2 is the largest among three samples.

For samples 1–3, the single grain growth inhibitor $Cr_3C_2$ is added to the WC–10Co cemented carbides. Many investigations have shown that the simultaneous use of the

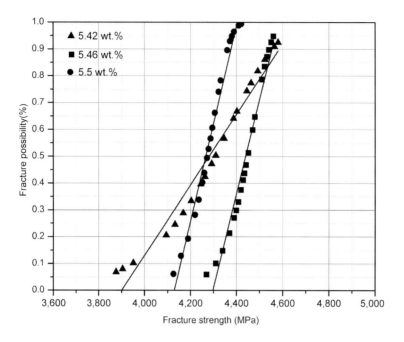

**Figure 10.5** Weibull graphics for WC–10 wt.% Co–0.5 wt.% Cr cemented carbides with different C contents. The figure is reproduced from Tian et al. (2019) with permission from Elsevier.

two types of grain growth inhibitors may have a stronger inhibitory effect than the single grain growth inhibitor alone. Following the same strategy for samples 1–3, we demonstrate that the mechanical properties and performance reliability of the WC–10 wt.% Co with the addition of two grain growth inhibitors ($Cr_3C_2$ and VC) can be further improved in comparison with sample 2. For that purpose, the phase equilibria of WC–10 wt.% Co with 0.4 wt.% V and WC–10 wt.% Co with 0.4 wt.% Cr + 0.2 wt.% V are calculated by using CSUTDCC1 thermodynamic database, as presented in Figure 10.6. The mechanical properties and Weibull parameters of the WC–10 wt.% Co with 0.4 wt.% Cr + 0.2 wt.% V are listed in Table 10.2 along with those from WC–10 wt.% Co with the single grain growth inhibitor 0.5 wt.% Cr or 0.4 wt.% V. It can be seen that the cemented carbides WC–10 wt.%Co-0.4 wt.% Cr –0.2 wt.% V has the best combination of HRA, TRS, KIC, and the Weibull parameter.

In this Section 10.2, for the ultrafine cemented carbides, we have demonstrated the strategy to (i) avoid or minimize segregation of the (Ta,W)C cubic phase by means of thermodynamic calculations and key experiment and (ii) to obtain a good combination of mechanical properties and performance reliability through a hybrid approach of thermodynamic calculations, Weibull distribution analysis, and calculation-guided key experiments. This strategy is equally valid for designing other ultrafine WC–Co cemented carbides.

The ultrafine WC–10 wt.% Co–0.6 wt.% $Cr_3C_2$ cemented carbide was developed by using the thermodynamic calculations in conjunction with key experiments. Figure 10.7a shows the SEM micrograph of a WC–10 wt.%Co–0.6 wt.%$Cr_3C_2$

**Table 10.2** Mechanical properties and Weibull parameter of WC–10 wt.% Co cemented carbides with different inhibitors.

| Alloy composition | HRA | TRS(MPa) | K$_{IC}$(MPa•m$^{1/2}$) | m |
|---|---|---|---|---|
| WC–10Co–0.5Cr | 92.8 | 4,380 | 15.5 | 52 |
| WC–10Co–0.4V | 93.2 | 4,120 | 14.1 | 37 |
| WC–10Co–0.2V–4Cr | 93.6 | 4,550 | 15.2 | 51 |

Tian et al. (2019).

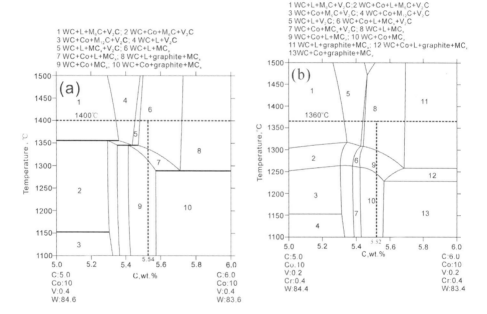

**Figure 10.6** Calculated phase equilibria near the sintering region: (a) 0.4 wt.% V; overall composition range covers 5C–10Co–0.4V–84.6W to 6C–10Co–0.4V–83.6W (wt.%). (b) 0.4 wt.% Cr–0.2 wt.% V; overall composition range covers 5C–10Co–0.2V–0.4Cr–84.4W to 6C–10Co–0.2V–4Cr–83.4W (wt.%). The figures are reproduced from Tian et al. (2019) with permission from Elsevier.

substrate. The preparation process of this substrate is the same as that shown in Figure 10.1. The grain size of this ultrafine substrate is 0.6 μm. It was HIP-sintered with a pressure of 6 MPa for the purpose of increasing the density. Subsequently, a Wickers hardness of about HV 1600 was obtained. This ultrafine cemented carbide substrate was coated with PVD TiAlSiN coating and used for the turning of high-temperature alloys, as shown in Figure 10.7b (www.achtecktool.com/en/).

## 10.3     Cemented Carbides with Composite Binder Phases of Co and $\gamma'$-Ni$_3$Al

It is well established that the microstructure (i.e., the morphology and grain size of WC, content, and composition of the binder phase as well as the interface

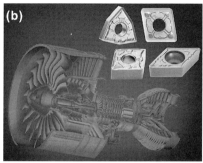

**Figure 10.7** (a) Ultrafine cemented carbide (grain size of WC: 0.6 μm); (b) inserts with ultrafine cemented carbide and PVD TiAlSiN coating for machining of high-temperature alloys in aviation engines. A black and white version of this figure will appear in some formats. For the colour version, please refer to the plate section.

structures between WC, the binder phase, and the other constitute phases) of cemented carbide determines its properties. The enhancement of WC–Co cemented carbide properties, such as oxidation and corrosion resistance as well as high-temperature strength and hardness, is possible by strengthening the binder phase. Since (Co) is the matrix for both cemented carbides and Co-based superalloys, the strategy for strengthening the (Co) in superalloys could be used as a reference to strengthen cemented carbides.

In cobalt-based superalloys, the ordered Fcc $L1_2$-type $\gamma'$-$Co_3$(Al,W) was found in the Al–Co–W system, which has led to the extensive research on $\gamma'$ reinforced Co-based superalloys (Sato et al., 2006). This new type of $\gamma$ (disordered Fcc) $+\gamma'$ Co-based superalloys has a better high-temperature strength than conventional Co-based alloy. The work of Sato et al. (2006) offers a new possibility for development of new WC–Co–Ni–Al cemented carbides that show excellent corrosion resistance and good high-temperature hardness. By an integrated approach of thermodynamic calculations and phase-field simulation as well as calculation-aided experiments, the main factors affecting the morphology and grain size of the Co–Ni–Al binder phase strengthened by the $\gamma'$ phase could be revealed. In the following, we will describe the procedure to design different phase compositions of the Co–Ni–Al binder phase that are responsive to the grain size and morphology of the $\gamma'$-strengthened composite binder phase. The decrease of the grain size of the Co–Ni–Al binder phase and the precipitation of $\gamma'$ phase lead to a rapid increase for both hardness and magnetic coercivity of the WC–Co–Ni–Al binder phase (Long et al., 2017a, 2017b). The coercivity is one magnetic property, the value of which is closely related to the structure of the alloy. For the present case, magnetic coercivity is associated with the amount of $\gamma'$ phase precipitation. Simultaneously, the mechanism of WC morphology evolution is clarified by phase-field simulation. Based on these microstructure designs, the cemented carbides with the composite binder phase and favored morphology of WC are developed, which possess the excellent combination of mechanical properties (TRS and HRA) and oxidation/corrosion resistance.

### 10.3.1  Optimization of Composition and Sintering Temperature

In order to select the best alloy composition prior to experiment, the driving force of $\gamma$-binder precipitation nucleation from liquid at 1,250°C was calculated, as shown in Figure 10.8a. The integrated driving force of precipitation from liquid increases with increasing Ni$_3$Al content, as shown in this diagram. High-nucleation driving force increases the nucleation rate during solidification. As a result, the grain size of binder phase decreases with the increase of grain number of binder phase.

In cemented carbides, the grain growth of WC follows Ostwald ripening. During the sintering process, large WC grains grow by consuming small grains. However, there are energy barriers to this process. When W and C atoms migrate from WC to the liquid binder, the barrier increases. On the other hand, when W and C precipitate into the Ostwald ripening, the barrier decreases. Therefore, it is suggested that a higher solid–liquid interface can increase the barrier of W or C atoms migration. In other words, the Co-rich binder with lower interfacial energy can promote the diffusion of W near the interface, thus increasing the solubility of W in the binder and accelerating the growth of WC grains. In contrast, Co-poor binder can increase driving force and solid–liquid interfacial energy to refine the grain of the binder phase and WC. It implies that the composition of the binder phase should be carefully selected.

The solid–liquid interfacial energies of WC–Co–Ni–Al with different Ni$_3$Al contents were calculated according to the model developed by Liu et al. (2020). According to this model, the solid–liquid interfacial energy can be estimated by considering chemical and structural contributions. In the stable W–C phase diagram, WC is formed via a peritectic reaction of $L + C = WC$ at 3,047 K. According to the metastable W–C phase diagram involving liquid and WC, the metastable congruent melting point for WC is 3,086 K. According to this metastable melting point of WC grains, the structural contribution to the interfacial energy is 0.31 J/m$^2$, while the chemical contribution can be calculated by using the thermodynamic database CSUTDCC1 (Peng et al., 2014). As indicated in Figure 10.8b, the interfacial energy increases with Ni$_3$Al contents. Based on Figure 10.8a,b, both high nucleation driving force and solid–liquid interfacial energy are beneficial for a smaller grain size, and accordingly we have designed three WC–50 (Co–Ni–Al) alloys containing 0, 11.4, and 15.2 wt.% Ni$_3$Al. In order to make the difference in binder grain size more obvious, we selected WC–50 wt.% (Co–Ni–Al) alloy with the same carbon content and coarse WC grains (9 µm) as the starting materials. Table 10.3 lists the chemical compositions and sintering schedules for the prepared WC–50 (Co–Ni–Al) alloys. In order to find the appropriate sintering temperature and carbon content for these three alloys, thermodynamic calculations are performed by using the database CSUTDCC1, as presented in Figure 10.8c. The calculated $\gamma + WC$ phase zone shows the different ranges of carbon content and sintering temperature for different Ni$_3$Al contents, and the blue dot at 1,350°C indicates the selected sintering temperature.

Following the same strategy used for WC–50 (Co–Ni–Al), we have designed six series of WC–24 (Co–Ni–Al) alloys. The carbon content was fixed at 4.64 wt.% by adding C or W powder to the selected samples. The chemical compositions and sintering schedules are shown in Table 10.4.

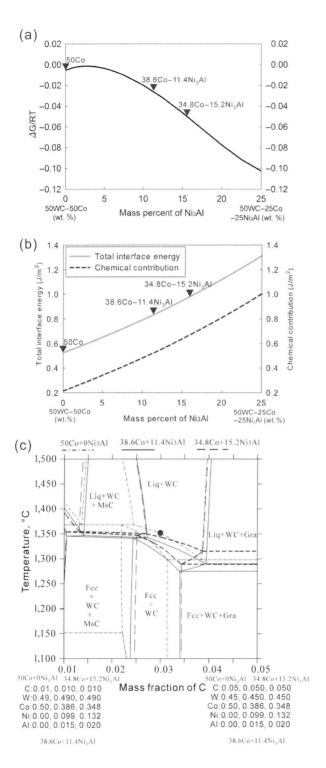

**Figure 10.8** (a) Calculated driving force for the binder phase precipitated from liquid in WC–50 (Co-Ni-Al) alloys at 1,250°C; (b) calculated solid–liquid interfacial energy of WC–50 (Co–Ni-Al) alloys, and (c) calculated phase equilibria in WC–50 (Co–Ni–Al) alloys. The overall composition ranges for three different binder phases (WC–50Co, WC–38.6Co–11.4Ni₃Al, and WC–34.8Co–15.2Ni₃Al) are indicated in the phase diagrams. The figures are reproduced from Long et al. (2017b) with permission from Elsevier.

**Table 10.3** Chemical compositions and sintering schedules of the prepared WC–50 (Co–Ni–Al) alloys.

| Sample | Chemical composition (wt.%) | | | | Sintering parameters | |
| | Co | Ni₃Al | C | W | Sintering holding time (h) | Sintering temperature (°C) |
| --- | --- | --- | --- | --- | --- | --- |
| 50Co | 50 | 0 | 3 | 47 | 1 | 1,350 |
| 11.4Ni₃Al | 38.6 | 11.4 | 3 | 47 | 1 | 1,350 |
| 15.2Ni₃Al | 34.8 | 15.2 | 3 | 47 | 1 | 1,350 |

Long et al. (2017b).

**Table 10.4** Chemical compositions and sintering schedules of the prepared WC–24 (Co–Ni–Al) alloys.

| Sample | Chemical composition (wt.%) | | | | Sintering parameters | |
| | Co | Ni₃Al | C | W | Time (h) | Temperature (°C) |
| --- | --- | --- | --- | --- | --- | --- |
| 1# | 24 | 0 | 4.64 | 71.36 | 1, 8, 20 | 1,450 |
| 2# | 20.5 | 3.5 | 4.64 | 71.36 | 1, 8 | 1,450 |
| 3# | 16.9 | 7.1 | 4.64 | 71.36 | 1, 8 | 1,450, 1,350 |
| 4# | 13.0 | 11.0 | 4.64 | 71.36 | 1, 8 | 1,450 |
| 5# | 8.9 | 15.1 | 4.64 | 71.36 | 1, 8 | 1,450 |
| 6# | 0 | 24 | 4.64 | 71.36 | 1, 8, 20 | 1,450 |

Long et al. (2017a).

## 10.3.2  Morphology Control of the Composite Binder Phases and WC Grains

Figure 10.9 shows the EBSD diagrams of the WC–38.6Co–11.4Ni₃Al (wt.%) and WC–34.8Co–15.2Ni₃Al (wt.%) alloys. The orientation of the binder phase is shown by the color inverse pole figure (IPF) diagram. A grain size of about 100 μm for the binder phase is obtained for the WC–38.6Co–11.4Ni₃Al alloy, as illustrated in Figure 10.9a. A noticeably reduced grain size of about 20 μm for the binder phase in the alloy WC–34.8Co–15.2Ni₃Al is observed as shown in Figure 10.9b. Through thermodynamic calculations, it is shown that the increase in the Ni₃Al content in the WC–50 (Co–Ni–Al) alloy can significantly reduce the grain size of the binder phase. The higher driving force for nucleation increases the nucleation rate during solidification, which leads to an increase in the number of grains in the binder phase and consequently results in a smaller grain size for the binder phase.

Figure 10.10a shows the typical $\gamma' + \gamma$ microstructure of the WC–38.6Co–11.4Ni₃Al alloy after vacuum sintering at 1,350°C for 1 hour. In order to find out whether the $\gamma'$ precipitate phase is coherent with the matrix phase as it is in the Co–Al–W alloy, a TEM image in Figure 10.10b and the selected-area electron diffraction

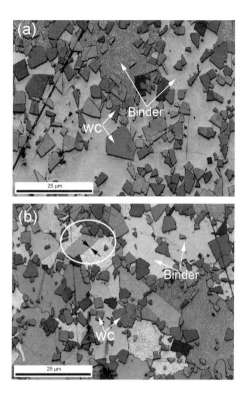

**Figure 10.9** EBSD maps of (a) WC–38.6Co–11.4Ni$_3$Al and (b) WC–34.8Co–15.2Ni$_3$Al. The alloys were sintered at 1,350°C for 1 hour under vacuum. The figures are reproduced from Long et al. (2017b) with permission from Elsevier. A black and white version of this figure will appear in some formats. For the colour version, please refer to the plate section.

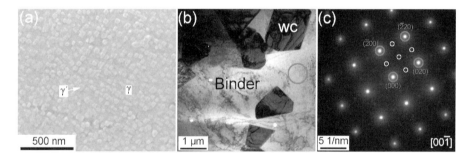

**Figure 10.10** Microstructure of the WC–38.6Co–11.4Ni$_3$Al alloy sintered for 1 hour at 1,350°C under vacuum: (a) FE-SEM image of the binder phase, (b) bright-field image of the microstructure between the binder phase and WC, and (c) the indexed SADP pattern marked by the red circles in (b). The figures are reproduced from Long et al. (2017b) with permission from Elsevier. A black and white version of this figure will appear in some formats. For the colour version, please refer to the plate section.

pattern (SADP) in Figure 10.10c are presented. Because the $d$ spacing of {0 2 0} in the $\gamma'$ phase is twice that of the $\gamma$ phase {0 2 0}, as shown in Figure 10.10c, the interface between ordered Fcc_L12 $\gamma'$ precipitates and the $\gamma$ matrix is coherent.

A phase-field-simulated microstructure of the binder phase with the composition 19.5 at.% Al, 10.5 at.% Co, and 70 at.% Ni aged at 1,373 K is shown in Figure 10.11 (Peng et al., 2020). In this phase-field model, the total energy is described by the sum of chemical energy density, interfacial energy density, and elastic energy density over a domain. As shown in Figure 10.11, in the early stage of the growth of the $\gamma'$ phase, $\gamma'$ precipitate has a spherical morphology. As the size of the precipitated phase increases, the elastic energy of the precipitated phase increases, and finally the shape of the precipitated phase become cubic. The cubic $\gamma'$ precipitates embedded in the (Co) matrix detected in this experiment are similar to the morphology observed in the Co–Al–W alloy. In addition, as the Ni$_3$Al content increases, the volume fraction of the $\gamma'$ phase increases. At the same time, this cuboidal shape of the coherent $\gamma'$ precipitation phase will hinder the sliding of dislocations in the $\gamma$ matrix, thereby helping to improve the high-temperature mechanical properties of the cemented carbide.

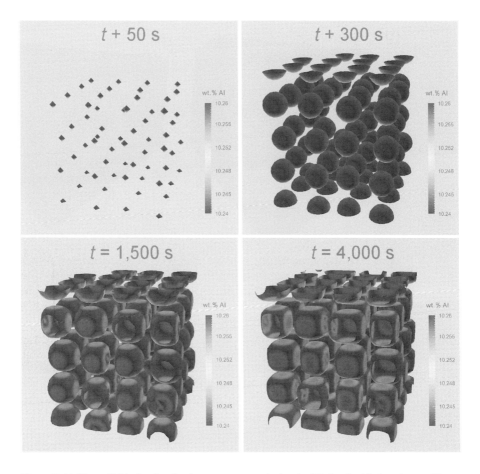

**Figure 10.11** Phase-field-simulated microstructure evolution in 3D for the binder phase with the composition 19.5 at.% Al, 10.5 at.% Co, and 70 at.% Ni aged at 1,373 K, showing the growth of $\gamma'$. The figures are reproduced from Peng et al. (2020) with permission from Elsevier. A black and white version of this figure will appear in some formats. For the colour version, please refer to the plate section.

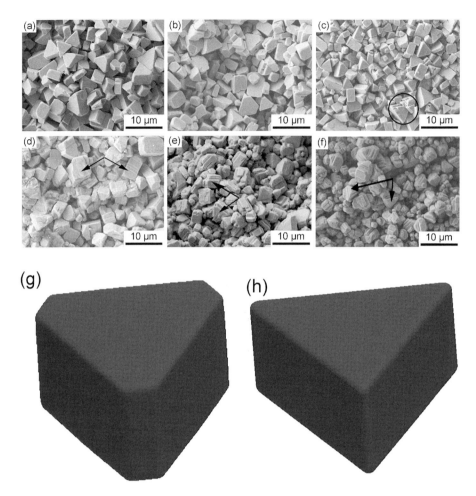

**Figure 10.12** Morphology (Long et al., 2017a) of WC grains extracted from WC–24 (Co–Ni–Al) alloys with different binder composites sintered at 1,450°C for 8 hours: (a) 0 Ni$_3$Al, (b) 3.5Ni$_3$Al, (c) 7.1Ni$_3$Al, (d) 11Ni$_3$Al, (e) 15.1Ni$_3$Al, and (f) 24Ni$_3$Al; and phase-field simulation (Li et al., 2021): (g) truncated trigonal prism, and (h) triangular prism. The figures are reproduced from Li et al. (2021); Long et al. (2017a), with permission from Elsevier.

Figure 10.12 shows the three-dimensional morphology of WC grains in the WC–24 (Co–Ni–Al) alloy after sintering at 1,450°C for 8 hours. For Co-rich binder samples (samples 1–3 in Figure 10.12a–c), the degree of faceting of WC grains decreases with the increase of the Ni$_3$Al content in the binder phase. In addition, the WC grains show a truncated triangular prism shape for Co-rich binder alloys. This morphology shows that the WC morphology is close to the equilibrium morphology. In the case of Ni$_3$Al-rich binder samples (samples 4–6, Figure 10.12d–f), the WC grains exhibit more obvious faceted steps. Furthermore, the WC shape becomes more irregular. At the same time, the WC grain size becomes smaller, in particular for fine WC grains. The triangular layer-by-layer growth morphology, which is shown by the black arrows in

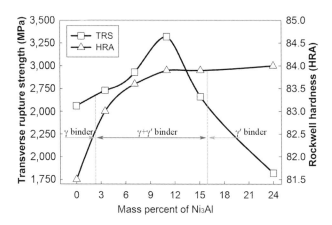

**Figure 10.13** Transverse rupture strength and Rockwell hardness of WC–24 (Co–Ni–Al) alloys. The figure is reproduced from Long et al. (2017a) with permission from Elsevier.

Figure 10.12d–f, provides direct evidence of anisotropic growth for WC grains in Ni$_3$Al-rich binder alloys (samples 5 and 6).

The phase-field model (Li et al., 2021), which considers the anisotropic interfacial energies (basal and prismatic facets) of WC grains, can reproduce both truncated trigonal prism and triangular prism, as shown in Figure 10.12g, h.

The preceding results show that with the increase of Ni$_3$Al content, the grain size of the binder phase, and the volume fraction of the $\gamma'$ phase change noticeably. In addition, the morphology and grain size of WC also change with the increase of Ni$_3$Al in the composite binder phase of Ni$_3$Al + Co. This means that properties of the WC–Co–Ni–Al cemented carbides can be enhanced through the microstructure design focusing on grain size, morphology, and volume fraction of WC and the binder phase. Figure 10.13 shows the HRA and TRS of WC–24 (Co–Ni–Al) alloys with different Ni$_3$Al contents. As shown in the diagram, the HRA exhibits a substantial increase up to about 11 wt.% Ni$_3$Al. There may be two reasons for this phenomenon. The main reason is WC grain refinement. On the other hand, the Co–Ni–Al binder can be hardened through the formation of the $\gamma'$ strengthening phase. In addition, the hardness of Ni$_3$Al is much higher than that of Co. It was noted that the HRA increases slightly due to the weakening of the $\gamma'$ phase strengthening effect, when the Ni$_3$Al content increases to 11 wt.% and beyond.

The TRS of the alloy containing 11 wt.% Ni$_3$Al reached the maximum value of 3,320 MPa, and then decreased significantly until the minimum value, which appeared in sample 6 with 24 wt.% Ni$_3$Al binder. On one hand, the strengthening effect of the $\gamma'$ phase may be the main reason for the increase of TRS. On the other hand, as the Ni$_3$Al content increases, the solubility of W in the Co–Ni–Al binder decreases, which may weaken bonding strength of the WC–binder interface. Therefore, under the combined effect of these two factors, the strength of the alloy first increases and then decreases.

It was found that the corrosion resistance to HNO$_3$ acid and electrochemical corrosion of WC–Co–Ni–Al alloys were enhanced with the increase of Al content.

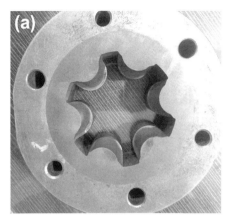

**Figure 10.14** WC–Co–Ni–Al-based cemented carbides: (a) hot heading die, (b) carbide rolls. A black and white version of this figure will appear in some formats. For the colour version, please refer to the plate section.

At the same time, Al can form a dense $Al_2O_3$ oxide film to prevent oxygen from further oxidizing the interior of the alloy, thereby improving the oxidation resistance of the alloy. Due to the outstanding properties, particularly good high-temperature properties, the aforementioned designed WC–Co–Ni–Al cemented carbides find wide applications in the fields of hot heading die and carbide rolls, as shown in Figure 10.14. By means of the hybrid approach of thermodynamic calculation and phase-field simulations in conjunction with calculations-guided experiment, Zhuzhou Cemented Carbide Group Corp. Ltd. of China has developed several cemented carbides with the composite binder phases (Chinese patents ZL201110371031.4 and ZL201310654779.4).

## 10.4    Gradient Cemented Carbides

As mentioned previously, cemented carbides have been used widely for machining, mining, cutting, drilling, and wear–resistant parts due to an excellent combination of hardness and toughness. In order to further improve the performance of the cemented carbide inserts, the surface of the cemented carbide is usually coated with a hard, thin film such as TiC, TiN, TiCN, TiAlN, $Al_2O_3$, or their multilayers by high-temperature chemical vapor deposition (CVD) or low-temperature physical vapor deposition (PVD) technique. In Chapter 11, the computational design of CVD and PVD hard-coatings will be described in detail. For CVD hard coating, cracks would be formed inside the coating unavoidably due to the large difference of thermal expansion coefficients between the substrate and hard coating. When cutting tools are employed in metal machining, the formed cracks propagate easily into the substrate to cause failure. To diminish or even avoid crack propagation, several attempts have been performed to make cemented carbides with a gradient distribution of structure or properties. One type of such gradient cemented carbides consists of WC and a second hard phase (NaCl-type cubic carbide or carbonitride) that are embedded in a cobalt-rich binder phase (Frykholm et al., 2001; Schwarzkopf et al., 1988; Zhang et al., 2011, 2013). This section will concentrate on the computational design of the gradient WC–Co–Ti(C,N)–TaC–NbC cemented carbide (Zhang et al., 2016). After sintering under the denitriding condition (i.e., the nitrogen partial pressure in the sintering atmosphere is below the nitrogen equilibrium partial pressure in the cemented carbides), this gradient layer in the cemented carbide is free of cubic carbide phase (Fcc) and enriched in binder phase (the so-called Fcc-free surface layer). This kind of surface can prevent crack propagation due to the tough feature (mainly resulting from the high volume fraction of the binder phase in this surface) and thus extend the service lifetime of the cemented carbides.

Gradient cemented carbides have several phases (WC, the binder phase, and cubic phase, etc.), some of which include multicomponents. Several factors, mainly including composition and sintering parameters (time, temperature, and atmosphere), affect the microstructure and properties of gradient cemented carbides. It would be extremely time consuming and costly to manipulate so many factors through pure experimental method. In the following subsections, we will demonstrate that prior to experiment, (1) diffusion modeling can predict measured concentration profiles and volume fractions of the phases for the given alloy composition, sintering temperature, and time as well as atmosphere and (2) a microstructure-based hardness model can be used to predict the hardness distributions from the surface to the core of the gradient cemented carbides with the input of microstructure parameters and the hardness values of the individual phases.

## 10.4.1 Computational Design of Gradient Microstructure

The formation of the gradient structure is a diffusion-controlled process during sintering of multicomponent cemented carbides. Under the denitriding atmosphere, the strong thermodynamic interactions between N and transition metals (Ti, Ta, and Nb) lead to the outward diffusion of N from the substrate and inward diffusion of these metals in the opposite direction. This complex diffusion phenomenon can be simulated by means of DICTRA software applied to the established thermodynamic CSUTDCC1 and atomic mobility CSUDDCC1 databases (Peng et al., 2014; Zhang et al., 2014). The related diffusion equations for such simulations are described in Chapter 6. The simulation of the gradient structure is based on the model for a long-range diffusion in a continuous matrix (the liquid binder phase) with dispersed phases of WC and the cubic phase. Because of the occurrence of the dispersed phases, the diffusion in the matrix is reduced. The labyrinth factor $\lambda(f)$, where $f$ is the volume fraction of the liquid matrix, was used to reduce the diffusion coefficient of the matrix (Frykholm et al., 2001):

$$D^n_{kj_{eff}} = \lambda(f)D^n_{kj}. \tag{10.2}$$

Frykholm et al. (2001, 2003) found that the simulation with $\lambda(f) = f$ could yield the better agreement with experiment data than that with $f^2$ (Engstrom et al., 1994; García and Prat, 2011; Turpin et al., 2005).

The first step for the computational design of the gradient cemented carbides is to optimize the alloy composition. The carbon content should be precisely controlled to ensure reasonable phase assemblage (no $M_6C$, $M_7C_3$ and graphite phases). For that purpose, the optimal alloy composition and sintering temperature are selected based on the thermodynamic calculations. Figure 10.15 presents the calculated phase equilibria close to the sintering region of alloys with different Ta and Nb additions by using the thermodynamic database of CSUTDCC1 (Peng et al., 2014). As shown in Figure 10.14a, there is a minor difference for C content in the sintering region for the alloys with Ta. In contrast, the large difference for C content is found in the case of the alloys with Nb. Based on these computed phase diagrams, the alloy compositions are selected, as listed in Table 10.5. The alloys are prepared by means of standard powder metallurgy methods. All the samples are sintered in vacuum at 1,450°C for 2 hours.

As a second step for the computational design of the gradient cemented carbides, diffusion simulations for the alloys sintered at 1,450°C for 2 hours are performed by using DICTRA software employing the CSUTDCC1 (Peng et al., 2014) and CSUDDCC1 databases (Zhang et al., 2014). Figure 10.16 shows the elemental concentrations and mole fractions of the phases as functions of the depth in the representative sample 2, indicating a reasonable agreement between prediction and experiment. As shown in Figure 10.16b, the binder phase is enriched inside the surface zone, where there is almost no cubic phase. It is this tough surface zone that prevents crack propagation from the CVD hard coating mainly due to the excellent toughness of the binder phase. The simulated phase fraction curve agrees with the SEM micrograph in Figure 10.16c, where an enrichment for the binder phase and a deficiency for the cubic phase are observed.

The preceding comprehensive comparisons indicate that major aspects of the experimental gradient microstructure can be captured through diffusion simulations for targeted alloys under the specified composition, sintering temperature, and time as well as atmosphere. Based on these predicted results, alloy composition and sintering schedules can be designed efficiently in order to obtain the expected gradient microstructure. The input files for the preceding diffusion simulations using DICTRA software are provided in Appendix A.

We have demonstrated the computational design of gradient cemented carbides under the denitriding condition. It should be mentioned that the strategy that we have described is equally valid for the design of the gradient cemented carbides under the nitriding condition (i.e., the nitrogen partial pressure in the sintering atmosphere is above the nitrogen equilibrium partial pressure in the cemented carbides). In fact, by just changing the nitrogen partial pressure, various gradient microstructure can be obtained (Li et al., 2020), as shown in Figure 10.17.

## 10.4.2    A Microstructure-Based Hardness Model for Gradient Cemented Carbides

In the preceding subsection for gradient cemented carbides, we have described the computational design for the selection of composition and microstructure evolution using thermodynamic calculations and diffusion simulation in steps 1 and 2,

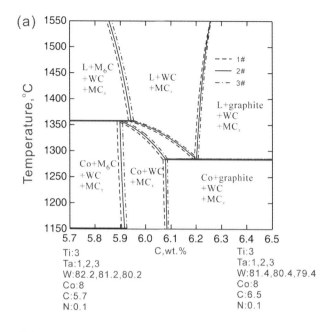

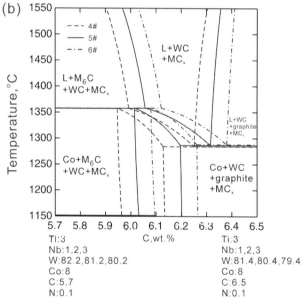

**Figure 10.15** Calculated phase equilibria close to the sintering region of alloys with different (a) Ta content (1#: 1 wt.% Ta; 2#: 2 wt.% Ta; 3#: 3 wt.% Ta) and (b) Nb content (4#: 1 wt.% Nb; 5#: 2 wt.% Nb; 6#: 3 wt.% Nb). The figures are reproduced from Zhang et al. (2016) with permission from Elsevier.

**Table 10.5** Compositions of the alloys with different Ta and Nb contents.

| Sample | Co | Ti | Ta | Nb | C | N | W |
|---|---|---|---|---|---|---|---|
| | | | Chemical composition (wt.%) | | | | |
| 1# | 8 | 3 | 1 | – | 6 | 0.1 | 81.9 |
| 2# | 8 | 3 | 2 | – | 6 | 0.1 | 80.9 |
| 3# | 8 | 3 | 3 | – | 6 | 0.1 | 79.9 |
| 4# | 8 | 3 | – | 1 | 6.07 | 0.1 | 81.83 |
| 5# | 8 | 3 | – | 2 | 6.14 | 0.1 | 80.76 |
| 6# | 8 | 3 | – | 3 | 6.20 | 0.1 | 79.7 |

Zhang et al. (2016).

respectively. The third step for computational design of the gradient cemented carbides is the model-predicted hardness as a function of depth from the surface to the core of the cemented carbides. The hardness distribution is crucial to crack propagation resistance, and its value is strongly related to the microstructure. From the surface to the core of gradient cemented carbides, a two-phase region composed of WC and the binder phase occurs firstly and then a three-phase region with the cubic phase. A microstructure-based hardness for cemented carbides with the three phases of WC, the binder, and cubic phase is given as follows (Pang et al., 2021):

$$H_{CC} = (H_{WC}V_{WC} + H_{Cubic}V_{Cubic})C_{cont} + H_{Co}[1 - (V_{WC} + V_{Cubic})C_{cont}], \quad (10.3)$$

where $C_{cont}$ means the contiguity of WC grains, and $V_{WC}$ and $V_{Cubic}$ the volume fractions of the WC and cubic phases, respectively. $H_{WC}$, $H_{Co}$, $H_{Cubic}$, and $H_{CC}$ are the hardness of WC, Co binder, cubic phase, and the alloy, respectively. For the two-phase region, where $V_{Cubic}$ is zero, (10.3) can still be used. The hardness for the individual phases is given by the Hall–Petch relationship:

$$H_{WC} = 13.5 + \frac{7.2}{\sqrt{d_{WC}}} \qquad (10.4a)$$

$$H_{Co} = 2.98 + \frac{3.9}{\sqrt{\lambda}} \qquad (10.4b)$$

$$H_{Cubic} = 16.5 + \frac{1.8}{\sqrt{d_{Cubic}}}, \qquad (10.4c)$$

where $d_{WC}$ is the mean grain size of WC phase, $\lambda$ the mean free path of the binder phase, and $d_{Cubic}$ the mean grain size of the cubic phase.

The contiguity $C_{cont}$ is approximately represented by an equation of the following form:

$$C_{cont} \cong 1 - \frac{V_{Co}}{(V_{Cubic}d_{WC}/d_{Cubic} + V_{WC})\lambda/d_{WC}}. \qquad (10.5)$$

The $\lambda$ for the binder phase is given by the following equations, and (10.6a) is for the two-phase region and (10.6b) for the three-phase region:

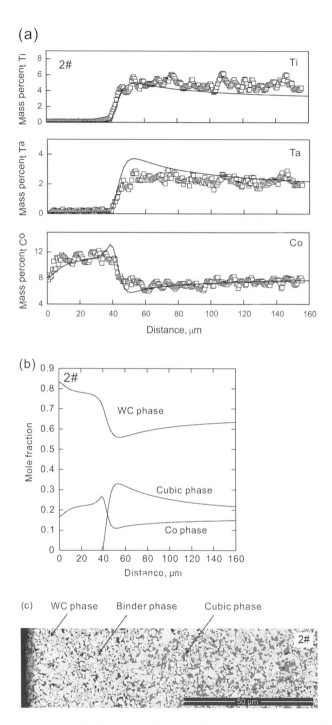

**Figure 10.16** (a) Measured and calculated elemental concentration depth profiles in sample 2 sintered at 1,450°C for 2 hours under vacuum; (b) calculated mole fractions of the phases; and (c) an SEM micrograph of the cross section for sample 2 sintered at 1,450°C for 2 hours under vacuum. The figures are reproduced from Zhang et al. (2016) with permission from Elsevier.

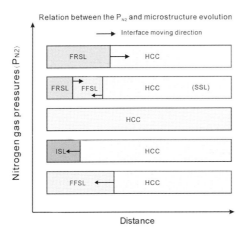

**Figure 10.17** Schematic illustration of microstructures against the nitrogen gas pressure. FFSL = Fcc-free surface layer; ISL = intermediate type of surface layer; HCC = homogeneous cemented carbides; SSL = sandwich surface layer, and FRSL = Fcc-rich surface layer. The figures are reproduced from Li et al. (2020) with permission from Elsevier.

$$\lambda = d \frac{V_{Co}}{(1 - V_{Co})(1 - C_{cont})} \tag{10.6a}$$

$$\lambda = \frac{V_{Co} d_{WC} d_{Cubic}}{(d_{WC} V_{Cubic} + d_{Cubic} V_{WC})(1 - C_{cont})}. \tag{10.6b}$$

According to these equations, the volume fractions of the phases, the grain sizes of WC, and the cubic phases are needed to predict the hardness. Both volume fractions and grain sizes can be obtained from experiments such as electron backscatter diffraction. On one hand, based on the thermodynamic and atomic mobility databases for cemented carbides (Peng et al., 2014; Zhang et al., 2014), the volume fractions of the phases can be calculated using DICTRA software. On the other hand, the grain sizes of WC and the cubic phase can be taken from microstructure simulations, such as the TC-Prisma calculation (Zhang and Du, 2017) and phase-field simulation.

Figure 10.18 presents the measured and calculated micro-indentation hardness as a function of distance from the surface for sample 2 sintered at 1,450°C for 2 hours under vacuum. As shown in the diagram, a good agreement between prediction and experiment is obtained.

Figure 10.19a shows an optical micrograph of gradient cemented carbide with a Co-rich layer on the surface. The surface has high toughness and is resistant to plastic deformation under external load. Figure 10.19b shows inserts (www.achtecktool.com .cn) with gradient cemented carbide substrate and CVD coating used for steel turning. Inserts with a thick Co-rich layer are mainly used for heavy turning while inserts with a thin Co-rich layer are good for finish-turning operations.

By means of the preceding hybrid approach of thermodynamic calculation, diffusion simulation and hardness predictions in conjunction with calculations-aided

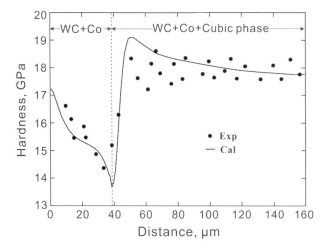

**Figure 10.18** Model-predicted hardness (Pang et al., 2021) against distance from the surface along with the experimental data (Zhang et al., 2016) for sample 2 (see Figure 10.16) sintered at 1,450°C for 2 hours under vacuum.

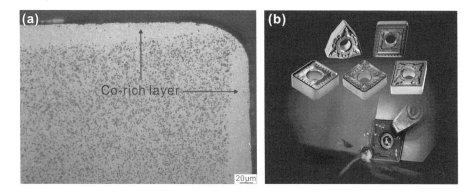

**Figure 10.19** (a) Gradient cemented carbide, and (b) Inserts with gradient cemented carbide substrate and CVD coating used for steel turning. A black and white version of this figure will appear in some formats. For the colour version, please refer to the plate section.

decisive experiment, Ganzhou Achteck Tool Technology Co., Ltd., in China has developed several gradient cemented carbides. The photo for one of the products is shown in Figure 10.19b.

# References

Bolognini, S., Mari, D., Viatte, T., and Benoit, W. (2001) Fracture toughness of coated TiCN–WC–Co cermets with graded composition. *International Journal of Refractory Metals and Hard Materials*, 19(4), 285–292.

Danzer, R. (1992) A general strength distribution function for brittle materials. *Journal of the European Ceramics Society*, 10(6), 461–472.

Engström, A., Höglund, L., and Ågren, J. (1994) Computer simulation of diffusion in multi-phase systems. *Metallurgical and Materials Transaction A*, 25(6), 1127–1134.

Frykholm, R., Ekroth, M., Jansson, B., Ågren, J., and Andrén, H. O. (2003) A new labyrinth factor for modelling the effect of binder volume fraction on gradient sintering of cemented carbides. *Acta Materialia*, 51(4), 1115–1121.

Frykholm, R., Ekroth, M., Jansson, B., Andrén, H. O., and Ågren, J. (2001) Effect of cubic phase composition on gradient zone formation in cemented carbides. *International Journal of Refractory Metals and Hard Materials*, 19(4), 527–538.

García, J., Ciprés, V. C., Blomqvist, A., and Kaplan, B. (2019) Cemented carbide microstructures: a review. *International Journal of Refractory Metals and Hard Materials*, 80, 40–68.

Garcia, J., and Prat, O. (2011) Experimental investigations and DICTRA simulations on formation of diffusion-controlled Fcc-rich surface layers on cemented carbides. *Applied Surface Science*, 257(21), 8894–8900.

Hunt, R. A., and McCartney, L. N. (1979) A new approach to Weibull's statistical theory of brittle fracture. *International Journal of Fracture*, 15(4), 365–375.

Lambrigger, M. (1999) Weibull master curves and fracture toughness testing, Part IV, dynamic fracture toughness of ferritic-martensitic steels in the DBTT-range. *Journal of Materials Science*, 34, 4457–4468.

Li, H., Du, Y., Long, J. Z., et al. (2021) 3D phase field modeling of the morphology of WC grains in WC–Co alloys: the role of interface anisotropy. *Computational Materials Science*, 196, 110526.

Li, N., Li, X., Zhang, W., and Du, Y. (2020) Relation between the nitrogen gas pressure and structure characteristics of WC–Ti(C,N)–Co graded cemented carbides. *Journal of Alloys and Compounds*, 831, 154764.

Li, N., Zhang, W., Du, Y., Xie, W., Wen, G., and Wang, S. (2015) A new approach to control the segregation of (Ta,W)C cubic phase in ultrafine WC–10Co–0.5Ta cemented carbides. *Scripta Materialia*, 100, 48–50.

Li, N., Zhang, W., Peng, Y., and Du, Y. (2016) Effect of the cubic phase distribution on ultrafine WC–10Co–0.5Cr–xTa cemented carbide. *Journal of the American Ceramics Society*, 99(3), 1047–1054.

Liu, Y., Zhang, C., Du, C., et al. (2020) CALTPP: a general program to calculate thermophysical properties. *Journal of Materials Science and Technology*, 42, 229–240.

Long, J., Li, K., Chen, F., et al. (2017a) Microstructure evolution of WC grains in WC–Co–Ni–Al alloys: effect of binder phase composition. *Journal of Alloys and Compounds*, 710, 338–348.

Long, J., Zhang, W., Wang, Y., et al. (2017b) A new type of WC–Co–Ni–Al cemented carbide: grain size and morphology of γ′-strengthened composite binder phase. *Scripta Materialia*, 126, 33–36.

Pang, M., Du, Y., Zhang, W.-B., Peng, Y.-B., and Zhou, P. (2021) A simplified hardness model for WC–Co–Cubic cemented carbides. *Journal of Mining and Metallurgy B*, 57(2), 253–259.

Peng, Y., Du, Y., Stratmann, M., et al. (2020) Precipitation of γ′ in the γ binder phase of WC–Al–Co–Ni cemented carbide: a phase-field study. *CALPHAD*, 68, 101717.

Peng, Y., Du, Y., Zhou, P., et al. (2014) CSUTDCC1 – a thermodynamic database for multi-component cemented carbides. *International Journal of Refractory Metals and Hard Materials*, 42, 57–70.

Sato, J., Omori, T., Oikawa, K., Ohnuma, I., Kainuma, R., and Ishida, K. (2006) Cobalt-base high-temperature alloys. *Science*, 312(5770), 90–91.

Schwarzkopf, M., Exner, H. E., Fischmeister, H. F., and Schintlmeister, W. (1988) Kinetics of compositional modification of (W,Ti)C–WC–Co alloy surfaces. *Materials Science and Engineering A*, 105–106, 225–231.

Tian, H., Chen, J., Zhu, G., Du, Y., and Peng, Y. (2019) Investigation of WC–Co alloy properties based on thermodynamic calculation and Weibull distribution. *Materials Science and Technology*, 35(18), 2269–2274.

Turpin, T., Dulcy, J., and Gantois, M. (2005) Carbon diffusion and phase transformations during gas carburizing of high-alloyed stainless steels: experimental study and theoretical modeling. *Metallurgical and Materials Transactions A*, 36(10), 2751–2760.

Weibull, W. (1951) A statistical distribution function of wide applicability. *Journal of Applied Mechanics*, 18(3), 293–297.

Zhang, C., and Du, Y. (2017) A novel thermodynamic model for obtaining solid-liquid interfacial energies. *Metallurgical and Materials Transactions A*, 48(12), 5766–5770.

Zhang, W., Du, Y., Chen, W., et al. (2014) CSUDDCC1 – a diffusion database for multi-component cemented carbides. *International Journal of Refractory Metals and Hard Materials*, 43, 164–180.

Zhang, W., Du, Y., and Peng, Y. (2016) Effect of TaC and NbC addition on the microstructure and hardness in graded cemented carbides: simulations and experiments. *Ceramics International*, 42(1), 428–435.

Zhang, W. B., Du, Y., Peng, Y., Xie, W., Wen, G., and Wang, S. (2013) Experimental investigation and simulation of the effect of Ti and N contents on the formation of Fcc-free surface layers in WC–Ti(C,N)–Co cemented carbides. *International Journal of Refractory Metals and Hard Materials*, 41, 638–647.

Zhang, W. B., Sha, C. S., Du, Y., Wen, G. H., Xie, W., and Wang, S. Q. (2011) Computer simulations and verification of gradient zone formation in cemented carbides. *Acta Metallurgica Sinica*, 47(10), 1307–1314.

# 11 Case Studies on Hard Coating Design

## Contents

## 11.1  Introduction to Cutting Tools and Hard Coatings

As the core of manufacturing by machining, cutting tools are applied to remove material from the workpiece utilizing a shear deformation. Cutting tool materials must be harder than the workpiece and need to withstand the heat generated during the metal-cutting process. Different types of cutting tool materials are produced with a variety of properties and performance capabilities. Typical cutting tool materials include high-speed steel, cemented carbide, ceramics, cubic boron nitrides, and polycrystalline diamonds. High-speed steel, which was developed from tool steel, has a good strength–toughness combination and is mainly used to make cutting tools with complex shapes, preserving an excellent impact resistance. However, the cutting efficiency of the high-speed steel is restricted by a quick drop of high-temperature hardness above 550°C. Cemented carbide tools, which were initially developed in the 1930s, possess higher thermal conductivity and hardness compared to high-speed steel. The cutting speed of carbides can be three to five times faster than that of high-speed steel. The advantage of ceramics, such as $Si_3N_4$ and $Al_2O_3$, is their excellent high-temperature hardness. However, ceramic tools are rarely used in the cutting process with impact and collision because of their intrinsic brittleness. The cubic boron nitride (c–BN), which is synthesized via a phase transition from hexagonal boron nitride (h–BN) under high temperature and high pressure, can maintain high hardness and excellent wear resistance even at 1,500°C. However, c–BN has the

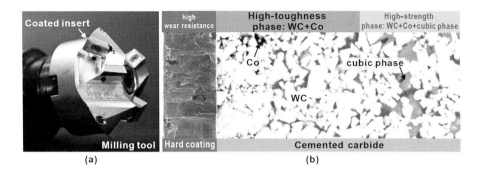

**Figure 11.1** (a) A cutting tool with indexable coated inserts; (b) microstructure of a nitride hard coating onto a gradient cemented carbide.

drawback of low toughness. As the hardest material, diamond has a high thermal conductivity and a low coefficient of friction. The wear resistance of polycrystalline diamond tools, which are sintered from diamond micropowder and a metal bonded phase, can reach 500 times that of cemented carbide. The brittleness of diamond cutters, however, limits their range of application. Also, diamond reacts with Fe and cannot be used to cut steel-based materials. In the past five years, the output value of cutting tools continues to rise in the global market, where carbide tools account for half of the output value, owing to a proper strength–toughness matching and moderate cost (Bobzin, 2017).

In order to increase the hardness and abrasive resistance, a thin coating can be deposited onto the surface of these cutting tools. Such a coating can overcome the imperfection and broaden the application range of individual tool materials. The coated carbide tools, as shown in Figure 11.1, effectively combine the high strength and toughness of the substrate with the high hardness of the coating, and therefore yield an excellent cutting performance. Nowadays, about 85% of cemented carbide tools are coated. The thickness of the protective coating deposited on the cutting tool is from several micrometers to tens of micrometers, and the nanoindentation hardness of the coating is generally above 25 GPa. Hard coatings are classified into the following categories: transition metal nitrides, transition metal carbides, transition metal borides, metal oxides, and carbon-based materials (e.g., diamond-like carbon, DLC, tetrahedral amorphous carbon, ta–C, and diamond). Among these coatings, the transition metal nitride has a complex chemical bond feature (a mixture of covalent and ionic bonds). It thus integrates high thermal and electrical conductivities of the metal as well as a high melting point and high hardness of the ceramic. At present, the most commonly used hard coating systems are TiN, TiC, TiCN, CrN, TiAlN, CrAlN, $Al_2O_3$ single layers, and their combinations, such as multilayer coatings and nanocomposite coatings.

In this chapter, we will demonstrate the design of two types of hard coatings, TiAlN-based physical vapor deposition (PVD) hard coating and TiCN chemical vapor deposition (CVD) hard coating, via a variety of calculations complemented with key experiments. For the PVD TiAlN coating, we will show how to adjust the formation of the metastable phase, select the appropriate deposition temperature, and manipulate the microstructure to obtain desired mechanical properties. The deposition of TiAlN/ TiN and TiAlN/ZrN multilayers guided with first-principles calculations will also be briefly described.

In case of CVD TiCN hard coating, we will show that the computed CVD phase diagram can accurately describe the experimentally observed phases and their compositions at a specified temperature, total pressure, as well as the partial pressures of various gases. Subsequently, the computational fluid dynamics (CFD) can provide reliable temperature fields, velocities, and distributions of various gases inside the CVD reactor. Key experiments based on these thermodynamic and CFD calculations can then be performed to deposit the TiCN hard coating efficiently.

Ancillary material for this chapter is available online at Cambridge University Press. This is the related PDF-file with complete instructions used in the through-process modeling of CVD moderate temperature MT-Ti(C,N) hard coating (Qiu et al., 2019) for the software ANSYS Fluent 16.1.

## 11.2     PVD Hard Coating

Physical vapor deposition is a well-established technique to synthesize hard coatings in industry. It utilizes thermal evaporation of a substance or energetic particles to bombard the surface of a target, so-called sputtering. This causes the target to be particularized into gas phases and/or ions, which are ultimately deposited onto the substrate under a vacuum or a low-pressure gas (or plasma) environment, as indicated in Figure 11.2. The PVD used for tool coatings mainly includes two basic categories of evaporation and sputtering. In order to obtain high purity, high density, and a large deposition rate and avoid the reaction of molecules in the air with the source material particles or colliding with each other, the PVD process requires a certain degree of vacuum $\left( <10^{-2}\,\text{Pa} \right)$.

### 11.2.1     Cathodic Arc Evaporation and Magnetron Sputtering

In this subsection, a very brief introduction about widely used cathodic arc evaporation and magnetron sputtering techniques is presented. The interested reader could refer to the comprehensive book by Martin (2010) for more details on the two techniques. Cathodic arc evaporation deposition technology uses arc discharge to evaporate cathode material (i.e., target). The discovery of cathodic arc discharge can be traced back to the study of "electricity" by Joseph Priestley in 1766, who discovered the corrosion pit left by the cathodic arc spot in the arc experiment (Anders, 2008). The application of substrate biasing technology for the sake of increasing adhesion between films and substrates and cathodic arc filtering technology aiming at reducing growth defects (Takikawa and Tanoue, 2007) has given more technical content to cathodic arc evaporation deposition. Today, cathodic arc evaporation

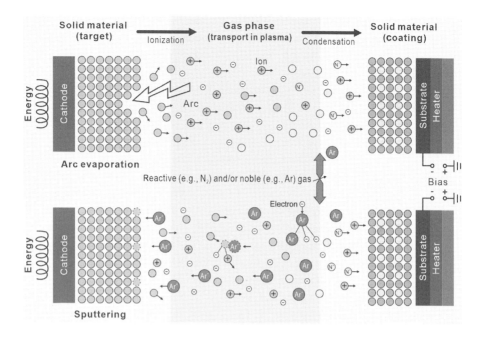

**Figure 11.2** A schematic diagram of deposition processes involving arc evaporation and sputtering for nitride coatings. The light gray circles in the coating symbolize nitrogen atoms in the metal nitride, while the dark gray circles stand for the metallic material from the target (e.g. Ti and Al) in this example.

deposition technology has been widely used for the deposition of hard coatings such as TiN, TiAlN, TiSiN, TiAlSiN, CrAlN, and CrAlSiN.

The sputtering method relies on the characteristic that the charged particles, which are accelerated under an electric field, have specified kinetic energies and are thus attracted to the sputtered substance (target). When the incident particle energy is high enough, the target materials will be sputtered out during the colliding process. These sputtered particles have certain kinetic energies and move toward the substrate to achieve film deposition, as shown in Figure 11.2. The sputtering method generally uses plasma as an ion source for bombarding the target. There are many ways to generate plasma, and the gas discharge method is commonly used for coating deposition. After filling pressure (0.1–1 Pa) of inert gas (usually Ar) under a base vacuum of $10^{-4}$ Pa, a high-voltage glow discharge is applied between the two electrodes to cause inert gas ionization. An abnormal glow discharge is a form of discharge, which is often used in thin-film sputtering or other thin-film preparation methods because it can provide a large area and relatively uniform distribution of plasma. This is a benefit for achieving uniform sputtering and thin-film deposition over a large area. However, the sputtered target particles have no directivity, leading to a low ionization rate. Consequently, many sputtered particles cannot reach the surface of the substrate, reducing the deposition rate. Direct current sputtering is the basic sputter deposition configuration. Ar is usually used as the working gas. Ar ions in the plasma near the cathode are led to the cathode target under the electric field, and the surface sputtering occurs. However, the direct current mode can only be used to sputter a conductive target. For targets that are not electrically conductive, radio-frequency or pulsed sputtering can be employed.

The introduction of a magnetic field in the sputtering apparatus (for example, placing a permanent magnet or an electromagnetic coil behind the target) can reduce the gas pressure during the sputtering process and effectively increase the sputtering efficiency and deposition rate under the same current and pressure conditions. The magnetic field prolongs the movement path of electrons in the plasma, increasing the probability of electrons participating in atomic collisions and ionization processes, thereby increasing the ionization rate of the gas near the surface of the target and contributing to the enhancement of film deposition rate. The advantage of a magnetron sputtering is the high-density plasma generated in a low-pressure environment. Such an environment reduces the energy loss due to physical collisions or charge-exchange collisions during the accelerated movement of ions in the plasma into the target. By mixing an appropriate amount of active gas such as $N_2$, $O_2$, and $CH_4$ during sputtering, the metal particles and atoms (or ions) originating from the metallic target can react with decomposed (N, O, C) from the active gas and thus form the slowly growing nitride, oxide, or carbide in the coating during deposition. By controlling the partial pressures of the reactive gas components during the deposition process, the deposited product could be a solid solution or a compound or even a mixture of several phases.

As described earlier, PVD is far from the equilibrium process, and it involves complex thermodynamic and kinetic phenomena. It will be extremely time consuming and costly if only experimental approaches are utilized to develop new protective hard coatings. Under the guidance of a variety of calculations, selected key experiments can be performed highly efficiently so that the cycle for the development of hard coatings can be reduced significantly. In the following, we will demonstrate the establishment of a metastable TiN–AlN phase diagram by taking into account of both thermodynamic and kinetic aspects (Liu et al., 2019). The metastable phase diagram can provide vital information for the composition of hard coating and deposition temperature. To improve properties of protective hard coatings, the as-deposited coatings are usually subjected to subsequent heat treatments. For example, the age hardening of TiAlN-based coatings at about 900°C for a few hours leads to a noticeable increase in hardness (Mayrhofer et al., 2003). Such an increase in hardness is beneficial for wear resistance during cutting. Thus, the microstructural evolution of TiAlZrN coating during age hardening will be described by using an elastic-chemical phase-field model (Attari et al., 2019). These microstructure simulations can provide a deep understanding on how the morphology changes through a slight change of composition and heat treatment schedule (such as temperature and time), paving a novel way to achieve desired properties according to microstructure design. Multilayer coatings usually possess better properties than individual single layers. To guide the experiment for the deposition of multilayer coatings, we will briefly describe how first-principles calculations can provide a clear insight into the effect of the modulation ratio on the interface structure of TiAlN/TiN and TiAlN/ZrN multilayers (Xu et al., 2017).

## 11.2.2 Metastable Phase Formation and TiN–AlN Phase Diagrams

Cathodic arc evaporation and magnetron sputtering generally lead to the formation of a metastable phase. In the field of metal cutting and forming, the TiN–AlN system has been regarded as a benchmark hard coating. The metastable ternary Fcc–TiAlN phase was introduced through the addition of Al to TiN to improve the oxidation resistance of TiN coating (Münz, 1986), which is oxidized at about 600°C.

Figure 11.3a shows the stable TiN–AlN pseudobinary phase diagram (Zeng and Schmid-Fetzer, 1997) and Figure 11.3b the metastable TiN–AlN phase formation diagram under different deposition temperatures and deposition rates (Liu et al., 2019). First-principles calculations can be performed to calculate the thermodynamic quantities of metastable forms. For the example of AlN, the Gibbs energy difference between the stable hexagonal form of AlN (Hcp–AlN), which is more precisely the hexagonal wurtzite structure of w–AlN, and its metastable structure Fcc–A1N, which is more precisely the cubic zincblende structure of c–AlN (Lin and Bristowe, 2007; Wang et al., 2012a, 2012b; Zhang and Veprek, 2007) and elastic properties of nitrides (Holec et al., 2013; Wang et al., 2014, 2015) are important. These calculated thermodynamic properties are key inputs for thermodynamic modeling of these metastable phase diagrams by means of CALPHAD approach, and the computed elastic properties are needed for an elastochemical phase-field simulation of the microstructural evolution of hard coatings during heat treatment.

Currently, two computational approaches are used to establish metastable phase diagrams in nitride-based PVD systems. One is based on energetic (thermodynamic) consideration. First-principles calculation, the CALPHAD method, or their combination belong to this category. The other relies on both energetic and kinetic (such as surface diffusion) considerations. Using the CALPHAD approach, one can calculate the metastable solubility limits during thin-film deposition according to Gibbs energy versus composition diagrams (Saunders and Miodownik, 1987). Several CALPHAD-type calculations (Liu et al., 2019; Spencer, 2001; Spencer and Holleck, 1989) reported that the maximal solubility values of Al in metastable Fcc–TiAlN are within the range of 60–72 mol% AlN, as shown in Figure 11.3b. For example, taking the underlying Gibbs energy functions of the stable phase diagram in Figure 11.3a, the intersection of the $G$(Fcc) and $G$(Hcp) curves at 0°C occurs at $x = 0.68$ or 68 mol% AlN in the TiN–AlN system. Thus, the "metastable TiN–AlN phase diagram," which is defined by excluding the formation of a two-phase structure at 0°C, shows single-phase Fcc for $0 < x \leq 0.68$ and single-phase Hcp for $0.68 \leq x < 1$. That is in reasonable agreement with the elaborate metastable phase formation diagram in Figure 11.3b only at the lowest temperature.

Also based on energetic data, a few first-principles calculations (Euchner and Mayrhofer, 2015; Hans et al., 2017; Holec et al., 2010; Hugosson et al., 2003; Mayrhofer et al., 2006), even including Fcc/Hcp interfacial energy (Hans et al.,

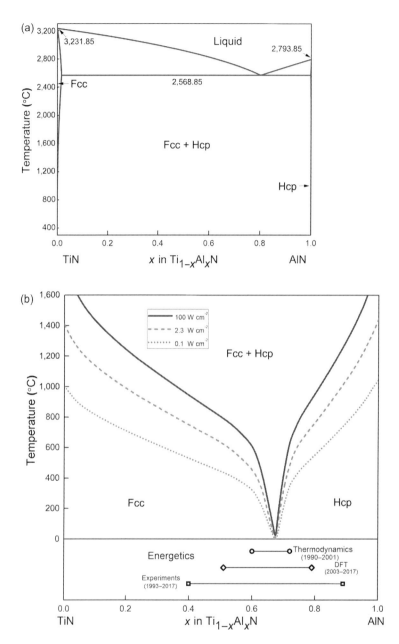

**Figure 11.3** (a) Stable TiN–AlN pseudobinary phase diagram (Liu et al., 2019; Zeng and Schmid-Fetzer, 1997); (b) metastable TiN–AlN phase formation diagram (Liu et al., 2019) under different deposition temperatures and power density rates. The maximal solubility limit values of Al in metastable Fcc–TiAlN from thermodynamic calculations, first-principles calculations, and experiments are also indicated in (b). The figures are reproduced from Liu et al. (2019) with permission from Elsevier.

2017), predict a wider range of maximal solubility from 50–79 mol% AlN. It is clear that the computational approach considering the energetic aspect alone cannot describe the experimentally observed maximal solubility values (40–90 mol% AlN) of Al in metastable Fcc–TiAlN, as indicated in Figure 11.3b for the temperature range $0-550°C$.

In view of this drawback associated with the modeling solely based on energetic consideration, a model that describes the formation of metastable phase(s) during thin-film growth according to information on both thermodynamics and kinetics was recently developed by Chang and coauthors (Chang et al., 2015, 2016). It was clarified that the surface diffusion during the deposition of PVD hard coating governs the metastable phase formation, and the diffusion distance of a single atom can be described as follows (Einstein, 1908):

$$X = \sqrt{2D_s t}, \tag{11.1}$$

in which $X$ is the diffusion distance, $D_s$ the surface diffusivity, and $t$ the diffusion time. Later on, the relationship to describe the surface diffusion, deposition rate, and diffusion distance during deposition was described by Cantor and Cahn, (1976):

$$X = \sqrt{2v\frac{a}{r_D} \cdot a} \cdot \exp\left(-\frac{Q_s}{2k_B T}\right), \tag{11.2}$$

where $a$ is the individual jump distance, $Q_s$ the atomic activation energy for surface diffusion, $v$ the atomic vibrational frequency (Ohring, 2001), $r_D$ the deposition rate, $k_B$ the Boltzmann constant, and $T$ the substrate temperature in Kelvin. Equation (11.2) was revised into the following description considering the critical temperature ($T_c$) and diffusion distance ($X_c$) (Chang et al., 2015):

$$X_c = \sqrt{2v\frac{a}{r_D} \cdot a} \cdot \exp\left(-\frac{Q_s}{2k_B T_c}\right). \tag{11.3}$$

The composition dependence of $X_c$ can be obtained using (11.3) with all the parameters determined by the comparative theoretical and experimental data (Chang et al., 2015, 2016).

Generally speaking, one key combinatorial deposition experiment along with the CALPHAD and first-principles calculations is needed to establish a complete metastable phase formation diagram for PVD hard coating. One can establish the relationship between the solubility and the critical temperature toward constructing the metastable phase formation diagrams for the sputtered thin films. The details on the phase formation diagrams modeling can be found in the literature (Chang et al., 2015, 2016). The metastable TiN–AlN phase formation diagram constructed with the preceding equations and shown in Figure 11.3b agrees well with the experimental data. Such phase (formation) diagrams play a key role in designing the experiment to deposit thin films. Higher Al solubility in Fcc–TiN results in better properties of nitride thin films. However, the formation of the Hcp phase deteriorates their properties drastically. Therefore, the maximum Al solubility is of great interest, which can be directly read from Figure 11.3b. In order to obtain the maximum solubility of Al in Fcc–TiN, one can use Figure 11.3b as guidance to select appropriate setup parameters (i.e., temperature and power density) during magnetron sputtering. This is definitely more efficient than the trial-and-error method.

## 11.2.3    Spinodal Decomposition

In the preceding subsection, we described the establishment of the metastable TiN–AlN phase formation diagram based on which the composition of the as-deposited hard coating can be obtained under specified experimental conditions. The properties of as-deposited protective hard coatings can be manipulated by subsequent heat treatments. Age hardening is one of the important experimental methods to enhance the hardness of as-deposited hard coatings (Chen et al., 2008; Hörling et al., 2002). Such an increase of hardness originates from the coherency strain because of the precipitation of c–TiN-rich and c–AlN-rich nanometer-size domains from c–TiAlN solid solution via spinodal decomposition (Mayrhofer et al., 2003). Here "c" stands for cubic, as detailed previously for c–BN before. The nanostructure due to spinodal decomposition is metastable, and under high-temperature conditions such as high-speed machining, the thermally activated diffusion could lead to the coarsening of the microstructure. In addition, the metastable cubic c–AlN-rich zincblende phase will transform into its stable hexagonal wurtzite-type AlN (w–AlN) when the temperature continues to increase. Such a phase transformation causes a dramatic loss of performance since a molar volume expansion of about 26% is associated with this transformation. Consequently, extensive research work has been performed to understand the mechanisms behind age hardening for c–TiAlN-based coatings and the corresponding microstructure evolution.

The phase-field method can be utilized to simulate the microstructure evolution during the age hardening of the hard coatings. Since the differences between lattice parameters and elastic constants of the different binary nitrides (Holec et al., 2013; Wang et al., 2014, 2015) are significant, the elastic contribution to the microstructure evolution should be included in phase-field simulations. In addition, the elastic anisotropy of different nitrides is expected to make a noticeable contribution to the morphology pattern of the microstructure. Integrating the Cahn–Hilliard formulation (Cahn, 1961) for chemical spinodal decomposition and linear microelasticity theory, Attari et al. (2019) performed a phase-field simulation for $Ti_{1-x-y}Al_xZr_yN$ alloys within $x = 0.25 - 0.7$ and $y = 0.05$ or $0.24$ at $1,200°C$, where $x$ and $y$ are the site fractions of Al and Zr in the metal sublattice (Ti,Al,Zr), respectively. Depending on the composition and operating conditions, the phase-field simulation for such a quaternary hard coating demonstrates the product of $c-AlN + c-Ti(Zr)N$ or $c-AlN + c-ZrN + c-Ti(Zr)N$ after age hardening and a final phase transition of c–AlN to w–AlN. For details about this phase-field simulation, the interested reader may refer to Attari et al. (2019). Figure 11.4 shows the experimental microstructure of $Ti_{0.3}Al_{0.46}Zr_{0.24}N$ at $1,200°C$ for 2 hours along with the phase-field simulated microstructure. The simulated microstructure and energy-dispersive X-ray

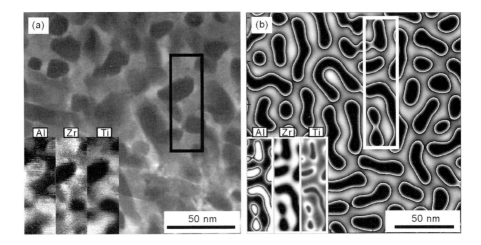

**Figure 11.4** (a) STEM micrograph of $Ti_{0.3}Al_{0.46}Zr_{0.24}N$ coating annealed at 1,200°C for 2 hours. Reproduced from Lind et al. (2014) with permission from Elsevier. (b) Phase-field simulation from Attari et al. (2019) with permission from Elsevier. The regions marked by rectangles in (a) and (b) are shown in the left-bottom inserts, indicating the corresponding elemental maps.

spectroscopy (EDX) elemental maps agree well with the experimental ones. The good agreement between simulation and experiment (Attari et al., 2019) demonstrates that microstructure design based on simulations can provide a rigorous understanding on how morphology patterns evolve during heat treatment of hard coatings for the sake of achieving desired properties.

## 11.2.4   Multilayer Hard Coating

PVD TiAlN hard coatings with excellent mechanical and thermal properties are widely used in the manufacturing industry (Mayrhofer et al., 2003; PalDey and Deevi, 2003). However, due to the widespread use of advanced cutting techniques (e.g., high-speed and dry cutting), the tool edge temperature may go beyond 1,000°C, which exceeds the working temperature (about 850°C) of the currently used TiAlN coatings. Combining the advantages of individual single layers, a multilayer architecture to improve the mechanical and thermal properties of TiAlN has attracted extensive attention (Barnett and Shinn, 1994; Holleck and Schier, 1995). Rotating the substrate at a specific rate to collect substance from different target materials is a straightforward way to prepare multilayer coatings, as shown in Figure 11.5. The modulation period is the repeating total thickness of the multilayers, whereas the modulation ratio describes the thickness ratio of the repeating single layers. The compositionally modulated multilayers prepared by alternating deposition of two kinds of nitrides can obtain a synergistic improvement for hardness, toughness, and thermal stability (Buchinger et al., 2019; Madan et al., 1997; Pei et al., 2019).

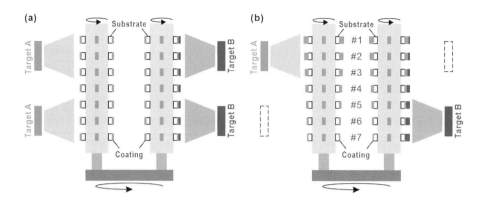

**Figure 11.5** The schematics of the deposition for multilayered coatings with various (a) modulation periods and (b) modulation ratios.

Recently, theoretical calculations (such as CALPHAD, first-principles, molecular dynamics, etc.) as materials design tools to accelerate the development of hard coatings have been widely recognized (Holec et al., 2017). For nanomultilayer coatings, the modulation period largely determines their structure and properties. Consequently, much theoretical work has been done to find the critical modulation period of the nanomultilayer coating for the optimal design of nanomultilayer geometry (Chawla et al., 2013; Ivashchenko et al., 2015; Marten et al., 2010; Stampfl and Freeman, 2012; Wang et al., 2016). The first-principles calculations by Chawla et al. (2014) show that the critical modulation period of the AlN/TiAlN system is influenced by the chemical composition and interface dislocation density of TiAlN. Also, the differences in lattice parameters and electronic structures of the different nitrides can affect the stability of the coherent interface (Ivashchenko et al., 2014). In addition to the modulation period, the interfacial structure and mechanical properties of multilayer coatings are also closely related to corresponding modulation ratios. Li et al. (2007) demonstrated that the TiAlN/ZrN coating with a bilayer thickness of 6.5 nm has the highest hardness of $\sim 30$ GPa when the modulation ratio $t_{TiAlN} : t_{ZrN}$ ($t_{TiAlN}$ and $t_{TiN}$ are the thicknesses of TiAlN and TiN sublayers, respectively) is 3:2. In this subsection, we will demonstrate the achievement of a high hardness for the multilayer TiAlN/TiN and TiAlN/ZrN coatings through the manipulation of the modulation ratio using the first-principles calculations in conjunction with key experiments (Xu et al., 2017).

Firstly, the energy difference ($\Delta E$) between the TiAlN–MeN (Me = Ti or Zr) structure and the single-layered one was calculated through the first-principles method. For simplification, the energy difference is approximated as the enthalpy difference at zero Kelvin. As depicted in Figure 11.6a, $\Delta E$ of TiAlN(001)–TiN(001) supercells with different $t_{TiAlN} : t_{TiN}$ ratios are almost constant at zero, with a slightly negative $\Delta E = -0.004$ eV/atom. The negative $\Delta E$ shows that the thermodynamic stability of the coherent interface in TiAlN(001)/TiN(001) is given for all ratios of TiAlN and TiN layers. In contrast, the positive $\Delta E$ with a large variation in TiAlN (001)/ZrN(001) suggests the thermodynamic instability regardless of the ratios for

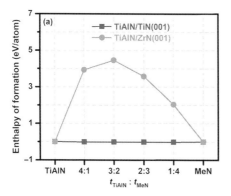 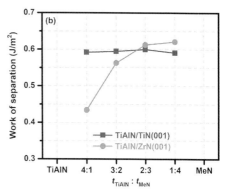

**Figure 11.6** First-principles calculated (a) enthalpy of formation and (b) work of separation for cubic TiAlN/TiN(100) and TiAlN/ZrN(100) interface structures. The figures are reproduced from Xu et al. (2017) with permission from Elsevier.

$t_{TiAlN} : t_{ZrN}$. The TiAlN/ZrN with $t_{TiAlN} : t_{ZrN} = 1 : 4$ shows the lowest $\Delta E$, indicating the highest possibility for achieving coherent interfaces.

The work of separation ($W_{sep}$) for the interfaces in the TiAlN/MeN (Me = Ti or Zr) supercells was also computed using first-principles calculations. The $W_{sep}$ means the work required to separate the coherent interface reversibly into two free surfaces corresponding to pure nitrides and was calculated from the energy difference per unit area corresponding to the introduction of the interface in comparison to two separate slabs (Lin and Bristowe, 2007).

TiAlN(100)/TiN(100) with different modulation ratios reveals a similar $W_{sep}$ value, as shown in Figure 11.6b, indicating the comparable interface strength. The $W_{sep}$ of TiAlN/ZrN exhibits a continuous increase with the decrease of $t_{TiAlN} : t_{ZrN}$. The highest $W_{sep}$ of the supercell occurs at $t_{TiAlN} : t_{ZrN} = 1 : 4$, corresponding also to the lowest $\Delta E$ among all the TiAlN/ZrN supercells, suggests that the coherent growth with this modulation ratio is more likely to occur.

Secondly, according to the preceding calculations, a series of TiAlN/TiN and TiAlN/ZrN multilayers (3~5 μm) with different modulation ratios were prepared by cathodic arc evaporation and investigated regarding interfacial structures. Ti$_{50}$Al$_{50}$ and Me (Me = Ti or Zr) targets were mounted at different heights in the deposition chamber, as indicated in Figure 11.5b. A series of nanomultilayers with a comparable modulation period but different modulation ratios can be obtained by mounting the substrates at different heights indicated by #1 to #7. Transmission electron microscopy (TEM) results confirm the epitaxial growth with coherent interfaces for all TiAlN/TiN multilayers. The bright-field image, as shown in Figure 11.7a (left panel), reveals the well-defined nanoscale layered structure with alternate bright and dark contrasts. The scanning transmission electron microscope (STEM) high-angle annular dark field (HAADF) micrograph of an individual columnar grain, as shown in Figure 11.7a (right panel), indicates that TiAlN sublayers are coherent to TiN sublayers. These experimental results agree well with the calculations, where the coherent interface is

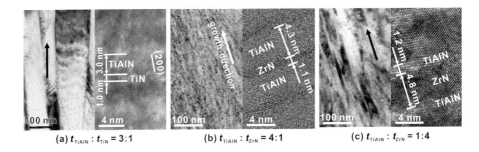

**(a)** $t_{\text{TiAlN}} : t_{\text{TiN}} = 3{:}1$    **(b)** $t_{\text{TiAlN}} : t_{\text{ZrN}} = 4{:}1$    **(c)** $t_{\text{TiAlN}} : t_{\text{ZrN}} = 1{:}4$

**Figure 11.7** TEM bright-field and lattice-resolved STEM HAADF images comparing three different multilayers: (a) TiAlN/TiN coating with a modulation period of 4 nm and $t_{\text{TiAlN}} : t_{\text{TiN}} = 3 : 1$, showing the columnar grain growth with coherent interfaces; (b) TiAlN/ZrN coating with $t_{\text{TiAlN}} : t_{\text{ZrN}} = 4 : 1$, indicating fine grain size and incoherent interfaces; (c) TiAlN/ZrN coating with $t_{\text{TiAlN}} : t_{\text{ZrN}} = 1 : 4$, presenting slender columnar crystals with coherent interfaces. The figures are reproduced from Xu et al. (2017) with permission from Elsevier.

stable for various modulation ratios. The calculated $\Delta E$ and $W_{sep}$ for the TiAlN/ZrN interface structures imply that coherent growth in the TiAlN/ZrN supercell with $t_{\text{TiAlN}} : t_{\text{ZrN}} = 1 : 4$ is more likely to occur. The TEM results for TiAlN/ZrN multilayers are shown in Figure 11.7b, in which the multilayer with a higher modulation ratio $(t_{\text{TiAlN}} : t_{\text{ZrN}} = 4 : 1)$ exhibits nonepitaxial growth. The grain growth of TiAlN is periodically interrupted by ZrN sublayers, resulting in a very fine grain size with incoherent interfaces (Figure 11.7b). When the modulation ratio changes to 1:4, the TiAlN/ZrN multilayer indeed can obtain coherent interlayer interfaces with fine columnar grain morphology as presented in Figure 11.7c.

Thirdly, the hardness and elastic modulus of the TiAlN/TiN and TiAlN/ZrN hard coatings are measured by nanoindentation. Figure 11.8 presents the hardness and elastic modulus of these multilayer coatings as a function of the modulation ratio. The hardnesses of the TiAlN, TiN, and ZrN monolithic coatings are $28.7 \pm 0.5$, $27.1 \pm 0.6$, and $27.4 \pm 0.4$ GPa, respectively. The coherent interface strengthening effect of TiAlN/TiN multilayer coatings with different modulation ratios leads to a higher hardness than that of TiAlN or TiN coating. Among them, TiAlN/TiN-5# coating shows the highest hardness value of $32.4 \pm 0.6$ GPa among the deposited coatings. According to the Koehler model (Koehler, 1970), the maximum hardness increase of the multilayer coating is related to the difference in shear modulus of the different constituent materials. Since the shear modulus of TiAlN and TiN is close to each other, the hardness enhancement of TiAlN/TiN multilayer coatings is not apparent. For TiAlN/ZrN multilayers, the hardness reveals a more significant correlation with the modulation ratio. The hardness of the TiAlN/ZrN-1# coating with incoherent interfaces is $27.4 \pm 0.8$ GPa between the value of the TiAlN and ZrN, which is consistent with the mixing rule. As the modulation ratio decreases, the hardness of the TiAlN/ZrN-2#, 3#, and 4# coatings decreases to $23.9–25.0$ GPa due to their lower degree of crystallization. The crystallinity of multilayers increases

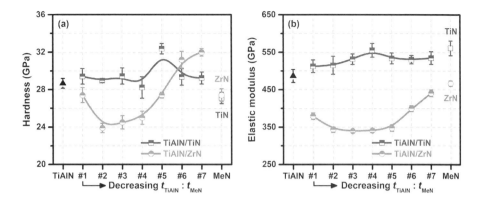

**Figure 11.8** Nanoindentation (a) hardness and (b) elastic modulus of TiAlN/TiN and TiAlN/ZrN multilayer coatings with various modulation ratios. The figures are reproduced from Xu et al. (2017) with permission from Elsevier.

with the further decrease of the modulation ratio, and thus improves the coating hardness. Notably, the strengthening effect from coherent interfaces makes the hardness of TiAlN/ZrN-6# and 7# coatings increase significantly. The TiAlN/ZrN-7# multilayer coating exhibits the highest hardness of H = 32.0 ± 0.4 GPa.

The elastic modulus of TiAlN/TiN and TiAlN/ZrN multilayers, as depicted in Figure 11.8b, shows a similar trend as hardness. Owing to their consistent epitaxial growth with different modulation ratios, TiAlN/TiN multilayer coatings obtain comparable elastic modulus in the range of 475–550 GPa. First a decrease and then an increase in modulus of elasticity can be detected for TiAlN/ZrN multilayers. The incoherent interface structure with very fine crystalline grains results in the decrease of elastic modulus for TiAlN/ZrN-2# to 4#. Also, the epitaxial growth with well-developed columnar grain growth leads to the rise of elastic modulus for TiAlN/ZrN with a lower modulation ratio. The highest elastic modulus (440.6 ± 8.9 GPa) was achieved by the TiAlN/ZrN-7# coating with a modulation ratio of 1:4. The coherent interfaces are unstable for the TiAlN/ZrN system due to a significant lattice mismatch between TiAlN and ZrN. The decrease of the modulation ratio is beneficial to the formation of coherent interfaces and thus improves the hardness and elastic modulus of TiAlN/ZrN.

The case study for the TiAlN/TiN and TiAlN/ZrN multilayers demonstrates that an efficient approach combining first-principles calculations and key experiments can be used to design TiAlN-based multilayered coatings efficiently.

## 11.3 CVD Hard Coating

### 11.3.1 Experimental Setup

CVD could be defined as a deposition of solid on a heated surface from a gas phase because of chemical reactions among gas species. It is one of the vapor-transfer processes where the deposition relies on atoms or molecules or their combination. Simply speaking, nucleation from the (supersaturated) gas mixture occurs only at the substrate surface, usually the hottest part in the CVD reactor. While the solid layer grows, it is usually very close to thermodynamic equilibrium at the gas–solid interface. This explains the huge significance of thermodynamic modeling and calculations in the multicomponent gas + solid two-phase equilibrium.

Figure 11.9 (Pierson, 1999) shows a schematic of CVD deposition process. CVD coatings are widely used in the fields of metal cutting, oil exploration, electrical, optoelectrical, optical, mechanics, chemistry, and so on. A brief summary for state-of-the-art CVD protective hard coating is listed in Table 11.1. TiN, TiC, moderate temperature MT-Ti(C,N), TiCN, $Al_2O_3$, TiAlN, $TiB_2$, and diamond are the most widely used CVD hard coatings for cutting tools. A typical schematic diagram for

**Table 11.1** Summary for CVD hard coatings as cutting tools.

| Year | Hard coating | Process | Substrate of cutting tool |
|------|--------------|---------|---------------------------|
| 1968 | TiC, TiN | CVD | Cemented carbide |
| 1973 | TiCN, TiC + $Al_2O_3$ | CVD | Cemented carbide |
| 1981 | TiC + $Al_2O_3$ + TiN, Al-O-N | CVD | Cemented carbide |
| 1982 | TiCN | MT-CVD | Cemented carbide |
| 1986 | Diamond, c–BN | CVD | Cemented carbide |
| 1990 | TiN, TiCN, TiC | PCVD | High-speed steel |
| 1993 | TiN + TiCN + TiN | CVD | Cemented carbide |
| 1996 | Thick-fiber-like TiCN | MT-CVD[*] | Cemented carbide |
| 1996 | $CN_x$ | CVD | High-speed steel |
| 2003 | $TiN/\alpha\text{-}Si_3N_4{}^{+}$ | PCVD[*] | Cemented carbide |
| 2008 | TiAlN | CVD | Cemented carbide |
| 2013 | TiSiCN | CVD | Cemented carbide |

[*] MT-CVD and PCVD mean moderate temperature and plasma-enhanced chemical vapor deposition, respectively. $^{+}\alpha$ means amorphous.

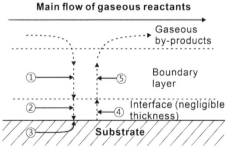

**Main flow of gaseous reactants**

Gaseous by-products

Boundary layer

Interface (negligible thickness)

Substrate

1. Diffusion of reactants through boundary layer
2. Reactants adsorption on substrate
3. Chemical reaction occurs
4. Desorption of adsorbed species
5. By-products diffusion out

**Figure 11.9** Sequence of phenomena during the CVD coating deposition process (Pierson, 1999).

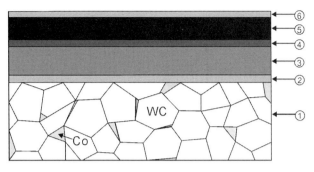

① Cemented carbide substrate

② TiN: Co diffusion barrier

③ MT–Ti(C,N): Wear resistance

④ TiAlOCN: Bonding layer for MT–Ti(C,N) and Al₂O₃

⑤ Al₂O₃: Oxidation resistance

⑥TiN: Crater wear resistance and easy identification of wear

**Figure 11.10** A schematic of CVD multilayered hard coating.

multilayered CVD hard coating is shown in Figure 11.10. Each individual layer has its own function. Currently, most of the CVD hard coatings utilized for cutting tools are usually developed by the costly and time-consuming trial-and-error method. In the following subsection, a through-process modeling for CVD MT–Ti(C,N) hard coating will be demonstrated step by step.

## 11.3.2 Through-Process Modeling of CVD MT–Ti(C,N) Hard Coating

Figure 11.11 shows a flowchart for the through-process modeling of CVD moderate temperature MT–Ti(C,N) hard-coating, which combines thermodynamic calculations, computational fluid dynamics (CFD) simulation, and decisive experiments to develop CVD-hard coatings efficiently. Thermodynamic calculations can predict the phase assemblages and compositions of CVD coatings deposited at different temperatures, pressures, and gas ratios. The thermodynamic calculations employ the Thermo-Calc software (Andersson et al., 2002) applied to thermodynamic databases of gases and solution phases. CFD simulations can predict the distributions of temperature, pressure, gas velocity and concentration, and deposition rate for a specific process in the CVD reactor. Under the guidance of both thermodynamic calculations and CFD simulations, key experiments can be carried out by selecting appropriate deposition parameters. Furthermore, the measured chemical compositions, phases, and deposition rates of CVD coatings give feedback to the predicted results for the sake of refining the simulations. If large differences exist between experiment and calculation, the thermodynamic parameters and/or the CFD models, meshes and setups can be modified in order to obtain a reasonable agreement between calculation and experiment. In practice, there could be a few such cycles.

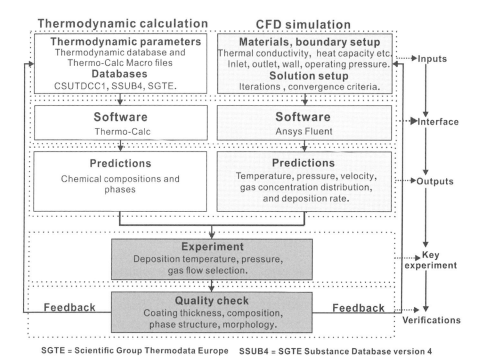

SGTE = Scientific Group Thermodata Europe    SSUB4 = SGTE Substance Database version 4
CSUTDCC1 = Central South University Thermodynamic Database for Cemented Carbide version 1

**Figure 11.11** A flowchart for the development of CVD multilayered coating through a through-process modeling along with experiment.

Firstly, thermodynamic calculations are performed for MT–Ti(C,N) hard coatings. Many deposition parameters influence the compositions and phase assemblage of MT–Ti(C,N) coatings. Thermodynamic calculations can provide appropriate values for these parameters. Through the minimization of the Gibbs energy for the system (Deng et al., 2012), thermodynamic simulations were performed for the CVD MT–Ti (C,N) coatings using Thermo-Calc software (Andersson et al., 2002) applied to the Ti–C–N–H–Cl system including the $TiCl_4$–$CH_3CN$–$N_2$–$H_2$ initial gas phase and various solid phases. The general principles of such CALPHAD calculations are provided in Chapter 5. In the simplest case, one may write

$$\min G = \min \left\{ \sum_{i=1}^{s} n_i^{cond} \Delta G_{m,j}^{\theta}(cond.) + \sum_{i=s+1}^{N} n_i \left[ \Delta G_{m,i}^{\theta}(gas) + RT \ln p + RT \ln \frac{n_i}{\sum\limits_{j=s+1}^{N} n_j} \right] \right\}.$$

(11.4)

in which $s$ is the total number of the stoichiometric condensed phases considered and all kinds of gas species in the system are counted from $s + 1$ to $N$. The total pressure is $p$, $n_i^{cond}$ is the amount of the $i^{th}$ condensed phase and $n_i$ is the amount of the $i^{th}$ gaseous species. In the case of solid solution phases, such as the most important

**Table 11.2** Solid phases and gaseous species considered in thermodynamic calculations of the Ti–C–H–C1 system (Qiu et al., 2019); Hcp–Ti, Bcc–Ti, Fcc–Ti(C,N), and Ti$_2$(N,C) are solid-solution phases.

| Solid phases | Gaseous species |
|---|---|
| Hcp–Ti, Bcc–Ti, Fcc–Ti(C,N)), Ti$_2$(N,C), C, TiCl$_2$, TiCl$_3$, TiH$_2$ | H, H$_2$, N$_2$, HCl, TiCl$_4$, Cl, Cl$_2$, Ti, Ti$_2$, TiCl, TiCl$_2$, TiCl$_3$, TiCl$_6$, C, C$_2$, C$_3$, C$_4$, C$_5$, C$_{60}$, CH, CH$_2$, CH$_3$, CH$_4$, C$_2$H, C$_2$H$_2$, C$_2$H$_3$, C$_2$H$_3$N, C$_2$H$_4$, C$_2$H$_5$, C$_2$H$_6$, C$_3$H, C$_3$H$_3$, C$_3$H$_4$, C$_3$H$_6$, C$_3$H$_8$, C$_4$H, C$_4$H$_{10}$, C$_2$H$_2$, C$_4$H$_4$, C$_4$H$_6$, C$_4$H$_8$, C$_6$H$_6$, CCl, CHCl, CH$_2$Cl, CH$_3$Cl, CCl$_2$, CHCl$_2$, CH$_2$Cl$_2$, CCl$_3$, CHCl$_3$, CCl$_4$, C$_2$Cl, C$_2$ClH, C$_2$ClH$_3$, C$_2$ClH$_5$, C$_2$Cl$_2$, C$_2$Cl$_2$H, C$_2$Cl$_2$H$_2$, C$_2$Cl$_2$H$_4$, C$_2$Cl$_3$, C$_2$Cl$_3$H, C$_2$Cl$_3$H$_3$, C$_2$Cl$_4$, C$_2$Cl$_4$H$_2$, C$_2$Cl$_5$, C$_2$Cl$_5$H, C$_2$Cl$_6$, C$_6$ClH$_5$, CClN, CHN, CN, CN$_2$, C$_2$N, C$_2$N$_2$, C$_3$HN, C$_3$N, C$_4$N, C$_4$N$_2$, C$_5$HN, C$_5$N, C$_6$N, C$_6$N$_2$, C$_9$N, NH, N$_3$H, NH$_2$, NH$_3$, N$_2$H$_4$, N$_2$H$_2$, N$_2$H$_4$, N, TiN, N$_3$ |

Fcc–Ti(C,N) phase treated here, the more general equations for Gibbs energy in Chapter 5 apply.

The thermodynamic descriptions for the Ti–C–N system were taken from the work of Frisk et al. (2004). Furthermore, the thermodynamic parameters for the gaseous species were adopted from the SSUB4 database, which was developed by Scientific Group Thermodata Europe (SGTE). The solid phases and gaseous species considered in thermodynamic calculations are shown in Table 11.2. The conditions for each thermodynamic calculation are those at a given temperature and pressure. A defined initial gas phase composition of TiCl$_4$–CH$_3$CN–N$_2$–H$_2$ is imposed onto the substrate so that under equilibrium the solid phase(s) and a more complex gas are formed. The H$_2$ is typically the carrier gas and the balance in initial gas composition, e.g. TiCl$_4$ = 2.27%, CH$_3$CN = 0.76%, N$_2$ = 24.24%, and H$_2$ = 72.73% (in vol %). The MT–Ti(C,N) coatings may be single-phase Fcc–Ti(C,N), single-phase Fcc–Ti(C,N) saturated with graphite, single phase Ti$_2$(**N**,C), two-phase Fcc–Ti(C,N) + Ti$_2$(**N**,C), or two-phase Fcc–Ti(C,N) + Hcp–Ti. Generally speaking, the pure or dominant single-phase Fcc–Ti(C,N) is preferable for MT–Ti(C,N) coatings. Nevertheless, the graphite phase will not be formed during the deposition process of the coating due to the sluggish kinetics. Thus the graphite phase could be suspended during thermodynamic calculations to ensure the existence of the single-phase Fcc–Ti(C,N).

The effects of deposition temperature, pressure, and mole fraction of CH$_3$CN on the chemical compositions of the MT–Ti(C,N) coating are shown in Figure 11.12a–c. When the deposition temperature increases from 900 to 1,200 K, as shown in Figure 11.12a, C content increases continuously and the variation of N shows an opposite tendency to that of C. The compositions of MT–Ti(C,N) coatings are almost unaffected by the deposition pressure within the range of 8–20 kPa, as shown in Figure 11.12b. CH$_3$CN influences the chemical compositions significantly when its mole fraction is less than 0.04 (Figure 11.12c). With a higher mole fraction of CH$_3$CN ($>$0.04), the coating compositions are constant and the constitution of

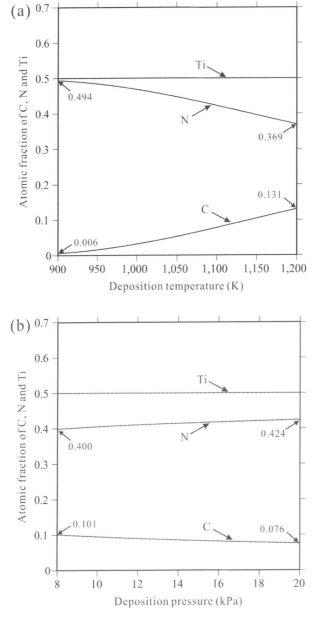

**Figure 11.12** Calculated chemical compositions of the single-phase Fcc–Ti(C,N) in MT–Ti(C,N) coatings depending on (a) deposition temperature; (b) deposition pressure; (c) the mole fraction of CH$_3$CN; (d) the MT–Ti(C,N) coating with different C contents as a function of deposition temperature and pressure. The basis conditions for thermodynamic calculations are $p = 12$ kPa, $T = 1{,}123$ K, TiCl$_4 = 2.27\%$, CH$_3$CN $= 0.76\%$, N$_2 = 24.24\%$, H$_2 = 72.73\%$, and they are varied as follows: (a) $T = 900$–$1{,}200$ K; (b) $P = 8$–$20$ kPa; (c) CH$_3$CN $= 0$–$20\%$, H$_2 =$ Balance; (d) iso-$y''_C$ curves, where $y''_C$ is the site fraction of C in the second sublattice of Ti$_1$(C,N)$_1$ used for the phase Fcc–Ti(C,N) within the pressure range $p = 2$–$20$ kPa and temperature range $T = 700$–$1{,}300$ K.

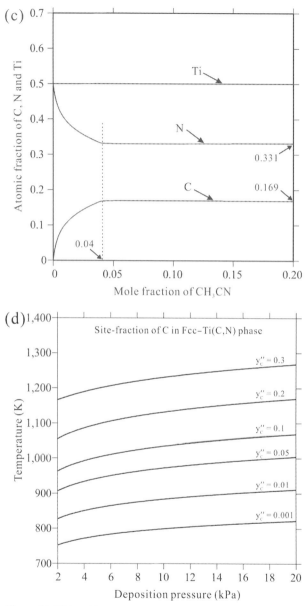

Figure 11.12 (cont.)

MT–Ti(C,N) coating is still in single-phase Fcc–Ti(C,N) due to the sluggish kinetics for the formation of graphite. Based on the CVD phase diagram, as indicated in Figure 11.12d, the single-phase Fcc–Ti(C,N) in MT–Ti(C,N) coatings can be designed with different C contents by selecting the deposition conditions.

Three MT–Ti(C,N) coatings are designed under the guidance of thermodynamic calculations through choosing deposition temperatures of 1,063, 1,103, and 1,153 K under a pressure of 12 kPa. The chemical compositions of the deposited coatings are

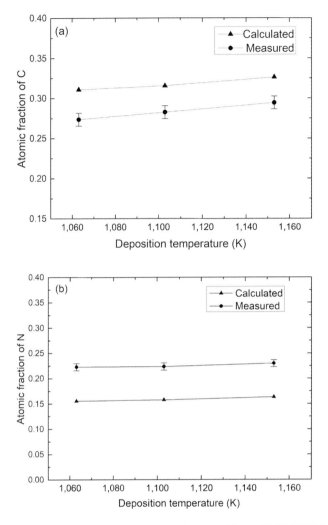

**Figure 11.13** Calculated and measured compositions of MT–Ti(C,N) coatings under the total pressure of 12 kPa. The initial gas composition is $TiCl_4 = 1.95\%$, $CH_3CN = 0.76\%$, $N_2 = 24.32\%$, $H_2 = 72.97\%$ for $T = 1,063$ K; $TiCl_4 = 1.98\%$, $CH_3CN = 0.76\%$, $N_2 = 24.31\%$, $H_2 = 72.95\%$ for $T = 1,103$ K; $TiCl_4 = 2.52\%$, $CH_3CN = 0.87\%$, $N_2 = 13.00\%$, $H_2 = 83.61\%$ for $T = 1,153$ K: (a) C; (b) N; (c) Ti. The diagrams are reproduced from Qiu et al. (2019) with permission from Elsevier.

measured and compared with calculated ones, as indicated in Figure 11.13. The variation tendency of the calculated composition of MT–Ti(C,N) coatings versus deposition temperature agrees reasonably with that of the experimental composition (Qiu et al., 2019).

Secondly, CFD simulations are carried out for the MT–Ti(C,N) coatings. The properties of the coating are not only influenced by phase assemblage but also affected by the flow dynamics of gas components, especially for an industrial-scale reactor with complex flow paths. CFD simulations can provide information on the distribution

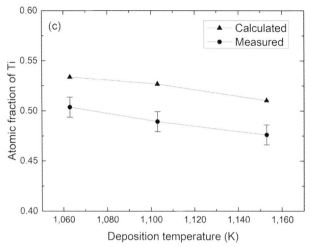

Figure 11.13 (cont.)

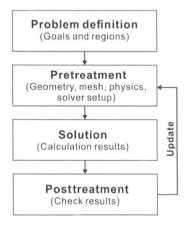

**Figure 11.14** Procedure for CFD simulations.

of temperature, gas concentration, and deposition rate during the deposition process. A procedure for CFD simulation is presented in Figure 11.14. Conservation equations of continuity, momentum, enthalpy, and gas species should be solved. For the steady-state gas flow, Favre-averaged equations (Favre, 1992) are written as follows (Ni et al., 2014).

Conservation equation for continuity:

$$\frac{\partial}{\partial x_i}(\rho \overline{u_i}) = S_m,\tag{11.5}$$

where $\rho$ is the gas density, $u_i$ are the spatial components of a gas velocity vector (ms$^{-1}$), and $S_m$ is a scalar source term accounting for the mass consumption or production due to reactions. The average of any quantity, such as $u_i$, is denoted by a bar, $\overline{u_i}$.

Conservation equation for momentum:

$$\frac{\partial}{\partial x_i}\left(\rho \overline{u_i u_j}\right) = -\frac{\partial \overline{p}}{\partial x_j} + \frac{\partial}{\partial x_j}\mu\left(\frac{\partial \overline{u_i}}{\partial x_j} + \frac{\partial \overline{u_j}}{\partial x_i}\right) - \frac{\partial}{\partial x_i}\left(\rho \overline{u_i'' u_j''}\right) + F_s, \qquad (11.6)$$

where $p$ is the pressure (Pa), and $\mu$ the dynamic viscosity. The fluctuation of a variable $u_i$ is denoted by $u_i''$. $F_s$ is a sink/source term of momentum corresponding to mass deposition on the surface.

Conservation equation for enthalpy:

$$\frac{\partial}{\partial x_i}\left(\rho \overline{u_i} \overline{h}\right) = \frac{\partial}{\partial x_i}\left[-\rho \overline{u_i'' h''}\right], \qquad (11.7)$$

in which $h$ is the enthalpy of gas mixture (J kg$^{-1}$).

Conservation equation for gas species:

$$\nabla \cdot \left(\rho \overrightarrow{u_i} Y_i\right) = -\nabla \cdot \overrightarrow{J_i} + R_i, \qquad (11.8)$$

where $\overrightarrow{u_i}$ and $Y_i$ are the velocity vector and mass fraction of $i$th species, respectively. $\overrightarrow{J_i}$ is the diffusion flux of species $i$ caused by concentration gradient, and $R_i$ the net rate of production of species by chemical reaction.

$$\overrightarrow{J_i} = -\left(\rho D_{i,m} + \frac{\mu_t}{S_{Ct}}\right)\nabla Y_i, \qquad (11.9)$$

where $S_{Ct}$ and $D_{i,m}$ are the turbulent Schmidt number and the diffusion flux of species $i$, respectively; $\mu_t$ is the turbulent viscosity.

CFD calculations on the MT–Ti(C,N) coating are conducted by means of the commercial software ANSYS Fluent 16.1. In the CVD reactor, the deposition and preheating areas are schematically shown in Figure 11.15. In view of the limitation for computational resources, a simplified 1/4 model with the same size as the actual CVD

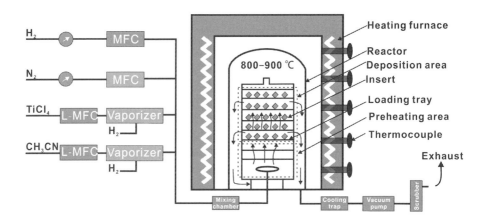

**Figure 11.15** A schematic of the CVD MT–Ti(C,N) coating deposition process. The diagram is reproduced from Qiu et al. (2019) with permission from Elsevier.

system was established, and only six layers of graphite spacers and trays are taken into consideration. Tetrahedral units are used during the meshing of the model to capture the changes of temperature and complex gas flow in the reactor. A total of mesh cells of about 5 million are utilized. As a simplification, the rotation of the feeding gas pipe and the inserts are neglected.

As an example, one of the deposition conditions ($T = 1,153$ K, $p = 12$ kPa, the initial gas mixture of $TiCl_4 = 2.52\%$, $CH_3CN = 0.87\%$, $N_2 = 13\%$, $H_2 = 83.61\%$) is selected for CFD simulation. An assumed overall chemical reaction was given as follows:

$$6TiCl_4(g) + 2CH_3CN(g) + 9H_2(g) = 6TiC_{0.67}N_{0.33}(s) + 24HCl(g), \qquad (11.10)$$

where $(g)$ and $(s)$ mean gas and solid phase, respectively. Note that compared to the thermodynamic calculations (Table 11.2), this is a simplified reduction to majority gas species. The full thermodynamic calculation enables checking if unexpected gas species or solid phases, beyond (11.10), become significant. The chemical reaction rate can be expressed in Arrhenius form:

**Table 11.3** Boundary conditions for the CFD simulations.

| Boundary | Boundary type | Boundary conditions |
|---|---|---|
| Inlet | Velocity inlet | 58.2 m/s, 1,123 K |
| Inlet $H_2$ mole fraction | – | 0.8361 |
| Inlet $N_2$ mole fraction | – | 0.1300 |
| Inlet $TiCl_4$ mole fraction | – | 0.0252 |
| Inlet $CH_3CN$ mole fraction | – | 0.0087 |
| Outlet | Pressure outlet | 1,153 K |
| Inner wall of tray | No-slip | Thermally coupled surface reaction |
| Outer wall of tray | No-slip | 1,153 K |
| Operating pressure | – | 12 kPa |

Qiu et al. (2019).

**Table 11.4** Thermophysical and the other parameters for the reactants and products (Qiu et al., 2019), unit mol stands for mol-formula.

| Parameters | $H_2$ | $N_2$ | $CH_3CN$ | $TiCl_4$ | $HCl$ | $TiC_{0.67}N_{0.33}$ |
|---|---|---|---|---|---|---|
| Specific heat $C_p$ (J mol$^{-1}$ K$^{-1}$) | 28.836 | 29.124 | 52.2 | 145.205 | 29.136 | 582.599 |
| Thermal conductivity (W m$^{-1}$ K$^{-1}$) | 0.17064 | 0.02475 | 0.1883 | 0.145 | 0.01456 | 21 |
| Viscosity (kg m$^{-1}$ s$^{-1}$) | $8.8 \cdot 10^{-6}$ | $1.76 \cdot 10^{-5}$ | $3.47 \cdot 10^{-4}$ | $8.43 \cdot 10^{-4}$ | $1.46 \cdot 10^{-5}$ | $2.13 \cdot 10^{-5}$ |
| Molecular weight (g mol$^{-1}$) | 2.02 | 28 | 41.05 | 189.692 | 36.4609 | 60.55 |
| Standard state enthalpy (J mol$^{-1}$) | 0 | 0 | 74,000 | −804,165 | −92,312 | −234,608 |
| Standard state entropy (J mol$^{-1}$ K$^{-1}$) | 130.68 | 191.609 | 243.4 | 252.379 | 186.901 | 26.529 |
| Reference temperature (K) | 298.15 | 298.15 | 298.15 | 298.15 | 298.15 | 298.15 |
| L-J characteristic length $\sigma$ (Å) | 2.827 | 3.798 | 5.399 | 5.747 | 3.339 | Not used |
| L-J energy parameter $\varepsilon/k_B$ | 59.7 | 71.4 | 416.717 | 483.269 | 344.7 | Not used |
| Density (g cm$^{-3}$) | – | – | – | – | – | 5.08 |

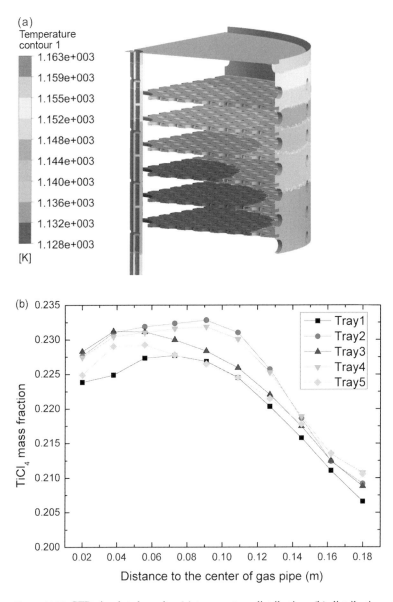

**Figure 11.16** CFD simulated results: (a) temperature distribution; (b) distribution of TiCl$_4$ mass fraction; (c) gas flow velocity and direction. Trays are numbered from bottom (1) to top (5). The diagram is reproduced from Qiu et al. (2019) with permission from Elsevier. A black and white version of this figure will appear in some formats. For the colour version, please refer to the plate section.

$$k = AT^\beta \exp(-E_a/RT), \tag{11.11}$$

where $k$ is the chemical reaction rate. $A(8.26 \times 10^{17})$, $\beta$ (5.72), and $E_a$ (436 kJ mol$^{-1}$) are the preexponential factor, temperature exponent, and activation energy (Wang et al., 2016), respectively, which are necessary inputs for CFD simulation to calculate the deposition rate.

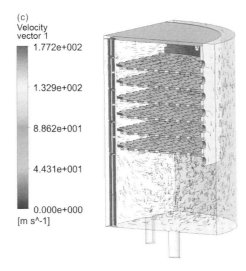

Figure 11.16 (cont.)

A steady-state and pressure-based solver is chosen. In order to improve the computation accuracy, the spatial discretization is set to be second-order upwind for conservation equations of continuity, momentum, enthalpy, and gas species transport. Moreover, to increase the convergence rate, a coupled scheme is used for pressure–velocity coupling. The convergence criteria are that the residuals should be less than $10^{-3}$, $10^{-3}$, and $10^{-6}$ for equations of species transport, continuity, and enthalpy, respectively.

The boundary conditions are shown in Table 11.3. The inputs for the materials setup of CFD simulation include thermal conductivity, viscosity, specific heat, density, molecular weight, standard state entropy and enthalpy at 298 K, and Lennard-Jones (L-J) parameters. The thermophysical parameters for the reactants and products are taken from (Chase, 1998; Poling et al., 2001; Yaws, 1999) and listed in Table 11.4. The input files used in the demonstrated CFD simulations using ANSYS Fluent 16.1 are provided in Appendix A.

The predicted results of temperature contour, mass fraction of $TiCl_4$, and gas flow velocity and direction are presented in Figure 11.16. The gradually reduced blue area shows a temperature increase from the bottom to the top trays (Figure 11.16a; see the color insert section). Furthermore, the temperature difference is within 6–10 K between the center and rim of the reactor. As a result of short distance to the heating furnace, the temperature of the rim location is higher. For the upper trays, higher temperatures are also demonstrated when a location with the same distance to the central gas pipe is taken into account. Because of the heat exchange with the adjacent cooling water located at the CVD system base, lower temperatures are shown on the bottom trays. The mass fraction variation versus distance for $TiCl_4$ is shown in Figure 11.16b. On the same tray, the mass fraction essentially decreases from center to rim. This could be caused by the earlier coating deposition at the center due to the earlier arrival of transported reactants. Besides, with a distance of 0.055–0.145 m to

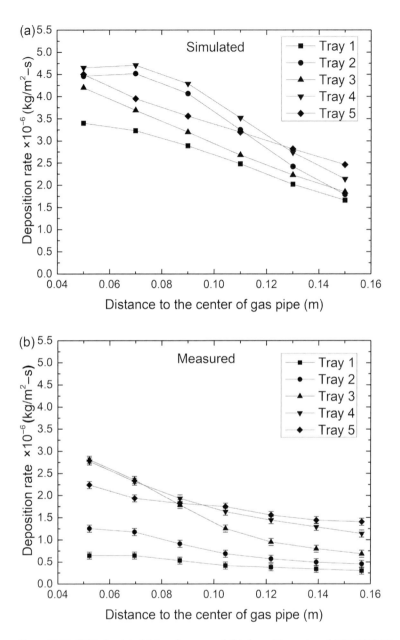

**Figure 11.17** CFD simulated (a) and measured (b) deposition rates of MT–Ti(C,N) coating deposited at 1,153 K. The diagram is reproduced from Qiu et al. (2019) with permission from Elsevier.

the central gas pipe, the mass fraction of $TiCl_4$ on the graphite trays 2 and 4 is higher than that of trays 1 and 3. The variation in concentration could be also caused by the viscosity change. The gas viscosity increases with temperature, leading to a lower gas flow rate and higher gas concentration due to the gas accumulation. Moreover, CFD

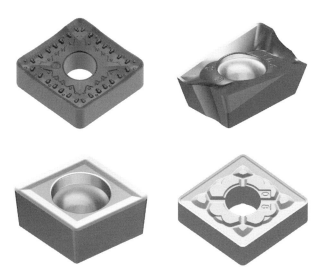

**Figure 11.18** CVD-coated tungsten carbide inserts developed via the presented hybrid approach of modeling and experiment. A black and white version of this figure will appear in some formats. For the colour version, please refer to the plate section.

simulation also provides information on the gas velocity and flow direction, as shown in Figure 11.16c.

The measured deposition rates of MT–Ti(C,N) coatings are compared with the simulated ones, as shown in Figure 11.17. From the center to rim location on the same tray, both the simulated (Figure 11.17a) and the measured (Figure 11.17b) deposition rates decrease. Although the temperature (Figure 11.16a) of the rim location is higher, the $TiCl_4$ mass fraction (Figure 11.16b) is lower than that at the central location, which has more significant impact on the deposition rate of the chemical reactions. At the rim location of the upper trays, higher deposition rates are obtained due to the combined contribution of higher temperature and concentration of reactants. With similar mass fractions of gas components, the higher deposition rate is obtained at higher temperature. The measured deposition rates are about 65% of the simulated ones so that scaling may be done.

Thirdly, the experimental deposition of CVD hard coating can be performed highly efficiently under the guidance from the previously discussed thermodynamic calculations and CFD simulations. This led to design optimization of the CVD MT–Ti(C,N) coating deposition process shown schematically in Figure 11.15. Gases $H_2$, $N_2$, $TiCl_4$, and $CH_3CN$ are mixed in the mixing chamber and flowed to the preheating chamber before entering the deposition area. One rotating gas pipe with lateral holes is located in the center of trays. The gas components flow to the central gas pipe from the preheating chamber and distribute to different trays through the lateral holes.

Finally, the gas components flow to the bottom of the reactor and are exhausted after cooling and neutralized with NaOH solution in the scrubber. In the industrial CVD reactor, there are 42 trays, and up to 10,000 inserts could be loaded depending on their sizes. The coating quality could be influenced by many factors, such as deposition parameters, fixture design, loading method, and quantity. Compared to simulation, the cost for running CVD coating experiments is high. One batch of CVD coating failure may cause an economic loss of at least $30,000 US in industry.

Guided by thermodynamic calculations and CFD simulations, CVD-coated cutting tools (Figure 11.18) with MT–Ti(C,N) coatings working as a wear-resistant layer were developed and produced at a large scale for tool companies in China (www .achtecktool.com).

# References

Anders, A. (2008) *Cathodic Arcs from Fractal Spots to Energetic Condensation*. New York: Springer.

Andersson, J.-O., Helander, T., Höglund, L., Shi, P. F., and Sundman, B. (2002) Thermo-Calc & DICTRA, computational tools for materials science. *CALPHAD*, 26(2), 273–312.

Attari, V., Cruzado, A., and Arroyave, R. (2019) Exploration of the microstructure space in TiAlZrN ultra-hard nanostructured coatings. *Acta Materialia*, 174, 459–476.

Barnett, S. A., and Shinn, M. (1994) Plastic and elastic properties of compositionally modulated thin films. *Annual Review of Materials Science*, 24(1), 481–511.

Bobzin, K. (2017) High-performance coatings for cutting tools. *CIRP Journal of Manufacturing Science and Technology*, 18, 1–9.

Buchinger, J., Koutná, N., Chen, Z., et al. (2019) Toughness enhancement in TiN/WN superlattice thin films. *Acta Materialia*, 172, 18–29.

Cahn, J. W. (1961) On spinodal decomposition. *Acta Metallurgica*, 9(9), 795–801.

Cantor, B., and Cahn, R. W. (1976) Metastable alloy phases by co-sputtering. *Acta Metallurgica*, 24, 845–852.

Chang, K. K., Music, D., to Baben, M., Lange, D., Bolvardi, H., and Schneider, J. M. (2016) Modeling of metastable phase formation diagrams for sputtered thin films. *Science and Technology of Advanced Materials*, 17(1), 210–219.

Chang, K. K., to Baben, M., Music, D., Lange, D., Bolvardi, H., and Schneider, J. M. (2015) Estimation of the activation energy for surface diffusion during metastable phase formation. *Acta Materialia*, 98, 135–140.

Chase, M. W. (1998) *NIST-JANAF Thermochemical Tables*, fourth edition. Woodbury: American Chemical Society and the American Institute of Physics for the National Institute of Standards and Technology.

Chawla, V., Holec, D., and Mayrhofer, P. H. (2013) Stabilization criteria for cubic AlN in TiN/ AlN and CrN/AlN bi-layer systems. *Journal of Physics D: Applied Physics*, 46(4), 045305.

Chawla, V., Holec, D., and Mayrhofer, P. H. (2014) The effect of interlayer composition and thickness on the stabilization of cubic AlN in AlN/Ti–Al–N superlattices. *Thin Solid Films*, 565, 94–100.

Chen, L., Du, Y., Mayrhofer, P. H., Wang, S. Q., and Li, J. (2008) The influence of age-hardening on turning and milling performance of Ti–Al–N coated inserts. *Surface and Coatings Technology*, 202(21), 5158–5161.

Deng, J. L., Cheng, L. F., Hong, Z. L., Su, K. H., and Zhang, L. T. (2012) Thermodynamics of the production of condensed phases in the chemical vapor deposition process of zirconium diboride with $ZrCl_4$–$BCl_3$–$H_2$ precursors. *Thin Solid Films*, 520, 2331–2335.

Einstein, A. (1908) Elementare Theorie der Brownschen Bewegung. *Zeitschrift für Elektrochemie und Angewandte Physikalische Chemie*, 14(17), 235–239.

Euchner, H., and Mayrhofer, P. H. (2015) Vacancy-dependent stability of cubic and wurtzite $Ti_{1-x}Al_xN$. *Surface and Coatings Technology*, 275, 214–218.

Favre, A. J. A. (1992) Formulation of the statistical equations of turbulent flows with variable density, in Gatski, T. B., Speziale, C. G., and Sarkar, S. (eds), *Studies in Turbulence*. New York: Springer, 324–341.

Frisk, K., Zackrisson, J., Jansson, B., and Markström, A. (2004) Experimental investigation of the equilibrium composition of titanium carbonitride and analysis using thermodynamic modeling. *Zeitschrift für Metallkunde*, 95(11), 987–992.

Hans, M., Music, D., Chen, Y.-T., et al. (2017) Crystallite size-dependent metastable phase formation of TiAlN coatings. *Scientific Reports*, 7(1), 16096.

Holec, D., Rovere, F., Mayrhofer, P. H., and Barna, P. B. (2010) Pressure-dependent stability of cubic and wurtzite phases within the TiN–AlN and CrN–AlN systems. *Scripta Materialia*, 62(6), 349–352.

Holec, D., Zhou, L. C., Rachbauer, R., and Mayrhofer, P. H. (2013) Alloying-related trends from first principles: An application to the Ti–Al–X–N system. *Journal of Applied Physics*, 113, 113510.

Holec, D., Zhou, L. C., Riedl, H., et al. (2017) Atomistic modeling-based design of novel materials. *Advanced Engineering Materials*, 19(4), 1600688.

Holleck, H., and Schier, V. (1995) Multilayer PVD coatings for wear protection. *Surface and Coatings Technology*, 76–77(0), 328–336.

Hörling, A., Hultman, L., Odén, M., Sjölén, J., and Karlsson, L. (2002) Thermal stability of arc evaporated high aluminum-content $Ti_{1-x}Al_xN$ thin films. *Journal of Vacuum Science and Technology A*, 20(5), 1815–1823.

Hugosson, H. W., Högberg, H., Algren, M., Rodmar, M., and Selinder, T. I. (2003) Theory of the effects of substitutions on the phase stabilities of $Ti_{1-x}Al_xN$. *Journal of Applied Physics*, 93(8), 4505–4511.

Ivashchenko, V. I., Veprek, S., Argon, A. S., et al. (2015) First-principles quantum molecular calculations of structural and mechanical properties of $TiN/SiN_x$ heterostructures, and the achievable hardness of the $nc\text{-}TiN/SiN_x$ nanocomposites. *Thin Solid Films*, 578, 83–92.

Ivashchenko, V. I., Veprek, S., Turchi, P. E. A., et al. (2014) First-principles molecular dynamics investigation of thermal and mechanical stability of the TiN(001)/AlN and ZrN (001)/AlN heterostructures. *Thin Solid Films*, 564, 284–293.

Koehler, J. S. (1970) Attempt to design a strong solid. *Physical Review B*, 2(2), 547–551.

Li, D. J., Cao, M., Deng, X. Y., Sun, X., Chang, W. H., and Lau, W. M. (2007) Multilayered coatings with alternate ZrN and TiAlN superlattices. *Applied Physics Letters*, 91(25), 251908.

Lin, Z. S., and Bristowe, P. D. (2007) Microscopic characteristics of the Ag(111)/ZnO(0001) interface present in optical coatings. *Physical Review B*, 75(20), 205423.

Lind, H., Pilemalm, R., Rögstrom, L., et al. (2014) High temperature phase decomposition in $Ti_xZr_yAl_zN$. *AIP Advances*, 4(12), 127147.

Liu, S. D., Chang, K. K., Mráz, S., et al. (2019) Modeling of metastable phase formation for sputtered $Ti_{1-x}Al_xN$ thin films. *Acta Materialia*, 165, 615–625.

Madan, A., Kim, I. W., Cheng, S. C., Yashar, P., Dravid, V. P., and Barnett, S. A. (1997) Stabilization of cubic AlN in epitaxial AlN/TiN superlattices. *Physical Review Letters*, 78(9), 1743–1746.

Marten, T., Isaev, E. I., Alling, B., Hultman, L., and Abrikosov, I. A. (2010) Single-monolayer $SiN_x$ embedded in TiN: a first-principles study. *Physical Review B*, 81(21), 212102.

Martin, P. M. (2010) *Handbook of Deposition Technologies for Films and Coatings: Science, Applications and Technology*. Norwich: William Andrew.

Mayrhofer, P. H., Hörling, A., Karlsson, L., et al. (2003) Self-organized nanostructures in the Ti–Al–N system. *Applied Physics Letters*, 83(10), 2049–2051.

Mayrhofer, P. H., Music, D., and Schneider, J. M. (2006) Influence of the Al distribution on the structure, elastic properties, and phase stability of supersaturated $Ti_{1-x}Al_xN$. *Journal of Applied Physics*, 100(9), 094906.

Münz, W. D. (1986) Titanium aluminum nitride films: a new alternative to TiN coatings. *Journal of Vacuum Science and Technology A*, 4(6), 2717–2725.

Ni, H. Y., Lu, S. J., and Chen, C. X. (2014) Modelling and simulation of silicon epitaxial growth in Siemens CVD reactor. *Journal of Crystal Growth*, 404, 89–99.

Ohring, M. (2001) *Materials Science of Thin Films*, second edition. Washington: Academic Press.

PalDey, S., and Deevi, S. C. (2003) Single layer and multilayer wear resistant coatings of (Ti, Al)N: a review. *Materials Science and Engineering A*, 342(1–2), 58–79.

Pei, F., Liu, H. J., Chen, L., Xu, Y. X., and Du, Y. (2019) Improved properties of TiAlN coating by combined Si-addition and multilayer architecture. *Journal of Alloys and Compounds*, 790, 909–916.

Pierson, H. O. (1999) *Handbook of Chemical Vapor Deposition (CVD)-Principles, Technology and Applications*, second edition. Norwich: Noyes Publications / William Andrew Publishing, LLC.

Poling, B. E., Prausnitz, J. M., and O'Connell, J. P. (2001) *The Properties of Gases and Liquids*, fifth edition. New York: McGraw-Hill Education.

Qiu, L. C., Du, Y., Wang, S. Q., et al. (2019) Through-process modeling and experimental verification of titanium carbonitride coating prepared by moderate temperature chemical vapor deposition. *Surface and Coatings Technology*, 359, 278–288.

Saunders, N., and Miodownik, A. P. (1987) Phase formation in co-deposited metallic alloy thin films. *Journal of Materials Science*, 22(2), 629–637.

Spencer, P. J. (2001) Computational thermochemistry: from its early CALPHAD days to a cost-effective role in materials development and processing. *CALPHAD*, 25(2), 163–174.

Spencer, P. J., and Holleck, H. (1989) Application of a thermochemical data-bank system to the calculation of metastable phase formation during PVD of carbide, nitride and boride coatings. *High Temperature Science*, 27, 295–309.

Stampfl, C., and Freeman, A. J. (2012) Structure and stability of transition metal nitride interfaces from first-principles: AlN/VN, AlN/TiN, and VN/TiN. *Applied Surface Science*, 258(15), 5638–5645.

Takikawa, H., and Tanoue, H. (2007) Review of cathodic arc deposition for preparing droplet-free thin films. *IEEE Transactions on Plasma Science*, 35(4), 992–999.

Wang, A. J., He, M. Z., Zhang, R., et al. (2015) Mechanical properties and spinodal decomposition of $Ti_xAl_{1-x-y}Zr_yN$ coatings. *Physics Letters A*, 379(36), 2037–2040.

Wang, A. J., Shang, S.-L., Du, Y., Chen, L., Wang, J. C., and Liu, Z. K. (2012a) Effects of pressure and vibration on the thermal decomposition of cubic $Ti_{1-x}Al_xN$, $Ti_{1-x}Zr_xN$ and $Zr_{1-x}Al_xN$ coatings: a first-principles study. *Journal of Materials Science*, 47(21), 7621–7627.

Wang, A. J., Shang, S.-L., He, M. Z., et al. (2014) Temperature-dependent elastic stiffness constants of Fcc-based metal nitrides from first-principles calculations. *Journal of Materials Science*, 49(1), 424–432.

Wang, A. J., Shang, S. L., Zhao, D. D., et al. (2012b) Structural, phonon and thermodynamic properties of Fcc-based metal nitrides from first-principles calculations. *CALPHAD*, 37, 126–131.

Wang, F., Abrikosov, I. A., Simak, S. I., Odén, M., Mücklich, F., and Tasnádi, F. (2016) Coherency effects on the mixing thermodynamics of cubic $Ti_{1-x}Al_xN/TiN$ (001) multilayers. *Physical Review B*, 93(17), 174201.

Xu, Y. X., Chen, L., Pei, F., Chang, K. K., and Du, Y. (2017) Effect of the modulation ratio on the interface structure of TiAlN/TiN and TiAlN/ZrN multilayers: first-principles and experimental investigations. *Acta Materialia*, 130, 281–288.

Yaws, C. L. (1999) *Chemical Properties Handbook*, first edition. New York: McGraw-Hill Education.

Zeng, K. J., and Schmid-Fetzer, R. (1997) *Thermodynamic Modeling and Applications of the Ti–Al–N Phase Diagram, Thermodynamics of Alloy Formation, Proceedings of a Symposium Held at the Minerals, Metals & Materials Society Annual Meeting*. Orlando: Minerals, Metals & Materials Society.

Zhang, R. F., and Veprek, S. (2007) Metastable phases and spinodal decomposition in $Ti_{1-x}Al_xN$ system studied by ab initio and thermodynamic modeling, a comparison with the $TiN–Si_3N_4$ system. *Materials Science and Engineering, A*, 448(1), 111–119.

# 12 Case Studies on Energy Materials Design

## Contents

With the rapid development of modern industries, the demand for energy and fuels is increasing. The fossil fuels are finite, and their products of combustion are not environmentally friendly, thus new energy storage/conversion materials are being developed. In the automobile industry, for example, new energy sources including hydrogen-powered fuel cell vehicles and lithium battery electric vehicles have emerged and grown rapidly. One of the key issues of this energy revolution is to find suitable materials for hydrogen storage or electrodes in Li–ion batteries (LIBs). Moreover, LIBs are an important part of portable electronic devices. In this chapter, we will show the strategy for understanding or designing hydrogen storage materials, such as $LiBH_4$, and electrode materials of LIBs by the application of multiscale computational methods. These methods not only can calculate material properties directly and figure out the mechanisms behind the performance of materials theoretically but also serve as a tool to screen the target materials based on high-throughput calculations. It can be expected that the combination of experiments and calculations and the integration of multiscale computational methods will become a common method in the development of new materials, including energy materials.

## 12.1 Case Study for Design of Hydrogen Storage Materials

### 12.1.1 Overview of Hydrogen Storage Materials

The widespread adoption of hydrogen as a fuel for vehicles depends critically on the material ability to store hydrogen on-board at high gravimetric and volumetric densities and on the ability to extract/insert it at sufficiently rapid rates. The traditional hydrogen storage method is to use a high-pressure gas cylinder to store gaseous hydrogen or a cryogenic container for liquid hydrogen at extremely low temperatures. Both methods are unlikely to satisfy the targets for the on-board application and may have the risk of explosion. According to the targets for on-board hydrogen storage systems issued by the US Department of Energy (see Table 12.1), hydrogen storage materials must fulfill certain essential criteria: (1) high gravimetric and volumetric hydrogen capacities; (2) suitable thermodynamic properties to ensure reversible hydrogen sorption/desorption under moderate delivery temperature ($-40$ to $+85°C$) and delivery pressure (5–12 bar); (3) rapid kinetics for hydrogenation and

**Table 12.1** Some targets for on-board hydrogen storage systems issued by the US Department of Energy (DOE).[a]

| Storage parameter | Units | 2020 | 2025 | Ultimate |
|---|---|---|---|---|
| **System gravimetric capacity** | | | | |
| Usable, specific-energy from $H_2$ (net useful energy/max system mass[b]) | kWh/kg (kg $H_2$/kg system) | 1.5 (0.045) | 1.8 (0.055) | 2.2 (0.065) |
| **System volumetric capacity** | | | | |
| Usable energy density from $H_2$ (net useful energy/max system volume[b]) | kWh/L (kg $H_2$/L system) | 1.0 (0.030) | 1.3 (0.040) | 1.7 (0.050) |
| **Cost** | | | | |
| Storage system cost | $/kWh net ($/kg $H_2$) | 10 (333) | 9 (300) | 8 (266) |
| Fuel cost[c] | $/GGE at pump | 4 | 4 | 4 |

[a] More details given at www.energy.gov/eere/fuelcells/doe-technical-targets-onboard-hydrogen-storage-light-duty-vehicles.
[b] Energy (kWh) relates to usable quantity of hydrogen deliverable to a fuel cell system.
[c] GGE is the gasoline gallon equivalent of a hybrid vehicle.

dehydrogenation; and (4) a low cost. In the past three decades, various hydrogen storage materials have been investigated, including metal hydrides, complex light metal hydrides, and sorbents.

Metal hydrides are formed by the reaction of $H_2$ with a metal or metal alloy (M): $(x/2)H_2 + M \leftrightarrow MHx$. The metal hydrides can be classified into two types according to the crystal structure of M after they are hydrogenated: interstitial hydride and structural hydride. For the *interstitial hydrides*, H atoms are inserted in the interstitial sites of the host metal or alloy, thus the crystal structure of the metal or alloy does not change topologically. Examples of interstitial hydrides include $Mg_2NiH_4$, $LaNi_5H_6$, $VH_2$, $ZrH_2$, and PdH. Due to involvement of heavy transition metal or rare earth metal elements, the gravimetric capacity of interstitial hydrides is generally too low. The *structural hydrides* include $MgH_2$ and $AlH_3$. Unlike interstitial hydrides, a new crystal structure is formed when hydrogen is incorporated in Mg or Al. The hydrogen capacities of $MgH_2$ and $AlH_3$ are higher than those of interstitial hydrides because of the light weight of Mg and Al. However, since the bonding between H and metal is too strong (Mg–H bond) or too weak (Al–H bond), the thermodynamics of $MgH_2$ and $AlH_3$ are unfavorable for easy hydrogen insertion/removal.

*Complex light metal hydrides* are composed of light metal (e.g., Li, Na, Mg, K, and Ca) and complex anions (such as $[BH_4]^-$, $[AlH_4]^-$, and $[NH_2]^-$). In the anionic complex, hydrogen atoms are covalently bonded to central atoms. Examples of typical complex metal hydrides are sodium alanate ($NaAlH_4$), lithium amide ($LiNH_2$), and lithium borohydride ($LiBH_4$). Due to the high hydrogen weight and volume content, complex light metal hydrides have attracted great interest in the field of hydrogen storage. The common disadvantages of complex metal hydrides are the high dehydrogenation temperature and slow hydrogenation/dehydrogenation kinetics.

Another approach to store hydrogen in solid-state form is to use the physisorptive method, in which hydrogen molecules ($H_2$) are attracted by porous lightweight materials. Most carbon- or boron-based nanomaterials (such as fullerenes, nanotubes, and graphene) and metal-organic frameworks can absorb $H_2$ with a high gravimetric density, but these sorbents can only capture $H_2$ at very low temperatures. This is due to the weak van der Waals interaction between molecular hydrogen and the sorbents. Thus the goal to make sorbents for a practical use is to strengthen the binding between $H_2$ and the sorbent.

## 12.1.2    Complex Light Metal Hydride $LiBH_4$

In this section, we take complex light metal hydride $LiBH_4$, lithium tetrahydridoborate, with an extremely high hydrogen capacity, as an example to show how computational methods help to understand the hydrogenation/dehydrogenation process and guide the design to meet the goals for on-board application.

### 12.1.2.1    Overview of $LiBH_4$ Properties

$LiBH_4$, which consists of $BH_4^-$ anion and $Li^+$ cation, has a theoretical volumetric and gravimetric hydrogen capacity of 0.120 kg $H_2$/L and 0.185 kg $H_2$/kg (18.5 wt.% $H_2$), respectively. If it decomposes into LiH and B through the reaction of $LiBH_4 \rightarrow LiH + B + H_2$, it still yields 13.8 wt.% hydrogen, satisfying the requirement for capacity in Table 12.1. $LiBH_4$ shows three phases: an orthorhombic structure with a space group of *Pnma* below 383 K, a hexagonal structure with a space group of $P6_3mc$ from 383–550 K, and a liquid phase at higher temperatures. Due to the strong covalent bond formed between H and B in the $BH_4^-$ anion, the hydrogen desorption temperature of $LiBH_4$ is too high (about 677 K) and the decomposition kinetics is relatively slow. Several attempts have been carried out to manipulate the thermodynamics and hydrogen release kinetics of $LiBH_4$, such as using multiphase composite, adding catalyst and doping elements. It was reported that $LiBH_4$ could be destabilized via reactions with other metal hydrides or oxides. For example, the mixture of $LiBH_4$ with $(1/2)MgH_2$ forms a destabilized reversible hydrogen storage material system that releases hydrogen gas at 543 K (Vajo et al., 2005). Züttel et al. (2003) reported that $LiBH_4$ releases hydrogen gas at around 573 K by mixing it with $SiO_2$ powder, lowering the hydrogen desorption temperature for $LiBH_4$ by approximately 100 K. The hydrogen desorption properties of the systems mixing $LiBH_4$ with Al, Mg, Ti, Sc, V, Cr, Zn, Zr, etc., by ball milling, have also been studied (Li et al., 2007; Yang et al., 2007).

The thermodynamics of hydrogenation/dehydrogenation reaction is one of the most fundamental properties for a hydrogen storage material. The equilibrium relationship between the operating temperature and pressure is given by the van't Hoff equation. This equation describes equilibrium between gas phase and solid phases without considering surface energy effects:

$$\ln \frac{p_{eq}}{1\ \text{bar}} = \frac{\Delta H^\circ}{RT} - \frac{\Delta S^\circ}{R}, \qquad (12.1)$$

where $p_{eq}$ is the equilibrium pressure of $H_2$. $R$ is the gas constant, and $T$ the absolute temperature. $\Delta H^\circ$ and $\Delta S^\circ$ represent the standard enthalpy and entropy of reaction for a given hydrogenation reaction, such as $Mg + H_2 = MgH_2$. This reaction is exothermic with $\Delta H^\circ < 0$, and the exponential increase of $p_{eq}$ with $T$ is mainly governed by $\Delta H^\circ$. Small variations in $\Delta H^\circ$ can significantly modify the hydrogenation/dehydrogenation conditions. From a thermodynamic point of view, one can adjust hydrogen desorption temperature by mixing the hydrogen storage material with reactive compounds. If the mixture shows different dehydrogenation reaction pathways compared to that of $LiBH_4$, the values of $\Delta H^\circ$ and $\Delta S^\circ$ are altered, and thus the thermodynamics of hydrogenation/dehydrogenation is changed accordingly.

The kinetics of hydrogen sorption/desorption is also an important indicator for hydrogen storage materials. In interstitial hydrides, hydrogen occupies the interstitial site. In this case, hydrogen desorption is very simple, namely, rapid diffusion of the hydrogen atom through the hydride and recombination of molecular hydrogen at the hydride surface. On the contrary, hydrogen reaction in $LiBH_4$ is generally more complicated than that in conventional metal hydrides, since the other constituent elements including Li and B besides H are also involved at various sorption/desorption steps. Desorption of $LiBH_4$ involves breaking the B–H bond, diffusion of hydrogen and/or other constituent atoms in the hydride, and nucleation and growth of new phases. Every step may be the rate-limiting step for desorption.

Although the final products of decomposition of $LiBH_4$ are LiH, B, and $H_2$, the dehydrogenation of $LiBH_4$ may involve several intermediate steps and thus various byproducts are formed during the decomposition. Experimental research has found evidence of important reaction intermediates such as $Li_2B_{12}H_{12}$ and $Li_2B_{10}H_{10}$, which may play a role during the decomposition of $LiBH_4$ in the presence of $B_2H_6$ (diborane) (Friedrichs et al., 2010; Orimo et al., 2006).

## 12.1.2.2    Strategy for Understanding Dehydrogenation of $LiBH_4$

To understand the dehydrogenation mechanism of $LiBH_4$ and guide optimization of hydrogen sorption/desorption properties of $LiBH_4$ experimentally, comprehensive investigations on the thermodynamics and kinetics of hydrogen release process are highly required. Figure 12.1 is a schematic diagram for understanding hydrogen desorption in $LiBH_4$ from a computational point of view. The computation includes, but is not limited to, first-principles calculations, ab initio molecular dynamics (AIMD) simulation, and thermodynamic modeling. These modeling methods are used to understand the structural information, thermodynamic, and kinetic properties of the decomposition (products) of $LiBH_4$ at different time and space scales.

*First-Principles Calculations*

First-principles calculations are widely used in studying hydrogen storage materials. They can obtain various material properties, as stated in Chapter 2. For $LiBH_4$, if the space group and atomic positions of the ground state are known, the lattice parameter and internal atomic coordinates at 0 K can be obtained by first-principles calculation.

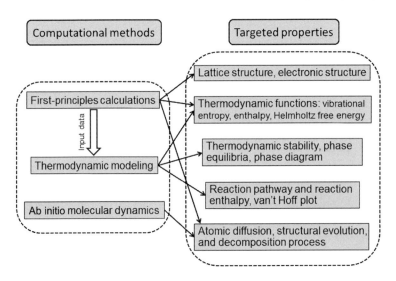

**Figure 12.1** Schematic diagram for understanding hydrogen desorption from $LiBH_4$ in terms of computational design.

Besides, first-principles can be used to predict the crystal structure of the ground state if the experimental crystal structure is in controversy or not known. For example, the ground structure of $LiBH_4$ is orthorhombic with space group *Pnma*, but first-principles calculations by Tekin et al. (2010) predicted a new stable orthogonal structure with *Pnma* symmetry, which is 9.66 kJ/mol lower in energy than the experimentally proposed structure. The electronic structure, such as electron density of states and charge distribution, is generally calculated by means of first-principles method in the literature in order to get chemical bond and charge transfer data. Thermodynamic properties are extremely important in hydrogen storage materials. In addition to the formation enthalpy at 0 K, finite temperature thermodynamic functions, such as vibrational energy, vibrational entropy, the Gibbs energy can be obtained by taking into account phonon calculation in first-principles calculations. If the Gibbs energy of $LiBH_4$ and its decomposition products are known, the enthalpy of dehydrogenation can be obtained (Frankcombe and Kroes, 2006). The calculated formation enthalpy and/or entropy data can be used as input parameter for CALPHAD modeling if experimental data are not available. The basic steps of $LiBH_4$ decomposition are accompanied with the breaking of B–H bond and diffusion of constituent elements. These basic steps can be viewed as formation and diffusion of native point defects. Thus investigation of formation and diffusion of defects in $LiBH_4$ using first-principles calculations could help to understand the initial stage of hydrogen release.

*Thermodynamic Modeling*

Thermodynamics is a cornerstone of materials science and engineering. CALPHAD is a powerful method that allows the prediction of equilibrium behavior of a multi-component system. The goal of CALPHAD approach is to obtain a description of the dependence of the Gibbs energy of all phases on temperature, pressure, and composition. All other properties, such as heat capacity and entropy, are consistently obtained according to thermodynamic laws as detailed in Chapter 5. From the Gibbs energies of all phases, the pressure–temperature–composition phase diagram can be determined as well. For $LiBH_4$, the Gibbs energy difference between $LiBH_4$ and its dehydrogenated species determines the reaction enthalpy, and the favorable reaction pathway has the lowest reaction enthalpy among all possible reactions. The reaction enthalpy controls the desorption temperature according to the van't Hoff equation.

*Ab Initio Molecular Dynamics*

Although first-principles calculations and thermodynamic modeling can provide plentiful information on $LiBH_4$ and its decomposition stage, such as reaction pathways and enthalpies, some structural and dynamic properties, especially for time-dependent structural evolution, are not available using those two computational methods. *Ab initio* molecular dynamics, which is a quantum mechanical method, can serve as a supplement to first-principles calculations and thermodynamic modeling. Based on the atom's trajectory of the AIMD simulation, the diffusion route and diffusion coefficient of atoms can be obtained. In addition, from the structural information in the AIMD time scope, the structural evolution in the sorption/desorption process can be illustrated.

### 12.1.2.3 Thermodynamics of LiBH$_4$

The thermodynamic properties of LiBH$_4$ were assessed by El Kharbachi et al., (2012) using the CALPHAD. A good description of the Gibbs energy function $G$(LiBH$_4$, $\varphi$, $T$) for each stable phase $\varphi$ of LiBH$_4$ is very important for understanding the dehydrogenation behavior of several systems containing this compound. The CALPHAD assessment of $G$(LiBH$_4$, $\varphi$, $T$), involving phases in decomposition of LiBH$_4$, is based on a critical examination of experimental data and first-principles calculated data (El Kharbachi et al., 2012). The evaluated phases include low- and high-temperature phases and the liquid phase of LiBH$_4$, as well as the possible intermediate compounds of Li$_2$B$_{12}$H$_{12}$ and Li$_2$B$_{10}$H$_{10}$, which are all considered strictly stoichiometric phases. The calculated molar heat capacity and enthalpy of various phase of LiBH$_4$ are in good agreement with the experimental data within the whole temperature range (see Figure 12.2). In the following, the possible thermal decomposition reactions of LiBH$_4$ are illustrated from the calculated $G$(LiBH$_4$, $\varphi$, $T$).

The $T$–$p$ phase diagram of the Li–B–H system can be calculated from $G$(LiBH$_4$, $\varphi$, $T$) and $G$ of the other condensed phases (intermediate compounds, LiH, B) and the well-known $G$(H$_2$, gas, $T$, $p$) for various temperatures and H$_2$ pressures $p$ as explained in Chapter 5. Figure 12.3 shows the calculated phase diagram of the Li–B–H system at fixed overall composition of LiBH$_4$. Phase transitions related to condensed phases without change in H content are independent of pressure and represented by the vertical lines, a reasonable assumption in the diagram range up to 100 bar. These lines are in perfect agreement with the experimental data. However, phase transformations involving H$_2$ evolution display the pressure dependence of the temperature of

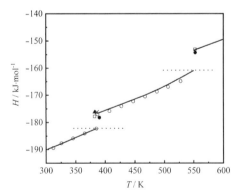

**Figure 12.2** Calculated enthalpy $H$ (the solid line) as a function of temperature for different phases of LiBH$_4$ in the respective ranges of stability. Open circles ○ show the experimental data, which are obtained by integration of experimental molar heat capacity data. Experimental enthalpies of phase transformations refer to the values shown at the dotted lines and are marked by □, ▲, ●, and ×. This figure is reproduced from El Kharbachi et al. (2012) with the permission from Elsevier.

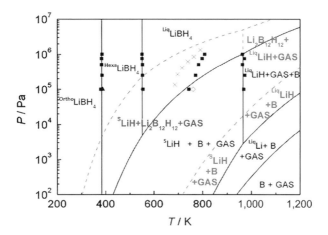

**Figure 12.3** Calculated $T$–$p$ phase diagram of the Li–B–H system at LiBH$_4$ composition. The dashed lines show the calculations considering also Li$_2$B$_{12}$H$_{12}$ and Li$_2$B$_{10}$H$_{10}$ compounds, while the solid lines exclude Li$_2$B$_{12}$H$_{12}$ and Li$_2$B$_{10}$H$_{10}$ in the calculations. The symbols ■, ▲, ◊, and × are experimental values of phase transitions. This figure is reproduced from El Kharbachi et al. (2012) with permission from Elsevier.

transformation, and the experimental data are not in agreement, as discussed later. The solid lines in Figure 12.3 show the constricted equilibria, considering only LiH and B as the decomposition products of LiBH$_4$. However, the truly stable CALPHAD phase diagram calculation includes the competition of all possible phases, also the intermediates of Li$_2$B$_{12}$H$_{12}$ and Li$_2$B$_{10}$H$_{10}$ without constriction, i.e., without excluding any phase during the Gibbs energy minimization of the system. That result, shown by the dashed lines in Figure 12.3, demonstrates that the stable range of LiBH$_4$ is narrowed and Li$_2$B$_{12}$H$_{12}$ is stable in a wide range of temperature and pressures, together with LiH and gas. Interestingly, the potential intermediate Li$_2$B$_{10}$H$_{10}$ phase does not show up in the stable phase diagram. Unfortunately, El Kharbachi et al. did not follow the standard convention to plot the stable phase boundaries in the form of solid lines (the dashed lines in Figure 12.3) and the metastable, i.e., constricted, phase boundaries dashed. That would apply to all the solid lines inside the stable three-phase region Li$_2$B$_{12}$H$_{12}$ + LiH + gas.

It is known that the phase transformations do not always occur along the equilibrium phase boundaries because the formation of one or more phases may be kinetically hindered. In order to discuss the thermal decomposition reactions of LiBH$_4$ in more detail, the following four possible decompositions of LiBH$_4$ were considered by El Kharbachi et al. (2012), and they have calculated the van't Hoff plot lines for these four decomposition reactions as shown in Figure 12.4:

①    LiBH$_4$ → $(1/12)$Li$_2$B$_{12}$H$_{12}$ + $(5/6)$LiH + $(13/12)$H$_2$
②    LiBH$_4$ → $(1/10)$Li$_2$B$_{10}$H$_{10}$ + $(4/5)$LiH + $(11/10)$H$_2$
③    LiBH$_4$ → LiH + B + $(3/2)$H$_2$
④    LiBH$_4$ → Li + B + 2H$_2$

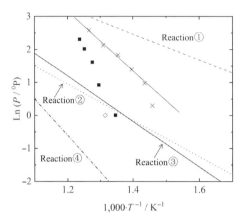

**Figure 12.4** Calculated van't Hoff plot for four possible LiBH$_4$ decomposition reactions. Experimental values are marked by ■, ◊ and ×. This figure is re-designed by adding the reaction numbers and otherwise taken unchanged from (El Kharbachi et al., 2012) with permission from Elsevier.

It is emphasized that no new information is gained in the van't Hoff plot compared to the phase boundaries from the stable calculation and all metastable CALPHAD phase diagram calculations because the same thermodynamic data are used. Just the abscissa in Figure 12.3 is transformed to inverse temperature, and the important information on phase stability regions is lost. Also the experimental data shown in the range 600–900 K are the same. The thin solid line through the data marked by (×) is just a guide for the eye.

It is impossible to read from the van't Hoff plot which of the intermediate phases is more stable in competition with other product phases. This is because the "stability" of product phases is already assumed by arbitrarily writing down a reaction equation. For example, the dotted line for reaction ② in Figure 12.4, involving Li$_2$B$_{10}$H$_{10}$, is not shown in Figure 12.3, but a part of it would appear in another metastable phase diagram calculation if only the more stable phase Li$_2$B$_{12}$H$_{12}$ is excluded. That phase diagram was given in the appendix by El Kharbachi et al. (2012) and is shown as revised Figure 12.5 here. It reveals that along the line corresponding to reaction ③, also seen as a solid line in the phase diagram shown in Figure 12.3, a five-phase equilibrium point, LiBH$_4$ + Li$_2$B$_{10}$H$_{10}$ + LiH + B + gas, metastable with respect to Li$_2$B$_{12}$H$_{12}$, occurs at about 10$^5$ Pa (1 bar). From that point, a phase region with Li$_2$B$_{10}$H$_{10}$ and two phase boundaries develops down to lower temperatures. Only the upper phase boundary, marked here by ②, but not the lower one, could be deduced from Figure 12.4. The crossing of the lines for reaction ② and reaction ③ in the van't Hoff plot can only be related to this metastable five-phase equilibrium point after the CALPHAD phase diagram calculation with minimizing the Gibbs energy of the system has clearly shown which phase (combinations) would be more stable at that state point.

The advice to solve this kind of problem is to start with the simple CALPHAD calculation of the stable phase diagram involving all known phases, LiBH$_4$,

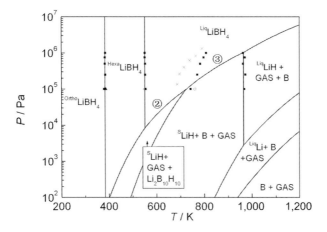

**Figure 12.5** Calculated LiBH$_4$ phase diagram excluding only the Li$_2$B$_{12}$H$_{12}$ compound from the calculations. The experimental data symbols are the same as those in Figure 12.3. This figure is redesigned by adding the reaction numbers ② and ③ and otherwise taken from El Kharbachi et al. (2012) with permission from Elsevier.

Li$_2$B$_{12}$H$_{12}$, Li$_2$B$_{10}$H$_{10}$, LiH, B, and gas. In the second calculation, the "next stable Li-phase" adjoining the single-phase LiBH$_4$ region and identified as Li$_2$B$_{12}$H$_{12}$ should be excluded. In this diagram, given here as Figure 12.5, the "next-next stable phase" adjoining the now metastable extended, single-phase LiBH$_4$ region is identified as Li$_2$B$_{10}$H$_{10}$ (at $p < 1$ bar and lower $T$) and as LiH (at $p > 1$ bar and higher $T$). In the third calculation, this "next-next stable phase" Li$_2$B$_{10}$H$_{10}$ is also excluded together with Li$_2$B$_{12}$H$_{12}$, shown in Figure 12.3 with "solid lines only." One may proceed with excluding also LiH, resulting in a phase diagram with a threefold supersaturated region of LiBH$_4$, limited by a part of the line for reaction ④, adjoining the metastable extension of the Li $+$ B $+$ gas region. In the B $+$ gas region, all the Li should be in the gas phase, therefore $p$ in the phase diagrams is not the partial pressure of H$_2$ but the total pressure. Using the stable and the sequential metastable phase diagrams, the pressure or temperature difference between the phase boundaries limiting the existence of LiBH$_4$ may be taken as a measure for the thermodynamic driving force to overcome the kinetic barriers of the LiBH$_4$ decomposition reaction. In this way, the possible intermediate phases can be considered one by one.

For the discussion, one may view the experimental data in the temperature range 650–850 K as a signal for LiBH$_4$ decomposition by monitoring the pressure at constant $T$ or by monitoring the H$_2$ evolution during heating. It is obvious that all data above 1 bar are between the calculated equilibrium phase boundaries, above the metastable reaction ③ in Figure 12.5 and below the (stable) reaction ① in Figure 12.3. Therefore, on heating, the stable equilibrium decomposition (boundary ①) appears kinetically hindered since significant superheating (more generally supersaturation) beyond the phase region with Li$_2$B$_{12}$H$_{12}$ is required. Also, the necessary superheating increases at lower $T$, being consistent with an increased kinetic barrier.

However, there is no thermodynamic driving force for the formation of LiH or $Li_2B_{10}H_{10}$ since these phase regions in Figure 12.5 are not touched, except for the two data points at 1 bar. These statements, however, assume the correctness and reliability of the thermodynamic data behind the phase diagrams. Assessed parameters for $G(Li_2B_{12}H_{12}, T)$ are based only on *ab initio* calculations, and a note of warning was already given by El Kharbachi et al. (2012).

They have also pointed out that equilibrium conditions were not fulfilled during these experiments, another reason for discrepancy of the cited experimental data with the calculations. Data are often obtained in so-called *pcT* measurements, starting with $LiBH_4$ at very low pressure, and then, at preset $T$, allowing decomposition under controlled constant hydrogen outflow from the sample. The pressure first increases rapidly and then reaches a plateau. The plateau value extrapolated to zero flow rate is assumed as equilibrium decomposition pressure, as shown by Mauron et al. (2008), and their data are denoted by the symbol $\times$ in the diagrams. The beauty of the phase diagram approach is that this reaction path by releasing the pressure at constant $T$, which is different from the heating path assumed earlier, can also be clearly discussed. The experiment at fixed 790 K (Mauron et al., 2008) started at a pressure well below the $10^2$ Pa in Figures 12.3 and 12.5. At this state, the $LiBH_4$ sample is highly supersaturated with respect to both LiH and $Li_2B_{12}H_{12}$. During the initial short period of strong pressure increase, both phase regions are passed, so it is not surprising that traces of both phases may be found as occasionally observed. Along the path taken at the lowest temperature of 697 K, the phase region of $Li_2B_{10}H_{10}$ is also briefly passed in Figure 12.5. The experimental error, however, is larger at the low temperature (Mauron et al., 2008). The phase amounts formed would depend on the height of kinetic barrier, currently unknown, and the changing thermodynamic driving forces during the pressure increase. These driving forces could be obtained quantitatively from CALPHAD calculations. Such future studies could reveal the thermodynamics and kinetics of the rate-limiting step for the decomposition.

In conclusion, it is emphasized that the fundamental advantage of the approach exemplified in this section – compared to just reading the van't Hoff plot – is that in these phase diagrams the regions of stability or supersaturation with respect to well-defined phase selections are clearly visible and that different reaction paths can be discussed. Simple classical chemical reaction equations, such as used in the van't Hoff plot, show a very special case of the phase equilibria or invariant reactions occurring in phase diagrams. The superiority of using (calculated stable/metastable) phase diagrams to understand reactions and transformations – compared to chemical reaction equation approach – becomes overwhelming if nonstoichiometric phases with a distinct solution range are involved, as discussed by Schmid-Fetzer (2014). For example, the solubility of B in liquid Li significantly increases above ~900 K.

As mentioned before, doping is one effective method to adjust dehydrogenation performance of $LiBH_4$. Thermodynamic modeling can also be used to investigate the effect of dopants (M) on the thermodynamic properties of $LiBH_4$. Lee et al. ( 2010) studied the effect of M = Mg, Ca, and Zn on stability of $LiBH_4$ through the CALPHAD method. The Gibbs energy of the solid solution phase of M in $LiBH_4$ in

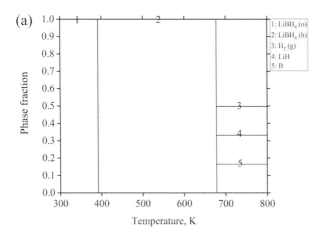

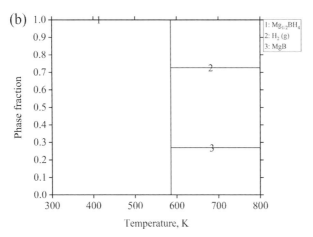

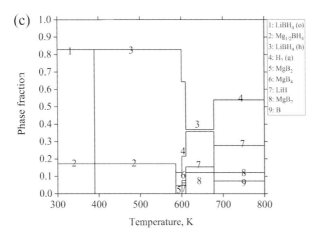

**Figure 12.6** Predicted mole fractions of equilibrium phases of the $(Li_{1-x}Mg_{x/2})BH_4$ system as a function of temperature: (a) $LiBH_4$; (b) $Mg_{1/2}BH_4$; and (c) $(Li_{9/11}Mg_{1/11})BH_4$. This figure is reproduced from Lee et al. (2010) with permission from Elsevier.

the thermodynamic modeling is obtained through experimental data or first-principles calculation. The ionic sublattice model of $(Li^{+1}, M^{+2}, Va)_1(BH_4^{-1})_1$ is adopted for the metal substituted $LiBH_4$ phase. It is found that all the three additives reduce the decomposition temperature of $LiBH_4$, but Ca has a significant effect on $H_2$ releasing temperature (about 110 K) due to the formation of highly stable $CaB_6$ in the Li–Ca–B–H system. This theoretic calculation gives an example how thermodynamic modeling guides screening of appropriate additives to $LiBH_4$ to improve its dehydrogenation properties.

Figure 12.6 shows mole fractions of equilibrium phases in the $(Li_{1-x}Mg_{x/2})BH_4$ system as a function of temperature. Their first-principles calculations show a high positive enthalpy of mixing and thus negligible equilibrium solubility of Mg in $LiBH_4$ at room temperature. Thereby, addition of Mg into $LiBH_4$ does not result in the formation of solid solution but in the two-phase equilibrium at room temperature. Decomposition reaction starts at 677 K for $LiBH_4$ and at 586 K for $Mg_{1/2}BH_4$, presumably at 1 bar, according to Lee et al. (2010).

### 12.1.2.4   Point Defects in LiBH$_4$: Understand the Dehydrogenation of LiBH$_4$

In essence, the thermal decomposition of $LiBH_4$ involves the breaking of H–B bonds, the diffusion of H and other constituent elements, as well as the nucleation and growth of new phases. From a microscopic point of view, the decomposition process inside the $LiBH_4$ phase could be viewed as formation and migration of point defects. Thus, it is worth investigating defect properties to understand the hydrogen release process. First-principles calculation is widely used as a powerful technique for exploring defect-related properties, including defect formation energy, migration barrier of defect, and diffusion coefficient (Mantina et al., 2009; Wang et al., 2018). Research was done on native defects in complex hydrides of $NaAlH_4$, $LiAlH_4$, $LiNH_2$, $LiBH_4$, and $Li_4BN_3H_{10}$ by first-principles calculations in order to understand the decomposition process and find the rate-limiting steps for the dehydrogenation (Hoang and Van de Walle, 2012; Wang et al., 2011; Wilson-Short et al., 2009). It should be noted that complex metal hydrides are electrical insulators with a large band gap, thus defects in $LiBH_4$ may be in a charged state, just like defects in semiconductors. In order to investigate charged defects computationally, two aspects should be taken into account. One is the formation energy, which is not only related to atomic chemical potentials but also to chemical potential of electrons, as shown in Section 2.3.4. Different atomic chemical potentials represent different experimental conditions. The other aspect is that the global charge neutrality must be maintained.

Figure 12.7 shows first-principles calculated formation energies of H-related vacancies ($V_H$) and interstitials ($H_i$) in all possible charge states in $LiBH_4$, which were carried out by Hoang and Van de Walle (2009). The atomic chemical potentials of Li, B, and H are determined by assuming equilibrium among solid $LiBH_4$, LiH, and B, shown as the (metastable) solid line in Figure 12.3 and correspond to the equilibrium of reaction ③. Figure 12.7 reveals that H vacancies and interstitials are all in the

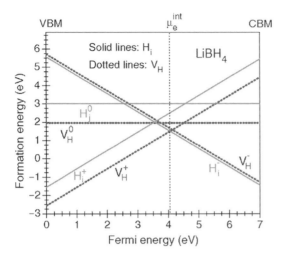

**Figure 12.7** Calculated formation energy of H-related defects in LiBH$_4$, assuming that LiBH$_4$, LiH, and B are in equilibrium with each other. The vertical dotted line is the intrinsic Fermi level, which is determined by the condition of charge neutrality. This figure is reproduced from Hoang and Van de Walle (2009) with permission from the American Physical Society.

charged state, because their formation energies are lower than the corresponding neutral H vacancy and interstitial. The defects with the lowest formation energy are the combination of $V_H^+$ and $H_i^-$: the positively charged H vacancy $(V_H^+)$ is formed by removing one H atom and an extra electron from a BH$_4$ unit and placing it as negatively charged H interstitial $(H_i^-)$ initially near the center of the void formed by Li atoms and BH$_4$ units. The vertical dash line in Figure 12.7 shows the intrinsic Fermi level $(\mu_e^{int} = 4.06\ \mathrm{eV})$, where the formation energies and hence approximately the concentrations of intrinsic point defects $V_H^+$ and $H_i^-$ are equal.

The energetically favorable H-related defects in LiBH$_4$ are in charged state, and hence their formation energies are related to the Fermi level. If other defects and impurities are absent, the Fermi level would be fixed at the position shown in Figure 12.7, $\mu_e^{int} = 4.06\ \mathrm{eV}$, in which positively and negatively charged defects occur in equal numbers and thus charge neutrality is obeyed. If electrically active defects are incorporated in LiBH$_4$ and their concentrations exceed those of native defects, the Fermi level of the system will be changed accordingly. Consequently, the formation energies of some H-related defects will be reduced. This opens a strategy to find catalysts to accelerate hydrogen desorption kinetics in complex hydrides by controlling the position of the Fermi level.

Hoang and Van de Walle (2009) calculated formation energies for seven dopants, namely Ti, Zr, Fe, Ni, Zn, Pd, and Pt in LiBH$_4$. They considered three possible sites for the dopants, i.e., the Li site, the B site, and interstitial site. Their calculations show that all the impurities are predominantly on the B site, and the charged defect states

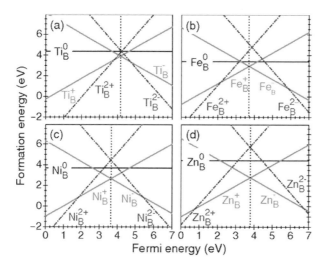

**Figure 12.8** Calculated formation energies for (a) Ti, (b) Fe, (c) Ni, and (d) Zn impurities in various charge states $-2$, $-1$, $0$, $+1$, and $+2$ on the B site in LiBH$_4$. The dotted vertical line is the transition level between charge states $+1$ and $-1$ of the impurity. This figure is reproduced from Hoang and Van de Walle (2009) with permission from the American Physical Society.

are, again, energetically more favorable than the neutral ones, as indicated in Figure 12.8. The transition level $\varepsilon(q_1/q_2)$ is defined as the Fermi-level position, where the formation energies of charge states $q_1$ and $q_2$ are equal. Their calculated transition levels $\varepsilon(+1/-1)$ for Ti$_B$, Fe$_B$, Ni$_B$, and Zn$_B$ in LiBH$_4$ are 4.16, 3.71, 3.65, and 3.79 eV, respectively. Note that the intrinsic Fermi level of LiBH$_4$ is 4.06 eV. We assume that the concentration of transition metal impurities is larger than that of the hydrogen-related defects. Thus, the Fermi level of the doped system will shift. The shift values $\Delta\varepsilon = \varepsilon(+1/-1) - \mu_e^{int}$ caused by Ti$_B$, Fe$_B$, Ni$_B$, and Zn$_B$ are $+0.1$, $-0.35$, $-0.41$ and $-0.27$ eV, respectively. A direct outcome caused by the shift of the Fermi level is the change in the formation energy of H-related defects. For example, Ti$_B$ results in the decrease of the formation energy of H$_i^-$ by 0.1 eV. The reduction of formation energy of H-related defects is in turn responsible for the improvement of thermodynamics in the dehydrogenation of LiBH$_4$. Thus, from a defect point of view, it is possible to find out appropriate catalysts to reduce the decomposition temperature of LiBH$_4$.

It is also useful to look at the atomic structures associated with defects. This implies not only to obtain the geometric structures around the defect, such as bond length and bond angle, but also to find the clue of forming intermediate products in the decomposition of complex hydrides. For instance, first-principles calculations on native defects in ammonia borane show that the creation of H-related defects results in highly reactive species of $[BH_2]^+$, $[BH_4]^-$, and $[NH_2]^-$ (Wang et al., 2017). The reactive species could form some complex structures observed in experiments, such as NH$_3$BH$_2$–NH$_2$BH$_3$ complexes and diammoniate of diborane (DADB, $[(NH_3)_2BH_2^+]BH_4^-$).

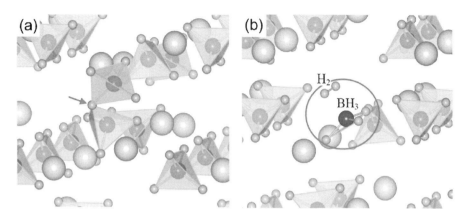

**Figure 12.9** Local structures induced by H-related defects in LiBH$_4$: (a) $V_H^+$, (b) $H_i^+$. Large spheres are Li atoms, medium spheres B atoms, and small spheres H atoms. This figure is reproduced from Hoang and Van de Walle (2009) with permission from the American Physical Society.

Coming back to LiBH$_4$, the creation of $V_H^+$ leads to the formation of a BH$_3$–H–BH$_3$ complex, namely, two BH$_4$ units sharing a common H atom marked by the arrow in Figure 12.9a. The creation of $H_i^0$ and $H_i^+$ results in a BH$_3$ unit plus one H$_2$ molecule, as shown in Figure 12.9b. Both neutral H vacancy $V_H^0$ and negatively charged H vacancy $V_H^-$ lead to the formation of a BH$_3$ unit. The BH$_3$ units observed in the case of $V_H^-$ and $H_i^+$ can combine and form diborane B$_2$H$_6$, which may be emitted during the dehydrogenation process. This could account for the unrecoverable loss of boron reported in the dehydrogenation experiment.

Diffusion of constituent elements is equally important in the dehydrogenation of LiBH$_4$. From the formation energy, one knows that the dominant H-related defects are $V_H^+$, $H_i^+$, $V_H^-$, and $H_i^+$. The calculated migration barriers using the climbing-image nudged elastic band method for $H_i^+$, $H_i^-$, $V_H^+$, and $V_H^-$ are 0.65, 0.41, 0.91, and 1.32 eV, respectively. The energy barriers for H vacancy are relatively high because the diffusion of these defects involves the breaking of B–H bonds. The activation energy (i.e., the sum of formation energy and migration barrier) for self-diffusion of $H_i^+$, $H_i^-$, $V_H^+$, and $V_H^-$ are, respectively, 3.43, 1.65, 2.68, and 2.52 eV. Thus, the mass transport of H is mediated by $H_i^-$, the negatively charged H interstitial.

### 12.1.2.5 Structural Evolution and Diffusivity of LiBH$_4$

The uncatalyzed decomposition of LiBH$_4$ at $\sim$677 K and 1 bar is above the melting temperature. Thus, it is important to characterize the liquid to understand the behavior of LiBH$_4$ during hydrogen desorption. A difficulty to characterize the liquid state of a complex hydride by experiments is that the study of light species such as hydrogen requires costly equipment (for example, neutron diffraction method). On the contrary, molecular dynamics (MD) provides a detailed description of atomic trajectories as a function of time, and it has been used to investigate diffusion kinetics and structural and vibrational properties of liquid and amorphous phases.

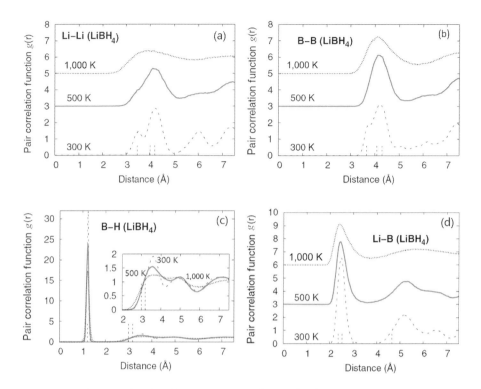

**Figure 12.10** Pair correlation functions, $g(r)$, versus distance in $LiBH_4$. Dashed vertical lines on the abscissa indicate the first few neighbor peaks in the relaxed structure at 0 K. This figure is reproduced from Farrell and Wolverton (2012) with permission from the American Physical Society.

Ab initio molecular dynamics has been used to study $LiBH_4$ and other complex hydrides (Farrell and Wolverton, 2012). In each run, the system is equilibrated for 1 ps (2,000 time steps), then data are collected over an additional 4 ps (8,000 time steps). The calculated time averaged pair distribution function of Li–Li, B–B, B–H, and Li–B in $LiBH_4$ at 300, 500, and 1,000 K are shown in Figure 12.10. At 300 K, the cation pair correlation of Li–Li displays obvious peaks in Figure 12.10a, corresponding to the first few neighbor shells in the solid. Due to the vibrations of Li ions around their equilibrium position, the peaks have a width. With increase of temperature, the peaks become less prominent and eventually approach a uniform distribution after the nearest-neighbor distance at 1,000 K, indicating that the cation order disappears. The pair correlation functions of central atoms in the $BH_4$ anion (i.e., B–B) in Figure 12.10b show similar trends as for the cations. In Figure 12.10c, there is a very sharp peak around 1.2 Å for the pair correlations between H and the central B atom at all temperatures. This is an indication that the anionic complexes $[BH_4]^-$ remain intact even at high temperatures. At longer distances (shown as an inset in Figure 12.10c), the order of the $[BH_4]^-$ anion decreases with increasing temperature. The broad peaks at long ranges are likely the results of atomic vibrations and diffusive motions of the anionic units relative to one another. Figure 12.10d shows the ordering between the

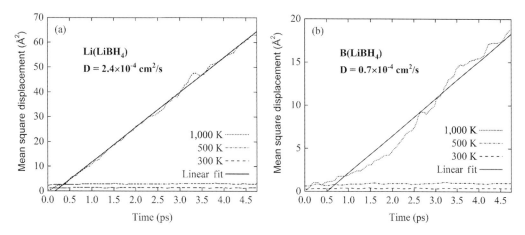

**Figure 12.11** Mean square displacements at three temperatures for (a) Li and (b) B in LiBH$_4$. This figure is reproduced from Farrell and Wolverton (2012) with permission from the American Physical Society.

anionic complexes $[BH_4]^-$ and Li cations, which is indicated by the pair correlation function between B and Li. It is revealed that the cation–anion ordering at the first neighbor range stays consistent across temperatures, but at longer ranges loss of order occurs as temperature increases.

In addition to the aforementioned time-averaged structural information, the diffusive motion of each species was also examined by Farrell and Wolverton (2012) as an important kinetic aspect for the decomposition of LiBH$_4$. Figure 12.11 shows the mean square displacement (MSD) of Li and B. The MSDs of Li and B show a nearly flat line at 300 and 500 K over the simulation time, indicating that little bulk motion of atoms occurs with negligible MSD increase. At 1,000 K, the MSD curves show clearly increasing MSDs developing, indicating the occurrence of diffusion. From the slope of a linear fit to the MSD data versus time, the diffusivities of both Li and B are obtained. The diffusivity of B at 1,000 K compares reasonably well with experimental data.

## 12.2 Case Study for Design of Li–Ion Batteries

### 12.2.1 Overview of Li–Ion Batteries

LIBs are embedded in our daily life. To meet the ever-increasing demands for portable electronic devices and electric vehicles or hybrid electric vehicles, extensive research has been conducted to develop the materials of LIBs toward the goal of high voltage, high energy/power density, long cycle life, superior safety, low cost, and environmental friendliness. LIBs consist of cathode, anode, electrolyte, and separator. The cathode and anode are immersed in the electrolyte and separated by a separator. Li ions transfer between the cathode and anode through an electrochemical reaction, as

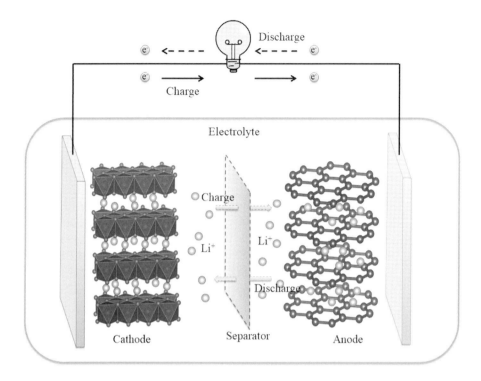

**Figure 12.12** Schematic diagram of a lithium–ion cell.

shown in Figure 12.12. It is worth noting that multiple processes occur during the electrochemical reaction, such as charge/mass transfer, interface formation, structural change, phase transformation, etc., that directly link to the performance of a specific LIB system. To design advanced LIBs to meet the requirements of portable devices and electric vehicles, it is important to establish the composition–structure–property relationship for electrode materials. An excellent overview of the materials science aspects of advanced batteries, especially LIBs, is given by Huggins (2009). This book also indicates some of the groundbreaking work on the thermodynamics, phase diagrams, and electrochemistry of LIBs made by Huggins's group starting already in the 1970s.

Currently, computational approaches have become powerful tools to guide and accelerate the development of new materials including LIBs (He et al., 2019; Meng and Arroyo-de Dompablo, 2009). First-principles calculations can predict the fundamental properties of electrode materials at the atomic and electronic scales (Ceder et al., 2011). Meanwhile, the results from the first-principles calculations can be used as the input for the CALPHAD approach. Within the CALPHAD framework, the phase equilibria and properties (thermodynamic, electrochemical, and physical properties) can be predicted, as shown for the Li–Si and Li–Si–H systems by Liang et al. (2017). That is essential to acquire the composition–structure–property relationships and understand the process of the battery during the charge/discharge process.

With this motivation, the present case study aims to describe a procedure that can be used to design the comprehensive performance of LIBs from the aspect of the composition–structure–property relationships. It should help researchers involved in the development process to select appropriate materials and optimize synthesis conditions by providing information for relevant issues.

## 12.2.2 Relationship among Phase Diagram, Thermodynamics, and Electrochemical Properties

The key materials of LIBs, cathode and anode, are generally multicomponent multiphase systems, and Li is always one component. Their phase diagrams are consistently obtained by CALPHAD calculations after the Gibbs energy of each stable or metastable phase is obtained by CALPHAD assessment as explained in Chapter 5. The Gibbs energy is assessed as the integral quantity of the phase $\varphi$, $G^\varphi(x_i, T)$, as a function of temperature $T$ and composition, where $x_i$ stands for the mole fraction of component $(i)$. From $G^\varphi(x_i, T)$, the partial thermodynamic quantities also are calculated, and the most important ones for LIBs are the chemical potential, $\mu_i$, and chemical activity, $a_i$. Both are a property of the component $i$ in phase $\varphi$ (or the multiphase equilibrium), and the component Li is the most important one. For the activity, a reference state must always be defined, such as

$$\mu_{Li} = G_{Li}^{0,ref} + RT \ln a_{Li}^\varphi, \tag{12.2}$$

where $G_{Li}^{0,ref} = \mu_{Li}^{0,ref}$ is the molar Gibbs energy of pure Li in the reference phase. In the simplest setup of the electrochemical cell in Figure 12.12, one envisages the anode as pure Li, the reference half-cell in this case. As the cathode a material is taken that would like to react with Li, such as silicon containing almost no Li, and that is the fully charged state. Since the electrolyte lets only $Li^+$ ions pass, an electromotive force (EMF) builds up because the activity of Li in the reference half-cell is unity but almost zero on the "cathode" side. The EMF value $E$ is a voltage and related to the chemical potential difference of Li by the Nernst equation:

$$\mu_{Li} - \mu_{Li}^{0,ref} = -zFE, \tag{12.3}$$

where $F$ is Faraday's constant and $z$ the number of electrons transferred ($z = 1$ for Li). This voltage is also denoted as equilibrium open-circuit voltage (OCV). By connecting the electrodes with a consumer, the electric current will flow and simultaneously the $Li^+$ ions passing to the cathode will react, in the present example by forming $Li_pSi_q$ stoichiometric compounds. Whenever a new compound is formed, the values of $a_{Li}$ and $\mu_{Li}$ in the multiphase cathode in our simple setup will rise and the voltage will drop, staying constant in any two-phase region because the chemical potential is then constant. This discharging process ends when the Li-richest compound, $Li_{17}Si_4$, was formed because $Li + Li_{17}Si_4$ are in chemical equilibrium, and $a_{Li} = 1$ in both half-cells. The charging process will reverse the chemical reaction. This setup with pure Li in one half-cell is actually used in basic research as discussed in detail for the Li–Si system, including the kinetic implications leading to measured voltages that may be different from the stepwise ideal equilibrium case during lithiation and delithiation, involving also amorphous Li–Si alloys (Liang et al., 2017). Note that the terminology of anode/cathode in electrochemistry can be confusing and that the anode is always the negatively charged electrode as simply tested by a voltmeter (Huggins, 2009).

In practical LIBs, pure Li is not used as negative electrode (anode) material to avoid short-circuit due to unfavorable Li-crystal growth in charging cycles. For example, the still dominating low-cost material is graphite, and during charging, single- or multi-phase Li–C (intercalation) compound(s) may be formed (Drüe et al., 2013; Kozlov et al., 2013). This case of insertion reaction of Li into materials with layer-type crystal structures is also called intercalation reaction. In that case, $a_{\text{Li}}^{\text{anode}} \neq 1$ occurs also in the anode, and the EMF $E$ is obtained by using (12.2) and (12.3) for both electrodes. For $E$, the pure Li half-cell standard $G_{\text{Li}}^{0,ref} = \mu_{\text{Li}}^{0,ref}$ used for both sides cancels out by taking the difference of these equations, resulting in the measurable cell voltage $E$ in this general form of the Nernst equation (Huggins, 2009):

$$E = -\frac{RT}{zF} \ln \frac{a_{\text{Li}}^{\text{cathode}}}{a_{\text{Li}}^{\text{anode}}}. \tag{12.4}$$

The electric charge capacity (in Coulomb, SI-unit = As) of an intercalation electrode is equal to $nF$, where $n$ is the amount of Li$^+$ cations (in mol) that can be extracted from or inserted to the electrode during battery operation. The theoretical specific capacity $C_M$ refers to the mass of active electrode materials by dividing with $M$, the molecular mass of the electrode (factor 3.6 is from conversion to the unit mAh/g):

$$C_M = \frac{nF}{3.6M}. \tag{12.5}$$

Besides capacity, energy density is also an important practical parameter for battery performance. The maximum theoretical specific energy $E_M$ (or energy density in unit mWh/g) is simply obtained by the product of $C_M$ with the average operating voltage, $E$ (unit V):

$$E_M = C_M E. \tag{12.6}$$

The energy density may also be given per electrode volume (in mWh/L). In addition to advancing the anode material, researchers have also made significant progress in developing advanced positive electrode (cathode) materials, which are typically complex Li–oxides, such as the first commercially introduced LiCoO$_2$, which is still broadly used in laptop batteries.

## 12.2.3     Li–Mn–O Spinel Cathode Material

The lithium manganese oxide (LMO) spinel Li$_x$Mn$_2$O$_4$ $(0 \leqslant x \leqslant 2)$ has attracted extensive attention due to its superior properties, such as nontoxicity and low cost. However, a wide range of spinel solid solution in LMO spinel can be directly sintered and lead to different properties of identical overall composition during the battery

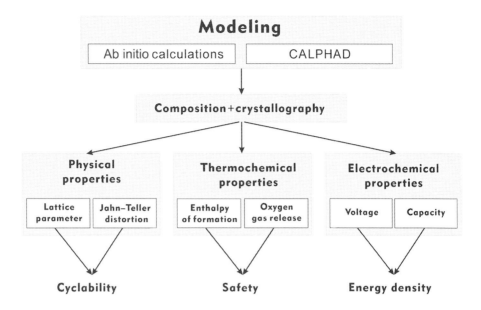

**Figure 12.13** The flowchart of the high-throughput computational framework to evaluate the performance of the spinel compounds in Li–ion batteries systematically. This figure is reproduced from Zhang et al. (2018) with permission from the American Chemical Society.

operation. This makes it extremely difficult to understand the intrinsic properties and evaluate battery performance. This is a major hurdle to select the best composition and structure with optimized cell performance for the various applications. Exclusive experimental studies of such a complex material are extremely time consuming and expensive.

In this case study, a high-throughput computational framework combining first-principles calculations and the CALPHAD approach is demonstrated to describe the composition–structure–property relationships under sintered and battery states of spinel cathodes systematically (Zhang et al., 2018). The phase diagram and related properties are calculated by Thermo-Calc software. Depending on composition and structure, various properties (physical, thermochemical, and electrochemical) are quantitatively predicted. This provides a general guide to evaluate and design the performance (cyclability, safety, and energy density) of LMO spinel cathodes with wide composition ranges, as shown in Figure 12.13.

### 12.2.3.1   Phase Diagrams

After the thermodynamic modeling is established, the phase diagrams of the Li–Mn–O system can be constructed to identify the phase relationship. Information on the phase equilibria and thermodynamic properties of the Li–Mn–O system is of great importance. During charge/discharge cycles of the Li–Mn–O cathode, different phases could appear, as clearly deduced from isothermal sections of the Li–Mn–O phase diagram shown in Figure 12.14. Assume an LMO cathode composition is in the shaded region

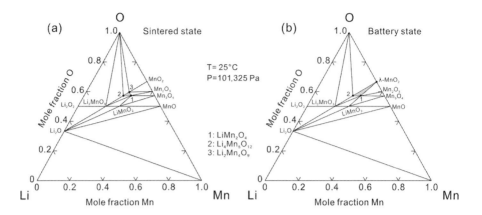

**Figure 12.14** Calculated phase diagrams of the Li–Mn–O system for (a) "sintered state" and (b) "battery state" (MnO₂ suspended) at room temperature. The shaded regions of the phase diagram represent the spinel single phase. This figure is reproduced from Zhang et al. (2018) with permission from the American Chemical Society.

(1–2–3) in Figure 12.14a. During the first charge cycle, Li is extracted from LMO, and the overall composition of the cathode must vary along a straight line from the pure Li corner toward the Mn–O binary edge. That results from the material balance, assuming that O or Mn cannot leave the cathode and the Mn/O ratio is constant. This straight line passes through various three- and two-phase equilibrium regions and, under the assumption of local equilibria, identifies the various phases that may form during the charging reactions along this "ideal" reaction path. Quite a number of binary, ternary, and nonstoichiometric phases in the Li–Mn–O system may be formed, and some of them are detrimental to the electrochemical performance of LIBs. It is obvious from this application of the phase diagram that, for example, shifting the initial cathode composition just slightly may have a significant impact on the phase regions traversed by the reaction path. This directly provides valuable information for selected experimental studies. In the next cycle upon discharging, the reaction path is simply reversed if the Mn/O ratio of the cathode is unchanged. However, the battery operation temperature, assumed to be 25°C also in the phase diagram, is much lower than the sintering temperature of the cathode. Therefore, some phases may not form during discharging/charging cycles, and that is quantitatively reflected in the calculated metastable phase diagrams by excluding (suspending) the phase(s) not forming due to any kinetic barrier.

Figure 12.14a,b shows the calculated phase diagrams in the Li–Mn–O system for "sintered" and "battery" states at room temperature, respectively. The phase diagram of "sintered state" is considered a stable state, while the "battery state" is a metastable phase diagram calculated with MnO₂ suspended. Note that the stable MnO₂ phase differs from the less stable phase λ-MnO₂ which crystallizes in the spinel structure. Therefore, the single-phase region of spinel grows from the triangular region 1–2–3 in Figure 12.14a to 1–2–λ–MnO₂ in Figure 12.14b without the necessity to mark point 3, now located on the oxygen-rich edge of the spinel region. The "sintered" and "battery" states can be used to predict the intrinsic properties for the freshly made batteries and electrochemical behavior during the battery operation, respectively.

### 12.2.3.2  Evaluation of Cyclability

It seems impossible to evaluate the cyclability in the work of computational modeling, but it can be assessed through indirect ways. In the spinel Li–Mn–O system, two physical properties are associated with cyclability. One is the lattice parameter, and the other is the extent of the detrimental Jahn–Teller distortion in the material.

For the spinel $Li_xMn_2O_4$, oxygen ions, which are at $32e$ sites, are arranged in a cubic close-packed arrangement, lithium ions are located at $8a$ tetrahedral sites, and manganese ions occupy half of the octahedral positions, which are $16d$ sites. The other half of the octahedral void consists of $16c$ sites, which reside halfway between neighboring tetrahedral sites and serve as a channel for migration of Li. Figure 12.15 shows in the left part the calculated and experimental lattice parameters of Li excess $Li_{1+y}Mn_{2-y}O_4$ ($0 \leq y \leq 0.33$, where $y$ indicates the excess amount of Li on the $16d$ sites). Note that the maximum of lattice parameters ($y = 0$, center of Figure 12.15) corresponds to the composition point 1 ($LiMn_2O_4$) in Figure 12.14. Now recall the reaction path from pure Li through point 1 to see that, in the "battery state," the path also matches the long solid-solution range from point 1 to binary $\lambda$-$MnO_2$. The composition point 2 ($Li_4Mn_5O_{12}$) in Figure 12.14b corresponds to the

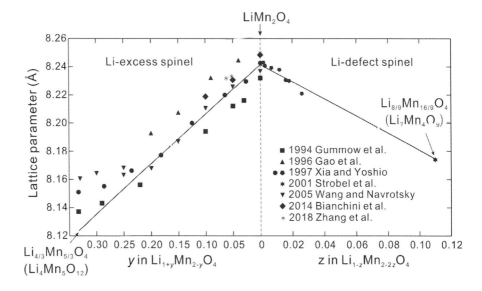

**Figure 12.15** Calculated lattice parameters of $Li_{1+y}Mn_{2-y}O_4 (0 \leq y \leq 0.33)$ and $Li_{1-z}Mn_{2-2z}O_4 (0 \leq z \leq 0.11)$ spinel for the Li-excess and Li-defect spinel in the "sintered state," respectively. This figure is reproduced from Zhang et al. (2018) with permission from the American Chemical Society.

minimum of lattice parameters ($y = 1/3$, the left edge of Figure 12.15). Starting from this point 2, any small extraction of Li leads the reaction path into the two-phase region spinel $+O_2$ and consequently release of oxygen gas.

Nevertheless, starting from point 1 toward point 2, the spinel electrodes with smaller lattice parameter can increase the structural stability and consequently improve cycling efficiency. Such spinel compounds with smaller lattice parameters exhibit better rechargeability, as shown by Xia and Yoshio (1997) in their study on the effect of lattice parameter decrease on the electrochemical performance of spinel $LiMn_2O_4$. They argued that the reduced lattice parameters are usually accompanied by shorter bond lengths, which contribute to the structural stability of the electrode material. On the other hand, the "battery state" phase diagram in Figure 12.14b reveals that the Li variation during charging within the single-phase spinel region decreases if instead of $LiMn_2O_4$, point 1, a spinel composition toward point 2 (a smaller lattice parameter) is used. Because of the growing range of oxygen gas release, a trade-off between capacity and cyclability needs to be considered.

The right part of Figure 12.15 shows that the lattice parameters decrease with Li deficiency along $Li_{1-z}Mn_{2-2z}O_4$ ($0 \leq z \leq 0.11$, where $z$ indicates the amount of vacancies on the $8a$ sites) and ends at the right edge ($z = 1/9$) corresponding to the composition point 3 ($Li_2Mn_4O_9$) in Figure 12.14. The spinel in the "sintered state" does not persist in the "battery state."

In the Li-excess range of $Li_{1+y}Mn_{2-y}O_4$ with $0 < y < 0.2$, the lattice parameter decreases linearly with the increase of excess Li. This reaction can be attributed to replacement of some $Mn^{3+}$ ions by $Li^+$ ions and partial $Mn^{3+}$ oxidation to $Mn^{4+}$ (ionic radius: $Mn^{4+} < Mn^{3+} < Li^+$). For $y > 0.2$, oxygen-defect-free $Li_{1+y}Mn_{2-y}O_4$ is difficult to sinter. Therefore, more $Mn^{3+}$ and less $Mn^{4+}$ must exist to balance the electroneutrality, and consequently this leads to an increase of the lattice parameter. For the Li-deficient spinel $Li_{1-z}Mn_{2-2z}O_4$, the lattice parameter decreases with increase of $z$ due to higher vacancy concentration on the tetrahedral and octahedral sites.

In addition to the lattice parameter, the Jahn–Teller distortion is also important for cyclability. It describes the geometrical distortion of ions resulting from certain electron configurations, which in turn breaks the symmetry of the LMO spinel crystal structure. Due to the existence of the orbital degeneracy of $Mn^{3+}$ ions, the Jahn–Teller distortion always occurs in spinel $Li_xMn_2O_4$. This distortion is one of the most important reasons of capacity fading for spinel cathode materials in LIBs. Intuitively, the Jahn–Teller distortion could be inhibited by decreasing the $Mn^{3+}$ amount. One way to prohibit the Jahn–Teller distortion and increase cyclability of LMO is to introduce excess Li ions in the $Mn^{3+}$ sites. However, this does not explain the fact that rapid charge/discharge (high current) leads to faster capacity fading. This hints at a kinetic issue that must be taken into consideration. With high-current operation, nonequilibrium state can cause the local Jahn-Teller distortion. Therefore, the tolerance, i.e., how much diversion from the equilibrium state is allowed before capacity fading occurs, was studied.

Figures 12.16 (a) and (b) show the CALPHAD-calculated room temperature voltage-composition profiles of $Li_xMn_2O_4$ ($0 \leq x \leq 2$) and $Li_xMn_{1.85}O_4$

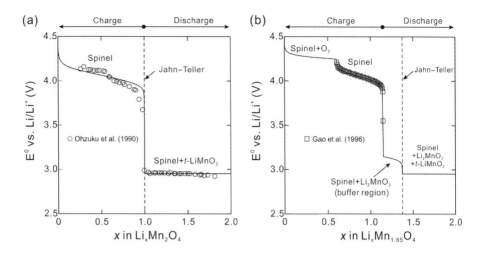

**Figure 12.16** Calculated open cell voltages and phase transitions of (a) $Li/Li_xMn_2O_4(0 \leq x \leq 2)$, (b) $Li/Li_xMn_{1.85}O_4(0 \leq x \leq 2)$ at room temperature using (12.4) with $a_{Li}^{anode} = 1$. This figure is reproduced from Zhang et al. (2018) with permission from the American Chemical Society.

$(0 \leq x \leq 2)$, measured against pure Li, respectively. Note that any variation of $x$ reflects the reaction path discussed in the text passing through point 1 ((a), for $Li_xMn_2O_4$) or halfway between points 1 and 2 ((b), for $Li_xMn_{1.85}O_4$). The calculated results reflect the phase diagram in the "battery state" with continuous change of $E$ in the single-phase spinel and a step to the lower value that remains constant along the tie line in the two-phase region spinel $+ t$-$LiMnO_2$ in (a). The slightly shifted reaction path in (b) reflects exactly the narrower single-phase spinel region from the phase diagram because of the lower Mn/O ratio. Also, the adjoining "buffer region" corresponds to the two-phase region spinel $+ Li_2MnO_3$, where the reaction path crosses the tie lines, resulting in the continuous drop of $E$.

It can be seen from Figure 12.16 that the calculated results for $E$ agree well with the measured data from the literature. $LiMn_2O_4$ can serve as a compound for both Li extraction and insertion. For $Li_xMn_2O_4$ with $0 \leq x \leq 1$, Li–ions on the $8a$ sites undergo lithiation/delithiation around 4 V in the spinel single-phase region. When the spinel $Li_xMn_2O_4$ transforms into tetragonal $Li_xMn_2O_4$ ($t$-$LiMnO_2$) due to the Jahn–Teller distortion of $Mn^{3+}$ ion, the voltage drops suddenly to 2.95 V at $x = 1$ and keeps this value in the range of $1 \leq x \leq 2$ because of the exceptional direction of the tie line along this path, as explained previously.

That is not the case for the $Li_xMn_{1.85}O_4$ with $0 \leq x \leq 2$, which shows more complicated electrochemical behavior during battery operation. When $x$ is in the range of $0.60 \leq x \leq 1.15$, the smooth voltage plateau around 4 V appears in the spinel single-phase region. When the composition of Li is within $1.15 \leq x \leq 1.375$, $Li_xMn_{1.85}O_4$ enters the two-phase region (spinel $+ Li_2MnO_3$). The voltage plateau in this two-phase "buffer" region is around 3.1 V. During further discharge to $1.375 \leq x \leq 2$, the composition is located in the three-phase region (spinel $+ Li_2MnO_3 + t\text{-}LiMnO_2$), where $\mu_{Li}$ must be constant, resulting in a flat plateau at 2.95 V. However, when $Li_xMn_{1.85}O_4$ is overcharged to $0 \leq x \leq 0.60$, it enters the two-phase region of spinel $+ O_2$ and may release oxygen gas. Overdischarging beyond $x \geq 2$ would lead to more drastic phase changes since, even after consumption of $t\text{-}LiMnO_2$, both spinel cathodes are still not in equilibrium with Li, as seen in Figure 12.14.

### 12.2.3.3    Evaluation of Safety

The enthalpy of formation is a key factor to evaluate the safety of LIBs because it is the main factor for the thermodynamic stability of electrode materials. The calculated and measured enthalpies of formation per mole of atoms of $Li_{1+y}Mn_{2-y}O_4 (0 \leq y \leq 0.33)$ and $Li_{1-w}Mn_2O_4 (0 \leq w \leq 1)$ spinel compounds are shown in Figure 12.17, and they are in agreement with each other. During the delithiation of the $LiMn_2O_4$ compound, the increasing (less negative) enthalpies of formation of $Li_{1-w}Mn_2O_4$ spinel result in decreasing phase stability. With increasing Li content, the enthalpies of formation of the $Li_{1+y}Mn_{2-y}O_4$ spinel increase only slightly, indicating that the thermodynamic stability of $Li_{1+y}Mn_{2-y}O_4$ is also slightly decreased compared to the most stable center point at $LiMn_2O_4$.

As shown in Figure 12.16b, the LMO spinel releases oxygen gas when the cell voltage exceeds the stability window of the electrodes during overcharge. Oxygen release inside the cells may lead to thermal runaway and combustion of the electrolyte. This oxygen release when being overcharged also applies to all of the $Li_{1+y}Mn_{2-y}O_4$ spinel phase if the ratio $Mn/O < 1:2$, i.e., for $(0 < y \leq 0.33)$. Only for $LiMn_2O_4 (y = 0)$ does the reaction path (Li – point $1 - \lambda\text{-}MnO_2$) in the phase diagram

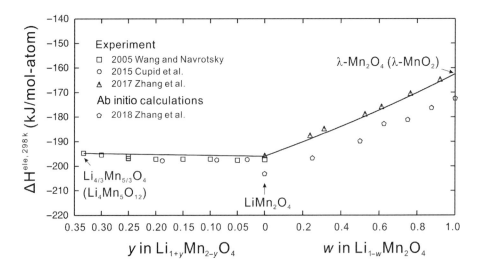

**Figure 12.17** Calculated enthalpies of formation per mol-atom for $Li_{1-w}Mn_2O_4 (0 \leq w \leq 1)$ and $Li_{1+y}Mn_{2-y}O_4 (0 \leq y \leq 0.33)$ spinel in the "battery state." This figure is reproduced from Li et al. (2019) with permission from Elsevier.

of Figure 12.14b not enter the two-phase region spinel + gas. Otherwise this "gas release" region is entered once the reaction path for delithiation of $Li_{1+y}Mn_{2-y}O_4$ compositions passes through the line $\lambda$-$MnO_2 - Li_4Mn_5O_{12}$ (point 2) in Figure 12.14b. Once this line is crossed, additional $Li^+$ ions and electrons can only be removed via the concomitant reduction of $O^{2-}$ to $0.5\ O_2$, leading to release of oxygen gas. Increasing of $y$ (reducing the ratio Mn/O in the spinel) exacerbates the oxygen gas release, especially at a high overcharging.

### 12.2.3.4 Evaluation of Energy Density

Voltage, capacity, and energy density are the key electrochemical properties of electrode materials in portable electronic devices and vehicles applications. The theoretical energy density can be calculated from the voltage and capacity via (12.6). The calculated theoretical capacities of $Li_{1+y}Mn_{2-y}O_4 (0 \leq y \leq 0.33)$ and $Li_{1-z}Mn_{2-2z}O_4 (0 \leq z \leq 0.11)$ solid solutions in the "sintered state," covering the composition range 2–1–3, are displayed in Figure 12.18. Why is the capacity zero at the terminal points $(2 = Li_4Mn_5O_{12}$ and $3 = Li_{12}Mn_4O_9)$? That is because the attempt to charge any spinel along the line 2–3 results immediately in oxygen gas release since the reaction path enters the spinel + gas region. This unsafe region does not count for the capacity. The calculated results in Figure 12.18 are slightly higher than the experimental data because the practical capacity cannot achieve the maximum theoretical value. The theoretical capacity is also proportional to the amount of $Mn^{3+}$, which is given at the right ordinate of Figure 12.18, due to the $Mn^{3+}/Mn^{4+}$ redox couples. The capacity decreases with the reduced ratio Mn/O due to increase of excess lithium $y$ in Li-rich spinel $Li_{1+y}Mn_{2-y}O_4$ or vacancy content $z$ in Li-defect spinel $Li_{1-z}Mn_{2-2z}O_4$.

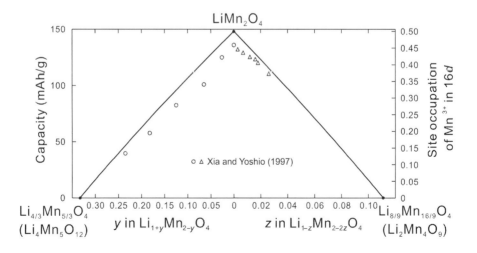

**Figure 12.18** Calculated theoretical capacity of $Li_{1+y}Mn_{2-y}O_4 (0 \leq y \leq 0.33)$ and $Li_{1-z}Mn_{2-2z}O_4 (0 \leq z \leq 0.11)$ solid solutions in the "sintered state." This figure is reproduced from Zhang et al. (2018) with permission from the American Chemical Society.

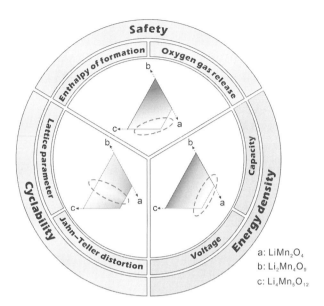

**Figure 12.19** A schematic diagram to evaluate the key factors (energy density, cyclability, and safety) determining battery performance. Energy density is evaluated based on cell voltage and capacity. Cyclability is evaluated based on electrochemical stability (the lattice parameter) and suppression of the Jahn–Teller distortion (the geometrical distortion of the $MnO_6$ octahedron). Safety is evaluated based on thermodynamic stability (enthalpy of formation) and less oxygen gas release. Favorable compositions for each property are represented with the dashed ellipse. This figure is provided by courtesy of Professor Weibin Zhang at Shandong University of China.

### 12.2.3.5    Optimization of the Composition Based on Comprehensive Consideration

This case study demonstrates a high-throughput computational framework to systematically describe the composition–structure–property relationships and evaluate the performance of the spinel Li–Mn–O compounds as a positive electrode for battery applications. Based on the previous calculations, the key factors (cyclability, safety, and energy density) must be taken into consideration to select the suitable 4 V spinel cathode material for specific applications, as shown in Figure 12.19. The cyclability is evaluated by considering the physical properties to find the compounds with higher electrochemical stability as well as suppression of the Jahn–Teller distortion. The safety is evaluated based on the thermochemical properties to select the compounds with a higher thermodynamic stability and less oxygen gas release. The energy density can be evaluated according to the electrochemical properties of cell voltage and capacity.

While the LMO spinel with the highest energy density is located close to the composition 1 in the solid-solution range 2–1–3 in Figure 12.14a, a trade-off between the other goals must be made. Moving with composition toward points 2 or 3 decreases the energy density, and it may also decrease safety and increase cyclability to different extents depending on the exact move taken as summarized in

Figure 12.19 derived from Zhang et al. (2018). This high-throughput computational framework employs combinatorial approaches to reveal complex phenomena and select the ideal composition of Li–Mn–O cathodes high-efficiently according to the practical application.

## References

Ceder, G., Hautier, G., Jain, A., and Ong, S. P. (2011) Recharging lithium battery research with first-principles methods. *MRS Bulletin*, 36(3), 185–191.

Drüe, M., Seyring, M., Kozlov, A., Song, X., Schmid-Fetzer, R., and Rettenmayr, M. (2013) Thermodynamic stability of $Li_2C_2$ and $LiC_6$. *Journal of Alloys and Compounds*, 575, 403–407.

El Kharbachi, A., Pinatel, E., Nuta, I., and Baricco, M. (2012) A thermodynamic assessment of $LiBH_4$. *CALPHAD*, 39, 80–90.

Farrell, D. E., and Wolverton, C. (2012) Structure and diffusion in liquid complex hydrides via ab initio molecular dynamics. *Physical Review B*, 86(17), 174203.

Frankcombe, T. J., and Kroes, G. J. (2006) Quasiharmonic approximation applied to $LiBH_4$ and its decomposition products. *Physical Review B*, 73(17), 174302.

Friedrichs, O., Remhof, A., Hwang, S. J., and Züttel, A. (2010) Role of $Li_2B_{12}H_{12}$ for the formation and decomposition of $LiBH_4$. *Chemistry of Materials*, 22(10), 3265–3268.

He, Q., Yu, B., Li, Z., and Zhao, Y. (2019) Density functional theory for battery materials. *Energy & Environmental Materials*, 2(4), 264–279.

Hoang, K., and Van de Walle, C. G. (2009) Hydrogen-related defects and the role of metal additives in the kinetics of complex hydrides: a first-principles study. *Physical Review B*, 80 (21), 214109.

Hoang, K., and Van de Walle, C. G. (2012) Mechanism for the decomposition of lithium borohydride. *International Journal of Hydrogen Energy*, 37(7), 5825–5832.

Huggins, R. A. (2009) Introductory material, in Huggins, R. A. (ed), *Advanced Batteries: Materials Science Aspects*. Boston: Springer US, 1–23.

Kozlov, A., Seyring, M., Drüe, M., Rettenmayr, M., and Schmid-Fetzer, R. (2013) The Li–C phase equilibria. *International Journal of Materials Research*, 104(11), 1066–1078.

Lee, S. H., Manga, V. R., and Liu, Z. K. (2010) Effect of Mg, Ca, and Zn on stability of $LiBH_4$ through computational thermodynamics. *International Journal of Hydrogen Energy*, 35(13), 6812–6821.

Li, H. W., Orimo, S., Nakamori, Y., et al. (2007) Materials designing of metal borohydrides: viewpoints from thermodynamical stabilities. *Journal of Alloys Compounds*, 446–447, 315–318.

Li, N., Li, D., Zhang, W., et al. (2019) Development and application of phase diagrams for Li–ion batteries using CALPHAD approach. *Progress in Natural Science*, 29(3), 265–276.

Liang, S. M., Taubert, F., Kozlov, A., Seidel, J., Mertens, F., and Schmid-Fetzer, R. (2017) Thermodynamics of Li–Si and Li–Si–H phase diagrams applied to hydrogen absorption and Li–ion batteries. *Intermetallics*, 81, 32–46.

Mantina, M., Wang, Y., Chen, L. Q., Liu, Z. K., and Wolverton, C. (2009) First principles impurity diffusion coefficients. *Acta Materialia*, 57(14), 4102–4108.

Mauron, P., Buchter, F., Friedrichs, O., et al. (2008) Stability and reversibility of $LiBH_4$. *Journal of Physical Chemistry B*, 112(3), 906–910.

Meng, Y. S., and Arroyo-de Dompablo, M. E. (2009) First principles computational materials design for energy storage materials in lithium ion batteries. *Energy & Environmental Science*, 2(6), 589–609.

Orimo, S.-I., Nakamori, Y., Ohba, N., et al. (2006) Experimental studies on intermediate compound of $LiBH_4$. *Applied Physics Letters*, 89(2), 021920.

Schmid-Fetzer, R. (2014) Phase diagrams: the beginning of wisdom. *Journal of Phase Equilibria and Diffusion*, 35(6), 735–760.

Tekin, A., Caputo, R., and Züttel, A. (2010) First-principles determination of the ground-state structure of $LiBH_4$. *Physical Review Letters*, 104(21), 215501.

Vajo, J. J., Skeith, S. L., and Mertens, F. (2005) Reversible storage of hydrogen in destabilized $LiBH_4$. *Journal of Physical Chemistry B*, 109(9), 3719–3722.

Wang, J., Du, Y., and Sun, L. (2018) Understanding of hydrogen desorption mechanism from defect point of view. *National Science Review*, 5(3), 318–320.

Wang, J., Du, Y., Xu, H., et al. (2011) Native defects in $LiNH_2$: a first-principles study. *Physical Review B*, 84(2), 024107.

Wang, J., Freysoldt, C., Du, Y., and Sun, L. (2017) First-principles study of intrinsic defects in ammonia borane. *Journal of Physical Chemistry C*, 121(41), 22680–22689.

Wilson-Short, G. B., Janotti, A., Hoang, K., Peles, A., and Van de Walle, C. G. (2009) First-principles study of the formation and migration of native defects in $NaAlH_4$. *Physical Review B*, 80(22), 224102.

Xia, Y., and Yoshio, M. (1997) Optimization of spinel $Li_{1+x}Mn_{2-y}O_4$ as a 4 V Li-cell cathode in terms of a Li–Mn–O phase diagram. *Journal of the Electrochemical Society*, 144(12), 4186–4194.

Yang, J., Sudik, A., and Wolverton, C. (2007) Destabilizing $LiBH_4$ with a metal (M = Mg, Al, Ti, V, Cr, or Sc) or metal hydride ($MH_2$ = $MgH_2$, $TiH_2$, or $CaH_2$). *Journal of Physical Chemistry C*, 111(51), 19134–19140.

Zhang, W., Cupid, D. M., Gotcu, P., et al. (2018) High-throughput description of infinite composition–structure–property–performance relationships of lithium–manganese oxide spinel cathodes. *Chemistry of Materials*, 30(7), 2287–2298.

Züttel, A., Wenger, P., Rentsch, S., Sudan, P., Mauron, P., and Emmenegger, C. (2003) $LiBH_4$ a new hydrogen storage material. *Journal of Power Sources*, 118(1), 1–7.

# 13 Summary and Future Development of Materials Design

## Contents

Computational design of engineering materials is an emerging multidiscipline of computational materials science, mechanics of materials, computational mathematics, physics, and materials characterization, among others. The purpose of this book is twofold. On one hand, we have presented broad and concise introductions to common computational methods currently used in materials design so that the interested reader can understand fundamentals associated with materials design and related fields. On the other hand, case studies for computational design of representative engineering materials (steels, light alloys, superalloys, cemented carbides, PVD and CVD hard coating, as well as lithium batteries and hydrogen storage materials) were demonstrated via step-by-step procedures mainly using the fundamentals described in this book.

In this chapter, we briefly summarize the main content of the book, highlight computational designs of some other engineering materials and processes not covered in the preceding chapters, and discuss future orientations and several challenges for materials design.

## 13.1    Brief Summary of This Book

Starting with a linkage of the advance of human civilization with materials and definitions of a few terms widely used in computational design of materials, Chapter 1 briefly introduces the past and state-of-the-art development of computational design of engineering materials as well as the scope and structure of the book. Particularly, two flowcharts (one is through-process simulation and experiment of Al alloys during solidification, homogenization, rolling, and age strengthening, and the other is about three major stages for the development of engineering materials) are described in some detail to demonstrate a general framework about materials design to the reader.

In Chapter 2, a brief introduction is presented about two main atomistic simulation methods, first-principles calculation based on density functional theory and molecular dynamics, focusing on these methods' theoretical basis. Subsequently, we demonstrate how to obtain some basic materials properties, such as lattice parameters, finite-temperature thermodynamic properties, mechanical properties, and point defect properties by first-principles calculations. After that, one case study is shown for designing Mg–Li alloys as ultralightweight applications, based on first-principles calculations.

Chapter 3 describes a few methods that can simulate the microstructure evolution at mesoscale. Modeling at the mesoscale is appropriate, as length scales intermediate between the atomic and continuum scales and can bridge between atomistic structures and macroscopic properties of materials. Among the mesoscale simulation methods, the phase-field method and the cellular automaton method are the most important. Consequently, the chapter is mainly focused on the fundamentals of the two methods, their applications, and integration with some other methods such as first-principles calculation, CALPHAD, and machine learning. One case study for design of high-energy-density polymer nanocomposites is briefly presented mainly using the phase-field method.

In Chapter 4, firstly, basic concepts and equations of continuum mechanics are briefly recalled for the sake of understanding fundamental equations of the crystal plasticity finite element method. Then, representative mechanical constitutive laws of crystal plasticity, including the dislocation-based constitutive model and the constitutive model for displacive transformation, are briefly described, followed by a short introduction to the finite element method and procedure for crystal plasticity finite element simulation. Subsequently, one case study of plastic deformation-induced surface roughening in Al polycrystals is demonstrated to show important features of the crystal plasticity finite element method.

Chapter 5 first presents an overview of the CALPHAD method, intended as a practical guide for readers without prior knowledge but also including useful advice for advanced readers. It covers origins, development, and principles of CALPHAD along with a brief introduction to commercial and open-source CALPHAD-based software. Thermodynamic modeling is first introduced for simple phases with fixed composition, followed by solution phases (substitutional solution, gas phase, associate solution, quasichemical model, and sublattice model for solid-solution phases). The effects of both pressure and magnetic contributions on thermodynamic functions and phase stability are also briefly described. Important features of thermodynamic data-bases for multicomponent and multiphase systems are highlighted, which are indispensable for efficient design of engineering materials. In the final sections, design applications for Al and Mg alloys are demonstrated, using solely thermodynamic database or extended CALPHAD-type databases, concluding with one case study showing the CALPHAD design of Al alloys with a high resistance to hot tearing.

Chapter 6 begins by describing Fick's laws on diffusion, diffusion mechanism, and diffusion in binary, ternary, and multicomponent phases, and a variety of computational methods to calculate and/or estimate diffusivity and atomic mobilities. Secondly, a few other important thermophysical properties are very briefly introduced, including interfacial energy, viscosity, volume, and thermal conductivity. Subsequently, the procedure to establish thermophysical databases is described from the materials design point of view. One case study for simulation of the precipitation and age hardening in AA6005 aluminum alloys is demonstrated mainly using thermophysical properties as input to show their importance for materials design.

Part two of the book begins with Chapter 7, in which a step-by-step material design for two representative steels, S53 high strength steel and AISI H13 hot-work tool steel, are demonstrated. In the case of S53 steel, both high strength and excellent corrosion resistance are needed. For that purpose, to screen the composition range with a high strength, the cross-plot of the thermodynamic driving force for precipitated phases versus the coarsening rate constant was established. The corrosion resistance was computationally designed by analyzing the multicomponent thermodynamic effects to maximize Cr partitioning in the spinel oxide. These calculations and subsequent simulation-guided experiments have reduced the development cycle time for S53 steel significantly compared to the traditional trial-and-error approach. For the case study of AISI H13 tool steel, firstly precipitations of carbides were simulated using a thermo-kinetic software. Secondly, simulated microstructural parameters were coupled with structure–property models to predict yield stress, flow stress curve, and creep properties. Subsequently, those simulated properties were coupled with a finite element method (FEM) to predict the relaxation of internal stresses and the evolution of deformations at the macroscopic scale. Finally, these simulated mechanical properties were coupled with FEM simulations to simulate the stress relaxation during tempering of a thick-walled workpiece made of AISI H13. These two case studies demonstrate the significant role of a variety of simulation methods for materials design.

In Chapter 8, selected case studies for material design of Al and Mg light alloys demonstrate the various methods. For Al alloys, there are two examples. The first example shows the solidification simulation of cast alloy A356, in which a dedicated microsegregation modeling is presented. The second example is about elaborate heat treatment simulation and precipitation kinetics. In both examples, thermodynamic databases and extended CALPHAD-type databases (atomic mobility and kinetic databases) are utilized. In the first set of examples for Mg alloys, the selection of alloy composition and optimized heat treatment schedules for new creep-resistant Mg alloys, AZ series alloys and Mg–Al–Sn–based cast alloys are demonstrated using both thermodynamic and extended CALPHAD-type databases by means of the PanSolidification and PanPrecipitation software modules. The final case study in Chapter 8 addresses biomedical applications. The CALPHAD-based design strategy is detailed, used for the metallurgical part of the development process that culminated in a recent bioresorbable Mg alloy stent to cure coronary artery disease.

The design of advanced superalloys requires comprehensive understanding of multicomponent and multiphase materials and multiphysics, as well as concurrent consideration of multiple descriptors at different length scales. In Chapter 9, taking the single-crystal Ni-based superalloys and NiFe-based superalloys for advanced ultra-supercritical (A-USC) boilers in power plants as examples, we demonstrated a few aspects for material design of superalloys, covering the selection of alloy composition, property prediction, control of the microstructure, and the structure–property correlation. The employed computational methods include thermodynamic calculations, property prediction model, the multi-objective optimization algorithm, and machine learning.

In Chapter 10, three types of cemented carbides (ultrafine cemented carbide, WC–Co–Ni–Al cemented carbide, and gradient cemented carbide) are designed according to several simulations coupled with experiments. In the case of ultrafine cemented carbide, thermodynamic calculations were utilized to select the optimal alloy composition and sintering temperature in order to avoid or minimize the segregation of (Ta, W)C cubic phase in cemented carbides. Such a segregation leads to an inhomogeneous microstructure and deteriorates the mechanical properties of the cemented carbides. In addition, the optimal mechanical properties were obtained via comprehensive control of two inhibitors (VC and $Cr_3C_2$), and the selected sintering temperature and alloy compositions, which are based on thermodynamic calculations and a Weibull plot of fracture strength. For the WC–Co–Ni–Al cemented carbides, the calculated Co–Ni–Al phase diagram and solid–liquid interfacial energy were employed to optimize the composition of the Co–Ni–Al binder phase and sintering temperature. The shape and size of the WC hard phase were controlled according to phase-field simulation and key microstructure characterization. It was shown that the best trade-off between transverse rupture strength and Rockwell hardness can be obtained through the aforementioned calculations and calculation-guided experiments. The customization of microstructure in gradient cemented carbides is a formidable challenge. In this case, the thermodynamic calculations and diffusion simulations for multicomponent and multiphase cemented carbides were performed to select alloy compositions and sintering schedules (temperature, time, and atmosphere) and provide the most important microstructure features (such as phase assemblage and fractions) prior to the key experiment. The microstructure-based hardness model was then developed to predict the hardness distribution from the surface to the core of the gradient cemented carbides. In comparison with time-consuming and costly experimental approach, this simulation-driven materials design leads to the development of actual industrial products within three years.

In Chapter 11, we demonstrated the design of two types of hard coatings, TiAlN-based PVD hard coating and TiCN CVD hard coating, via computations and experiments. In the case of TiAlN PVD coating, it was described how to adjust the formation of the metastable phase, select the appropriate deposition temperature, and manipulate the microstructure to obtain desired mechanical properties mainly through first-principles calculations and thermodynamic calculations for metastable phases. The deposition of TiAlN/TiN and TiAlN/ZrN multilayer guided by first-principles calculations was also briefly described. For the TiCN CVD hard coating, it was indicated that the computed CVD phase diagram can accurately describe the phases and their compositions under the given temperature and total pressure, as well as the partial pressures of various gases. Subsequently, computational fluid dynamics (CFD) can provide reliable temperature field, velocity, and distributions of various gases inside the CVD reactor. After that, key experiments based on the preceding thermodynamic and CFD calculations were conducted to deposit the TiCN hard coating efficiently. These simulation-driven designs for the hard coatings have found industrial applications in a very short time, just two years, much shorter compared to the development time via a purely experimental approach.

Chapter 12 focuses on energy storage/conversion materials. Two case studies for design of cathode material of lithium batteries and hydrogen storage materials were demonstrated. We show how to use the hybrid approach of CALPHAD modeling, first-principles calculations, and ab initio molecular dynamics to establish the relationship among composition, structure, and performance in these energy materials. For the design of hydrogen storage materials, the major aspects of thermodynamics, the phase diagram, and the hydrogen release process of the complex metal hydride $LiBH_4$ are considered. In the case of design for Li–ion batteries, thermodynamic properties, phase diagrams, cell voltage, and diffusion of Li–ion in Li–MnO spinel cathode material are taken into consideration.

Following this summary of the main content of the book in the previous chapters, we will now highlight the computational design of some other materials and processes and then discuss future orientations and key challenges for materials design.

## 13.2 Highlighting Computational Design of Other Engineering Materials and Processes

In this book, we have demonstrated the computational design of several engineering materials by means of step-by-step procedures. The strategy described in this book is equally valid for the computational design of other engineering materials and processes, such as $Mo_2BC$ thin film; $Cu_3Sn$ interconnect material; the Linz–Donawitz (LD) converter process model, which uses a basic oxygen furnace for slag, metal, and gas; and slag as recycled materials. The computational design of these four examples will be highlighted in the following subsections.

## 13.2.1    Highlighting the Design of $Mo_2BC$ Thin Film

In Chapter 11, we have mentioned that wear-resistant coatings such as TiAlN-based materials are widely applied in cutting and forming tools due to their high stiffness and hardness. However, they usually suffer crack initiation and growth as well as subsequent cracks, leading to failure of their protective function and the reduction of tool lifetime. This brittle deformation behavior mainly originates from the intrinsically low ductility of the ceramic materials. Therefore, it is of great interest to search for materials with intrinsic moderate ductility and high stiffness. Selection criteria of such candidates are as follows: (1) a bulk modulus to shear modulus ratio ($B/G$) that is above 1.75, indicating that the materials are ductile; (2) a positive Cauchy pressure value ($c_{12} - c_{44}$, where $c_{12}$ and $c_{44}$ are the longitudinal compression and shear modulus, respectively), which is a second indicator of moderate ductility (Pettifor, 1992); (3) a congruent melting point, which supports the high-stiffness notion; (4) a good phase stability; and (5) a high Young's modulus. All the information can be well predicted using first-principles calculations and the CALPHAD approach. Through computational study on a series of ceramic materials, $Mo_2BC$ (an unusually stiff and moderately ductile compound) is found to meet all the criteria (Emmerlich et al., 2009). The computational design strategy for this hard coating is summarized in an iterative fashion, as shown in Figure 13.1. $Mo_2BC$ exhibits a nanolaminated structure with a $B/G$ ratio of 1.75 and a Cauchy pressure of +43 GPa. For comparison, the typical $B/G$ ratio and Cauchy pressure of TiAlN are 1.44 and −40 GPa, respectively. $Mo_2BC$ melts congruently above 2,500°C, while TiAlN possesses a phase separation below 1,000°C. The calculated Young's modulus for $Mo_2BC$ is 470 GPa, which is comparable with that of TiAlN ($400 \sim 500$ GPa). Based on the theoretical results, the $Mo_2BC$ thin film was synthesized by magnetron sputtering, and the elastic properties were characterized by means of nanoindentation. The hardness of $Mo_2BC$ was determined to be $29 \pm 2$ GPa, which is close to that of TiAlN. No indication of brittle fracture was observed for $Mo_2BC$ thin film. Consequently, the synthesized $Mo_2BC$ thin film has found its application as a protective layer in industry.

## 13.2.2    Highlighting the Design of Nanocurvature-Enhanced Fabrication of $Cu_3Sn$

Low-electromigration interconnect materials, such as $Cu_3Sn$, are vital for lead-free microelectronic devices. They must be fabricated at temperature as low as possible to minimize potential interfering with surrounding electronic components in three-dimensional integrated circuits (3D ICs). However, the near–room temperature fabrication of $Cu_3Sn$ is substantially challenging, because the reaction of Cu and Sn

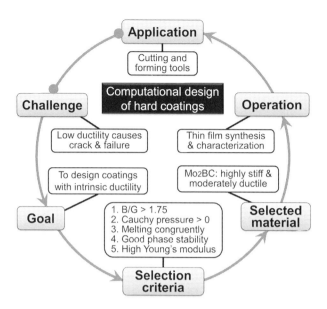

**Figure 13.1** Sketch of a computational design strategy for $Mo_2BC$ as an industrial protective thin film with high stiffness and moderate ductility.

produces preferably the intermetallic $Cu_6Sn_5$ compound with inferior properties. A breakthrough has been recently achieved by a conceptual design using dealloying and the nanocurvature effect on nucleation combined with a precursor ternary alloy design. Significant formation of $Cu_3Sn$ at both 70 and 55°C in 3 hours was achieved by Song et al. (2021) through dealloying of the tailored precursor alloy $Al_{67}Cu_{18}Sn_{15}$ (at.%) versus 20–100 hours at 120°C by conventional methods. $Cu_3Sn$ is formed just 10 minutes after the initial formation of $Cu_6Sn_5$ at 70°C, which is two orders of magnitude faster than the $Cu/Cu_6Sn_5$ diffusion-couple approach at 120°C. Dealloying was performed in 10% HCl solution, and phase formation/transformation was monitored by in situ synchrotron X-ray diffraction (XRD). The energy barrier to the formation of a $Cu_3Sn$ nucleus on nano islands of $Cu_6Sn_5$ was assessed in detail versus dealloying temperature and phase compositions using the CALPHAD method by exploiting the initial driving force (DF) concept and DF-plateau values. The dealloying process is summarized in Figure 13.2, and its three stages are explained in this subsection.

The as-cast $Al_{67}Cu_{18}Sn_{15}$ consists of a three-phase equilibrium structure $Al_2Cu + Bct(Sn) + Fcc(Al)$ in agreement with the experimental work and CALPHAD assessment of Mirkovic et al. (2008). In stage *I*, the HCl-etching of $Al_{67}Cu_{18}Sn_{15}$ first releases the Fcc(Al) phase, virtually pure Al, from this initial three-phase structure $Al_2Cu + Bct(Sn) + Fcc(Al)$. The alloy composition moves along the straight line toward the composition of $Cu_6Sn_5$ as indicated in Figure 13.2 until the two-phase equilibrium line of $Al_2Cu + Bct(Sn)$ is reached. At this point, the acid etches Al no longer from free Al but from $Al_2Cu$, and the dealloying path veers

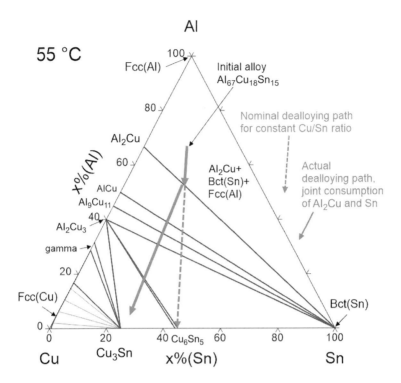

**Figure 13.2** Dealloying pathways of $Al_{67}Cu_{18}Sn_{15}$ based on Song et al. (2021), superimposed on the calculated isothermal section of the Al–Cu–Sn system at 55°C using the CALPHAD description of Mirkovic et al. (2008). The nominal pathway must follow a straight line extending from the top corner, representing pure Al, because of the constant Cu/Sn ratio due to the mass balance. However, after consumption of free Al in the Fcc(Al) phase, the actual pathway continues from the two-phase equilibrium line $Al_2Cu + Bct(Sn)$ with increasing Cu/Sn ratio due to the joint consumption of $Al_2Cu$ and Sn, shown schematically by the kink in the continuation of the solid arrow. A black and white version of this figure will appear in some formats. For the colour version, please refer to the plate section.

off to a more Cu-rich direction, sketched only schematically in Figure 13.2. Now nucleation becomes the dominant issue, as discussed later for the relevant binary system Cu–Sn. $Al_2Cu$ is left behind as a source for Cu to react with the abundant Sn in the system.

In stage *II* we first have a coverage of the abundant Bct(Sn) by freshly produced Fcc(Cu) from dealloying of $Al_2Cu$ in HCl solution. The freshly Cu layer is easily saturated by in-diffusion of Sn from the abundance of Sn in the system. To understand which of the competing phases ($Cu_3Sn$ or $Cu_6Sn_5$) will form in stage *II*, the concept of initial driving force, denoted as *DF*, need to be introduced here because the small nucleus of the new phase is formed from a supersaturated solution phase. The corresponding phase-type reactions are (13.1) and (13.2):

$$Fcc(Cu) + Bct(Sn) \rightarrow Cu_6Sn_5 \tag{13.1}$$

$$Fcc(Cu) + Bct(Sn) \rightarrow Cu_3Sn \tag{13.2}$$

The calculated plateau value of DF at any temperature between 38 and 150°C for reaction (13.1) is larger than that for reaction (13.2), which is consistent with experimental observations of initial $Cu_6Sn_5$ formation. However, the Fcc(Cu)/$Cu_6Sn_5$ interface still provides a driving force for the formation of $Cu_3Sn$, leading to stage *III* with reaction (13.3):

$$Fcc(Cu) + Cu_6Sn_5 \rightarrow Cu_3Sn \tag{13.3}$$

The total energy barrier for the formation of the $Cu_3Sn$ nucleus also involves the specific interfacial energy for the Fcc(Cu)/$Cu_3Sn$ and Fcc(Cu)/$Cu_6Sn_5$ interfaces, which was calculated based on the generalized broken bond (GBB) model in the CALPHAD approach using the Pandat software (Cao et al., 2009), as detailed in Section 13.3.2. In the tailored dealloying process, the Fcc(Cu)/$Cu_6Sn_5$ interface is not planar but characterized by the formation of $Cu_6Sn_5$ nanocaps. This nanocurvature effect reduces the energy barrier for nucleation of $Cu_3Sn$. The same barrier was also calculated from the thermodynamic data. The conceptual design and comprehensive predictions from computational thermodynamics are confirmed by the nanocurvature-assisted fabrication of $Cu_3Sn$ during the dealloying process (Song et al., 2021).

### 13.2.3    Highlighting the Design of Steel Production Process

In the LD converter process for steel production, pure oxygen is blown on a molten carbon-rich iron bath for refining purposes. Carbon is oxidized to gas bubbles containing CO and $CO_2$. Other dissolved elements, such as Si and Mn, but also part of the molten iron, are oxidized, forming a slag phase. Several reaction zones can be identified: (1) gas–metal hot spot reactor, (2) slag reactor, (3) metal–slag reactor, and (4) metal bath reactor. A simple cell model was developed (Modigell et al., 2008) and relationships were deduced that define the mass and energy transfer between the cells (reaction zones) and thus the model parameters. The kinetic and transport limitations of the process are modeled by the interlink of the reaction zones, and the zones themselves are treated assuming thermochemical equilibrium. The dissolution of fluxes, especially lime, was implemented into the model regulating lime participation in reactions leading to the formation of slag. Furthermore, the addition and melting of steel scrap were modeled depending on scrap temperature and carbon concentration.

The aforementioned process can be modeled using the modeling tool called SimuSage, which is a combination of the thermochemical programmer library ChemApp and a library of additional graphical components for Borland's programming environment Delphi. The equilibrium calculations within each reaction zone using ChemApp are performed by the same Gibbs energy minimization code as in the interactive software ChemSage (now FactSage). The thermodynamic data comprise the gas phase (60 species), the nonideal liquid Fe-rich metal phase, the nonideal liquid slag, and a large number of potentially additional occurring phases (Modigell et al., 2008).

This process modeling tool simulates decarburization as a function of time, and the simulation results are in agreement with experimental data and the behavior of other elements dissolved in metal and slag. Energy losses from the process are calculated throughout the simulation. The predicted temperatures of the hot spot and the metal bath also show a good agreement with real process data, despite the simplicity of the model structure employed. This process model enables users to predict the effects of controllable process parameters by the onsite engineer, such as the mass of the initial metal bath, the time profile of the oxygen blow rate, the mass of scrap, and lime input.

## 13.2.4    Highlighting the Design of Slag as Recycled Material

It is remarkable that the strategy described in this book for materials design works well not only for small pieces of materials such as thin film and coating but also for macroscale large-size products such as the industrial waste. The term "industrial waste" refers to waste generated from industrial production. Three types of industrial wastes, i.e., industrial solid waste, industrial effluent, and industrial exhaust gases, have gained global visibility due to their pollution to Earth's air, water resources, soil, and so on. Among these three types of wastes, industrial solid waste has long been considered for reuse or recycling as raw material for construction purposes instead of discharging it in landfills. This valorization is both economically and ecologically friendly.

Industrial solid waste, such as blast furnace slag or steel slag and construction or cement raw materials, may have similar chemical compositions. However, even for a fixed composition, they may be composed of different phases (i.e., mineralogical compositions), which results in quite different properties, and this uncertainty may limit their utilization and valorization. This complexity has been mainly tackled by the trial-and-error approach and leaves much room for improvement using computational thermodynamics.

The core quaternary system of slags is $CaO - SiO_2 - Al_2O_3 - MgO$, disregarding the minor $Fe_2O_3$. The cooling condition of the liquid slag strongly influences the formation of various compound phases. While the thermodynamic solidification simulation using the Scheil assumptions (see Chapters 5 and 8) has been widely used for decades for alloys, the first published application to slags was found in the year 2007 (Durinck et al., 2007). Durinck and his colleagues studied $CaO - SiO_2 - MgO$ slags with two distinct compositions (slag A and slag B), which were molten at $1,640°C$ and cooled to room temperature with 1 K/minute. The Scheil assumption of no diffusion in the solid state reveals a better agreement of predicted phase formation with experiments compared to the simulation under the complete equilibrium assumption for the MgO-richer slag B. For slag A, both simulations perform similarly because of the less complex solidification path. For applications to industrial slags, Durinck et al. (2007) emphasized that more work was needed to determine the actual cooling path, control the slag composition accurately, and develop the thermodynamic database.

Another important example is the conversion of steel slag into cement. Figure 13.3 presents a schematic flowchart in which the computation strategy follows a series of steps using the production of converter slag as an example.

Converter slag that contains V is used in industry to produce cements. The key point for this study is that the enrichment and extraction of V from slag are performed first by the formation of a $FeV_2O_4$ magnetic phase, followed by magnetic separation

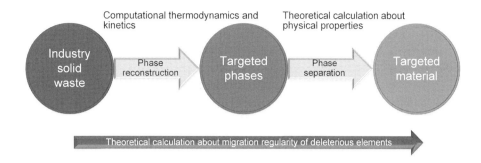

**Figure 13.3** The schematic flowchart of utilization of industrial solid waste.

before its use as raw material for the production of cement. The possibility and limitations of the $FeV_2O_4$ spinel phase reconstruction are calculated thermodynamically, and the rate of the reaction is calculated kinetically. The remaining phases in the slag after the formation of spinel phase must be reconstructed again for its preparation for use in the production of cement. The spinel phases must be magnetically separated with respect to their physical properties. The complex composition of the slag involves the dissolution of many different elements into the $FeV_2O_4$ spinel phase. Thus, the magnetic properties of the $FeV_2O_4$ formation after phase reconstruction must also be determined by computational calculation. Also, density functional theory (DFT) calculations on mechanical property in related thaumasite and ettringite mineral crystals and doping behaviors of Mn ions in clinker phases have been performed to support such efforts (Tao et al., 2018).

For more examples, the reader is referred to section 2.11, "Recycling Processes," in the recent *Handbook of Software Solutions for ICME* (Schmitz and Prahl, 2016).

## 13.3 Future Orientations and Challenges for Computational Design of Engineering Materials

### 13.3.1 General Aspects of Computational Design of Engineering Materials, ICME, MGI, and CDMD

Currently, computational design of engineering materials is emerging as a compelling and powerful discipline for new materials design. It may be considered a subfield of integrated computational materials engineering (ICME) (ICME, 2008), which is the broader field including modeling of existing and new materials. In that sense, it is related to the Materials Genome Initiative (MGI) (MGI, 2011), which is also a subfield of ICME because of its focus on new materials and the acceleration of its development cycle by employing fundamental scientific databases (the materials genome database), as pointed out by Xiong and Olson (2016). This has shown significant advantages in comparison to the traditional trial-and-error approach, and the significant reduction of development cycle time for new materials has been proved. The central paradigm was given in the three-link chain model by Olson (1997). A graphical extension of that

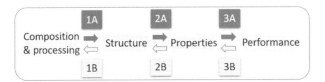

**Figure 13.4** Sketch of the four cornerstones of materials science and engineering and the links (1A, 2A, and 3A) of the materials-centered perspective, as well as the links (3B, 2B, and 1B) of a user-centered perspective.

model is given in Figure 13.4. The original term "processing" is extended by "composition" because the chemical composition of the material is at the same level of importance compared to processing (heat treatment temperature–time profiles, thermomechanical stress/pressure, and other applied external fields). Thus, "composition and processing" comprise the tools of the materials engineer forming the first of the four cornerstones in Figure 13.4. Each two of these adjacent cornerstones form one of the three links in the original paradigm (Olson, 1997). Figure 13.4 shows the direct links between these cornerstones. From the (traditional) materials-centered perspective, starting at the tools, these are 1A, 2A, and 3A, leading to the performance of the device or part. However, adopting a user-centered perspective, one starts from this goal backward along 3B, 2B, and 1B to find the appropriate tools.

For a long time, the link 1A was studied quite independently from the link 2A in the materials science community. On the other hand, traditional materials engineers may start from property compilations and design the device through link 3A. It is also clear that, in this A direction, a single experiment will provide a single "data point" on the performance side. That approach is not feasible in the B direction, because at each link a multitude of possibilities is encountered. Thus, the traditional trial-and-error method comprises a larger number of trials in the A direction, with the hope that the performance goal will be met. This is why computational simulation tools for each of the links are indispensable for predictive capability, allowing the simulation to enlighten both directions with limited effort and limited key experiments for calibration/validation of the underlying models. The integration results in the power of ICME along the complete chain in either direction.

In view of the need for establishing more quantitative relationships among these four cornerstones in materials science and engineering as well as advancing highly efficient product design methods for industry, several future orientations and challenges for computational design of engineering materials are suggested and discussed in the following subsection.

## 13.3.2 Advancement of Models and Approaches for More Quantitative Simulation in Materials Design

### 13.3.2.1 Heterointerface and Homointerface Thermodynamics

In order to better describe important phenomena in engineering materials, a deeper understanding about properties along interfaces in thin film or bulk materials is prerequisite. We need to distinguish the heterointerface between two different phases, $\alpha/\beta$, and the homointerface, $\alpha/\alpha$, such as in grains differing mainly by orientation of the polycrystals of the same phase.

For heterointerfaces the classical Gibbs model (dividing surface, zero volume) and Guggenheim model (homogeneous surface region, "interphase" finite volume) are established and the derived Gibbs adsorption equation is frequently used in physical chemistry and surface science (Láng, 2015). Some established concepts on interfacial energy have been discussed in Section 6.3, especially on surface ($\alpha =$ gas) energy (or surface tension) and for liquid–solid interfaces. Here we focus on the interface between solid–solid phases. The nearest-neighbor broken-bond model for the interfacial energy suggested by Becker in 1938 (Becker, 1938) has been further developed and combined with the CALPHAD approach for successful applications in metal/oxide and alloy systems (Nishizawa et al., 2001; Sonderegger and Kozeschnik, 2010). That generalized model is also implemented in the PanPrecipitation module of the software package Pandat to provide the composition and temperature-dependent values of interfacial energy (Cao et al., 2009). Its seamless integration for precipitation simulation in multicomponent alloys is attractive; however, due to the basically simple assumptions in the model (Sonderegger and Kozeschnik, 2010), the model predictions still require validation/calibration in specific material systems. Thus, there is still room for improvement.

Considering hetero- and homointerfaces in polycrystalline materials, grain growth during processing or service is another of the key issues determining performance. For example, grain growth inhibitors, such as VC and $Cr_3C_2$, are added into WC–Co cemented carbides to retard the grain growth of the WC phase. To design the amount and type of grain growth inhibitors effectively, reliable knowledge of both bulk and interfacial thermodynamics is needed.

More recently, the term "complexion" has been established to describe an interphase or grain-boundary phase more precisely (Cantwell et al., 2014). The complexion is defined as the interfacial material that is in thermodynamic equilibrium with the abutting phase(s) but may be chemically and structurally distinct from any of the abutting bulk phases. The complexion is treated as a separate thermodynamic phase with a finite thickness, typically 0.2–2 nm. Even though surface complexions and heterophase boundary complexions are also possible, the main focus is on grain-boundary complexions existing in the homointerface of single-phase polycrystalline material, often as impurity containing intergranular films (Cantwell et al., 2014; Raabe et al., 2014). Similar to a phase, transitions may occur in such a complexion. They can be first-order or continuous phase-like transitions and lead to abrupt changes for transport and physical properties, thereby influencing overall materials properties, such as stress–strain, creep, fatigue, corrosion, strength, embrittlement, and electrical and thermal conductivities (Cantwell et al., 2014; Raabe et al., 2014).

To understand the mechanisms behind these phenomena, physically sound interfacial thermodynamic models for description of grain-boundary complexion transitions and construction of grain-boundary complexion diagrams are needed.

In comparison with the systematic theory and/or principle about thermodynamics of bulk phases, the theory about many aspects of the interface is far from well established. This is also the case for interfacial thermodynamics. Attempts in the field of interfacial thermodynamics have been performed on several binary and ternary systems. In particular, grain-boundary complexion diagrams (including both grain-boundary adsorption complexion diagrams with well-defined transition lines and grain boundary diagrams that predict trends for average general grain boundaries) were established to describe grain-boundary behaviors as functions of bulk compositions (Zhou and Luo, 2015; Zhou et al., 2016). However, grain-boundary complexion transitions in almost all the multicomponent systems remain largely unexplored. Grain-boundary diagrams in multicomponent alloys can describe interactions among multiple alloying elements at grain boundaries, providing a novel strategy to manipulate grain boundaries. Within the framework of grain-boundary engineering, there is a need for integration of interfacial thermodynamic model with atomistic simulations to supply important interfacial data such as grain-boundary energy, mobility, diffusivity, cohesive strength, and sliding resistance for materials design. A long-term scientific objective is to extend the current CALPHAD method to an interface and develop interfacial complexion diagrams as one of new tools for materials design.

In some recent publications, the terms "nanothermodynamics", "nano-CALPHAD", or "nanophase diagram" may be found. However, in our opinion, these terms are disapproved, or at least to be taken with caution. On closer inspection, according to our experience, the size effects in the so-called "nanothermodynamics" always boil down to heterointerface effects. Often simple geometries are considered, such as a spherical phase with small radius, alternatively the specific surface area (surface per volume of a phase) (Kaptay, 2012) to define the size (or length), which is not a state variable in Gibbs thermodynamics. The additional Gibbs energy contribution to a phase with a large specific surface area, or small size, in contact with another phase is derived from the well-known curvature dependent Gibbs–Thomson effect on heterointerfacial energy. While these contributions can become dominant and certainly need to be further studied, it is considered distracting to invent a new terminology.

In addition, surface melting of macrocrystals and melting of nanocrystals should be taken into consideration to develop the thermodynamic models for a nanosize system (Vegh and Kaptay, 2018). Currently, thermodynamic databases for bulk phases are well established for many systems and need to be amended by clearly separated additional contributions due to nanosize phases. It is emphasized that "size" as a parameter must be accompanied by the complete geometry and embedment of phases in the system. For example, in the reassessment of Ag–Au systems including size effects, isotropic spherical particles in equilibrium with a gas are considered, thus only the solid–gas and liquid–gas interfaces (Park and Lee, 2008). A different result would certainly be obtained if the spherical particles of the same size would touch so that the solid–liquid interface comes into play. That consideration also demonstrates that we cannot have a unique alloy "nanophase diagram" by just specifying the "size."

Nonetheless, it is expected that nanotechnology will significantly benefit from further development and combination of bulk and interface thermodynamics.

### 13.3.2.2   Thermodynamics under External Fields

In order to manipulate desired structure and properties of materials that could not be obtained by traditional materials processes, external fields such as magnetic and electric fields can be utilized. An early indication was given by Chuang et al. (1986), who demonstrated by CALPHAD-type calculations that the Curie temperature in Fcc Fe–Ni alloys splits at a tricritical point at 735 K and 47.6 at.% Ni, forming a miscibility gap in the Fcc $(\gamma)$ phase, $\gamma_1$(paramagnetic) $+ \gamma_2$(ferromagnetic), at lower temperature. Even in a moderate external magnetic field, these two phases with different compositions will react differently in a heat treatment designed in that gap. The research work in these areas currently focuses on experimental aspects. New theories and/or approaches that can describe the effect of these external fields on microstructure and properties deserve extensive investigations.

For a strong external magnetic field, a remarkable impact on phase transition may be observed. For example, the $\alpha/\gamma$ transition temperature of pure Fe increases by a few Kelvin in a field of some 2–3 Tesla, such as in typical medical magnetic resonance imaging (MRI) systems. In that example, and also in Fe–Si alloys, only paramagnetic phases are considered. A simple explicit equation is found for the contribution of the external magnetic field to the Gibbs energy of $\alpha$ and $\gamma$ phases to calculate the phase diagram (Gao et al., 2006). This contribution is given by the following equation with molar volume, magnetization, and external magnetic field flux density ($B_{ext}$) as variables:

$$dG_m^{ext\ \ mag} = -V_m M dB_{ext}. \tag{13.4}$$

For paramagnetic phases, the integration is explicitly solved because magnetization and external magnetic fields are taken to be proportional (Gao et al., 2006).

However, for ferromagnetic phases, where the impact of the external field is also much stronger, the magnetization cannot be calculated from that simple Curie–Weiss law. Currently, the Weiss molecular field theory (Cullity and Graham, 2009) is usually employed to determine the implicit relation between the magnetization of a ferromagnetic phase and the external magnetic field. That involves numerical calculations, without an explicit solution of (13.4). It is the basis for many related investigations under the application of an external magnetic field. So far, there are only limited investigations (Mitsui et al., 2013; Zeng et al., 2022) on the thermodynamic behavior of engineering materials even under the application of a constant magnetic field. The external magnetic field is a truly additional state variable in Gibbs thermodynamics, increasing the degrees of freedom of system, leading to four-phase equilibrium in a binary system even under constant pressure. It is highly expected that the thermodynamics, including the effect from the external magnetic field, will extend the field of computational materials science. The utilization of other external fields such as an electric field along with related new models and simulation efforts on these fields will provide new opportunities for materials design.

### 13.3.2.3    More Quantitative Phase-Field Models

Although process–structure and structure–property linkage models have been the main focus for materials design for the past half-century, predictive physics-based models are needed. For example, the phase-field (PF) method has been one effective process–structure approach to simulate the mesoscale microstructure evolution during processes. There is still a lot of room to improve the predictive power of this method.

First, more quantitative PF models should be developed. Though various PF models accounting for different phenomenon have been proposed, quantitative modeling of microstructure evolution remains challenging. One main concern is that interface width is an adjustable parameter that can be set to physically unrealistic values, resulting in a loss of detail and unphysical interactions between different interfaces. To achieve a more quantitative PF model, Plapp has pointed out that the "antitrapping current" term, the standard "relaxational equation of motion" for the orientation field, and the current procedure to "incorporate fluctuations" into the PF approach should be improved (Plapp, 2011).

Second, measurable input parameters needed for PF models should be provided. A drawback of the current PF methods is related to the large number of material parameters involved in the evolution equations. Each of these parameters impacts the predicted microstructure evolution and must therefore be carefully calibrated for the targeted problem. However, determining the requisite parameters can be time consuming, and for quantitative models, some of the necessary material properties may not be known. For this purpose, the scientific databases of thermodynamics, kinetics, and thermophysical properties (interfacial energy, anisotropy, and so on) need to be further improved.

Third, more progress in integrating the PF method with different scale simulation methods should be achieved. The PF method has almost become the method of choice for simulation of microstructure evolution under different driving forces in materials science and engineering and can be the link between macroscale and atomic scale, and bridge the gaps between composition/processing and structure of materials. We need to further improve and develop the description methods at all levels of the phase transformation process and develop effective multiscale cross-level coupling simulation methods, beyond current PF, to bridge all the gaps among a variety of simulation methods.

Fourth, more efficient algorithms for solving the PF equations should be developed. One of the drawbacks of current PF method is the significant computational effort required. However, the techniques of adaptive gridding and parallel computing including both the central processing unit (CPU) and graphics processing unit (GPU) have resulted in assuaging this drawback to some extent. Massive calculation is still the bottleneck restricting the large-scale PF simulation and its industry applications.

The authors argue that in the future, more applications beyond microstructures description, such as materials design, properties prediction, and multiscale modeling, will be conquered by the PF method.

## 13.3.3    Databases and Materials Informatics

### 13.3.3.1    Scientific Databases

It is indisputable that simulation results will be reliable only if the parameters are accurate (garbage in, garbage out). Since almost all of the engineering materials are multicomponent and multiphase, their computational design relies heavily on the quality of available scientific databases (Li et al., 2018), which are also called genome databases (Xiong and Olson, 2016). Among a variety of databases, the scientific databases, which are related to physically sound models, are the most important ones. These databases can design alloy composition, optimize heat treatment schedules, describe microstructural evolution, and even predict mechanical and other properties if they are armed with materials design software and/or a data mining algorithm. Representative scientific databases include thermodynamic and diffusivity databases. Nowadays, thermodynamic databases for many engineering materials are well established even though there is still room for improvement. By contrast, for most thermophysical properties, such as diffusivity, interfacial energy, viscosity, and thermal conductivity, databases are less well established. Consequently, more dedicated research on establishing high-quality thermophysical property databases should be promoted.

In addition, many technologically important metastable phases are not included in current scientific databases. Specifically, these are not stable at any temperature/ composition as opposed to metastable extension of stable phases, often well described by suspension of some stable phases in CALPHAD calculations. The properties of metastable phases could be obtained by a hybrid approach of experiment, atomistic calculations, and the CALPHAD approach. It is expected that the inclusion of metastable phases in scientific databases will further increase the power of computational design for engineering materials.

In addition to the aforementioned thermodynamic and thermophysical databases, other types of databases including optical properties (such as refraction and reflectivity) and plasticity properties (such as formation and migration energies for dislocations) in the mechanistic constitutive laws for materials properties should be initiated. Currently, the investigation for these databases is still in its infancy. Furthermore, the successful and long-term utilization of scientific databases for materials design requires extensive efforts on their standardization, updating, and management. The interested reader can refer to Lukas et al. (2007) and Schmid-Fetzer et al. (2007) for more details on this topic, such as thermodynamic databases.

### 13.3.3.2    Materials Informatics

Perhaps the most severe challenge for an effective computational design of engineering materials is to establish the quantitative and/or predictive relationships among the four cornerstones of materials science and engineering shown in Figure 13.4. At present, the combined use of extensive experiment and a variety of computational tools is employed to understand these complex interactions, which govern the final performance of an engineering system. The current situation is that the rate for materials discovery is still slow even with advanced computational and experimental tools. Recently, materials design driven by materials informatics has drawn attention, e.g., by learning from biological systems (Rajan, 2008). Materials informatics applies the ideas of informatics to materials science and engineering. Informatics is one research field in which information science, statistics, computational mathematics, and system thinking are integrated to examine the structure and behavior of information. Such an integration may provide one new way to access and explore plentiful information. The information in materials science and engineering is mainly presented in data, which include various experimental data and the data resulting from simulation and/or calculations. The main carrier for materials informatics is machine learning (ML). ML is an interdisciplinary discipline of statistics, approximation theory, convex analysis, and algorithm theory. It uses large amounts or even an ocean of data to optimize models and make reasonable predictions for the targeted system. The procedure for ML usually includes the definition of the problem, data collection, establishment of a model, evaluation, and result analysis. Armed with the theory from computer science and statistics, a machine learning model can be established through continuous evaluation and correction. The data mining method within the framework of ML can search important information hidden by a large amount of data through algorithms, build a prediction model for material behavior, and then achieve rapid optimization of the targeted system. Consequently, the machine learning model may predict behavior and properties of materials satisfactorily in a much shorter time than currently used materials design approaches. Wen et al. (2019) formulated a materials design strategy combining an ML model and decisive experimental feedback to search for high-entropy alloys with a targeted hardness value in the Al–Co–Cr–Cu–Fe–Ni system. Their strategy shows a promise for designing other multicomponent materials, such as bulk metallic glasses and superalloys, to obtain desired structure or properties in a very short time. It is expected that the collective integration of ML with various simulations tools and calculations-guided experiments will provide one new strategy for materials design.

## 13.3.4     Enhanced Simulation Software Packages

Computational design of engineering materials relies heavily on combination and simultaneous or consecutive uses of a variety of software packages. Usually, one individual package is targeted for one of the four scales (nano-, micro-, meso-, and macroscales), the sizes of which were described in Chapter 1. Further improvement of these simulation packages is needed for more efficient materials design and rapid development of new materials and processes, such as additive manufacturing (AM). The complex phenomena (rapid solidification, solid-state phase transformation, and so on) as well as the interaction and control of process parameters, which happen in a very short period, make AM a tremendous challenge for currently available simulation packages. According to the analysis by Schmitz and Prahl (2016), the simulation packages on the macroscopic scale of the manufacturing processes are quite successful in industry, reaching a high technology readiness level (TRL) of 8–9 (Moorhouse,

2002). TRL 8 means that the final stable version of the software is released to the end-users, and TRL 9 implies that the software is under continuous development and improvement. Despite the great success for the macroscopic scale simulation tools, some work remains to be done on these macroscopic simulation packages. Most recently, the integration of crystal plasticity with the finite element method or phase-field method has demonstrated its great potential for describing extremely complex structures, their micromechanical behaviors, and a process-structure-property relation during AM and processes far from equilibrium.

In contrast to the well-established macroscopic scale simulation tools, there is a long way to go for continuum mesoscale models and discrete models including electron/atomistic/mesoscopic scales. The integration and compatibility between these types of simulations tools are the major challenge for future development of computational design of engineering materials. Schmitz and Prahl (2016) have compiled globally available software solutions for ICME. For more details, the interested reader can refer to their very comprehensive and recent handbook.

Another issue associated with simulation software packages is the functions of software platform, data interaction mode and manual operation level, which is usually time consuming and labor intensive and requires considerable experience. Therefore, the focus of future development work will be on the automation and "intelligence" of material design software. Under this background, calculation of thermophysical properties (CALTPP) software came into being (Du et al., 2020; Liu et al., 2020; Wen, 2020). CALTPP is an intelligent software for thermophysical property calculation. It is a bold attempt for the development of future material design software. Adhering to this idea, more such software is expected in the near future.

### 13.3.5 Concurrent Design of Materials and Products

Generally speaking, most universities and research institutes, but only a limited number of companies worldwide, are performing materials design, focusing on establishing the relations among process, microstructure, and properties. On the contrary, product design is the main focus for almost all of the companies in the world. Nowadays, materials design and product design are usually performed separately. In order to shorten significantly the development time of a new material from idea to product, it is clear that materials design should be performed concurrently with product design. Another argument for concurrent design of materials and product (CDMP) is that future markets will not relate to microstructure and their properties but rather to functionality and performance directly. CDMP relies on continuous development of several aspects: predictive models and efficient simulations on multiple length and time scales, a variety of software and modeling tools and the related scientific databases, quantitative (even 3D) representations of multiscale structure, and powerful visualization and management of materials informatics. Currently, CDMP is in its infancy. Many changes and issues to be solved in this creative design strategy are discussed by McDowell et al. (2010). The interested reader can refer to their comprehensive book for more details. In the following, a few important challenges/issues associated with CDMP are briefly indicated.

For intelligent CDMP, more powerful models and tools are needed to track the evolution of materials and products along their entire production and service life cycle. To decrease simulation costs significantly, it is necessary to integrate or combine all individual modeling components into a single workflow so that all simulation tools are independent but mutually connected. Currently, multiscale bridging tools are limited, indicating the gaps and missing bridges among diverse modeling and simulation tools.

Another issue is that major sources of uncertainty in processing, microstructure, modeling, etc., should be taken into consideration in CDMP in a quantitative way. The uncertainties include natural uncertainty resulting from errors induced by processing and operating conditions, model parameter uncertainty due to insufficient or inaccurate data, systematic model uncertainty originating from approximations and simplifications associated with the model, and uncertainty propagation along the entire workflow for design. Techniques for characterizing and managing these uncertainties are greatly needed since the uncertainties can dominate the configuration of the design process and affect the range of acceptable solutions.

In Section 13.3.3, it has been stressed that scientific databases, including thermodynamic and thermophysical databases, have a decisive role for various simulations. In addition to these databases, it is identified that materials structure and property databases resulting from both experiment and simulation are also fundamental elements of materials informatics, especially for multicomponent and multiphase materials. As a result, the establishment of materials structure and property databases needs collaboration among experts for experiments, modeling, and informatics.

Another aspect that can significantly advance the development of CDMP is decisive experiments. Simulations and calculations cannot replace experiment completely. The concurrent design will be more efficient if it is combined with simulations predicting the conditions of key experiments, which are thus to be performed in a focused way. For some phenomena, such as far-from-equilibrium material processing (PVD, rapid solidification, additive manufacturing, etc.), mechanisms may be poorly understood using current theory and principles, and the model uncertainty is so high that experiments must be carried out. Recently, modern materials characterization techniques, such as SEM, aberration-corrected TEM, electron tomography, in situ heating, in situ biasing (applying electric fields to samples), and in situ tensile/compressive tests, have been widely used. In TEM methods as well as three-dimensional atom probe (3DAP), detailed quantitative descriptions of 3D microstructures and their dynamic evolution during heat treatments and deformation processes will be possible. It allows for predictions of materials properties as a function of processing parameters, including temperature–time profiles and composition. Most microstructural parameters obtained in simulations can now be well verified by SEM, TEM, 3DAP, and so on. For example, the quantitative 3D distribution of second-phase microparticles, grain size distribution, and grain orientations in samples with dimensions from several microns to several hundred microns can simultaneously be measured by the slice-and-view method (also called serial section tomography) in a dual-beam SEM with focused ion beam (FIB) and electron backscatter diffraction (EBSD) apparatuses (Burnett et al., 2016; Cao et al., 2012). Electron tomography in a TEM can realize a quantitative 3D imaging of particles in the nanoscale and determine their size, shape, distribution, and volume fraction (Malladi et al., 2014), whereas high-precision quantitative measurement of these parameters of nanoparticles can be performed using conventional TEM based on a sample thickness measurement by convergent beam electron diffraction (CBED) (Li et al., 2016) or electron energy-loss spectrometry (EELS). 3DAP not only can do the same thing but also can reveal local solute concentration in particles or boundaries (Li et al., 2016). Column-by-column identification of the chemical nature of atoms in a structure by Z-contrast scanning TEM with a high-resolution energy dispersive X-ray spectrometry (EDS) or EELS system is now widely used, and the obtained atomic resolution structures of particles and boundaries (Chen et al., 2020) facilitate the deep understanding about microstructure. Atomic-resolution high-angle annular dark field (HAADF) scanning transmission electron microscopy (STEM) combined with atomic-resolution energy-dispersive X-ray spectroscopy (EDS) can now precisely determine the structure of intermetallic phase in ternary Mg–Nd–Ag alloys (Zhao et al., 2020). Various in situ techniques in TEM further provide information about how the particles in materials nucleate, grow, transform, and deform during heating (Malladi et al., 2014), biasing, and/or deformation of the sample. It is highly expected that quantitative 3D microstructure will tremendously advance the concurrent design of materials and products.

As mentioned previously, CDMP includes many aspects: a wide array of simulation and software tools at various scales, scientific databases, structure and property databases, experiments, and various design exploration tools (such as materials informatics and decision-making tools). The detailed descriptions for design exploration tools were presented elsewhere (McDowell et al., 2010). It is necessary to develop system design methods and tools that can bridge or integrate this array of aspects into one whole framework system so that the individual work can be performed independently but while mutually connected. This makes it possible that the resource-sharing and complementary advantages are realized in the whole design chain to create the maximum synergy effects.

Last but not least, economy should be also considered in CDMP. Generally, the optimal material is a balance between meeting the product performance goals and minimizing the cost of the materials. Thus, the recycling and reuse of the materials can be considered in CDMP. For example, most recently, an increasing focus has been on the recycling of waste-cemented carbides for the sake of extracting WC powder and binder phase from the carbides as starting materials to make new cemented carbides.

## 13.3.6   ICME and MGI as Well as Their Correlations to CDMP

In Chapter 1, it was stated that, in addition to optimizing existing materials and processes, the goal of ICME is to optimize materials, processes, and component design long before components or products are fabricated, by linking materials models at multiple lengths and time scales into a holistic system. Similarly, the aim of MGI is to discover, develop, and manufacture advanced materials at least twice as fast as possible in comparison with the traditional trial-and-error approach by means of the integration of three platforms: high-throughput calculation, high-throughput experiments, and databases. Since the definition and content for both MGI and CDMP are almost identical, we could equally argue that MGI and CDMP are identical concepts within the framework of ICME and the development of both will bring revolutionary insight and tremendous opportunities for research and development of new materials.

## References

Becker, R. (1938) Die Keimbildung bei der Ausscheidung in metallischen Mischkristallen. *Annals of Physics (Oxford)*, 424(1–2), 128–140.

Burnett, T. L., Kelley, R., Winiarski, B., et al. (2016) Large volume serial section tomography by Xe Plasma FIB dual beam microscopy. *Ultramicroscopy*, 161, 119–129.

Cantwell, P. R., Tang, M., Dillon, S. J., Luo, J., Rohrer, G. S., and Harmer, M. P. (2014) Grain boundary complexions. *Acta Materialia*, 62, 1–48.

Cao, S. S., Pourbabak, S., and Schryvers, D. (2012) Quantitative 3-D morphologic and distributional study of $Ni_4Ti_3$ precipitates in a $Ni_{51}Ti_{49}$ single crystal alloy. *Scripta Materialia*, 66(9), 650–653.

Cao, W., Chen, S., Zhang, F., et al. (2009) PANDAT software with PanEngine, PanOptimizer and PanPrecipitation for multi-component phase diagram calculation and materials property simulation. *CALPHAD*, 33(2), 328–342.

Chen, H., Lu, J., Kong, Y., et al. (2020) Atomic scale investigation of the crystal structure and interfaces of the B′ precipitate in Al–Mg–Si alloys. *Acta Materialia*, 185, 193–203.

Chuang, Y., Chang, Y. A., Schmid, R., and Lin, J. (1986) Magnetic contributions to the thermodynamic functions of alloys and the phase equilibria of Fe–Ni system below 1200 K. *Metallurgical and Materials Transactions A*, 17(8), 1361–1372.

Cullity, B. D., and Graham, C. D. (2009) *Introduction to Magnetic Materials*, second edition. Indianapolis: John Wiley & Sons.

Du, C., Zheng, Z., Min, Q., et al. (2020) A novel approach to calculate diffusion matrix in ternary systems: application to Ag–Mg–Mn and Cu–Ni–Sn systems. *CALPHAD*, 68, 101708.

Durinck, D., Jones, P. T., Blanpain, B., Wollants, P., Mertens, G., and Elsen, J. (2007) Slag solidification modeling using the Scheil–Gulliver assumptions. *Journal of the American Ceramics Society*, 90(4), 1177–1185.

Emmerlich, J., Music, D., Braun, M., Fayek, P., Munnik, F., and Schneider, J. (2009) A proposal for an unusually stiff and moderately ductile hard coating material: $Mo_2BC$. *Journal of Physics D: Applied Physics*, 42(18), 185406.

Gao, M. C., Bennett, T. A., Rollett, A. D., and Laughlin, D. E. (2006) The effects of applied magnetic fields on the $\alpha/\gamma$ phase boundary in the Fe–Si system. *Journal of Physics D: Applied Physics*, 39(14), 2890–2896.

ICME (Committee on Integrated Computational Materials Engineering and National Materials Advisory Board) (2008) *Integrated Computational Materials Engineering: A Transformational Discipline for Improved Competitiveness and National Security*. Washington: National Academies Press, USA.

Kaptay, G. (2012) Nano-CALPHAD: extension of the CALPHAD method to systems with nano-phases and complexions. *Journal of Materials Science*, 47(24), 8320–8335.

Láng, G. G. (2015) Basic interfacial thermodynamics and related mathematical background. *ChemTexts*, 1(4), 1–17.

Li, B., Du, Y., Qiu, L., et al. (2018) Shallow talk about integrated computational materials engineering and Materials Genome Initiative: ideas and practice. *Materials China*, 37(7), 506–525.

Li, K., Idrissi, H., Sha, G., et al. (2016) Quantitative measurement for the microstructural parameters of nano-precipitates in Al–Mg–Si–Cu alloys. *Materials Characterization*, 118, 352–362.

Liu, P. W., Wang, Z., Xiao, Y. H., et al. (2020) Integration of phase-field model and crystal plasticity for the prediction of process-structure-property relation of additively manufactured metallic materials. *International Journal of Plasticity*, 128, 102670.

Liu, Y. L., Zhang, C., Du, C. F., et al. (2020) CALTPP: a general program to calculate thermophysical properties. *Journal of Materials Science and Technology*, 42, 229–240.

Lukas, H. L., Fries, S. G., and Sundman, B. (2007) *Computational Thermodynamics: The CALPHAD Method*. Cambridge: Cambridge University Press.

Malladi, S. K., Xu, Q., van Huis, M. A., et al. (2014) Real-time tomic scale imaging of nanostructural evolution in aluminum alloys. *Nano Letters*, 14(1), 384–389.

McDowell, D. L., Panchal, J. H., Choi, H.-J., Seepersad, C. C., Allen, J. K., and Mistree, F. (2010) *Integrated Design of Multiscale, Multifunctional Materials and Products*. Woburn: Butterworth-Heinemann.

MGI (2011) *Materials Genome Initiative for Global Competitiveness*. Washington: NSTC, National Science and Technology Council, Executive Office of the President, USA.

Mirković, D., Gröbner, J., and Schmid-Fetzer, R. (2008) Liquid demixing and microstructure formation in ternary Al–Sn–Cu alloys. *Materials Science and Engineering A* 487(1), 456–467.

Mitsui, Y., Oikawa, K., Koyama, K., and Watanabe, K. (2013) Thermodynamic assessment for the Bi–Mn binary phase diagram in high magnetic fields. *Journal of Alloys and Compounds*, 577(30), 315–319.

Modigell, M., Güthenke, A., Monheim, P., and Hack, K. (2008) IV.8 – Non-equilibrium modelling for the LD converter, in Hack, K. (ed), *The SGTE Casebook*, second edition. Sawston: Woodhead Publishing, 425–436.

Moorhouse, D. J. (2002) Detailed definitions and guidance for application of technology readiness levels. *Journal of Aircraft*, 39(1), 190–192.

Nishizawa, T., Ohnuma, I., and Ishida, K. (2001) Correlation between interfacial energy and phase diagram in ceramic-metal systems. *Journal of Phase Equilibria*, 22(3), 269–275.

Olson, G. B. (1997) Computational design of hierarchically structured materials. *Science*, 277(5330), 1237–1242.

Park, J. C., and Lee, J. H. (2008) Phase diagram reassessment of Ag–Au system including size effect. *CALPHAD*, 32(1), 135–141.

Pettifor, D. G. (1992) Theoretical predictions of structure and related properties of intermetallics. *Materials Science and Technology*, 8(4), 345–349.

Plapp, M. (2011) Remarks on some open problems in phase-field modelling of solidification. *Philosophical Magazine*, 91(1), 25–44.

Raabe, D., Herbig, M., Sandlöbes, S., et al. (2014) Grain boundary segregation engineering in metallic alloys: a pathway to the design of interfaces. *Current Opinion in Solid State and Materials Science*, 18(4), 253–261.

Rajan, K. (2008) Learning from systems biology: an "omics" approach to materials design. *JOM*, 60(3), 53–55.

Schmid-Fetzer, R., Andersson, D., Chevalier, P.-Y., et al. (2007) Assessment techniques, database design and software facilities for thermodynamics and diffusion. *CALPHAD*, 31(1), 38–52.

Schmitz, G. J., and Prahl, U. (2016) *Handbook of Software Solutions for ICME*. Indianapolis: John Wiley & Sons.

Sonderegger, B., and Kozeschnik, E. (2010) Interfacial energy of diffuse phase boundaries in the generalized broken-bond approach. *Metallurgical and Materials Transactions A*, 41(12), 3262–3269.

Song, T., Schmid-Fetzer, R., Yan, M., and Qian, M. (2021) Near room-temperature fabrication of $Cu_3Sn$ interconnect material: in-situ synchrotron X-ray diffraction characterization and thermodynamic assessments of its nucleation. *Acta Materialia*, 213, 116894.

Tao, Y., Zhang, W. Q., Shang, D. C., et al. (2018) Comprehending the occupying preference of manganese substitution in crystalline cement clinker phases: a theoretical study. *Cement and Concrete Research*, 109, 19–29.

Vegh, A., and Kaptay, G. (2018) Modelling surface melting of macro-crystals and melting of nano-crystals for the case of perfectly wetting liquids in one-component systems using lead as an example. *CALPHAD*, 63, 37–50.

Wen, C., Zhang, Y., Wang, C., et al. (2019) Machine learning assisted design of high entropy alloys with desired property. *Acta Materialia*, 170, 109–117.

Wen, S. Y., Du, Y., Liu, Y. L., et al. (2020) Atomic mobilities and diffusivities in Fcc_A1 Ni–Cr–V system: modeling and application. *CALPHAD*, 70, 101808.

Xiong, W., and Olson, G. B. (2016) Cybermaterials: materials by design and accelerated insertion of materials. *NPJ Computational Materials*, 2(1), 1–14.

Zeng, Y. P., Mittnacht, T., Werner, W., Du, Y., Schneider, D., and Nestler, B. (2022) Gibbs energy and phase-field modeling of ferromagnetic ferrite ($\alpha$)$\rightarrow$ paramagnetic austenite ($\gamma$) transformation in Fe–C alloys under an external magnetic field. *Acta Materialia*, 225, 11759.

Zhao, X. J., Li, Z. Q., Chen, H. W., Schmid-Fetzer, R., and Nie, J. F. (2020) On the equilibrium intermetallic phase in Mg–Nd–Ag alloys. *Metallurgical and Materials Transactions A*, 51(3), 1402–1415.

Zhou, N. X., Hu, T., and Luo, J. (2016) Grain boundary complexions in multicomponent alloys: challenges and opportunities. *Current Opinion in Solid State and Materials Science*, 20(5), 268–277.

Zhou, N. X., and Luo, J. (2015) Developing grain boundary diagrams for multicomponent alloys. *Acta Materialia*, 91, 202–216.

# Appendix A   Ancillary Materials

## Ancillary, Chapter 5

(a) Precipitation simulation and calculation of the isothermal transformation diagram [time–temperature–transformation (TTT) diagram] for the Ni–14 at.% Al alloys.
(b) Solidification simulation to obtain crack susceptibility index (CSI) for a range of Al–Cu–Mg alloys.

The READ-ME file describes how to download the free simulation software package "Pandat Trial/Education," including the modules PanPrecipitation and PanSolidification. It also provides a step-by-step instruction for using the related thermodynamic and mobility parameters (.tdb), kinetic precipitation model parameters (.kdb), and kinetic solidification parameters (.sdb) for the two examples and the simple batch files (.pbfx). These .pbfx examples are used for calculations of the TTT diagram of Ni–14Al in Figure 5.23, the solidification profiles, $T$ versus $(f_S)^{0.5}$, for Al–1Mg–$x$Cu (wt.%) alloys in Figure 5.23. Instructions to create Figure 5.25d, the CSI map at 25 K/s using the high-throughput calculation (HTC) function implemented in Pandat, are also included. A comprehensive description of the models and parameters used in the PanPrecipitation and PanSolidification modules can be found in "Pandat_Manual_2021.pdf."

The READ-ME file and input files that are needed for both simulations are available in the directory "\Ancillary-material-Chapt-5" with two subfolders: "\PanPrecipitation_Ni-14Al-TTT" and "\PanSolidification_Hot-Tearing."

## Ancillary, Chapter 6, #1

One C + + code to implement Sauer-Freise method for the sake of calculating the interdiffusion coefficients in a binary phase.

The READ-ME file, input file, and code are available in the directory "\Ancillary-material-Chapt-6" with the subfolder "\Chapt-6-Ancillary-1-Calculation-of-binary-diffusivity." The READ-ME file presents detailed instructions on how to input the experimental concentration profiles and annealing schedule (time and temperature) and run the code. This example demonstrates the calculation of interdiffusion coefficient for Bcc Mn–V alloys at 1,373 K for 1,216,800 seconds.

## Ancillary, Chapter 6, #2

One C + + code to implement the Matano–Kirkaldy method for calculating the interdiffusivities in a ternary phase.

The READ-ME file, input file, and code are available in the directory "\Ancillary-material-Chapt-6" with the subfolder "\Chapt-6-Ancillary-2-Calculation-of-ternary-diffusivity." The READ-ME file presents detailed instructions on how to input the experimental concentration profiles and annealing schedule (time and temperature) and run the code. This example demonstrates the calculation of interdiffusion coefficient for Fcc Ni–Cr–V alloys at 1,373 K for 586,200 seconds.

## Ancillary, Chapter 6, #3

The thermodynamic and atomic mobility parameters in the Ni–Si system, which are used to perform diffusion simulation in Figure 6.16 by means of DICTRA software.

The READ-ME file and input files needed for DICTRA simulation are available in the directory "\Ancillary-material-Chapt-6" with the subfolder "\Chapt-6-Ancillary-3-Growth of Stoichio-Compounds." The READ-ME file provides detailed instructions to input thermodynamic ("NiSi-thermodynamic.tdb") and kinetic parameters ("NiSi-kinetic.tdb") as well as a Marco file ("Ni5Si2-thickness.tcm"), and experimental data ("Ni5Si2p.exp") are described for the sake of simulating diffusion growth of stoichiometric phase $Ni_5Si_2$ in the Ni–Si system by means of DICTRA software. The DICTRA demo version can be downloaded free of charge from Thermo-Calc's website: www.thermocalc.se.

## Ancillary, Chapter 6, #4

Precipitation and age hardening in the AA6005 Al alloy.

The READ-ME file describes how to download the free simulation software package "Pandat Trial/Education," including the module "PanPrecipitation." It also provides step-by-step instructions for using the related thermodynamic (.tdb) and thermophysical parameters (.kdb) files for the two examples, "AA6005_yield_strength" and "AA6005_reheating" and the simple batch files (*.pbfx). These examples are used to calculate Figures 6.28–6.30 in Chapter 6. A comprehensive description of the models and parameters used in "PanPrecipitation" can be found in "Pandat_Manual_2021.pdf."

The READ-ME file and input files needed for "PanPrecipitation" simulation are available in the directory "\Ancillary-material-Chapt-6" with the subfolder "\Chapt-6-Ancillary-4-PanPrecipitation-in-Al-alloys."

## Ancillary, Chapter 8

Precipitation hardening in the 7xxx Al alloy and solidification kinetics in AT72 Mg alloy.

The READ-ME file describes how to download the free simulation software package "Pandat Trial/Education," including the modules "PanPrecipitation" and "PanSolidification." It also provides step-by-step instructions for using the related thermodynamic and mobility parameters (.tdb), kinetic precipitation model parameters (.kdb), and kinetic solidification parameters (.sdb) for the two examples and the simple batch files (*.pbfx). These pbfx examples are used for calculations related to Figures 8.4–8.6 for the ternary alloy Al–6.1Zn–2.3Mg (wt.%) and Figure 8.10 for the ternary Mg–7Al–2Sn (wt.%) alloy. A comprehensive description of the models and parameters used in the "PanPrecipitation" and "PanSolidification" modules can be found in "Pandat_Manual_2021.pdf."

The READ-ME file and input files needed for both simulations are available in the directory "\Ancillary-material-Chapt-8" with two subfolders: "\PanPrecipitation_AA7010" and "\PanSolidification_AT72."

## Ancillary, Chapter 9

One MATLAB code to implement the design of Ni–Fe based superalloys for advanced ultrasupercritical unit (A-USC) based on an artificial neutral network (ANN) model combined with a genetic algorithm (GA) is provided. To run this example, a license for MATLAB is required.

The READ-ME file, input files, and code are available in the directory "\Ancillary-material-Chapt-9" with the subfolder "Chapt-9-Ancillary-1-AUSC-Design." In this READ-ME file, detailed instructions about these files are compiled in three directories: "\Code_alloy_design_ANN_MATLAB," "\Code_alloy_design_ANN_doc," and "\Training data" were provided. The source code is in .m format, such as "BP_GA.m" in the first file. The source code enables the reader to understand directly the principle and procedure of calculation, set target parameters (expected properties and compositional range), and perform calculations in the MATLAB environment. One can simply view the source code provided in the second file with the help of Office software. In addition, an experimental dataset including 200 instances is available in the file of "Training data." Using these data, one is supported to repeat the work from Hu et al. (Hu, X., Wang, J., Wang, Y., Li, J., et al., 2018, Two-way design of alloys for advanced ultra supercritical plants based on machine learning. *Computational Materials Science,* 155, 331–339) and carry out other new cases.

## Ancillary, Chapter 10

The thermodynamic and atomic mobility parameters for the gradient cemented carbide, which are used to perform diffusion simulation in Figure 10.16b by means of DICTRA software.

The READ-ME file and input files needed for DICTRA simulation are available in the directory "\Ancillary-material-Chapt-10" with the subfolder "\Chapt-10-Ancillary-1-Gradient-structure." In this READ-ME file, detailed instructions are provided to input thermodynamic parameters (C-W-Co-Ti-N-thermodynamic.tdb) and atomic mobility parameters (C-W-Co-Ti-N-kinetic.tdb), together with the Marco file (Gradient-C-W-Co-Ti-N.dcm) to run the calculations. These instructions can be used to calculate mole fractions of WC, binder (Co), and cubic phases for the gradient cemented carbide ($Co_8Ti_3C_{5.95}N_{0.1}W_{82.95}$, in wt.%) sintered at 1,450°C for 1,200 seconds under vacuum by means of DICTRA software. In order to save the simulation time, 1,200 seconds of sintering time are employed to simulate the formation of gradient layer. In the real simulation for a real engineering alloy, the sintering time of 2 hours was used. The simulation software package "DICTRA (R version)" is used to run this example. If the reader uses different version of DICTRA software, sometimes it is necessary to modify some commands in the file "Gradient-C-W-Co-Ti-N.DCM" in order to perform diffusion simulation. Usually different versions of DICTRA software correspond to slightly different DCM files. The DICTRA demo version can be downloaded free of charge from Thermo-Calc's website: www.thermocalc.se/.

## Ancillary, Chapter 11

A tutorial guide file to provide step-by-step instructions for the CFD simulation of an industrial scale chemical vapor deposition (CVD) furnace.

The tutorial guide file ("READ-ME-Chapt-11-CFD-CVD-calculation-YDU-20221216.pdf") is in the download folder "\Ancillary-material-Chapt-11." To repeat this CFD calculation, the reader must have a license for the software ANSYS Fluent. The deposition condition ($T = 1,153$ K, $p = 12$ kPa, the initial gas mixture of $TiCl_4 = 2.52\%$, $CH_3CN = 0.87\%$, $N_2 = 13\%$, $H_2 = 83.61\%$) is selected for CFD simulation. An assumed overall chemical reaction was given as follows: $6TiCl_4(g) + 2CH_3CN(g) + 9H_2(g) = 6TiC_{0.67}N_{0.33}(s) + 24HCl(g)$. The boundary conditions, the solver, spatial discretization, and convergence criteria can be found in the tutorial file. The distributions of temperature, mass fraction of $TiCl_4$, and gas flow velocity and direction, as indicated in Figure 11.16, can be demonstrated after running the simulation.

**Figure 0.1** Five authors (Jianchuan Wang, Jincheng Wang, Rainer Schmid-Fetzer, Yong Du, and Shuhong Liu, from left to right) discussing the overall structure of the book in Changsha on September 21, 2018.

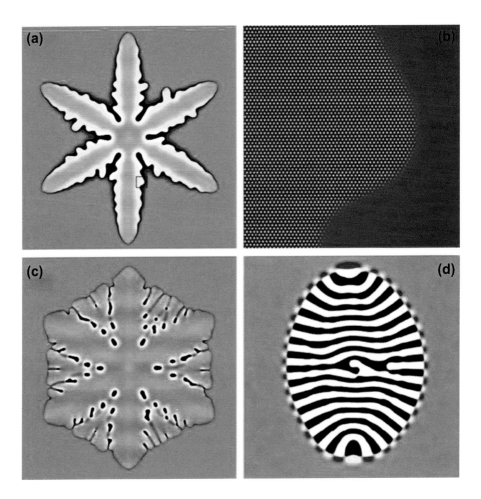

**Figure 3.5** Illustrative PFC simulations for solidification in binary alloys. (a) Number density difference map of a solutal dendrite. (b) Total number density map of the small square area shown in (a). (c) A more compact dendritic structure grown at a higher driving force. (d) Eutectic structure. This figure is reproduced from Tegze et al. (2009) with permission from Elsevier.

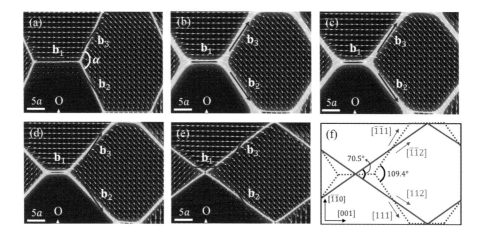

**Figure 3.6** Equilibrium dislocation structures within $(1\bar{1}0)$GB in (a) Nb, (b) Mo, (c) W, (d) Ta, (e) $\beta$ – Ti, together with (f) a schematic drawing of two special states of equilibrium dislocation structures in Bcc crystals. This figure is reproduced from Qiu et al. (2019) with permission from Elsevier.

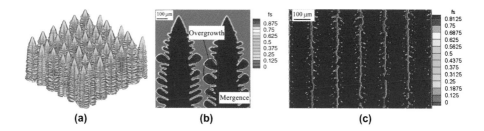

**Figure 3.11** Calculated dendritic microstructure of an Al-3.9 wt.% Cu-0.9 wt.% Mg alloy solidified directionally. The temperature gradient ahead of the S/L interface is 7 K mm$^{-1}$ and the cooling rate is 1 K s$^{-1}$. (a) The 3D dendrite tip morphology; (b) the 2D dendrite tip morphology; (c) the 2D morphology of the dendrites after solidification. This figure is reproduced from Zhang et al. (2012) with permission from Elsevier.

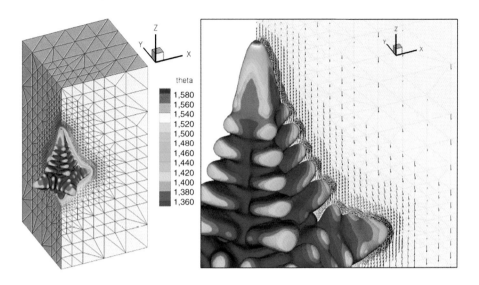

**Figure 3.12** Ni–Cu crystal growth with inlet flow from the top. Left: dendrite colored with interface velocity, mesh colored with temperature; right: flow passing by the upstream and perpendicular stream. This figure is reproduced from Tan and Zabaras (2007) with permission from Elsevier.

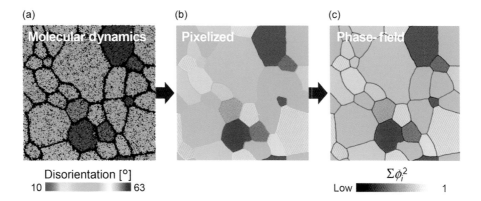

Figure 3.15 Creation of an initial structure for multi-PF grain growth simulations from MD data. (a) MD-generated solidified microstructure at time $t = 1{,}000$ ps. (b) Polycrystalline system created by discretizing the atomic configuration shown in (a) into 2D difference grid points. (c) Multi-PF interfacial profiles obtained through the relaxation simulation starting from (b). This figure is reproduced from Miyoshi et al. (2018) with permission from Elsevier.

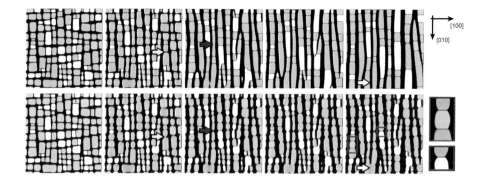

**Figure 3.17** Snapshots of the microstructure (colors indicate $\gamma'$ variants) in a $2.3 \times 2.3 \ \mu m^2$ periodic box at $t = 0$, 1, 5, 15, and 45 hours (from left to right) in the elastic (top) and elastoviscoplastic (bottom) cases. This figure is reproduced from Cottura et al. (2016) with permission from Elsevier.

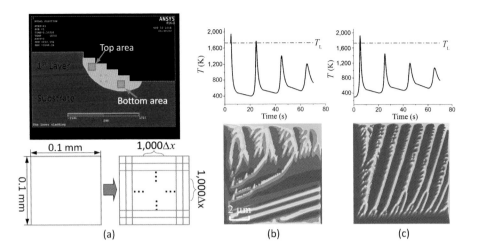

**Figure 3.19** PF modeling of microstructure evolution during additive manufacturing coupling with macro temperature field. (a) Schematics of melting pool and the interpolation, (b) the temperature history and microstructures in the top area, and (c) the temperature history and microstructures in the bottom area. (Li and Wang, 2021).

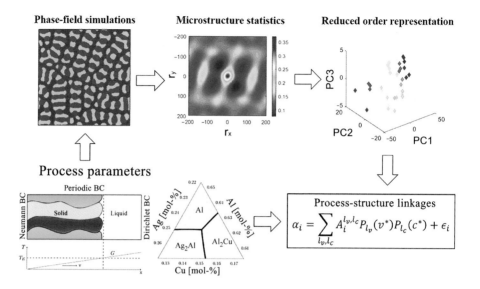

**Figure 3.21** Framework to establish process-structure linkages from microstructures simulated by PF models and machine learning. This figure is reproduced from Yabansu et al. (2017) with permission from Elsevier.

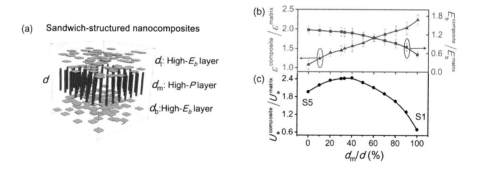

**Figure 3.22** (a) A designed sandwich microstructure based on PF simulations and (b) the corresponding effective dielectric permittivity, $\varepsilon$, and the breakdown strength, $E_b$, and (c) the energy density as a function of the fraction of the middle layer. This figure is reproduced from Shen et al. (2018) with permission from Wiley.

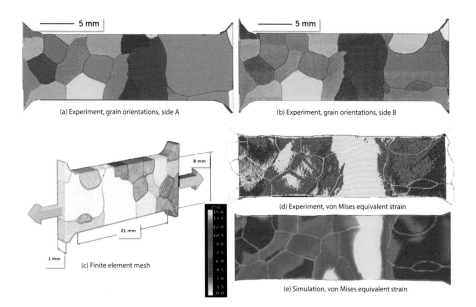

(a) Experiment, grain orientations, side A

(b) Experiment, grain orientations, side B

(c) Finite element mesh

(d) Experiment, von Mises equivalent strain

(e) Simulation, von Mises equivalent strain

**Figure 4.7** Plastic strain localization and deformation-induced surface roughening in a 3D aluminum polycrystal consisting of a few coarse grains (Roters et al., 2010; Zhao et al., 2008). (a) and (b) are the two sides of the view of the sample. (c) is the finite element model of uniaxial tension after mesh division, which is displayed based on the view of (b). (d) and (e) are the stress distribution obtained by experiment and simulation, respectively. The figures are reproduced from Roters et al. (2010); Zhao et al. (2008) with permission from Elsevier.

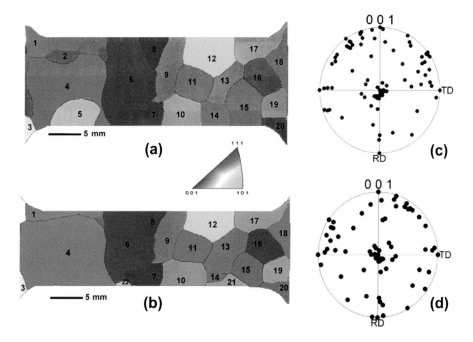

**Figure 4.8** Surface grain profiles and orientation distribution obtained by EBSD (Zhao et al., 2008). Grain shape on top (a) and bottom (b) surface and overall orientation distributions (c, d) on these two surfaces (a, b), respectively. The figures are reproduced from Zhao et al. (2008) with permission from Elsevier.

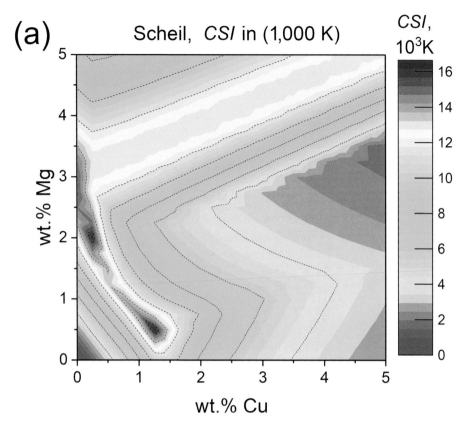

**Figure 5.25** Crack susceptibility maps for Al–Cu–Mg alloys: (a)–(e) calculated with PanSolidification and PanAl (www.computherm.com); (a) Scheil conditions with values of *CSI* from (5.60); (b) Scheil conditions with values of *CSI** from (5.61) normalized with $CSI_{max} = 16{,}672$ K; (c) simulation with $CR = 100$ K/s, (d) simulation with $CR = 25$ K/s; (e) equilibrium (lever rule) solidification conditions; (f) experimentally determined map (Pumphrey and Moore, 1948) showing composition ranges in which cracking is greatest in ring castings. (f) is reproduced from Liu and Kou (2017) with permission from Elsevier.

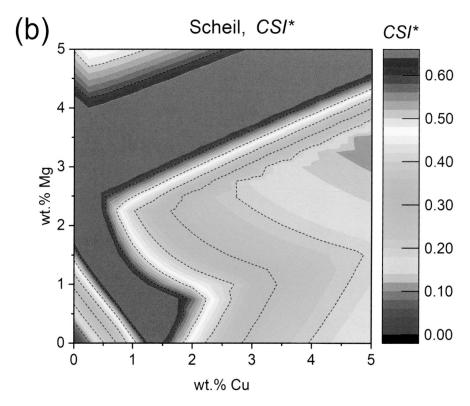

**Figure 5.25** (*cont.*)

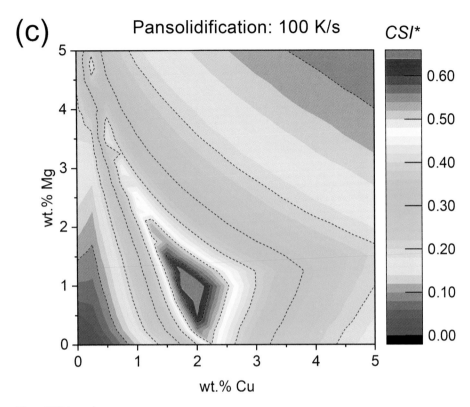

**Figure 5.25** (*cont.*)

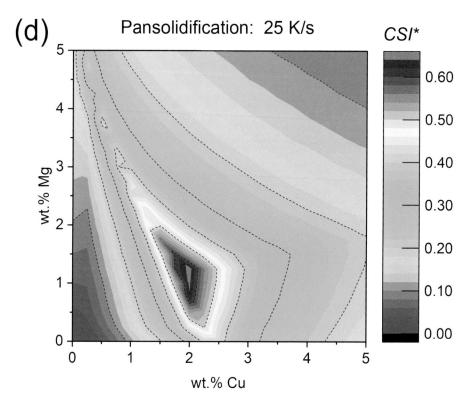

**Figure 5.25** (*cont.*)

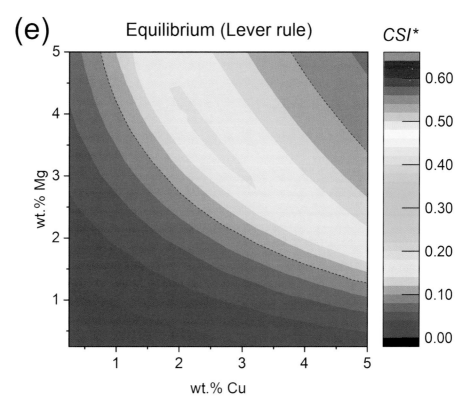

**Figure 5.25** (*cont.*)

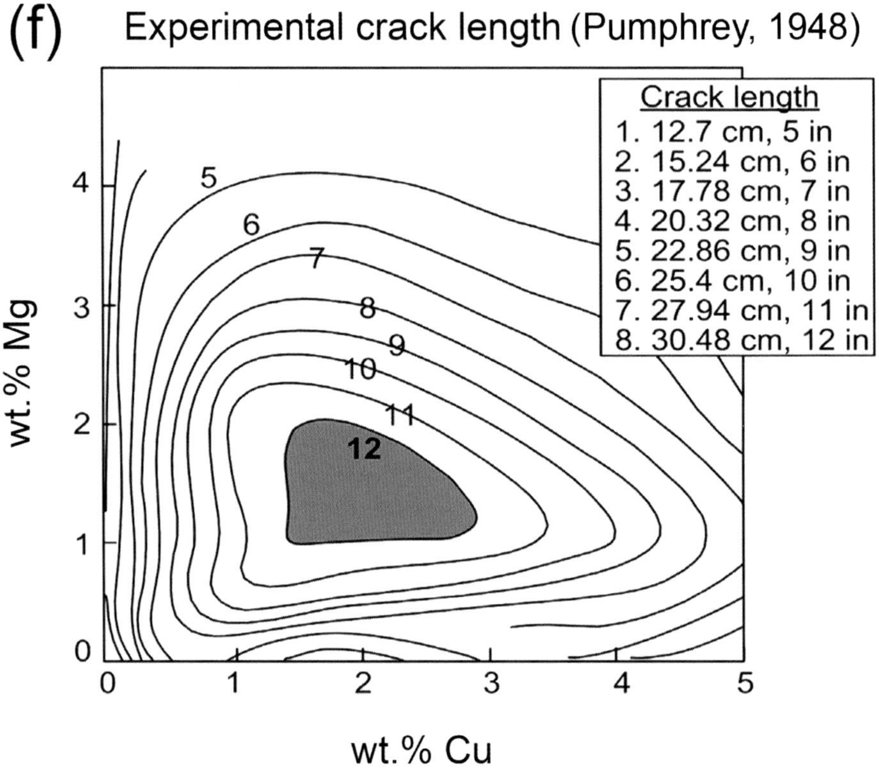
(f)
Experimental crack length (Pumphrey, 1948)
Crack length
1. 12.7 cm, 5 in
2. 15.24 cm, 6 in
3. 17.78 cm, 7 in
4. 20.32 cm, 8 in
5. 22.86 cm, 9 in
6. 25.4 cm, 10 in
7. 27.94 cm, 11 in
8. 30.48 cm, 12 in
5
6
7
8
9
10
11
12
wt.% Mg
wt.% Cu
0
1
2
3
4
5

**Figure 5.25** *(cont.)*

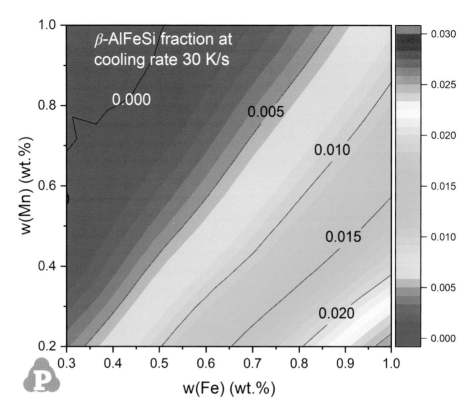

**Figure 8.2** Color map of $\beta$-AlFeSi phase fraction as function of the Fe and Mn compositions for Al–8Si–0.35Mg–$x$Fe–$y$Mn (wt.%) alloys at a fixed cooling rate of 30 K/s. Adapted from Zhang et al. (2020) with permission from Springer Nature.

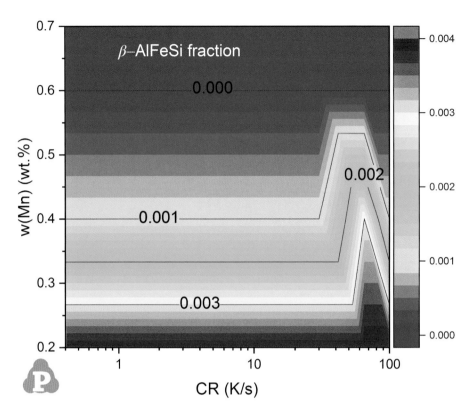

**Figure 8.3** Color map of the $\beta$-AlFeSi phase fraction as function of the Mn composition and the cooling rate (CR) for Al–8Si–0.35Mg–0.3Fe–$y$Mn (wt.%) alloys. Adapted from Zhang et al. (2020) with permission from Springer Nature.

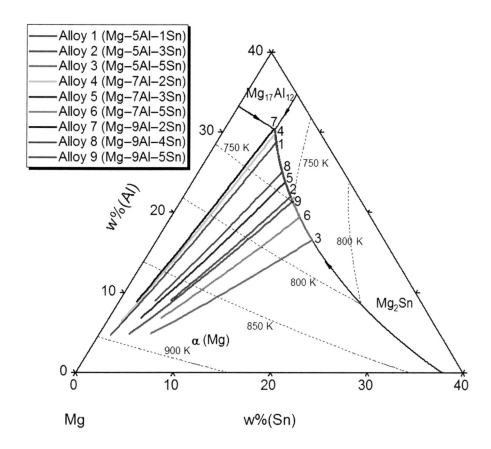

**Figure 8.9** Calculated Mg-rich corner of the Mg–Al–Sn liquidus projection and the solidification paths of the experimental AT alloys. Reprinted from Luo et al. (2012) with permission from Springer Nature.

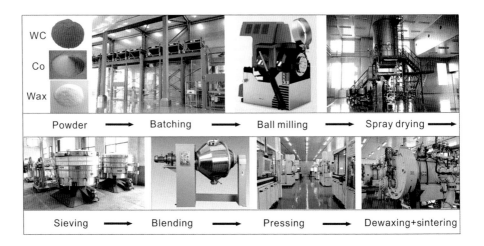

**Figure 10.1** Process illustrations for preparation of cemented carbide.

**Figure 10.3** EBSD maps of (a) sample 1, (b) sample 2, and (c) sample 3, where the WC phase is shown in the band comparison chart and the orientation of the Fcc phase is shown in the color inverse pole figure (IPF). The figures are reproduced from Li et al. (2015) with permission from Elsevier.

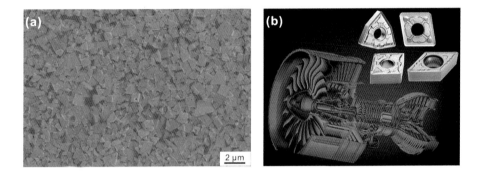

**Figure 10.7** (a) Ultrafine cemented carbide (grain size of WC: 0.6 μm); (b) inserts with ultrafine cemented carbide and PVD TiAlSiN coating for machining of high-temperature alloys in aviation engines.

**Figure 10.9** EBSD maps of (a) WC–38.6Co–11.4Ni$_3$Al and (b) WC–34.8Co–15.2Ni$_3$Al. The alloys were sintered at 1,350°C for 1 hour under vacuum. The figures are reproduced from Long et al. (2017b) with permission from Elsevier.

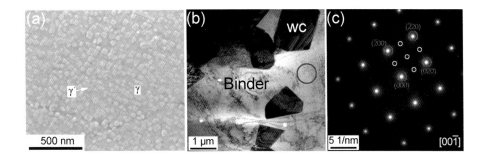

**Figure 10.10** Microstructure of the WC–38.6Co–11.4Ni$_3$Al alloy sintered for 1 hour at 1,350°C under vacuum: (a) FE-SEM image of the binder phase, (b) bright-field image of the microstructure between the binder phase and WC, and (c) the indexed SADP pattern marked by the red circles in (b). The figures are reproduced from Long et al. (2017b) with permission from Elsevier.

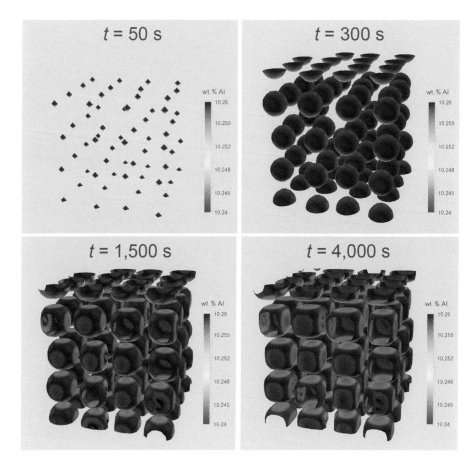

**Figure 10.11** Phase-field-simulated microstructure evolution in 3D for the binder phase with the composition 19.5 at.% Al, 10.5 at.% Co, and 70 at.% Ni aged at 1,373 K, showing the growth of $\gamma'$. The figures are reproduced from Peng et al. (2020) with permission from Elsevier.

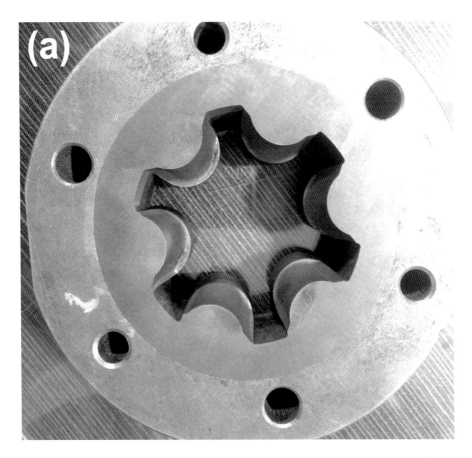
**Figure 10.14** WC–Co–Ni–Al-based cemented carbides: (a) hot heading die, (b) carbide rolls.

**Figure 10.14** (*cont.*)

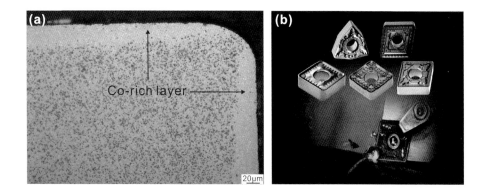

**Figure 10.19** (a) Gradient cemented carbide, and (b) Inserts with gradient cemented carbide substrate and CVD coating used for steel turning.

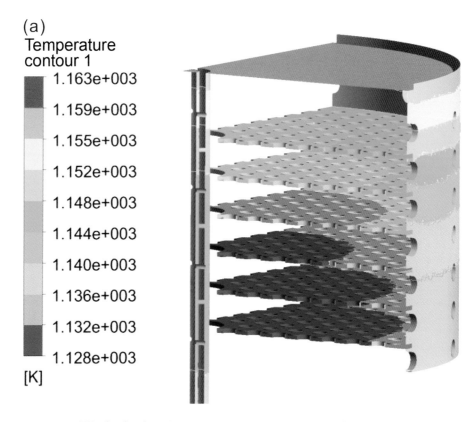

**Figure 11.16** CFD simulated results: (a) temperature distribution; (b) distribution of $TiCl_4$ mass fraction; (c) gas flow velocity and direction. Trays are numbered from bottom (1) to top (5). The diagram is reproduced from Qiu et al. (2019) with permission from Elsevier.

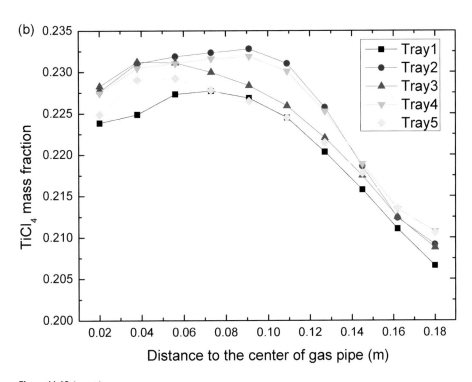

**Figure 11.16** (*cont.*)

(c)
Velocity
vector 1

1.772e+002

1.329e+002

8.862e+001

4.431e+001

0.000e+000
[m s^-1]

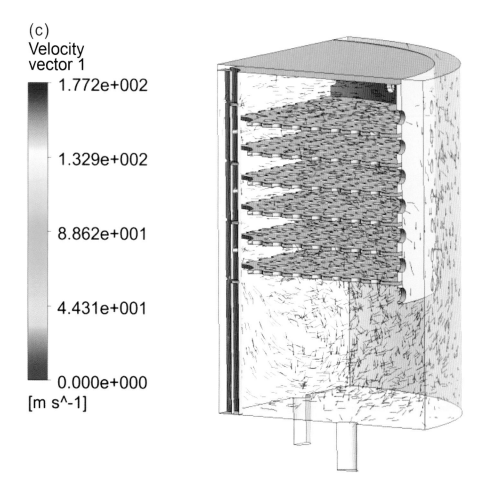

**Figure 11.16** (*cont.*)

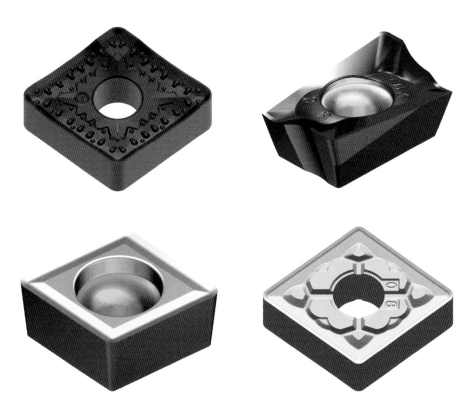

**Figure 11.18** CVD-coated tungsten carbide inserts developed via the presented hybrid approach of modeling and experiment.

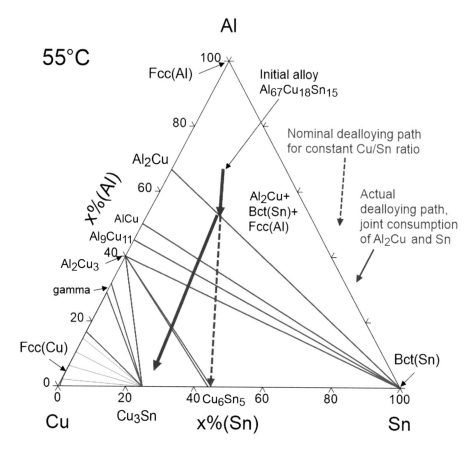

**Figure 13.2** Dealloying pathways of $Al_{67}Cu_{18}Sn_{15}$ based on Song et al. (2021), superimposed on the calculated isothermal section of the Al–Cu–Sn system at 55°C using the CALPHAD description of Mirkovic et al. (2008). The nominal pathway must follow a straight line extending from the top corner, representing pure Al, because of the constant Cu/Sn ratio due to the mass balance. However, after consumption of free Al in the Fcc(Al) phase, the actual pathway continues from the two-phase equilibrium line $Al_2Cu + Bct(Sn)$ with increasing Cu/Sn ratio due to the joint consumption of $Al_2Cu$ and Sn, shown schematically by the kink in the continuation of the solid arrow.

# Appendix B  Notations

## Appendix B1: Latin Symbols

| Symbol | Meaning | SI-unit |
|---|---|---|
| $a_i$ | Chemical activity of element $i$; see reference states in (5.33) and (5.34) | |
| $A_m$ | Interfacial area per mole of interface atoms; see (6.113) | $m^2\,mol^{-1}$ |
| $b$ | Magnitude of Burgers vector; see (4.18) | nm |
| $B$ | Bulk modulus; see (2.61) | GPa |
| $B_0$ | Bulk modulus at equilibrium volume; see (2.41)–(2.43) and (2.60) | GPa |
| $B_0{}'$ | Derivative of $B_0$ with respect to pressure; see (2.41) | |
| $c$ | Composition, i.e., atomic/molar fraction of component (solute); see (3.2) | |
| $c_L$ | Composition of liquid phase; see (3.20) | |
| $c_S$ | Composition of solid phase, see (3.20) | |
| $c_L^*$ | Composition of liquid phase at equilibrium state; see (3.41) and (3.42) | |
| $c_S^*$ | Composition of solid phase at equilibrium state; see (3.41) and (3.42) | |
| $c_s(x, t)$ | Local solute composition of the solid phase; see (3.21) | |
| $c_L(x, t)$ | Local solute composition of the liquid phase; see (3.21) | |
| $C$ | Concentration of particle; see (6.1) | $mol\,m^{-3}$ |
| $C_0$ | Concentration of vacancy; see (6.68) | $mol\,m^{-3}$ |
| $C_i$ | Concentration of species $i$; see (6.10) | $mol\,m^{-3}$ |
| $C_i^-$ | Initial concentration of a terminal alloy at left side of a diffusion couple; see (6.11) | $mol\,m^{-3}$ |
| $C_i^+$ | Initial concentration of a terminal alloy at right side of a diffusion couple; see (6.11) | $mol\,m^{-3}$ |
| $C_i^0$ | Concentration of element $i$ at the Matano plane; see (6.29) | $mol\,m^{-3}$ |
| $c_p$ | Molar heat capacity at constant pressure; see Table 5.2 | $J\,mol^{-1}K^{-1}$ |
| $C_p$ | Heat capacity at constant pressure; see (2.53) and Table 5.2 | $J\,K^{-1}$ |
| $C_V$ | Heat capacity at constant volume; see (2.51) and Table 5.2 | $J\,K^{-1}$ |
| $C_M$ | Specific capacity of Li–ion battery; see (12.5) | mAh/g |
| $C_{ij}$ | Elastic constants, $i, j = 1, 2, 3, 4, 5, 6$; see (2.55) | GPa |
| $d_{WC}$ | Grain size of WC particles; see (6.136) | m |
| $D$ | Diffusion coefficient or diffusivity; see (2.27) and (6.1) | $m^2\,s^{-1}$ |

(cont.)

| Symbol | Meaning | SI-unit |
|---|---|---|
| $D_{eff}$ | Effective diffusivity, weighted average of lattice diffusivity and grain-boundary diffusivity; see (6.57) | $m^2\ s^{-1}$ |
| $\tilde{D}_i^{eff}$ | Effective interdiffusion coefficient; see (6.28) | $m^2\ s^{-1}$ |
| $\tilde{\tilde{D}}_{i,L}^{eff}$ | Average effective interdiffusion coefficient on the left side of the Matano plane; see (6.30) | $m^2\ s^{-1}$ |
| $\tilde{\tilde{D}}_{i,R}^{eff}$ | Average effective interdiffusion coefficient on the right side of the Matano plane; see (6.31) | $m^2\ s^{-1}$ |
| $D_0$ | Diffusion prefactor; see (6.73) and (6.74) | $m^2\ s^{-1}$ |
| $D_{gb}$ | Grain-boundary diffusivity; see Figure 6.18 and (6.57) | $m^2\ s^{-1}$ |
| $D_v$ | Lattice or bulk diffusivity; see Figure 6.18 and (6.57) | $m^2\ s^{-1}$ |
| $D_i$ | Intrinsic diffusion coefficient of species $i$; see (6.5) | $m^2\ s^{-1}$ |
| $D_{kj}^n$ | Intrinsic diffusion coefficient of element $k$ in element $n$ with the influence of element $j$; see (6.39) and (6.40) | $m^2\ s^{-1}$ |
| $\tilde{D}_{kj}^n$ | Interdiffusion coefficient of species $n$; see (6.37) and (6.40) | $m^2\ s^{-1}$ |
| $D_A^A$ | Self-diffusion coefficient, i.e., the diffusion of species $A$ in a solid of element A; see Section 6.2.1 | $m^2\ s^{-1}$ |
| $D_A^*$ | Tracer diffusion coefficient; see Section 6.2.1 | $m^2\ s^{-1}$ |
| $D_A^{B*}$ | Impurity or solute $B*$ tracer diffusion coefficient in $A$; $B*$ is the radioisotope of $B$; see Section 6.2.1 | $m^2\ s^{-1}$ |
| $D_{AB}^{A*}$ | Tracer diffusion coefficients for $A*$ in a homogeneous binary $AB$ alloy; see Section 6.2.1 | $m^2\ s^{-1}$ |
| $\tilde{D}$ | Interdiffusion coefficients; see (6.5) and (6.6) | $m^2\ s^{-1}$ |
| $D_s$ | Surface diffusivity; see (11.1) | $m^2\ s^{-1}$ |
| $e$ | Elementary charge; $1.6021766 \times 10^{-19}$ | C |
| $E$ | Total energy; see (2.1) and (2.41)–(2.43) | J |
| $E$ | Open-circuit voltage of Li–ion cell; see (5.56), (12.3) and (12.4) | V |
| $E$ | Elastic modulus; see (7.7) | GPa |
| $E_0$ | Ground state energy at 0 K; see (2.41)–(2.43) | J |
| $E_M$ | Specific energy of a Li–ion cell; see (12.6) | mWh/g |
| $e^{str}$ | Elastic strain energy; see (7.1) | $J\ m^{-3}$ |
| $f$ | Fugacity of gas; see (5.36) | Pa |
| $f(c, \phi)$ | Free energy density; see (3.2) | $J\ m^{-3}$ |
| $f_{chem}$ | Chemical free energy density; see (3.2) | $J\ m^{-3}$ |
| $f_{doub}$ | Double-well potential density; see (3.3) | $J\ m^{-3}$ |
| $f_{grad}$ | Gradient energy density; see (3.3) | $J\ m^{-3}$ |
| $f^L$ | Free energy density of liquid phase; see (3.12) | $J\ m^{-3}$ |
| $f^S$ | Free energy density of solid phase; see (3.12) | $J\ m^{-3}$ |
| $f^\alpha$ | Molar phase fraction of $\alpha$ phase; see Section 5.2.2 | |
| $F$ | Faraday's constant, $F = 96,485.3$; see (5.56) and (12.3) | $C\ mol^{-1}$ |
| $F$ | Helmholtz energy; see Table 5.2 | J |
| $F$ | Free energy of a system; see (3.1) | J |
| $F_{bulk}$ | Bulk free energy; see (3.1) | J |
| $F_{int}$ | Interfacial energy; see (3.1); in other chapters, $\gamma$ is used | J |
| $F_{GB}$ | Bulk diffusion activation energy multiplier (typical value 0.5) for grain-boundary diffusion, see (6.109) | J |

(cont.)

| Symbol | Meaning | SI-unit |
|---|---|---|
| $F_{str}$ | Strain energy; see (3.1) | J |
| $F_{vib}$ | Vibrational free energy; see (2.48) and (2.49) | J |
| $G$ | Shear modulus; see (2.62) | GPa |
| $G$ | Gibbs energy; see Table 5.2 | J |
| $\Delta_f G_{Va}$ | Gibbs energy for formation of vacancy; see (6.68) | J mol$^{-1}$ |
| $G^\varphi$ | Molar Gibbs energy of phase $\varphi$; see Section 5.2.2 | J mol$^{-1}$ |
| $G_i^\varphi$ | Molar Gibbs energy of the pure element ($i$) in phase $\varphi$; see Section 5.2.2 | J mol$^{-1}$ |
| $G_i^{0,\varphi}$ | Molar Gibbs energy of the pure element ($i$) in phase $\varphi$ with standard element reference (SER); $G_i^{0,\varphi}(T) = G_i^\varphi(T) - H_i^{SER}$; see (5.1a) | J mol$^{-1}$ |
| $G_m^{IF}$ | Molar Gibbs energy of the interface phase; see (6.116) | J mol$^{-1}$ |
| $G_{mag}^\varphi(T)$ | Magnetic contribution to the Gibbs energy for phase $\varphi$; see (5.4) | J mol$^{-1}$ |
| $\Delta g^{ch}$ | Critical chemical driving force per unit volume needed for nucleation; see (7.1) | J m$^{-3}$ |
| $G_m^L$ | Molar Gibbs energy of pure liquid phase; see (6.113) | J mol$^{-1}$ |
| $^E G_m^{IF}$ | Equilibrium molar Gibbs energy for interfacial energy calculation; see (6.113) | J mol$^{-1}$ |
| $\Delta G_j^*$ | Activation energy of viscous flow of a pure component $j$; see (6.122) | J mol$^{-1}$ |
| $H$ | Enthalpy; see Table 5.1 | J |
| $H_{CC}$ | Hardness of cemented carbide; see (10.3) | GPa |
| $H_i^{SER}$ | Molar enthalpy of the element ($i$) at 298.15 K and 1 bar in its SER state, value unknown; see (5.1a) | J mol$^{-1}$ |
| $H_i^{SER}$ | Molar enthalpy of the element ($i$) at 298.15 K and 1 bar in its SER phase, value zero; see (5.3) | J mol$^{-1}$ |
| $\Delta_{mix} H$ | Enthalpy of mixing; see Section 5.3.2.1 | J mol$^{-1}$ |
| $\Delta_f H_{Va}$ | Enthalpy of formation for vacancy; see (6.68) | J mol$^{-1}$ |
| $H^{mig}$ | Enthalpy of migration for atom; see (6.70) | J |
| $\Delta H_m^{magn}$ | Magnetic enthalpy; see Section 5.3.1.2 and (6.127) | J mol$^{-1}$ |
| $\Delta H_{ass}$ | Enthalpy of mixing of the associated liquid alloy; see Section 5.3.2.3 and (6.122) | J mol$^{-1}$ |
| $h$ | Planck's constant; $h = 6.626070 \times 10^{-34}$ | J s |
| $I_{SSS}$ | Solid-solution strengthening index; see (9.1) | |
| $J$ | Diffusion flux of particle, see (6.1) | g m$^{-2}$ s$^{-1}$ |
| $J_k$ | Intrinsic diffusion flux of the component $k$; see (6.39) | g m$^{-2}$ s$^{-1}$ |
| $\tilde{J}_i$ | Interdiffusion flux of the component $i$; see (6.10) | g m$^{-2}$ s$^{-1}$ |
| $k_B$ | Boltzmann constant, $k_B = 1.380649 \times 10^{-23}$ | J K$^{-1}$ |
| $L$ | Latent heat of phase transformation; see (3.14) | J |
| $L(\mathbf{r} - \mathbf{r}')$ | Symmetric microscopic kinetic matrix; see (3.36) | |
| $L_{i,j}^{\nu,\varphi}$ | Redlich-Kister parameter for interaction of order $\nu$ between elements ($i$) and ($j$) in phase $\varphi$; see (5.27), also (6.135) | J mol$^{-1}$ |
| $m$ | Mass of electron, $m = 9.1093837 \times 10^{-31}$ | kg |
| $m_L$ | Slope of liquidus in phase diagram; see (3.46) | K |
| $M$ | Taylor factor; see (7.6) | |
| $M_c$ | Dynamic coefficient related to atomic mobility; see (3.4) | |

(*cont.*)

| Symbol | Meaning | SI-unit |
|---|---|---|
| $m_I$ | Mass of nucleus $I$; see (2.2) | kg |
| $M_i$ | Mass of particle or atom $i$; see (2.18) | kg |
| $M_{Integ}$ | Integration of interdiffusion fluxes within the left or right side of the Matano plane; see (6.30) | g m$^{-1}$ s$^{-1}$ |
| $M_B$ | Mobility of element $B$; see (6.98) | m$^2$ s$^{-1}$ mol J$^{-1}$ |
| $M_B^0$ | Mobility frequency factor of element $B$; see (6.98) | m$^2$ s$^{-1}$ |
| $M_0^{bulk}$ | Frequency factor in the bulk phase; see (6.109) | m$^2$ s$^{-1}$ |
| $M^{gd}$ | Mobility for diffusion in the grain boundary; see (6.109) | m$^2$ s$^{-1}$ |
| $N_i^\alpha$ | Number of sites on the $\alpha$ sublattice occupied by atom $i$; see (6.108) | |
| $N_{total}^\alpha$ | Total number of sites on the $\alpha$ sublattice; see (6.108) | |
| $N$ | Number of particles or atoms | |
| $N_A$ | Avogadro's constant, $N_A = 6.02214 \cdot 10^{23}$ | mol$^{-1}$ |
| $n(\mathbf{r})$ | Electron density; see (2.5) and (2.6) | |
| $\vec{n}$ | Interface normal vector; see (3.41) | |
| $n$ | Amount of a system; see Section 5.2.2 | mol |
| $n_i$ | Amount of component $i$ in a multicomponent system; see Section 5.2.2 | mol |
| $n_{max}$ | Maximum density of nuclei; see (3.37) | m$^{-3}$ |
| $p$ | Pressure | Pa |
| $p'$ | Probability for formation of vacancy around an impurity; see (6.75) | |
| $p_0$ | Reference pressure, usually 1 bar $= 10^5$ Pa | Pa |
| $P_f$ | Fracture probability; see (10.1) | |
| $Q_s$ | Atomic activation energy for surface diffusion; see (11.2) | J |
| $Q$ | Activation energy for diffusion; see (6.46) | J mol$^{-1}$ |
| $Q^{bulk}$ | Activation energy for diffusion in the bulk phase; see (6.109) | J mol$^{-1}$ |
| $Q_i^j$ | Activation energy of diffusion for element $i$ in pure $j$; see (6.96) | J mol$^{-1}$ |
| $Q_B^{dis}$ | Activation energy contribution of the disordered state; see (6.106) | J mol$^{-1}$ |
| $R$ | Gas constant, $R = 8.3145$ | J mol$^{-1}$K$^{-1}$ |
| $\langle R^2(t) \rangle$ | Mean square displacement (MSD); see (2.26) and (6.84) | m$^2$ |
| $R_{M/WC}$ | Interfacial (metal/WC) thermal resistance (ITR); see (6.136) | m$^2$ K W$^{-1}$ |
| $r$ | Distance | m |
| $r_D$ | Deposition rate; see (11.2) | m s$^{-1}$ |
| $r_{SE}$ | Effective hydrodynamic particle radius; see (6.119) | m |
| $S^{mig}$ | Entropy of migration for atom; see (6.70) | J K$^{-1}$ |
| $S$ | Entropy; see Tables 5.1 and 5.2 | J K$^{-1}$ |
| $S$ | Seebeck coefficient; see (6.139) | V K$^{-1}$ |
| $s_i$ | Step function; see (3.25) | |
| $\Delta_f S_{Va}$ | Entropy of formation for vacancy; see (6.68) | J mol$^{-1}$·K$^{-1}$ |
| $t$ | Time | s |
| $\Delta t$ | Time step or time interval | s |
| $T$ | Absolute temperature | K |
| $T_C$ | Curie temperature; see (5.5b) | K |
| $T_N$ | Néel temperature; see (5.5b) | K |

(cont.)

| Symbol | Meaning | SI-unit |
|--------|---------|---------|
| $T_m$ | Melting temperature; see (6.85) | K |
| $T^*$ | Critical magnetic temperature, $T_C$ or $T_N$; see (5.5b) | K |
| $T^*$ | Interface temperature; see (3.46) | K |
| $\Delta T$ | Undercooling below liquidus temperature; see (3.38) and (3.40) | K |
| $\Delta T_{mn}$ | Mean nucleation undercooling; see (3.37) | K |
| $U$ | Internal energy; see Tables 5.1 and 5.2; also used for potential or potential energy; see (2.17), (2.19), and (2.21) | J |
| $V$ | Volume; see Tables 5.1 and 5.2 | $m^3$ |
| $V_0$ | Equilibrium volume or Molar volume at a reference state, see (2.41)–(2.43) and (5.13) | $m^3$ or $m^3\ mol^{-1}$ |
| $V_i$ | Partial molar volume of species $i$; see (6.5) | $m^3\ mol^{-1}$ |
| $V_{cell}$ | Molar volume of the unit cell; see (6.124) | $m^3\ mol^{-1}$ |
| $V_m$ | Molar volume | $m^3\ mol^{-1}$ |
| $V^E$ | Atmospheric excess volume; see (6.129) | $m^3\ mol^{-1}$ |
| $V_{i:j}$ | Molar volume of a hypothetical compound $i_x j_y$; see (6.131) | $m^3\ mol^{-1}$ |
| $V_{i,k:j},\ V_{i:j,l}$ | Interaction parameters to describe the molar volume of intermetallic solution phase; see (6.131) | $m^3\ mol^{-1}$ |
| $V_{WC}$ | Volume fraction of WC phase; see (6.136) | |
| $V_n$ | Interface growth velocity along the normal direction; see (3.41) | $m\ s^{-1}$ |
| $v$ | Kirkendall velocity; see (6.7) | $m\ s^{-1}$ |
| $v(\Delta T)$ | Growth rate of dendrite tip as a function of undercooling; see (3.38) and (3.39) | $m\ s^{-1}$ |
| $W$ | Energy barrier height; see (3.3) | $m$ |
| $W_{mol}$ | Molar weight, see (6.123) | $kg\ mol^{-1}$ |
| $W_\mu$ | Athermal frictional work; see (7.2), and $W_\mu = V_m \cdot w_\mu$ | $J\ mol^{-1}$ |
| $w_\mu$ | Athermal frictional work; see (7.1) | $J\ m^{-3}$ |
| $w_{th}$ | Thermal work of interfacial friction; see (7.1) | $J\ m^{-3}$ |
| $x,\ X$ | Diffusion distance; see (6.11) and (11.1) | $m$ |
| $x$ | Position parameter, for one dimension | $m$ |
| $x_0$ | Matano plane; see (6.11) | $m$ |
| $x_B^{IF}$ | Effective interfacial composition; see (6.116) and (6.117) | |
| $x_i$ | Molar/atomic fraction of element (species) $i$; see Section 5.2.2 and (6.6) | |
| $x_i^\gamma$ | Molar/atomic fraction of element $i$ in the $\gamma$ phase; see Section 5.2.2 | |
| $x_i^0$ | Overall composition (atomic fraction) of element $i$ in a multiphase system; see Section 5.2.2 | |
| $Y$ | Young's modulus; see (2.63) | GPa |
| $Y_i$ | Normalized concentration of species $I$; see (6.14) | |
| $y_i$ | Molar fraction of species $i$ in a gas phase; see (5.35) | |
| $y_i^\alpha$ | Site fraction of species $i$ on the $\alpha$ sublattice; see (5.45) and (6.108) | |
| $z$ | Number of electrons transferred; see (12.3) | |
| $z,\ Z$ | Coordination number; see (5.41) and (6.66) | |

# Appendix B2: Greek and Special Symbols

| Symbol | Meaning | SI-unit |
|---|---|---|
| $\hbar$ | Reduced Planck constant, $\hbar = h/2\pi = 1.0545718 \times 10^{-34}$ | J s |
| $\alpha$ | Thermal expansion coefficient (linear, also CLE); see (6.91) and (6.125) | $K^{-1}$ |
| $\alpha_h$ | Thermal expansion coefficient (linear) of the $h$-lattice direction; see (2.54) | $K^{-1}$ |
| $\alpha_V$ | Thermal expansion coefficient (volumetric); see (2.52) and (6.137) | $K^{-1}$ |
| $\alpha_i$ | Thermal expansion coefficient of pure component $i$; see (6.138) | $K^{-1}$ |
| $\alpha^E$ | Excess thermal expansion of solution phase; see (6.138) | $K^{-1}$ |
| $\alpha(\theta)$ | Function of anisotropy; see (3.9) | |
| $\gamma$ | (Specific) interfacial energy; see (6.110) | $J\ m^{-2}$ |
| $\gamma_{LV}$ | Surface energy of liquid–vapor; see (6.115) | $J\ m^{-2}$ |
| $\gamma_s$ | Interfacial energy between austenite (Fcc) and martensite (Bct) nucleus; see (7.1) | $J\ m^{-2}$ |
| $\gamma_{SL}$ | Interfacial energy of solid–liquid; see (6.112) | $J\ m^{-2}$ |
| $\gamma_{SV}$ | Surface energy of solid–vapor; see (6.115) | $J\ m^{-2}$ |
| $\dot{\gamma}^\alpha$ | Slip rate on each slip system $\alpha$; see (4.15) | $m\ s^{-1}$ |
| $\Gamma$ | Jump rate of the diffusion species; see (6.69) | $s^{-1}$ |
| $\Gamma$ | Gibbs–Thomson coefficient; see (3.46)–(3.47) | K m |
| $\delta_{ik}$ | Kronecker delta ($\delta_{ik} = 1$ if $i = k$, otherwise $\delta_{ik} = 0$); see (6.105) | |
| $\delta$ | Grain boundary width; see Figure 6.18 | m |
| $\delta$ | Variation symbol; see (2.13) and (2.14) | |
| $\varepsilon_\phi$ | Gradient term coefficient for phase-field variable $\phi$; see (3.2) | |
| $\varepsilon_c$ | Gradient term coefficient for phase-field composition $c$; see (3.2) | |
| $\varepsilon_K$ | Degree of anisotropy of interfacial energy; see (3.10) | |
| $\varepsilon_H$ | Plastic strain in the hard phase region; see (7.7) | |
| $\varepsilon_S$ | Plastic strain in the soft phase region; see (7.7) | |
| $\varepsilon_{tot}$ | Sum of elastic and plastic strains; see (7.7) | |
| $\dot{\varepsilon}^{pl}$ | Plastic strain rate; see (7.6) and (7.15) | $s^{-1}$ |
| $\dot{\varepsilon}^{cr}$ | Creep strain rate; see (7.8) | $s^{-1}$ |
| $\dot{\varepsilon}^{el}$ | Elastic strain rate; see (7.15) | $s^{-1}$ |
| $\dot{\varepsilon}^{thr}$ | Thermal and transformation contributions to strain rate; see (7.15) | $s^{-1}$ |
| $\dot{\varepsilon}^{tp}$ | Transformation-induced plasticity strain rate; see (7.15) | $s^{-1}$ |
| $\dot{\varepsilon}^{t}$ | Total strain rate; see (7.15) | $s^{-1}$ |
| $\sigma_i$ | Component of stress tensor; see (2.55) | MPa |
| $\sigma_0$ | Scaling parameter; see (10.1) | MPa |
| $\sigma_H$ | Local stress of the hard phase; see (7.7) | MPa |
| $\sigma_S$ | Local stress of the soft phase; see (7.7) | MPa |
| $\sigma_{Fe}$ | Intrinsic lattice strength of pure Fe; see (7.5) | MPa |
| $\sigma_{ss}$ | Solid-solution hardening of matrix; see (7.5) | MPa |
| $\sigma_P$ | Precipitation hardening contribution; see (7.5) | MPa |
| $\sigma_{dis}$ | Dislocation strengthening contribution; see (7.5) | MPa |
| $\sigma_{sgb}$ | Subgrain strengthening contribution; see (7.5) | MPa |
| $\sigma_{yield}$ | Overall yield strength/stress; see (7.5) | MPa |
| $\eta$ | Viscosity; see (6.118) | Pa s |
| $\eta_0$ | Preexponential of viscosity; see (6.118) | Pa s |
| $\eta^E$ | Excess viscosity; see (6.121) | Pa s |

(*cont.*)

| Symbol | Meaning | SI-unit |
|---|---|---|
| $\kappa$ | Isothermal compressibility, $\kappa = 1/B$; see (5.10) | $\text{Pa}^{-1}$ |
| $\kappa$ | Thermal diffusion coefficient or thermal diffusivity; see (3.14) | $\text{m}^2\,\text{s}^{-1}$ |
| $\kappa_{corr}$ | True (corrected) value of thermal diffusivity; see (6.132) | $\text{m}^2\,\text{s}^{-1}$ |
| $\bar{\kappa}$ | Average interface curvature; see (3.48) | $\text{m}^{-1}$ |
| $\boldsymbol{m}^\alpha$ | Normal direction of the slip plane; see (4.15) | |
| $\boldsymbol{s}^\alpha$ | Slip direction; see (4.15) | |
| $\lambda$ | Thermal conductivity; see (3.45) and (6.133) | $\text{W}\,\text{m}^{-1}\,\text{K}^{-1}$ |
| $\lambda$ | Mean free path of the binder phase; see (10.4b) | m |
| $\lambda_p$ | Thermal conductivity of a pure substance; see (6.134) | $\text{W}\,\text{m}^{-1}\,\text{K}^{-1}$ |
| $\lambda^\varphi$ | Thermal conductivity of a solution phase $\varphi$; see (6.135) | $\text{W}\,\text{m}^{-1}\,\text{K}^{-1}$ |
| $\lambda_{\text{HM}}$ | Thermal conductivity of the two-phase hard metal; see (6.136) | $\text{W}\,\text{m}^{-1}\,\text{K}^{-1}$ |
| $\lambda_f$ | Fraction of energy, straining lattice; see (6.90) | |
| $\lambda(f)$ | Labyrinth factor; see (10.2) | |
| $\mu_i$ | Chemical potential of component $i$; see Table 5.2, (2.65), and (5.33) | $\text{J}\,\text{mol}^{-1}$ |
| $\mu_i^{0,ref}$ | Chemical potential of element ($i$) in its reference phase; see (5.33) and (12.2), $\mu_i^{0,ref} = G_i^{0,ref}$ | $\text{J}\,\text{mol}^{-1}$ |
| $\xi$ | Magnetic order parameter ($0 < \xi < 1$); see (6.101) | |
| $\xi_\phi$ | Nonconserved Gaussian noise field; see (3.5) | |
| $\bar{\xi}_m(\lambda_{ct})$ | Maximum volume fraction of martensite; see (7.10) | |
| $\dot{\xi}_k$ | Formation rate for the product phase $k$; see (7.11) | |
| $\rho$ | Mass density, general; see Section 2.4 | $\text{kg}\,\text{m}^{-3}$ |
| $\rho_{dis}$ | Dislocation density; see (7.6) | $\text{m}^{-2}$ |
| $\rho_A^m$ | Mass density of element A; see (6.93) | $\text{kg}\,\text{m}^{-3}$ |
| $\tau$ | Kinetic attachment time; see (3.5) | s |
| $\phi$ | Order parameter and phase-field variable; see Section 3.2.2 and (3.2) | |
| $\phi_i$ | Electron wave function; see (2.4) | |
| $\phi_{ij}^k(\lambda)$ | Linearly independent basis functions; see (6.23) | |
| $\varphi_{jk}$ | Thermodynamic factor; see (6.8) and (6.80) | |
| $\varphi_{\text{ave}}$ | Average thermodynamic factor; see (6.51) | |
| $\Phi$ | Level-set function; see (3.50) | |
| $\Phi_B$ | A temperature-dependent atomic mobility parameter; see (6.102) | $\text{J}\,\text{mol}^{-1}$ |
| $\Phi_B^i$ | Self- or impurity atomic mobility parameter for element B; see (6.102) | $\text{J}\,\text{mol}^{-1}$ |
| $^r\Phi_B^{i,j}$ | Binary Redlich–Kister interaction parameters for atomic mobility; see (6.102) | $\text{J}\,\text{mol}^{-1}$ |
| $^s\Phi_B^{i,j,k}$ | Ternary interaction parameters for atomic mobility; see (6.102) | $\text{J}\,\text{mol}^{-1}$ |
| $\chi(\mathbf{r}, t)$ | Single-site occupation probability function; see (3.36) | |
| $\omega$ | Atom's jump rate; see (6.69) | $\text{s}^{-1}$ |
| $\omega'$ | Atom's jump rate; see (6.75) | $\text{s}^{-1}$ |
| $\omega_j(\mathbf{q})$ | Frequency of the $j$th phonon mode at wave vector $\mathbf{q}$; see (2.48), (2.50), and (2.51) | $\text{s}^{-1}$ |
| $\Psi$ | Wave function; see (2.1) | |
| $\nu$ | Poisson ratio; see (2.64) | |
| $\nu_i'$ | Vibrational frequencies of the transition state; see (6.71) | $\text{s}^{-1}$ |
| $\nu_i$ | Vibrational frequencies of the initial state; see (6.71) | $\text{s}^{-1}$ |
| $\nu_0$ | Ratio of products of $\nu_i$ and that of $\nu_i'$ with an imaginary frequency excluded (keeps dimension); see (6.71) | $\text{s}^{-1}$ |

## Appendix B3: Derivative Operation Symbols, Vectors, and Tensors

| Symbol | Meaning | SI-unit |
|--------|---------|---------|
| $\hat{H}$ | Hamiltonian operator; see (2.1) | |
| $\nabla$ | Gradient operator; see (3.2) | |
| $\nabla^2$ | Laplace operator; see (2.2) | |
| $\otimes$ | Tensor product; see (4.15) | |
| $\mathbf{X}$ | An arbitrary vector; see (4.1) and (4.2) | m |
| $\mathbf{I}$ | Second-order unit tensor; see (4.4) | |
| $\mathbf{E}$ | Eulerian finite strain tensor; see (4.12) | |
| $\mathbf{E}^*$ | Lagrangian finite strain tensor; see (4.13) | |
| $\boldsymbol{F}_i$ | Force vector acting on particle (atom) $i$; see (2.17) | N |
| $\mathbf{F}$ | Total deformation gradient tensor; see (4.2) | |
| $\mathbf{F}_\mathrm{V}$ | Volumetric deformation component of $\mathbf{F}$; see (4.4) | |
| $\mathbf{F}_\mathrm{iso}$ | Isochoric deformation component of $\mathbf{F}$; see (4.5) | |
| $\mathbf{F}^e$ | Elastic deformation component of $\mathbf{F}$; see (4.14) | |
| $\det\mathbf{F}$ | Determinant of matrix F | |
| $\mathbf{F}^p$ | Plastic deformation component of $\mathbf{F}$; see (4.14) | |
| $r, R$ | Position vector; see Section 2.1 | m |
| $\mathbf{R}$ | Rotation tensor; see (4.8) | |
| $\sigma$ | Cauchy stress tensor; see (4.19) | MPa |
| $\Delta\varepsilon$ | Plastic strain vector due to crystal slip; see (4.15) | |

# Index